VOLUME 1 SOUTH
3rd Edition

Published in Great Britain by Pesda Press

Tan y Coed Canol, Ceunant,

Caernarfon, Gwynedd

LL55 4RN

Wales

Gary Latter

Third Edition 2024

Second Edition 2017

First Edition 2008

Reprinted with minor updates 2012

ISBN 978-1-917182-01-0

Maps by Bute Cartographics.

Printed and bound in Poland, www.hussarbooks.pl

CONTENTS

Dave MacLeod on the first ascent of Ring of Steall, Glen Nevis, one of the hardest routes in the book at F8c+. Photo Claire MacLeod.

"Climb the mountains and get their good tidings. Nature's peace will flow into you as sunshine flows into trees. The winds will blow their own freshness into you, and the storms their energy, while cares will drop off of you like autumn leaves."

– John Muir

SCOTTISH ROCK

The area covered by this book, the Highlands and Islands, lies entirely to the north of the Highland Boundary Fault. With its mountain landscapes, deep glens, lochs, rivers and hundreds of islands, it represents one of the most extensive and least populated semi-natural areas remaining in Western Europe. Scotland can also lay claim to the only true areas of 'wilderness' remaining in Britain, with vast tracts of uninhabited areas in the far North West, and the similarly wild and unspoilt high arctic plateaux of the Cairngorm massif. Often, by choosing your venue carefully, it is possible not just to avoid queues but to have whole mountains to yourself.

Within this incredibly varied setting can be found stunning examples of every sub-sport that rock climbing has evolved. In UK terms, we have the longest mountain routes (such as *The Long Climb* on Ben Nevis); the biggest sea cliffs (St John's Head, Hoy), which also harbour the only multi-day big wall route in the country – the 23 pitch *Longhope Route*; the steepest cliff (Sron Ulladale, Harris). But size isn't everything. In contrast, a myriad of miniature sport routes have appeared in recent years, together with a resurgence of interest in outcrop climbing in general. There has also been the opening up of some wonderfully situated bouldering venues, together with exquisite deep-water soloing on a few esoteric locations.

This book is intended as a celebration of the wealth and variety of great climbing that Scotland has to offer. The selection of routes should have something for everyone, from the athlete to the aesthete. Climbing in Scotland is about more than the rock alone; there is the magnificent and awe-inspiring scenery, the sense of history, the wide open spaces, the clean fresh air and the possibility of solitude.

USING THE GUIDE

All the areas covered are described as approaching from the south, where the majority of visitors originate. Similarly, the routes are also laid out in the order they are encountered from the approach. Each area has an introduction outlining the style of climbing, together with detailed up-to-date information on **Accommodation** and **Amenities** – in short everything the visitor requires to familiarise themselves with an area. Each cliff or crag is described in summary, together with specific **Access, Approach** and **Descent** details clearly laid out. In addition, maps and photo-diagrams illustrate further. Routes are given an overall technical grade alongside the adjectival grade, with the individual pitch grades incorporated within the description. The abbreviations **FA** and **FFA** refer to the first ascent and first free ascent respectively. **PA** refers to the number of points of aid used on the first ascent. There are very few routes containing aid in this book, though in some instances the use of a couple of points of aid may give a more consistent route, and an alternative grade is offered. On a few harder routes, usually unrepeated, rest points were used and this is mentioned in the hope of encouraging subsequent free ascents. **PR** and **PB** refer to peg runner and belay respectively; **F** and **R** to friends and rocks; **BB** and **LO** refer to bolt belays and lower-offs on the sport crags. **TIC** stands for Tourist Information Centre; **ATM** for Automated Teller Machine.

You should have the relevant Ordnance Survey 1:50,000 map and, particularly for the remoter mountain crags, knowledge of how to use a map and compass is assumed. The middle of the Cairngorms or the top of Ben Nevis is not the place to attempt to learn to navigate. The Grid Reference and Altitude refers to the base of the cliff or crag. The approach times quoted are intended as a general guide (racing greyhounds and ramblers/tortoises can make their own adaptations accordingly), along the lines of Naismith's Rule (4.5km per hour and one minute for every 10m of ascent).

I have attempted to consult as many active climbers in Scotland as possible to get a broad range of opinions and a consensus on grades and quality, but the final selection of routes is a personal one. For instance, not all the routes are on immaculate rock, with some of the older routes in the traditional character-building mode. Jim Bell's famous adage, *"Any fool can climb good rock. It takes craft and cunning to get up vegetatious schist or granite."* may be worth bearing in mind.

ACCOMMODATION

Information on a range of budget accommodation is included for each area, from camp sites and youth hostels to private bunkhouses. There are also a number of well-situated mountaineering club huts in all the main mountain areas. These are available for booking by members of Mountaineering Scotland, the BMC and affiliated clubs. In addition, Tourist Information Centres (**TICs**) are detailed at the start of each main area. These are a good source of information including local recommendations. More information is available from Visit Scotland (www.visitscotland.com).

EATING OUT

One important point worth bearing in mind, (especially for those used to continental and transatlantic hospital-ity) is that the majority of Scotland still lurks in the dark ages when it comes to the service industry. Most pubs only serve food over a short period at lunchtime; often 12 – 2pm, and more importantly, the majority of pubs and hotels stop serving food at 9pm, some at 8pm even! I've had the misfortune to turn up at a restaurant/pub in Skye (in July, the height of the tourist season) to be informed *"We're not serving food: the chef's on his lunch"* – unbelievable. Establishments that are particularly good and worth seeking out are highlighted within the introductory section of each relevant chapter.

ACCESS

The **Land Reform (Scotland) Act 2003** gives statutory access rights to most land and inland water. These rights exist only if exercised responsibly by respecting the privacy, safety and livelihoods of others and by looking after the environment. The Scottish Outdoor Access Code (www.outdooraccess-scotland.scot) provides detailed guidance on the responsibilities of those exercising access rights and those managing land and water. The Access Code notes that access rights do not apply to motor vehicles and provides guidance on parking.

- Take responsibility for your own actions.
- Respect people's privacy and peace of mind.
- Help land managers and others to work safely and effectively.
- Care for your environment.
- Keep your dog under proper control.
- Take extra care if you are organising an event or running a business.

WILD CAMPING

In the rural areas it is often possible to camp at the side of the road. If in doubt, ask permission locally from farmers and crofters. Remember to remove all your litter, all trace of your tent pitch and not to cause any pollution. It should almost always be possible to camp in the hills, except perhaps in some areas during the stalking season.

CARAVANS

Those wishing to bring caravans please don't – go to the Lakes, the Borders or some other rolling hills well away from the Highlands and Islands. Even better, stay at home and play tiddlywinks or golf, or take up macramé or embroidery or some other suitably sedate pastime. Alternatively, travel under the cover of darkness, preferably at 3am on a Sunday morning.

BIRDS

Some of the sea cliffs are affected by nesting seabirds and should be avoided during the nesting season of April – July inclusive. Almost all birds, their nests and eggs are protected. The proliferation of guano on such cliffs makes it in the climber's interest to choose another venue. In particular, some popular routes, such as the *Old Man of Stoer* and *Hoy*, have the occasional fulmar nest on ledges, and it is definitely in the climber's interest to avoid close encounters, as they have the nasty habit of vomiting semi-digested fish oil onto uninvited visitors. It should still, however, be possible to climb these routes during the nesting season. In the unlikely event of coming across birds of prey (especially peregrine falcons, golden or white tailed sea eagles – all Schedule 1 birds) choose another route or cliff. It is an offence, under the Wildlife and Countryside Act 1981, to disturb any Schedule 1 bird, with an unlimited fine, up to 6 months imprisonment or both. Their continued existence is surely more important than another tick in the guidebook? Information on current restrictions is available from Mountaineering Scotland (01738 493942; www.mountaineering.scot).

SEASONAL RESTRICTIONS

The grouse shooting season is from 12 August (the 'glorious' twelfth) – 10 December and deer from 1 July – 20 October for stags and 21 October – 15 February for hinds. More information at www.outdooraccess-scotland. scot. There are few crags or cliffs included where access problems have been encountered in the past. A caring, responsible attitude towards parking, litter, conservation and a polite approach to landowners should ensure that the present situation continues. If any difficulties are encountered, contact the access officer at access@ mountaineering.scot. No commercial stalking takes place on National Trust for Scotland properties (such as Glen Coe and Torridon), ensuring access at all times.

DIRECTIONS

All directions (left and right) are given for climbers facing the crag, except in descent. Any ambiguous descriptions also include a compass point, but if you don't know your left from your right, chances are you won't have a clue where the North Pole lies.

CONSERVATION

Try to adopt a minimum impact approach at all times, leaving the place as you would like to find it. Approaches to some of the cliffs can be greatly aided by the use of bicycles. Their use should be restricted to solid paths such as private and forest roads or rights of way, not soft paths and open hillsides where considerable erosion can occur. Where there is a substantial time-saving advantage, such information is included in the approach information. Where repeated abseils from trees is the norm (such as on Creag Dhubh), slings and karabiners or maillons have been left in place, and their use is encouraged to prevent ringing of the bark, leading to the eventual demise of the trees. Always park with consideration for others, and avoid damage to fences and walls. And of course, as the country code stresses, avoid 'interfering' with animals (Aberdonians and Rick Campbell take note!). Do not leave any litter, including food scraps, finger tape, chalk wrappers and cigarette ends and remove any left by others. Bury or burn toilet paper. Scratching arrows or names at the base of routes can clearly be viewed in a modern light as nothing short of vandalism. Established markings are mentioned to aid identification, and it is hoped no further additions will be thought necessary. Many of the areas covered are within National Scenic Areas (NSA), National Nature Reserves (NNR) and Sites of Special Scientific Interest (SSSI), controlling development and ensuring the retention and preservation of the natural environment.

STYLE, PEGS & IN SITU PROTECTION

"Ethics change the experience for others, style only changes your own personal experience."
The use of chalk is no longer a burning issue. Nevertheless, its use should be kept to a minimum, hopefully only on extremes. Chalk has been spotted on descent routes (I kid you not) and on VDiffs, such as *Agag's Groove*, which must be an ultimate low point. Hold improvement is unacceptable on natural rock. If you can't climb a piece of rock with the holds available, leave it unclimbed rather than resort to the hammer and chisel. The use of hammered nuts should be discouraged, as their rapid deterioration soon blocks the placement possibilities for subsequent ascentionists.
Whilst the style a route is climbed in is a personal one,

I feel obliged to make a few comments. The use of 'rest points' (i.e. aid) and prior top-roping should be reserved for routes that are pushing new frontiers. It is true to say that such tactics percolate downwards. Try to give the rock a fighting chance, and approach the route on its own terms, in accordance with local practices. The majority of active pioneers in Scotland have attempted to push standards, and many very audacious leads have been achieved on-sight or ground up.

QUALITY ASSESSMENT

I had originally intended to adopt the Farquhar rating system, with its two extremes of PS and FB, but as hopefully there are no 'pure sh≈@‡' routes herein (unless included for historical interest, or to aid in crag descriptions) and masses of '#µ©k*≈g brilliant' routes, I have decided to opt for the conventional star rating system, with three star routes being of truly outstanding quality. As the climbing in Scotland is clearly superior to anything south of the border, a few exceptional routes have the honour of four stars. These are absolute 'must do's' that would rate amongst the best anywhere on the planet, such is their undeniable brilliance. On a few isolated routes, a wire brush symbol denotes that the route may require prior cleaning in its present state, and the stars assume the route is in a clean state. These are routes which were originally climbed following cleaning on abseil, but at the time of writing have not had much repeat traffic, and may require a quick abseil with a wire brush prior to an ascent.

CLIMATE

"They'll all be doing them when the sun comes out."
– Don Whillans.
The Highlands and Islands are dominated by the prevailing southwesterly winds, bringing moist and usually mild air from the Atlantic. In addition, many of the Atlantic depressions pass close to or over Scotland. *"It always rains up there"* is a commonly held myth. It is easy at first sight to confuse a map of annual rainfall with that of a relief map, for the two are closely linked. The wettest belt extends from the Cowal peninsula

(south and west of Arrochar) in a broad band as far as the hills just south of Torridon. In the mountains an annual precipitation of between 200–300cm and more is the norm, these dreich figures dropping markedly to 150–200cm on the coastal fringe. Within this broad belt there is much variation. As an example, at Dundonnell at the head of Little Loch Broom the annual rainfall is 195cm; 10km south it is 250cm, and 10km further north in Ullapool the average is 150cm.

The coastal promontories, especially in the north, and the Outer Hebrides receive only 130–170cm. Similarly, low ground around the Cairngorms and the eastern edge of the Central Highlands (such as Craig a Barns and The Pass of Ballater) benefit from the rain shadow effect of the hills further west (80–100cm). The higher ground in the Cairngorms receive around half the precipitation than the hills just in from the main Atlantic seaboard, with an average of 225cm recorded on Cairn Gorm summit. Lying in the centre of the country, their climate is more continental, with warmer summers than on the coasts. Many districts in the north and east have, on average over the four summer months from May–August, a total rainfall of less than 40cm, comparing favourably with the drier parts of England. Throughout the country the driest and sunniest period is from May to the end of June, the next driest from mid-September to mid-October.

In the Outer Hebrides gales are recorded on over 40 days of the year, and in the Northern Isles this figure is even greater, though most of these occur in the winter. Prolonged spells of strong wind are uncommon between May and August. Especially in the Western Isles and along the west coast, May is the sunniest month, closely followed by June. April is sunnier than the popular holiday months of July and August. The temperatures on the west coast and the islands are generally a couple of degrees cooler than inland, with the Northern Isles a couple of degrees cooler again. Finally, in midsummer there is no complete darkness in the north of Scotland, with Shetland receiving about 4 hours more daylight (including twilight) than London.

TIDAL INFORMATION

In general, the tide ebbs and flows twice daily. As a rough guide, the tide takes 6 hours to come in, spends a half an hour 'on the turn', then 6 hours to recede, before repeating the same process. Spring tides occur after a new and full moon, and have the greatest amplitude. Tide tables are published annually for specific areas and are available from yacht chandlers and in many newsagents, or from harbour offices. Also available online at www.tidetimes.org.uk. Tidal predictions up to 7 days in advance also available at www.tidetimes.co.uk.

WEATHER INFORMATION

Radio: BBC Radio Scotland broadcasts an outdoor conditions forecast daily at 18.25 weekdays and on Saturdays and Sundays during the 07.00 and 19.00 news bulletins, in addition to general forecasts at the end of each news bulletin.

TV: BBC Reporting Scotland broadcasts a good general forecast daily at 18.55 weekdays, on Saturdays and Sundays during the 17.00 and 18.50 news bulletins.

Online: www.bbc.co.uk/weather provides a detailed weather forecast including details for the week ahead, coastal forecasts and tide tables. In addition to the above www.metoffice.gov.uk provides daily mountain forecasts, as does www.mwis.org.uk, which issues a detailed 3-day forecast daily at approximately 16.30. Another popular option is the Norwegian website www.yr.no.

WEE BASTARDS

Little biting creatures, which the vast majority of tourist-orientated brochures and guidebooks fail to mention, can make a massive difference to one's stay in the Highlands and Islands. Of the thirty-four species of biting midge found in Scotland, only four or five species bite humans. By far the worst and most prevalent, accounting for more than 90 percent of all bites to humans is the female of the species *Culicoides impunctatus*, or the Highland Midge. This voracious creature first makes its appearance around the end of May and can persist until the end of September in a mild summer, with early June through to August being the worst periods. They are particularly active on

still, cloudy or overcast days, especially twilight (which lasts throughout the night in Scotland in summer). Wind speeds above a slight breeze force them to seek shelter. Mosquitoes are less of a problem, though the cleg (or horsefly) feeds mainly during warm bright days. Finally, sheep or deer ticks, small black or brown round-bodied members of the genus *arachnid* rest on vegetation, awaiting a host. The tick sinks its head into the victim's flesh, until it eventually swells up and drops off. Ticks in the UK regularly carry Lyme Disease, a potentially serious infection, so they should be removed as soon as possible. Remove with a tick removal tool or tweezers and apply antiseptic cream. If flu-like symptoms persist after a tick bite, you should see a doctor immediately. Avoid ticks by keeping your arms and legs covered if possible; wear light-coloured fabrics so you can see them easily; spray clothing and shoes with the repellent permethrin, such as Lifesystems EX4, available from outdoor retailers; read about correct tick removal and always carry a tick remover. For more information: www.lymediseaseuk.com; www.nhs.uk; www.mountaineering.scot. There are many insect repellents commercially available. Although most contain varying concentrations of diethyl toluamide (DEET), non-toxic alternatives are available such as Smidge; Mosi Guard Natural and Autan Protection Plus are particularly good against ticks.

MOUNTAIN RESCUE

In the event of a serious accident requiring medical attention, call 999 or 112 and ask for Police and Mountain Rescue/Coast Guard. Give concise information about the nature of the injuries, and the exact location, including a six-figure grid reference or the name of the route if possible. Try to leave someone with the casualty, who should be made as comfortable as possible, if injuries allow. If unconscious, be sure to place in the recovery position, ensuring the airway is clear.

If you need to call assistance but cannot make voice calls due to poor mobile phone reception, you can contact the emergency services using a text from your mobile but only if you have already registered with the emergency SMS text service. To register, text the word 'register' to 999. You will get a reply and should then follow the instructions you are sent. More information at www.emergencysms.net.

In a few instances Mountain Rescue posts (containing a stretcher and basic rescue kit) are located in the hills. More information at www.scottishmountainrescue.org.

GRADES

Routes are graded for on-sight ground up ascents, and the climber is assumed to be fully equipped with a wide range of protection devices. On some of the hardest routes skyhooks may be found useful. It goes without saying that people should make their own judgement regarding any in situ equipment encountered including fixed abseil points, all of which will rapidly deteriorate through exposure to the elements. I have tried to be as consistent as possible, though minor regional variations may occur. Any crucial runner information, especially relating to obscure gadgets or hidden or hard to place protection has been included where known. Where a route has only received an ascent after extensive top-rope practice this headpointed ascent has been highlighted within the first ascent details where known, in order to record such prior familiarisation.

ACKNOWLEDGEMENTS

Special thanks to my wife Karen, who has helped throughout on this 3rd edition. In particular, special thanks to Jules Lines for extensive feedback. Others who have helped by providing route descriptions, feedback or photos are: Nathan Adam, Miguel Alonso, Zoe Anderson, Ken Applegate, Ivan Bialy, Paul Brian, Rick Campbell, Robin Campbell, Andy Corbe, Angus Davidson, Dave Douglas, Robert Durran, Dave Fowler, Davy Gunn, Peter Herd, Billy Hood, Mike Hutton, Murdoch Jamieson, Callum Johnson, Will Jones, Cassim Ladha, Zelin Liu, Seonaid MacDonald, Gavin McColl, Eilidh Milne, Colin Moody, Dan Moore, Willis Morris, Andy Nelson, Dominic Oughton, Tim Rankin, Dave Riley, Gemma Robertson, Ali Rose, Rafael Salazar, Huw Scott, John Sharples, Hugh Simons, Jamie & Morag Skelton, Martin Stephens, David Thompson, Matthew Thomson and Graham Tyldesley.

DISCLAIMER

GEOLOGY

by Dr Darren McAulay & Gary Latter

"For its size Scotland has the most varied geology and natural landscapes of any country on the planet."
- Alan McKirdy and Roger Crofts, *Scotland: The Creation of its Natural Landscape,* Scottish Natural Heritage, 1999

Scotland hosts a wealth of rock types and topography, each of which provides unique characteristics and opportunities to the climber. The intricate geological history has been at the forefront of study; indeed many of the major earth moving mechanisms (e.g. thrust and fault movements) have been understood by analysing relief and rock outcrops throughout Scotland.

Rock climbers cannot escape the fact that geology plays a major role in the formation of their playground and many have more than a passing interest in the subject. An in-depth description of Scottish geology can be found elsewhere, but the following notes give an insight into how rock type influences not only the formation of our vertical arena but affects frictional qualities, hold type/ shape and quality, and the provision of opportunity for the placement of protection.

A BRIEF GEOLOGICAL HISTORY

The age of the Earth is estimated to be in the region of 4,500 million years. Scotland's oldest rock, the Lewisian gneiss, is dated at 2,300 – 2,600 million years, being some of the oldest rocks to outcrop in Europe. The gneiss originally formed part of a huge landmass that included Scandinavia, Greenland and Northeast America. This landmass was uplifted and then eroded to give an undulating landscape that can be seen in the ancient topography of the NW mainland and its namesake of the Isle of Lewis. Around 1,100 million years ago, the gneiss was covered by around 5,000 metres of red sandstone called the Torridonian which was deposited by rivers flowing from the north. Following this and further to the south in more marine waters, sediments were being heated and folded (metamorphism) to produce the Moine schists.

Around 570 million years ago, the sea level rose and a series of sandstones, siltstones and limestones formed on the Torridonian/Lewisian basement – called the Cambrian. The next event, which had major consequences, started around 450 million years ago and involved the collision of two landmasses over 130 million years. The result was the Caledonian mountain belt, believed to be of Himalayan stature, which produced the Dalradian schists, which together with the Moine schists make up most of the Highlands. Aside from the massive folding of the rocks during this time was the series of northeast-southwest trending faults that came into existence. The Highland Boundary Fault, which runs through Arran, Loch Lomond and on to Stonehaven now forms the dividing line between the Scottish Highlands to the north and the Central Lowland Belt to the south. During this period heating allowed molten rock to rise up into the schists, sometimes reaching the surface forming volcanoes (Ben Nevis and Glen Coe) but generally solidifying as granite deeper in the crust (Cairngorms, Starav, Rannoch). At this point Scotland was still connected to a landmass that included North-east America and Scandinavia.

After the Caledonian shake-up a further series of sandstones were produced; the Old Red Sandstones, which are no longer represented in the Highlands but still occur in Orkney and NE Scotland. The period between 360 – 100 million years ago produced various rock types including more sandstones, limestones, coal measures and shales (source of oil and gas) but apart from in the Central Belt these rocks are not widespread and are of little interest to the climber. However, the Great Glen Fault (now filled by Loch Ness) was active at various times in this period causing a lateral tearing of the earth (as the San Andreas Fault does today in California). It has been shown that the area to the north of the Great Glen has moved 100km relative to the south. Major rock forming events were still not finished though, as around 60 million years ago great movement occurred when North America broke away from the European continent. Apart from the obvious geographical (and some say fortunate!) consequence, the earth movements created substantial heating with the result that huge volcanoes were formed along the NW margins of Scotland (a line from Arran to Skye to St Kilda). Subsequent extensive erosion by elements such as changes in climate and periods of glaciation have reduced the mountains and volcanoes to the remnants left today. Fortuitously, these remnants offer a western coastline that is 260 miles from north to south but over 2,000 miles in outline, with 550 Hebridean islands, 40 mainland sea lochs and hundreds of mountains and cliffs to enjoy.

ROCK TYPES

Lewisian Gneiss – Ancient rocks that have undergone such extreme heating and pressure that all trace of original structure has been lost. The Lewisian rocks originally formed part of a great continent that was worn down to a low lying, gentle landscape, subsequently covered (in Scotland) by the Torridonian sandstones. Typically forms rounded and rocky outcrops (due to the lack of jointing) with exceptional friction when weathered, because of the coarse-grained nature and streaked texture that permits the formation of flat, banded holds and rough pockets. It occurs on Lewis and Harris and the Barra Isles, on almost all the worthwhile crags and cliffs in the Gairloch district, including the remote Carn Mor, and in bands in Sutherland, including around Sheigra.

Torridonian Sandstone – Transported by the action of large rivers on the margins of a mountainous continent to the north, the Torridonian deposits generally consist of coarse-grained, red sandstones that contain frequent pebbly bands. The Torridonian exhibits its original bedding, since it has escaped the mountain building forces imposed on many other rock types in Scotland. Good friction and quick drying, the holds often have a rounded feel and friends/cams ensure peace of mind in the sometimes flared cracks and horizontal breaks. The outcrops around Torridon, and almost all the cliffs and crags in Coigach and Assynt, are composed of this rock.

Cambrian Quartzite – Deposited on top of the Torridonian sandstones but with a southeast tilt, this quartzite is the result of heating of quartz-rich sandstones. They are brittle and well jointed which tends towards sharp flat holds. This rock outcrops mainly as resistant summit caps on the hills of Assynt and Sutherland where frost-shattering forces produce abundant sharp, blocky screes. Further south the extensive Coire Mhic Fhearchair on Beinn Eighe and the Bonaidh Dhonn overlooking Loch Maree are two of the finest examples.

Moinian/Dalradian Schists – These are the predominant rock types of the Highlands, the result of metamorphism (intense folding, alteration by heat and pressure and even melting) caused by the meeting of two crustal plates. They formed a chain of mountains of Himalayan stature as huge folds and thrusts occurred, sometimes causing older rocks to rise up over younger, with the culminating edge being the Moine Thrust which runs from Skye to the north coast near Durness. The metamorphism of original sandstones, mudstones and possibly igneous rocks caused a slaty cleavage to develop which cross cuts the original bedding planes. Hence, the schist crags vary enormously depending on

the local variations in the folding, and can even vary from route to route. At Creag Dhubh, for instance, the majority of the holds are of a horizontal, sometimes sloping nature whereas The Cobbler exhibits a variety of holds from slopers, to jugs and pockets, to quartz knobs. One notable property of the schists is that they offer poor friction when wet or lichenous, particularly the mica-schists in the Southern Highlands. Other locations include Glen Nevis, Craig a Barns, and venues as diverse as Glen Ogle and the alpine Aonach Beag.

Old Red Sandstone – Not a common rock type in terms of climbing (and geographic distribution) in Scotland, with the exception of the Northern Isles and the Old Man of Hoy. It is generally well bedded which results in many horizontal breaks and cracks (often rounded or flared). The friction is good where the rock is clean, the quality is generally impeccable on the dark russet coloured walls and it is at its best on some of the upper sections on Rora Head.

Jurassic Sandstone – Although most of the rocks from this period tend to be soft and friable, some climbing can be found. A notable exception is the quartz sandstone of Suidhe Biorach (Skye) which like the Cambrian quartzite has a tendency to produce sharp, flat edges and, where eroded by the sea, has numerous pockets.

Granite – The granites can be found all over the Highlands, formed when large amounts of molten rock rose up through the folds and faults of the Caledonian mountains and solidified. The quality of the granite can depend on the rate at which it cooled. Rapid cooling producing a fine grain such as at Binnein Shuas and Dirc Mhor, whilst slower cooling produces a coarser grain as in the Cairngorms. Typical granite holds consist of cracks and flakes of a rounded nature, which are the product of blocky fractures formed during cooling. In addition ice-scoured slabs can be noted for their absence of holds and scarcity of protection (Etive, Arran and some of the cliffs in the Cairngorms). Gullies, formed by preferential weathering of intrusions and minor faults, tend to be

Finger jamming on Cairngorm granite.

very loose, formed of a fragile substance more closely resembling Weetabix than rock. Many routes follow lines produced by similar intrusions including the quartz vein on *Swastika*, and pegmatite band on *Ardverikie Wall*.

Andesite & Rhyolite – Forming similarly to the granites above, but in areas where the molten rock was able to penetrate the surface, are a range of lavas and granites. Most notably in the Ben Nevis and Glen Coe areas, huge blocks of schists were undermined by molten rock, which caused a subsequent collapse and outpouring of lavas. Although the lavas have long since been eroded away, leaving mostly schist and granite, the collapse (estimated at 1,600m in Glen Coe) resulted in some rocks (mainly rhyolites and andesites) being dropped to a level that was not removed. The rhyolites and andesites are very resistant to erosion compared to the surrounding schists and so make up the higher parts of the mountains. The Ben Nevis complex experienced at least four episodes during the collapse and subsequent granite intrusions and the differential cooling has resulted in a variety of coarseness, the area around the Carn Mor Dearg arête being fine grained as the result of a smaller, later subsidence. The summit block of the Ben consists of a resistant andesite that is the core of the collapse. The main cliffs still rest on schists that are 600m above sea level. Everywhere on the cliffs there is an inward sloping trend that results in good ledges/holds. Both rhyolite and andesite are fine grained, and fracture to give good sharp holds, though some of the best climbing in Glen Coe is on wonderful compact rough bubbly brown-pink rhyolite, at its best when water washed and well weathered. A number of dykes, mainly of porphorytic microdiorite have eroded to give features such as the distinctive *Ossian's Cave* and the descent route down from the Terrace Face, both on Aonach Dubh.

Dolerite & Basalt – Huge basaltic lava fields covered much of the area of what is now the western coast of Scotland. They poured out from huge volcanoes that have long since been reduced to their roots but many layers built up over time. Dolerite is a fine-grained lava and where it was intruded between layers (as a sill) it often cooled in the famed columnar form seen at Fingal's Cave (Staffa) and Kilt Rock (Skye). Basalt is generally unreliable for climbing, but the harder dolerite can form excellent cracks and vertical columns giving sustained jamming, its long parallel cracks being well protected with an ample supply of cams.

Gabbro – This coarse rock is famed for its roughness and great friction. The origin of gabbro is associated with volcanic activity but unlike the dolerite and basalt lavas it solidified deep within the root of such volcanoes. Solidification was slow which accounts for the coarseness of gabbro. Subsequent pressuring caused fissures within the volcanic complexes, with the result that much of the original rock was crosscut by smaller basalt intrusions (dykes) that radiate out from the centres; *The Snake* on the Eastern Buttress of Sron na Ciche follows the line of one such prominent dyke. These basalt dykes are fine-grained and less resistant than the gabbros and

although weathering often removes the dykes to form gullies and ledges, caution must be taken when climbing as the rock can be brittle and has poor friction when wet. Gabbro on the other hand retains good friction even in the rain, which is fortunate considering their location on Skye and Ardnamurchan.

In addition to the above, there are lesser outcroppings of other rocks, such as slate, limestone, greywacke and conglomerate throughout the Highlands and Islands, though (with the exception of the conglomerate crags of The Camel and Sarclet) none of the venues are deemed worthy of consideration from a climbing point of view to date. The rapid pace of exploration in recent years, together with the acceptance of the worthiness of rock quality in other parts of the UK, will undoubtedly reveal other venues.

Finally, it should be noted that erosion of the mountains and crags is continually taking place. Notable examples in recent years include two huge rock falls (1995 and 2000) on the right wall of *Parallel B Gully* in Lochnagar, the top corner of *The Giant* and the third pitch of *Cougar* on Creag an Dubh-loch, large fresh rock fall scars around the base of *Collie's Route* on Skye's Sron na Ciche, the repositioning of huge chokestones in *The Chasm* on Buachaille Etive Mor, and the collapse of the natural arch that joined the top of the Old Man of Hoy to the mainland cliff as recently as just over one hundred years ago.

This roughly kidney-shaped island sits in the Firth of Clyde. It is the most southerly (and the most accessible) of the Scottish islands, and a popular tourist destination. It straddles the Highlands and Lowlands, as the Highland Boundary Fault passes through its centre – the 'String Road' (B880) between Brodick and Blackwaterfoot. Arran is often described as 'Scotland in miniature', displaying a wide range of rock types and landforms in such a small (32 x 16km) area. Despite the geological diversity, all the quality climbing is concentrated on the granite cliffs in the mountains to the north-east. A number of outcrops and sea-cliffs have also been climbed on but they do not compare with the bigger cliffs, and a wet day would be better spent exploring the many cafes and pubs. It should be noted that, on occasion, *"The climbing is anything but straightforward yet can be addictive. Arran is all wrong in the same way as a camel is a horse designed by a committee. Whoever thought up this place did so in a distinctly mischievous frame of mind."* – Bill Skidmore, Arran this Summer, 1978 SMCJ

"With the advent of climbing wall training, Arran seems even more baffling and inscrutable"
– Andrew Fraser, 1999

Access: A regular car ferry (Caledonian MacBrayne at the Ferry Terminal, Ardrossan ☎ 0800 066 5000 www.calmac.co.uk) runs from Ardrossan – Brodick, the main (only) town on the island, taking 55 minutes. There is a regular 7am ferry to the island, making day trips eminently possible (there is also a late ferry back on a Friday). The two popular camping spots with climbers and hillwalkers, Glen Rosa and Glen Sannox, can be reached by taxi; Sannox also has a regular bus service. There is also a summer crossing from Claonaig (Kintyre) – Lochranza at the north of the island. There is a frequent train service from Glasgow to Ardrossan, reaching Arran just two hours from the city.

> Access to all the cliffs on **Beinn Tarsuinn**, **A'Chir**, **Goatfell** and **Cir Mhor**: From the pier at Brodick head north up the main (A841) road for 1.3 miles/2km. Continue straight on along the B880 west for 0.1 mile/0.2km then turn right up a single track road for 0.8 mile/1.3km to limited parking at the end of the tarmac road overlooking the campsite.

Amenities: TIC The Pier, Brodick (☎ 01770 303774; www.visitscotland.com). Many cafes and pubs offering lunches, especially around Brodick. Numerous places in Brodick do bike hire, including: Arran Bike Hire (07825 160668; www.arranbikehire.com) & Auchrannie Resort (01770 302234; www.auchrannie.co.uk). Motorhome stopover facilities at Auchrannie Resort.

Accommodation: The most popular camping spot for climbers and hillwalkers is the Glen Rosa Campsite (07985 566 004) at the end of the single track road at the bottom end of Glen Rosa. No showers. Unfortunately, this usually becomes overrun with drunken weegies* at Bank Holidays. Also beware the midges! Numerous quiet isolated camping spots can be found much further up Glen Rosa, beyond the bridge over the Garbh Allt, the best in the flat ground just beyond where the path splits (due west of the summit of Goatfell, beneath the Rosa Slabs). A fine less-frequented camping spot can also be found near the bottom of Glen Sannox (5 – 10 minutes walk from road), on either side of the track, the best spots in the trees 100m beyond the navigation beacons.

Youth Hostel: at the north end of the island, at Lochranza (☎ 01770 830631; www.hostellingscotland.org.uk). **Bunkhouse:** Kilmory Haven Bunkhouse (☎ 01770 870345; www.kilmoryhaven.com). There are more hotels, guest houses, B&Bs and self-catering establishments than you could shake a stick at. Call **TIC** (above).

A bicycle may be useful, to make the long approach up Glen Rosa more bearable and a much more feasible day trip from the mainland. Mountain bikes can be taken at least as far as the bridge over the Garbh Allt, though the track is pretty rough and hard going. The corner of the Garbh Allt in Glen Rosa takes exactly the same time to reach by taxi or bike.

*weegies = natives of Scotland's largest conurbation

For those who go to the hills to get away from people, ignore the 'No Camping' notices (erected by the campsite owner!)

BEINN TARSUINN
(TRANSVERSE MOUNTAIN)
The next hill north of Beinn Nuis, connected by a grassy ridge.

NR 962 412 **Alt:** 570m
MEADOW FACE
2hr 15min

An extensive clean 200m high gradually steepening face, which takes its name from the prominent small area of lush grass below the base. Despite its sunny aspect, many of the routes can take a couple of days to dry after heavy rain, due to a fair bit of vegetation on ledges.

Approach: From the small car park by the Glen Rosa campsite, follow the track up the glen for 2.2km to cross the bridge over the Garbh Allt. Follow the path steeply up the right (north) side of the burn. After 600m, take the right path fork and follow it over the first small hill (Cnoc Breac). From the small col beyond, contour into Coire a' Bhradain and continue NW, roughly following the line of the burn into Ealta Choire.
Descent: Down the right side of the cliff, by a well-worn path which runs underneath Consolation Tor then drop back down into the coire.

① The Blinder ★★   140m E1 5b

FA Bill Skidmore & Jim Crawford (2 PA) 21 August 1971;
FFA Jim Perrin & Ian Nightingale August 1976

The striking corner line left of the imposing left edge of the main slabs. The top two pitches call upon old-fashioned chimney techniques and feel as hard as anything on the lower pitches.

1 35m 5b Gain the crack from a small ledge and follow it direct until it is possible to traverse left to a belay on a grass ledge.

2 20m 5a Regain the main corner and cross a small roof then continue past a grass clump to belay in a small recess.

3 30m 5a Climb the groove past a grass section and up to an undercut bulge, which leads to a good stance in a deep chimney.

4 25m 4c Climb the chimney via a chokestone to exit onto the floor of great square recess. It is also possible to avoid the main challenge by climbing a line of flakes (poorly protected) on the wall left of the chimney.

5 30m 4c Enter a tight slot in the right hand corner and struggle to a boulder bulge, moving onto the left wall above this. Continue left up easy groove to a large flake belay at top.

Shauna Clarke on the superb long single pitch Snakes and Ladders (page 20). Photo Dan Moore

2 **Brobdingnag** ★★   203m E2 5c
FA Ian Duckworth, John Fraser & Gordon Smith (4 PA) April 1975

Follows the great left edge of the cliff. Start on a sloping grass ledge to the right of *The Blinder.*

1   **30m 5b** Traverse right to a small corner. Climb the corner and continue straight up to a large grass ledge.

2   **18m 5b** Climb a grassy groove to an often wet overhang, turn it on the left and belay below twin cracks.

3   **18m 5a** Climb the cracks to an overhang and step right to a shelf below a shallow chimney. Climb the chimney to a belay below a thin crack in wall.

4   **45m 5b** Climb the crack and swing left to gain a ramp. Move right then left on good holds and follow a corner to belay below a small cave.

5   **12m** Climb up left easily through a remarkable rock arch to belay in a deep hole.

6   **20m 5a** Climb a loose slab to a ledge then directly up a good jamming crack and exit right to a large stance.

7   **15m 5a** Follow the left-hand crack and transfer to the right-hand crack to pull over a jammed block

8   **45m 4c** Climb a corner and slabs to finish.

3 **Blundecral** ★★   115m E3 5c
FA Graham Little & Kev Howett 5 August 1995

This varied and interesting route climbs a line on the wall between *Brobdingnag* and *Brachistochrone* taking the obvious break through the band of overhangs at the end of the long roof running left from *Brachistochrone.* Start on a vegetated ledge at a bay to the left of the first chimney of *Brachistochrone* gained by scrambling up the groove to below the chimney then traversing left.

1   **25m 4c** Climb a flake then move right to an obvious groove (which runs parallel to the *Brachistochrone* chimney). Ascend the groove or wall to right then move left to belay at a pointed turf ledge.

2   **25m 5c** Follow the line of a thin diagonal crack up and left to a left-trending ramp which leads to the base of a right-facing corner. Climb this then step left to grasp a huge block/flake. From its top make a difficult step right to gain a rock ramp and belay.

3   **15m 5c** Climb the diagonal undercling to reach a hidden left-trending groove. Ascend this for 3m then traverse back right across the wall to gain the obvious thin rock ramp. Move right to belay at a small turf ledge. A spectacular pitch.

4   **25m 5c** Climb the fine diagonal rock ramp above to step left onto a continuation ramp. Go up this to a knobbly vein on the wall above. Pull up on to a shelf and move left up this to gain the obvious flake crack, which is climbed to a ledge above. Belay on the right. A bold pitch.

5   **25m 4c** (for the jump!) Walk right along the ledge until a mauvais pas is reached. Jump down on to a grass ledge and grab an enormous flake. Ascend this then blocks to belay on *Brachistochrone* (at the end of the difficult climbing on this route). Scramble up a grassy groove then traverse off right to clear the crag.

3a **Blundecral True Finish** ★★   30m E5 6a
FA Robin McAllister & Dave McGimpsey (headpointed) 4 August 1996

A significantly harder finish, bold and poorly protected on the crux traverse. From the belay at the top of pitch 4 move left and climb a steep slab on tiny holds to a wide horizontal break. Move left along this break until a step up onto a narrow sloping ramp can be made. Move left again (joining the *Blunderbuss Finish*) to reach and climb a flake crack leading to a narrow chimney. Ascend the chimney taking either fork to reach a wide ledge. Belay well back on a large spike. Either continue up the *Blunderbuss Finish* or traverse right to quit the crag.

3b **The Blunderbuss Finish** ★★   215m HVS 5a
FA Andrew Fraser & Robin McAllister 31 August 1995

A finely situated left finish, extending the route to the top of the crag and missing out the crux section of the original route.

1-3   **65m** Climb the first three pitches of *Blundecral.*

4   **25m 5a** Climb the diagonal rock ramp, as for *Blundecral,* step left onto a continuation ramp and follow it for 7m to take a spectacularly-situated belay on the ramp.

5   **25m 4b** Continue up the ramp for 3m then move up

onto a higher ramp. Follow this leftwards until it leads to a wide grassy fault (*Brobdingnag*). Climb this to a cave belay beneath an arch at 7m.

6 25m 5a Quit *Brobdingnag* by jumping onto a grass ledge to the right of the belay. Follow this rightwards and down to its end then take the flake crack leading up to a narrow chimney, taking either fork to gain a wide ledge.

7 50m 4c Climb a slab at the extreme left of the upper face to its apex, move right then continue to climb the left-trending flake cracks above.

8 25m Pleasant climbing up walls and cracks leads to the top.

4 Brachistochrone ★★   240m E1 5b
FA Mike Galbraith & Alasdair 'Bugs' McKeith 18 September 1966;
FFA M.Lynch & Ed Cleasby 26 May 1974

"... that thought-provoking slit down the full height of the Meadow Face. To then hear it described as an 'interesting enough struggle' can make a man suspect the approach of old age or a leg-pull." – Bill Skidmore, 1972 SMCJ
The left of the two long parallel cracks with 'some remarkable situations' in the lower section. Outstanding climbing if climbed to the terrace and walk-off. Scramble up a groove to the base of the chimney.

1 45m 4c Climb the chimney then a flake overhang and the twin cracks above to a belay ledge beneath a huge roof.

2 15m 5b Climb twin overhanging cracks through the roof to a block belay in a chimney.

3 28m 4c Traverse left up a layback shelf then back right to a ledge. Climb a crack then a further crack past a chokestone to a ledge.

4 17m 5a Climb overhanging crack to recess. Swing down left to a steep slab and go up to a grass ledge and block belay. Walk off here, or:

5–7 90m Continue by grassy grooves, slabs, corners and flakes, crossing *Meadow Slabs* to the continuation of the crack in the upper slabs.

8 45m 4b Finish up the crack over three overlaps, turning the middle one on the right.

5 Bogle ★   220m E2 5c
FA Ian Rowe & Ian Dundas 22 July 1967;
FFA Jim Perrin & Ian Nightingale August 1976;
pitch 5 Varn. George Szuca & Andy Wren 27 May 1991

The right of the two long parallel crack lines splitting the front of the face with a disproportionately hard 6m of ferocious fist jamming on the crux pitch. Scramble up to belay at the base of the crack.

1+2 60m 4c Follow the general line of the crack to belay on a slab beneath the large overhang.

3 15m 5c Move up onto a pedestal then follow the 'strenuous, slimy' crux crack for 6m then traverse 3m right and up a parallel crack to belay on the ledge above.

4 25m 5b Continue up the crack.

5a 15m 4c Climb the slimy cave by a contorted through route then move left up a slab to belay 6m above.

5b 15m Ungradeable (XS). If, as reported, the through route is still blocked by turf and rubble, it can be bypassed by an intimidating stomach traverse left along a shallow wet sloping shelf.

6 30m Scramble up to the foot of the *Meadow Slabs*.

7 50m 4b Follow the continuation of the crack to a large summit block.

8 25m 4c Finish up two left cracks.

6 The Rake Direct ★   210m E2 5c
FA Bill Skidmore & Bob Richardson (2 PA) 18 August 1962;
Direct Start: George McEwan & Alastair Walker 21 May 1988

At the bottom right of the main face is a huge slab, bounded on the right with a large groove line. Start at the base of the slab.

1+2 90m 4b Climb directly up the centre of the slab to belay on a large grass ledge, beneath a slabby corner containing two small overlaps.

3 35m 5a Climb up and left through the overlaps then traverse left to another slabby corner. Climb this then traverse left to belay on top of a huge flake.

4 15m 5b Gain and climb the overhanging crack above to belay below a corner.

5 30m 5c Go up then round right into a corner, climb this with difficulty (often wet) to gain an obvious bay on the right. Follow a crack in the slab on the left and rock ledges to belay on a grass ledge.

6 **10m 4b** Climb the corner above to a grass ledge and PB in overhanging cave.

7 **15m** As for pitch 5 of *Bogle*, or from the top of pitch 5, traverse left for 7m, moving up onto a higher ledge to join *Brachistochrone* at the start of pitch 4.

8 **15m** Scramble up to the terrace.

7 Gulliver's Travels ★★   300m E2 5c

FA Andrew Fraser & Robin McAllister 20 September 1995

A fine adventure with spectacular positions making up for the relative shortage of new ground. It follows a natural rising traverse, using the leftward trend and ledged nature of the cliff. The longest extreme on the island. Follow the first 5 pitches of *The Rake Direct* to belay in *Bogle*.

6 **30m 5a** Move up to a ledge on the left and follow this left to the crack of *Brachistochrone*. Descend this for 3m past a chokestone then follow a ramp up left past an awkward break to take a spectacularly-situated belay on the ramp 7m beyond the break.

7 **30m 4b** Continue up the ramp for 3m then move up onto a higher ramp. Follow this leftwards until it leads to a wide grassy fault (*Brobdingnag*). Climb this to a cave beneath an arch at 7m then traverse left to a block belay overlooking *The Blinder*.

8 **40m 5c** A sensational and intimidating traverse leads to *The Blinder* (a rope looped over the top of the arch lowers the fall potential). Continue to a chimney which is climbed via a chokestone to exit on to the floor of a great square recess.

9 **30m XS** 4c Finish as for the last pitch of *The Blinder*.

8 The Curver ★★   130m E1 5b

FA Rab Carrington & Ian Fulton 14 June 1968

Good delicate slab climbing, though slow to dry. It follows a series of grooves up through the curving overlaps, above and roughly parallel to the original grassy line of *The Rake*. Start near two overlapping slabs, close to the end of pitch 1 of *Meadow Slabs*.

1 **25m** Climb a slab and groove to belay above a damp corner.

2 **25m** Climb the groove on the right for 9m, gain a higher slab on the right then follow the slab corner for 12m. Move out right to reach a small stance and belay.

3 **40m** Move leftwards to gain the base of a narrow slab gangway, and continue to a deep flake crack. Climb this, or the layback shelf on the right to a block belay on a large ledge.

4 **40m** Pull round the edge on the right on quartz holds then follow a line of holds across slabs to finish up a corner.

9 Snakes and Ladders ★★★   50m E2 5b

FA Colin Moody & Billy Hood 16 May 1998

A slab of perfect rock up right of *The Curver*. Easily seen from the slopes on the right, it contains a vertical heather crack on the right side. The pitch was gained by abseil from chokestones at the top. Traditionalists can alternatively gain the base by two pitches of grassy grooves. Start up a bulge on the left side and follow the crack to a downward-pointing flake. Traverse right for 5m then climb straight up on seams and pocks to the left of the heather crack. A F #2 protects the top section, a R #6 can be threaded just below it.

10 Meadow Slabs ★   60m Difficult

FA Geoffrey Curtis, Ken Moneypenny & E.Morrison 20 August 1944

Pleasant climbing up the left edge of the upper slabs. Start at the left end of The Terrace, beneath a chokestoned crack formed by a slab and the wall above. Climb the crack then descend obliquely left on turf ledges towards the base of the blunt skyline ridge. Cross a slab, beyond which will be seen the foot of a 3m chimney which marks the start of the climb proper.

1 **10m** Climb the little chimney which narrows to a crack then gives access to the crest.

2 **20m** Climb a short undercut chimney on the left then easier ground to a belay.

3 **30m** Go slightly right then slant left up a grassy crack to reach a branching chimney. Climb the left wall then above by a slanting crack. Gain a short arête to finish.

GOATFELL
(HILL OF THE WIND)

The highest, most prominent and by far the most popular hill on the island.

SOUTH SLABS   1hr 45min

NR 986 409 **Alt:** 420m

A prominent area of clean compact slabs high on the west flank of the hill. The rock is more continuous and cleaner than the main Rosa Slabs, offering far and away the best middle grade friction routes on the island. Generally the rock is climbable anywhere, though protection is often noticeably lacking.

Approach: From the campsite continue up the forestry track and the continuation path to the footbridge over the Garbh Allt after 2km. Cross the bridge and continue up the path close to the west bank of Glenrosa Water for a further 2km then cross the stream and slog steeply up the hillside on the right (300m of ascent) to the base of the slabs.

Descent: By a vague path down the right (south) edge of the slabs.

1 Trundle ★★   130m Severe 4a

FA W.Bailey & R.White 1 April 1964

Good climbing up the left side of the slabs. Start 12m left of the left-trending broken fault.

1 **30m 4a** Move up into a scoop, step right then climb up into narrow grooves and follow these up left to a prominent flake.

2 **30m 4a** Move up a short way then traverse right to a thread runner on an edge. Continue slightly leftwards by a fading layback crack then over bulges and

slabs to belay on a heather ledge.

3 35m 4a Traverse up rightwards across the fault to a mossy corner. Climb this then move right to climb a crack up the slab to a block belay.

4 35m 4a Traverse right along ledges for 5m then up rightwards past a large knob to finish.

2 Dogleg ★★   115m VS 4b

FA Jim Brumfitt & Alasdair 'Bugs' McKeith 21 June 1964

Similarly bold climbing to the neighbouring *Blank*. Start at the base of the left-trending broken fault.

1 50m 4b Climb the slab to a break in a thin overlap at 6m, surmount this then up a flake edge. Trend slightly rightwards up undulating slabs to gain a short flake groove, up this then traverse left to a heather groove and up this, moving up right to belay at base of a groove.

2 50m 4b Climb the easy flake crack in the groove then traverse left until a line of pockets leads up the fine slab direct into the curving right-facing corner. Step left and follow the rib left of the corner, leading

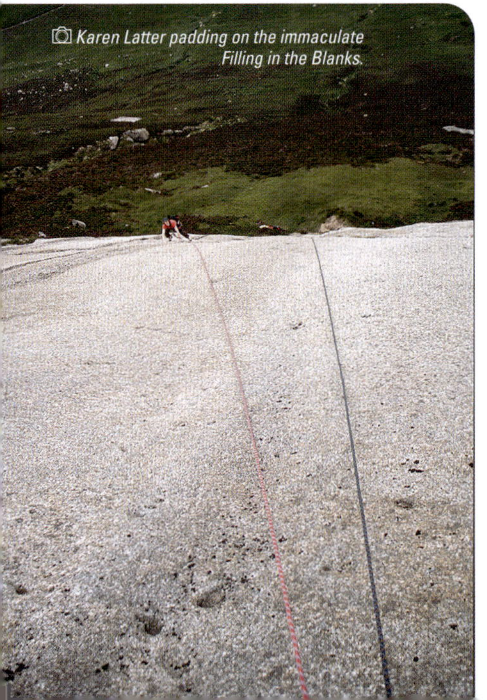

to easier upper slabs.

3 15m Scramble to finish.

3 Filling in the Blanks ★★★   120m HVS 4c

FA p1 Andy Nisbet 23 May 2004;
p2 Gary & Karen Latter 9 July 2005

Superb bold climbing plugging the obvious gap up the centre of the slabs.

1 50m 4b Climb the first pitch of *Blank* but instead of moving out right to the belay follow the lower dyke leftwards then follow a line of pockets trending leftwards up the slab to belay at the base of the easy flake groove on *Blank*.

2 60m 4c Climb easily to the break, step up right onto the slab above and move up leftwards to boldly climb the slightly protruding rib above. A superb bold pitch.

3 10m Scramble to finish.

4 Blank ★★★   110m VS 4b

FA Alasdair 'Bugs' McKeith & B.Kennelly 13 September 1963

Great climbing, taking a line near the centre of the slabs. Start 5m right of the left-trending broken fault.

1 30m 4b Climb the slab boldly, heading for a line of scoops and flakes then follow these trending slightly right to belay on a small gravel ledge in a heather groove (common to *Blankist*).

2 45m 4b Step left, move up then follow a wide vein running left across the slab until 5m short of a heather groove. Follow a line of pockets up a shallow scoop to gain and climb a flake groove. At its top traverse 3m right, surmount an overlap then climb straight up on pockets to a ledge.

3 35m 4b Climb directly up the centre of the big slab. A superb unprotected pitch.

5 Blankist ★★★★   110m HVS 4c

FA p1 Dave Bathgate & Jim Renny 21 June 1964;
Graham Little & Kev Howett 20 August 1995

The best route on the slabs taking a direct line up the right on perfect rock. Start 10m right of the left-trending broken fault.

1 **30m 4c** Climb straight up the holdless slab immediately right of a black streak to reach flakes. Ascend these to belay on a small gravel ledge in a heather groove on the right.

2 **25m 4c** Move out left and climb slightly leftwards up a line of perfect pockets moving left to gain an obvious long thin downward-pointing flake. Thin moves above this lead to a fine flake belay. A direct line right of this can also be taken to belay at a large flake crack further right.

3 **45m 4b** Climb straight up a bare slab to move right to gain and follow an obvious rib (overlooking a long corner to the right) then up easier-angled slabs to step right to belay at the base of a short banana-shaped groove. A memorable pitch – very bold, but steady.

4 **10m** Scramble up heather to finish.

A'CHIR (THE COMB)

This is the narrow rocky ridge (aligned south-north) midway between Beinn Tarsuinn and Cir Mhor.

1 A'Chir Ridge Traverse ★★★   1.5km Moderate

FA J.H.Gibson, T.F.S.Campbell, Willie Douglas, H.Fleming R.A.Robertson & Dr Leith 30 January 1892

An excellent ridge traverse with stunning views, usually traversed from south to north, saving the most difficult section to last. The rock is generally superb throughout with much variation possible, with very straightforward paths on the west flank avoiding most of the interesting sections. The best line keeps to the crest of the ridge as much as possible and is obvious and polished at a number of places where the route drops down into a succession of small cols on the ridge. About 300 metres north of the summit lies *'Le Mauvais Pas'*. The northern extremity of the ridge terminates in a long vertical drop down to a col. About 20m south of this, descend a short wall on the east flank (arrow scratched In rock above) to gain a fine incut dipping ledge (quite exposed) dropping down north towards the col. This peters out near the base, with the final 2 metre section to a good flat landing presenting the only real difficulties on the ridge. Either descend steeply (or jump!) to gain the col. The route continues much more easily to the Cir Mhor col, affording grand views of the Rosa Pinnacle on Cir Mhor. Continuing over Cir Mhor and Goatfell gives a fantastic day's ridge walking.

CIR MHOR (THE GREAT COMB)  2hr 30min

This distinctive pointed peak dominates the skyline at the head of the long boggy Glen Rosa and is far and away the most spectacular and distinctive of all the Arran hills. A selection of fluted buttresses lie on the south side of the mountain with the superb slabs and steep walls of the Rosa Pinnacle taking pride of place in the centre of the face. Despite the longest approach on the island, it is justifiably the most popular – well worth the effort!

Approaches: (A) Up the forestry track from the campsite and up the path to the footbridge over the Garbh Allt after 2km. Cross the bridge and continue close to the west bank of Glenrosa Water for a further 3km to where the stream bends round to the left and splits. Stay on the less obvious path which continues up the left bank for 350m. The way steepens on the path up the hillside into Fionn Choire. The cliff lies about 300m north-east from the path (half an hour from fork in path).

(B) A slightly shorter approach is possible from Glen Sannox. Park on the east side of the A841 at Sannox Bay. Follow the track west up Glen Sannox past the beacons to cross the stream after a kilometre. Continue up a boggy path close to the right (north) bank of the burn for about 3km to re-cross the burn. A steep path leads up the hillside (200m ascent) to The Saddle. Drop down right slightly then contour round rough ground for almost 1km to the base of the crags.

Descent: Down the right (west) side of Caliban's Buttress. Alternatively, for routes on the right side of the cliffs, descend the broad Sub Rosa Gully between the Rosa Pinnacle and Prospero Buttress, avoiding a short area of slabs low down by either of two forks, easiest by the left (east) fork.

ROSA PINNACLE

NR 972 428 **Alt:** 550m

The showpiece cliff on the island, with a range of excellent routes of almost every grade. Many of the routes criss-cross in the middle and upper sections of the cliff, and there is much scope for interchanging pitches.

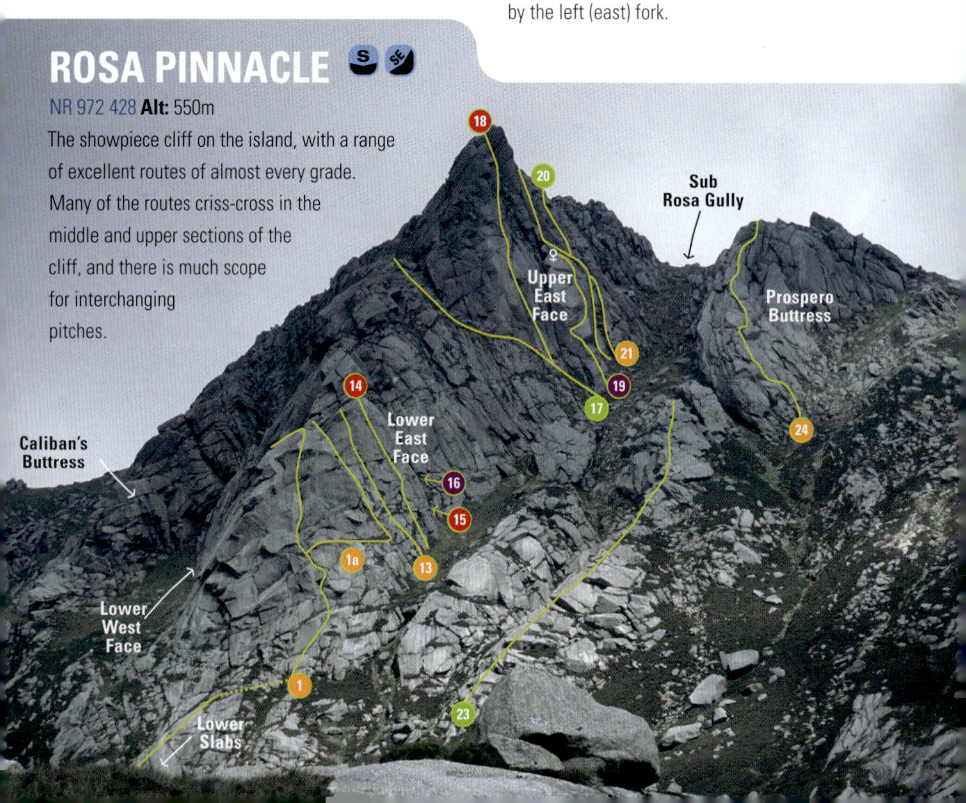

SOUTH FACE, ROSA PINNACLE

① South Ridge Direct ★★★★   330m VS 5a

FA J.F. 'Hamish' Hamilton & David Paterson September 1941

A very popular classic, with three harder pitches – the 'S', 'Y' and 'Layback' cracks. The ridge forms a steep nose rising out of a mass of vegetated slabs on the lower face. The first pitch of technical interest is the distinctive S-crack splitting the face of the nose. Start by scrambling up easy grass ledges right of the toe of the buttress.

1 **44m 4a** Follow cracked slabs on the left, leading to ledges on the right. Continue to a right-facing corner. Climb the layback flake on the right then move back left to belay at the top of the corner.

2 **6m** Step left and up easy corner to base of S-crack.

3 **25m 4c** The S-crack leads to a large block belay on a wide shelf. A fine pitch.

4 **10m 5a** The overhanging Y-cracks now loom above. Climb these strenuously (crux) to a belay just above.

5 **45m** Move up to a block-strewn terrace, step round a rib on the left then follow an easy diagonal traverse to belay in the far corner beneath the Layback Crack.

6 **25m 4b** Follow the Layback Crack until a vein runs out right. Follow it across the slab then climb a short easy flake crack to gain a large platform. The corner crack can also be followed in its entirety – *Lovat's Variation*, VS 4b.

7 **25m** Above lies the obvious Three-Tier Chimney on the left side of the crest of the ridge. Follow the chimney to gain the crest of the ridge.

8+9 **60m** Follow the crest of the ridge in two easy pitches to gain The Terrace beneath the Upper Pinnacle.

10 **40m** Start in a recess beneath the rightmost of two slabby right-facing corners. Climb either the corner or the slab on the right to gain a fault near the ridge crest. Follow this easily then the blocky chimney above to a long grass ledge.

11 **35m** Walk along the grass ledge to a little chimney. Climb the slab on its right to the ridge crest. Move round right onto the east face then make an exposed traverse right into an open flake-corner. Climb this to belay in a crevasse.

12 **15m** Continue easily up the crest to the top of the Upper Pinnacle.

①a South Ridge Original ★★   60m VS 4b

FA James Ramsay & party 1935

A worthy alternative, especially if there is a bottleneck of climbers on the S-crack, or to avoid the harder Y-crack.

1 **30m** Traverse right from the foot of the S-crack below the steep wall then climb up into a turfy recess beneath the large undercut corner.

2 **30m 4b** Climb out of the recess into the corner and climb it; then the cracks in the slabby left wall to the block-strewn terrace.

② Incus ★★   55m E6 6b

FA Gary Latter 27 August 2001

A series of pitches up the wall left of *South Ridge Direct*, the main centre of interest being the thin flange 15m left of the S-crack. Start on the front face just round right of *Anvil Recess Start* beneath a short groove.

1 **15m 5a** Climb the short undercut groove to layback onto a slab above then more easily up this. Climb easily right across a slab then up past a large flake to belay on a grass clod directly beneath the flange.

1a **30m 5a** A longer better pitch, occasionally climbed as an approach to the S-Crack. Start further right, beneath a left-facing groove. Enter and climb the wide slot in the groove to easier ground then continue by a steep hand crack and cracks just to its left to belay on grass clod above large flake.

2 **10m 6b** Move up the corner and step left across the slab to the flange. Climb this with interest (sustained and well protected) then use a one-finger pocket in the wall above to hand traverse left on a micro-granite vein to superb finishing holds. Belay on the shelf above.

3 **15m 5c** Walk right along the shelf to belay at the base of the curving corner. Climb this, moving up right at its top to a huge thread. Abseil off. Alternatively, climb up leftwards from the top of the corner to gain the block-strewn terrace on *South Ridge Direct*.

LOWER WEST FACE
The superb, generally immaculate slabs extending
leftwards up the slope from the toe of the South Face.
9
11
10
6a
4a
12
10
8
7
6
3
4
1
3a

3a Anvil Recess Start ★★★   40m HVS 5a

FA Bill Skidmore, Jim Crawford & John Madden 16 July 1964

Better than the parent route. Start at a block at the base of a left-slanting corner crack, directly beneath the huge inverted V-shaped recess on the front face of the ridge.

1 **15m 4b** Climb the corner crack to a spike belay.

2 **25m 5a** Pull round left to gain the slab above, and move up until it is possible to move right to the bottom of a large flake lodged under the right wall of the recess. Climb the flake and step left to the opposite wall then strenuously gain the slab above. Continue more easily by a crack on the right and over a grass tuft then move left to reach the cave belay on *Anvil*.

3 Anvil ★★   55m HVS 5a

FA Davy McKelvie & Bob Richardson 6 August 1960

Start at a wide crack a few metres right of *Hammer*.

1 **30m 5a** Follow the crack to just left of the corner of a small overlap. Descend 3m then traverse delicately right across the steep slab to gain holds on the slab edge which lead up through the overlap to a sloping platform. Continue up the narrow groove above to belay on a grass ledge.

2 **15m 4b** Move up grassy grooves left to belay in a cave.

3 **10m 5a** Break out right then follow an easy shelf to gain *South Ridge Direct*.

3b Variation Finish ★★   25m VS 4c

FA Bob Richardson & John Madden 1965

From the cave, traverse left and climb a rounded bulging groove to beneath an overhang. Right up a slab then climb a groove and further slabs to finish on *South Ridge Direct*.

4 Insertion ★★   105m E3 5c

FA Rab Carrington & Ian Fulton 15 June 1969

A very bold slab climb of its day. Start 1m right of *Anvil*.

1 **45m 5c** Climb straight up the steep slab to an overlap, cross this, continue up a further steep slab, trending right to belay a short way beneath the cave on *Anvil*.

2 **30m 4c** Traverse up and left beneath the large overhang then ascend the bulging rounded groove (as for *Anvil Variation Finish*) to beneath the overhang. Pull out right onto a slab then move easily left, crossing *South Ridge Direct* to belay below a steep slab.

3 **30m 4c** Move up left then traverse horizontally right below an overhanging wall. Continue right round an edge to a crack which leads to the large platform beneath the Three-Tier Chimney on *South Ridge Direct*.

4a Insertion Direct ★★   95m E5 5c

FA Mark Charlton & Kev Howett 6 June 1986

A very thin bold first pitch climbing direct up the slab. Start as for *Insertion*.

1 **50m 5c** Climb as for the normal route to over the overlap then straight up into a scoop in the centre of the slab. Ascend this then teeter rightwards to take a hanging belay under the big overhang at the base of the *"bulging rounded groove"* of *Anvil Variation Finish*.

2 **25m 4c** Gain and climb the groove to below a roof. Move out right up a slab to belay at its top below a bulging wall with a large pocket.

3 **20m 5c** Pull past the pocket and smaller ones above. Swing round the edge then climb to reach easier ground on *South Ridge Direct*.

5 Forge ★★   100m E5 5c

FA Jules Lines (on-sight solo) 28 July 2001

Bold sustained climbing.

1 **50m 5c** Start as for *Anvil*, but where that route goes right continue up the groove and flakes. When it becomes blank, move right onto the slab and up to pockets. Step right and continue up into a scoop with a tiny flake. Exit the scoop on the right and move up to a shallow-angled slab. Just before a bulging section, traverse down and left to reach *Hammer*.

2 **25m 5b** Follow a line of pockets up and right until they disappear then go straight up the slab to a horizontal flake-line. Move right to easy ground to belay below the slabby arête on the skyline.

3 25m 5c Using a short crack/pod on the left of the arête, make a thin move to gain the arête. Climb the arête on the right for 3m (easy) then make a hard move left onto a boss/smear and further hard moves up to small holds on the slab. Stretch right to good pockets in the very crest and climb direct to join *South Ridge Direct*.

6 Hammer ★★ 85m VS 4c

FA Donny Cameron & Dick Sim 6 August 1960

Good climbing, starting up the tapering slab just right of *West Flank Route*.

1 25m 4b Climb up onto the slab then move left to a crack and follow this and the small continuation groove to belay midway up the two-tiered chimney on *West Flank Route*.

2 20m 4c Descend a short way then traverse right and follow a good tufty crack to a large belay ledge.

3 40m 4b Follow the crack in the slab to reach a vein of micro-granite near the top. Delicately follow the vein right to a crack which leads to the base of the Layback Crack on *South Ridge Direct*.

6a Pocket Slab Variation ★★ 35m VS 4b

FA Colin Moody & Cynthia Grindley 9 July 2005

Apparently much better than the original. Instead of following the vein rightwards, follow a line of pockets up slightly left then back right. Finish straight up to join *South Ridge Direct*.

7 West Flank Route ★★★★ 155m E1 5b

FA Jim Crawford, John Madden, Bob Richardson & Bill Skidmore 3 August 1963

A classic route, taking the line of left-trending open chimneys and cracks. Low in the grade. Start beneath the prominent open chimney in the centre of the main slab.

1 35m 5a Struggle up the two-tiered chimney-like feature, (possible belay at 20m) to a ledge beneath a small overlap in the slab.

2 25m 4b Continue up the long left-slanting diagonal crack in the slab above to the large overlap. Move right round the corner to belay in a niche.

3 30m 5b Step up and left round an edge into a groove and crack, and follow these to gain a small spike and flake by a rounded layback. Take the right branch of the crack until a move left can be made across the slab. Climb to a horizontal crack then traverse left across the slab (crux) and climb up to a thread belay (on *Sou'wester Slabs*) at the top of a grassy groove.

4 35m 4c Ascend the corner above the belay, and step left to a ledge on the slab edge. Climb the edge to a large overhang, hand traverse 2m left to gain a wall and continue left round a bulge. Climb up easier ground to a belay.

5 30m Climb broken slabs to a short wall then a recess, to finish up a crack leading to The Terrace.

The two prominent chimney systems are *The Sickle* ★ HVS 4c, 5a, 4c and *The Iron Fist* ★ E1 4c, 5b respectively.

8 Vanishing Point ★★★ 100m E4 6a

FA Craig Macadam & Simon Steer 9 May 1985

Thin bold slab climbing. Start at the same point as *The Sickle*, at an easy flake system.

1 30m 6a Up the easy flake for 10m then up to flakes in the slab above. Traverse right with difficulty to gain the prominent flake crack. Continue up this with increasing interest, passing two prominent pockets to belay.

2 45m 4c Step right onto the slab and climb direct up this heading leftwards into the lower continuation of the *South Ridge* layback. Up this to belay at the base of the layback pitch.

3 25m 6a Up the corner for 4m then foot traverse left to a shallow groove in the arête. Up this then the thin bold slab above to finish on *Sou'wester Slabs*.

Starting up the crag classic
Sou'wester Slabs (page 31)

Hugh Simons padding up the committing The Rosetta Stone (page 33). Photo David Thompson

Karen Latter on the fine S-Crack on South Ridge Direct, (page 25) with the crux Y-cracks looming above.

Colin Dyer making short work of the crux Y-cracks on South Ridge Direct (page 25). Photo Billy Hood

9 Fourth Wall ★   125m HS 4b

FA Gordon Townend & Ken Moneypenny 5 August 1945

Good open climbing, with only the last pitch being of any difficulty. Start at the base of a long left-slanting groove, (level with the base of the vertical east wall on Caliban's Buttress).

1 **50m** Follow the groove then the rib on the left to a block belay just above the level of the right traverse on *Sou'wester Slabs*.

2 **50m** Continue direct by cracks and grooves to a huge plinth leaning against the face.

3 **25m 4b** Climb the initial section of the steep narrow chimney to a small ledge then traverse right, descending slightly for 6m until it is possible to climb the delicate and exposed wall. At the top a final awkward step right into an easy groove leads to The Terrace.

10 Sou'wester Slabs ★★★★   100m Very Difficult

FA Geoffrey Curtis, M.Hawkins, H.Hore & Gordon Townend 3 September 1944

The classic lower grade route of the island, consequently very popular.

1 **48m** Follow *Fourth Wall* to an obvious easy traverse line out right to belay at the base of a shallow left-facing groove.

2 **45m** Follow the groove to a large flake beneath twin parallel cracks. Continue up the cracks to the right edge of the slab. Step down to the lower slab on the right, and climb diagonally rightwards up a groove to belay at a PR under the great overhang.

3 **7m** Traverse easily right to join *South Ridge Direct* beneath the Three-Tier Chimney.

Either finish up the last 6 pitches of *South Ridge Direct* (175m), or continue traversing round right onto the east face then up on grassy ledges. Scramble down to the top of the vegetated line of *Old East* (Mod) which leads down diagonally rightwards into Sub Rosa Gully.

11 Arctic Way ★★   65m HVS 5a

FA Billy Hood & Colin Moody 19 June 1982

Good sustained climbing. Follow the lower section of *Fourth Wall* to belay at the base of the prominent left-facing corner crack, below and right of the huge plinth on *Fourth Wall*.

1 **40m 5a** Climb the corner then the continuation crack, crossing two bulges to belay below an overlap.

2 **25m 5a** Move left to a crack through the overlap and go left again to a bulge. Cross the bulge then left up the slab for just over a metre. Traverse right to climb a thin crack then trend left to a block belay at the top.

CALIBAN'S BUTTRESS  

NR 972 429 **Alt:** 630m

The buttress immediately left of the Rosa Pinnacle, higher up the slope, level with the base of *Sou'wester Slabs*.

12 Caliban's Creep ★★★   150m Very Difficult

FA Geoffrey Curtis & Gordon Townend 25 July 1943

Previously a renowned sandbag and *"the hardest Diff in the world"*, it may still prove problematic to those of ample girth. Start on the right of the toe of the buttress, which forms a square-cut overhang.

1 **25m** Climb diagonally leftwards up slabs to belay amongst a pile of boulders, beneath an overhang.

2 **25m** Traverse right across the wall on good holds, step down and continue round the edge then up an easy chimney to belay on the right on an area of slabs.

3 **15m** Continue up the slabby ridge crest to belay beneath the right edge of a short vertical wall.

4 **20m** *"What have we here? A man or a fish?"* Crawl through the narrow rock tunnel on the right (The Creep) then follow a narrowing ledge round right onto the east face and into a deep chimney on the right. Climb up this to belay on the floor of the great fissure above.

5+6 **65m** Escape from the fissure then about 60m of *"pleasant, carefree climbing near the right edge finishes the climb"*.

LOWER EAST FACE

The east flank of the South Ridge, between the corner pitch of *South Ridge Original* and *Old East*.
Approach: Up Sub Rosa Gully.

13 **The Crack** ★   **45m HVS 5a**

FA Ian Cranston, John Earl & Bob Hutchinson 29 May 1973

A fine looking flake crack forms a slim right-facing groove right of the corner pitch of *South Ridge Original*. Start at two parallel cracks in a protruding buttress of crumbly rock in Sub Rosa Gully.

1 **15m 5a** Follow the crumbly parallel cracks then the groove to belay on a grass ledge.

2 **30m 5a** Climb the groove on the left until level with the flake crack. Traverse left to gain the excellent flake crack, which leads to the block-strewn terrace on *South Ridge Direct*.

The following three routes lie right of *The Crack*. Abseil from in situ thread at the top. Alternatively, an easy traverse left leads to the belay at the top of the Y-crack on *South Ridge Direct*, where an abseil down *South Ridge Original* is possible.

14 **Ariel's Arête** ★★★   **40m E1 5b**

FA Martin Reynard & Dave Musgrove 6 May 1995

A superb pitch following flakes and cracks up the striking edge. Start at a messy bay just right of *The Crack*. Scramble up the bay then move left to a ledge. Climb an awkward little wall then pull into the flake system on the right. Follow this then an amazing thin crack up the edge to gain a rock ledge. Continue up the slab past large pockets to belay at a horizontal fault.

15 **Fox Among the Chickens** ★★★   **40m E4 6a**

FA Martin Reynard, Dave Musgrove & Pete Benson 6 May 1995

Right of *Ariel's Arête* is a prominent wall with an obvious pocket high up and three short flake-cracks low down. Start below the centre of the wall. Climb an undercut corner then move left and up to climb the right-most of twin flake-cracks. Gain and follow the wider right-slanting crack to its top. Make a move right then climb the wall directly, finishing up a short groove.

16 **Hardland** ★★★   **40m E4 6a**

FA Dan Honeymann & Tom De Gay (headpointed) 26 August 2001

A spectacular left finish to *Fox Among the Chickens*. From near the top of the right-slanting crack, break out left with difficulty to gain the huge pocket. Follow a line of pockets boldly leftwards with a final difficult move to gain the edge of the slab and the top.

17 **Old East**   **60m Moderate**

FA Gordon Townend & J.Jenkins August 1946

A useful approach or descent, following a vegetated curve cutting up diagonally leftwards below the base of the Upper Pinnacle, linking Sub Rosa Gully with The Terrace.

UPPER EAST FACE

A fine steep buttress rising out of the upper reaches of Sub Rosa Gully.

Approach: Up Sub Rosa Gully, or via a route on either the Lower West or Lower East Faces then descending the diagonal fault of *Old East*.

18 **Skydiver** ★★★   **70m E3 6a**

FA Graham Little & Colin Ritchie (4 PA) 1 August 1981; FFA with direct finish Colin MacLean & Andy Nisbet 7 May 1984

Excellent varied climbing with a short well protected crux. In the centre of the cliff is a slim groove splitting the arête. Scramble up (or down) *Old East* to a belay at the base of the arête.

1 **25m 5b** Climb the groove just left of the arête, moving right round the arête at 10m to a small ledge. Continue up the excellent twin cracks in the groove to a small sloping stance beneath the roof.

2 **45m 6a** Undercut round the roof and make contorted moves (crux) to gain the slim groove. Follow the groove and up past a large detached flake to a crack (possible belay – poor). Up the crack to a horizontal break then right to reach a flake. Climb the flake and continue easily on huge holds over the twin roofs above to a recess. Head rightwards up slabs to finish up a short overhanging corner.

19 The Sleeping Crack ★★★★   55m E7 6b
FA Dave MacLeod & Dan Honeyman 27 August 2001

The attractive intermittent crack in the huge projecting pillar at the right side of the cliff.

Gain the first crack and follow it with increasing difficulty until it is possible to escape left onto the slab and gain a huge thread. Traverse right and boldly climb a sloping shelf to regain the crack-line. Follow this with hard moves (crux), but excellent protection to an undercut flake. Follow this leftwards then up until beneath the final bulge. Climb rightwards up another flake then make a difficult move up left on finger pockets to gain the finishing slab. Pad easily up this to a flake belay on the terrace (large cams useful).

20 Labyrinth ★   120m Very Difficult
FA Geoffrey Curtis & Ken Moneypenny 26 September 1943

A popular old-fashioned classic, with all that that implies, based around the central chimney line. Start just below the bottom right corner of the monolithic wall beneath a rock alleyway or vent.

1 **20m** Enter the vent and ascend it, climbing behind a huge jammed block, to exit left. Continue up to gain a grass platform.

2 **10m** Climb the right-slanting groove to belay on a sloping grass ledge.

3 **15m** Traverse left on the horizontal ledge to gain a small undercut corner. Ascend this to belay in the groove above.

4 **30m** Ascend the chimney above, past a grass platform and two sets of chokestones to belay in 'The Eyrie' – the rocky recess beneath the base of the prow.

5 **35m** Traverse rightwards, gaining a little height, to an easy right-slanting grassy rake. Continue along this to belay at a cluster of blocks.

6 **10m** Drop behind the blocks to a recess overlooking Pinnacle Gully then follow the short steep chimney, finishing up a slab.

20a Labyrinth Direct Finish ★★   40m VS 4b
FA J.Stewart Orr & John MacLaurin May 1951

A much better though noticeably harder finish. From 'The Eyrie', climb easily direct up to the large corner. Climb the corner for 5m then traverse left along two horizontal faults with difficulty to the crest of the final pitch of *South Ridge Direct* and so to the top.

21 Easter Route ★★   95m HVS 4c
FA Ken Barber & Alfred Pigott Easter 1938

One of the hardest routes from the thirties, following the chimney line just up and right of *Labyrinth*. Start beneath the line of the chimney.

1 **15m** Scramble up to a block below the first chimney.

2 **20m** Follow the chimney to a grass ledge with blocks.

3 **10m** Continue up to a grass patch below a steeper chimney.

4 **12m 4c** Climb the steep chimney until a move right can be made onto a turf-capped ledge.

5 **10m 4c** Two shallow horizontal grooves cross the steep wall above. Using these, go left to the edge of the wall then climb direct to a grass ledge on the left.

6 **8m** Traverse left to the base of the steep curving crack on the right side of The Prow.

7 **20m 5a** Climb the crack with difficulty to a ledge then continue up the final chimney crack.

PINNACLE GULLY AREA

Pinnacle Gully bounds the upper side of the Upper East Face.

22 The Rosetta Stone ★★   12m E1 5a
FA James MacLay & William Naismith (top-rope) 1894;
FPA Robin Smith (solo) May 1957

This huge block lies at the top of *Pinnacle Gully*, on the west side of the Pinnacle summit. A fine friction problem, following the left edge of its west side. Descend the same line on all fives!

PROSPERO BUTTRESS

The discontinuous rib bounding the right edge of Sub Rosa Gully, steepening in its upper section.

23 Prospero's Prelude ★　　　　**120m Moderate**

FA Geoffrey Curtis & Ken Moneypenny 26 September 1943

A pleasant useful approach, either to *Prospero's Peril* or the Upper East Face. From near the watercourse in the right (east) branch, gain and follow slabby grooves in the crest.

24 Prospero's Peril ★★　　　　**122m HS 4b**

FA Gordon Townend & Geoffrey Curtis 25 July 1943

Varied and interesting climbing on good rock. Start right of the toe of the upper buttress.

1 **12m 4b** Bear slightly right and gain a slab. Climb the slab until a traverse right can be made into a crack.

2 **20m** Climb a narrow chimney.

3 **30m** Easy climbing, first up chimneys, and then a broad slab to a poor stance on its edge.

4 **15m** Continue up the slab to belay under a small overhang.

5 **15m** Traverse left for 3m. Climb a knee-width groove and then a slab.

6 **30m** Traverse left up a narrowing slab to a foothold on the vertical wall of the gully. The slab above is attained by two delicate and exposed steps to the right. The stance is 15m higher.

CIOCH NA H-OIGHE

(THE MAIDEN'S BREAST)

The distinctive elegant pointed peak of this magnificent mini mountain presents an exquisite panorama from the superbly positioned summit.

Access: From Brodick, follow the A841 north up the east side of the island for 8 miles/13km to park on the east side of the road at Sannox Bay.

Approach: Follow the track west up Glen Sannox past the beacons, to cross a burn after about 10 minutes (0.8km). Head left (south) by a small path following the general line of the Allt a' Chapuill (burn) up the hillside into Coire na Ciche (also known as The Devil's Punchbowl).

1 Midnight Ridge Direct ★　　　　**130m VS 4c**

FA Geoffrey Curtis & Gordon Townend 6 May 1944

Good airy climbing; the only worthwhile climbing on the mountain outwith The Bastion. The route follows the ridge formed by the south-east and east faces (directly below the top of Ledge 1 when viewed from the lip of the corrie). Start just right of a prominent gravelly fan on the right of the corrie lip.

1 **50m** Climb up to the base of the ridge by awkward vegetatious grooves, moving left to gain a thread belay in a little recess 4m right of the arête.

2 **15m 4c** Ascend the short strenuous overhanging scoop to a sloping ledge (crux) then climb a short wall to a grass ledge. Swing steeply left onto the undercut slab above the left end of the ledge and step left to a spike belay on the arête.

3 **25m 4b** Move up and left onto the vertical wall left of the belay and up this on good holds to a small ledge. Continue up a small corner to a grass ledge then over some ledges to climb a short left-facing layback groove leading to a large block belay.

4 **40m** Finish up the much easier pleasant narrow ridge.

Photo Andrew Fraser.

NS 000 439 **Alt:** 500m

THE BASTION E 1hr 30min

An isolated steep clean wall high up in the coire, harbour-ing a concentration of steep spectacular extremes.

Approach: A series of rightward slanting rakes, numbered 1–5 from right to left run across the generally broken and vegetated face. The central rake, Ledge 3 leads with care up to the base of The Bastion.

Descent: The easiest descent is to make two abseils down the line of *Klepht* from the bolt belay at the top. Alternatively, traverse right and carefully descend Ledge 3 back to the base of the cliff.

2 **Digitalis** ★★ 68m E2 5c

FA Graham Little & Bill Skidmore (3 PA & tension traverse) 4 April 1981; FFA lower pitches Pete Linning & Colin Ritchie 24 July1982; FFA top pitch Gary Latter & Alan Ramsay 25 May 1991

The roofed pillar at the left end of the cliff.

1 **18m 5c** Follow *Klepht* for 12m to a small ledge then traverse left across a slab to a small hold. Continue round the edge to belay in a grassy niche.

2 **15m 5a** Up a corner above a detached pillar to below a roof. Turn the roof on the right and up cracks trending left to a small ledge and belay.

3 **15m 5b** Ascend an awkward corner and then directly up to a small ledge below twin roofs.

4 **20m 5c** Climb the overhang directly above the belay to gain a groove. Move left and up easier rock to a heather ledge and flake runner. Continue up a slabby groove, trending right to a thread belay on a terrace. Walk off left to gain Ledge 4.

3 **Klepht Direct** ★★ 55m E2 5c

FA Andrew Maxfield & Bob Wilde (extensive aid) 25 May 1967; Direct Graham Little & Colin Ritchie (4 PA) 16 April 1981; FFA Craig Macadam & Andrew Fraser 10 August 1981

The prominent large open corner at the left end.

1 **27m 5c** Up the corner past a small ledge, move right and climb a corner directly to BB on a grass ledge.

2 **28m 5c** Follow the main crack past a recess until a smaller crack in left wall allows the final corner to be climbed to a grass ledge and good thread belay. Either walk and then scramble off left to gain Ledge 4, or move up right to abseil from the BB on *Armadillo*.

4 The Brigand ★★   60m E5 6b

FA Kev Howett, Graham Little & Lawrence Hughes 1996

The pillar midway between *Klepht* and *Armadillo*, spoilt by a band of disintegrating rock low down on the second pitch. Start a few metres down from *Armadillo* below a prominent left-facing double-roofed groove.

1 **20m 5c** Up the groove to swing out right onto a good flake near the top (good nuts in a small pocket just above). Move left on good holds on the lip to an undercling at the left end of the roof. Continue along the narrow sloping ledge to belay in a crack system.

2 **40m 6b** Climb the crack above the belay to a band of disintegrating rock. Gain a thin hanging flake groove in this band. Struggle up this to reach some pockets and step right back onto real rock again. Climb up to a diagonal flake corner, mantelshelf onto its top then climb pockets in the wall above. Palm left to gain a small flake which leads to a nasty exit onto a sloping ledge on the edge of the pillar. Climb the short open corner above the ledge, pulling out left at the top. A short slab and flake groove lead to a grassy ledge and the BB on *Armadillo*.

Three options exist: Abseil from the BB, finish up the last pitch of *Armadillo*, or drop down left and traverse clear of the face, as for *Klepht*.

5 Armadillo ★★   100m E3 6a

FA Bob Richardson & Bill Skidmore 14 July 1977 (3 PA);
FFA Craig Macadam & Simon Steer 29 May 1985

A fine route with a short well protected crux, taking the prominent open groove up and right of *Klepht*. Start below this.

1 **25m 6a** Climb the groove to a roof at about 20m. Make difficult moves round this to a ledge and belay.

2 **35m 5b** Up the groove above then break out right onto some pockets. The easy angled groove leads to a belay.

3 **40m 5b** From the top of the rake climb a series of short overhung corners to a grass ledge. Traverse left, go up then move right into a final easy groove.

6 Abraxas ★★★   105m E4 6a

FA Graham & Rob Little (12 PA) 26 May 1980;
FFA Craig Macadam & Derek Austin 1 June 1985

An excellent route in a fine situation. The traverse at the end of the second pitch is often wet. Start directly behind the right end of the prominent arched overhang in the centre of The Bastion.

1 **30m 6a** Traverse leftwards and slightly upwards to the left end of the arch. Move left around the edge to the base of a yellow roof-capped corner. Up this and pull right onto the wall. Move slightly left at a horizontal break then make a difficult move (crux) to gain a hidden finger pocket. Up from this to gain a rightward hand traverse line which leads to a belay just above.

2 **35m 6a** Climb a diagonal finger crack to a narrow sloping shelf. Gain this awkwardly and follow it to reach a vertical water-worn groove. Up this to its top and step left to a sloping ledge. Cross a narrow shelf on the left to gain a niche and BB.

3 **40m 5b** Up a crack on decomposing rock to an undercut flake. Pull left on good holds to reach a crack, up this to a heather ledge. Follow a grassy ramp and continue up further grass to belay well back.

7 The Great Escape ★★★   95m E7 6b

FA John Dunne & Andy Jack 21 June 2001; pitch 1 Jules Lines & Iain Small June 2015

A stunning line breaching the centre of the cliff.

1 **25m 6a** Climb a small V-groove 10m right of *Abraxas* to the roof, traverse left under the roof and pull up right onto the belay ledge. (Direct over the roof is a similar grade but on crumbly rock).

2 **40m 6b** From the left side of the ledge, make awkward moves up and right to gain a big thread in the base of the groove. Climb the groove until it dwindles before making committing moves left to reach some hidden pockets and low cams. Climb directly up the improbable wall following the obvious groove lines to a resting ledge. Traverse right for 3m to a letterbox pocket (and thread) for protection and move back onto the ledge. Climb

leftwards up the slab on friction holds until possible to gain the base of the incredible flake-crack. Follow this in a superb position to belay on *Tidemark*.

3 30m 6b Gain a narrow sloping shelf 3m above the belay, small wires in this and a C3 000 with a possibility of skyhooks up on the left. Make difficult and bold moves to gain the groove on the right leading to a large spike. Pull onto the slab above and veer slightly left and up this to its top. Easier climbing remains.

---

8 Rhino ★   **80m E2 5b**

FA Graham Little (2 PA, solo using back rope) 3 June 1979; FFA Graham Little & Bill Skidmore 1 July 1979

The prominent pointed flake high on the wall up and right of *Abraxas*. Start at a cairn.

1 20m 5b Gain and climb a 3m leftward-facing flake. Mantelshelf and go straight up to a bulge. Pull left to reach a small heather ledge. Climb up to a curious hold (PR above) and step right to a stance on a grass ledge.

2 20m 5a Climb straight up on good flakes then by a crack to enter the crux chimney crack (PR above). Climb the chimney and exit left round a chokestone to belay at the top of the flake.

3 30m 5b Walk right and climb the corner above a big block. Make a detour on the left wall then continue up the corner to a grass ledge and belays.

4 10m 4c Climb the overhanging corner with strenuous finish to a ledge and belay. A short scramble leads to the top end of Ledge 4.

---

9 Token Gesture ★★   **70m E5 6b**

FA Dave Cuthbertson & Kev Howett 5 June 1985

Climbs the wall right of *Rhino*. Start just right of this.

1 20m 6b Make a long reach to gain a hanging flake and climb the shallow scoop directly above then step right to good footholds. Climb a series of flakes diagonally left towards a vegetated flake (good holds). Step back right then up over a bulge and go left on huge hidden pockets, stepping down to a ledge and belay on *Rhino*.

2 20m 6b Traverse back out right on the pockets then up to a large jug at the base of the hanging

groove. Climb this to its top and step out right and make a difficult move onto the slab.

3 30m 5c Climb the wall above via a line of left trending pockets to gain a flake. Go rightwards up an easy ramp and ascend a small groove. Exit left with an awkward move onto a slab which leads right into the final corner of *Rhino*. Finish up this.

---

10 Tidemark ★★★   **75m HS 4b**

FA Andrew Maxfield & John Peacock 9 June 1960

A fine wildly exposed girdle taking the glaringly obvious (visible from the road!) curving shelf across the top third of the cliff. Start at a rounded flake belay well above the upper end of Ledge 3. This can be gained either by a serious scramble up and left from near the top end of Ledge 3 or by making a 55m abseil from a large flake-boulder at the top of the gully bounding the right end of the face.

1 30m 4a Traverse left across two slabs to a large split block. Climb over the block then walk across grass to an 'eye-hole' belay, atop the *Rhino* flake.

2 30m 4a Follow the magnificent exposed gangway to step left across the slab to a wide crack. Continue more easily up this to belay on a small ledge just left of an overhang. A stunning pitch.

3 15m 4b Climb the short flake crack above then traverse left into a heather groove which leads to Ledge 4.

---

11 Slipway ★   **45m HVS 5a**

FA Bill Skidmore & John Madden (1 PA) 7 June 1975

A good continuation to any of the routes on The Bastion. It lies on the short steep little buttress above Ledge 4, situated almost immediately above the finish of *Tidemark*. Start at a small open corner.

1 15m 4c Climb the right wall via a blocky flake and spike then up the left-trending ramp over a block to a small grass ledge on the left.

2 30m 5a Gain the ramp above with difficulty. (This can also be gained by a traverse left from the right end of the ledge, lowering the overall grade to VS). Follow the ramp for about a metre, step down then traverse left to cracks. Go up to a square cut recess and exit left to a large chokestone belay. Scramble to Ledge 5.

Punster's Crack, The Cobbler. Britta Dost climbing the second pitch. Photo Dave Cuthbertson, Cubby Images.

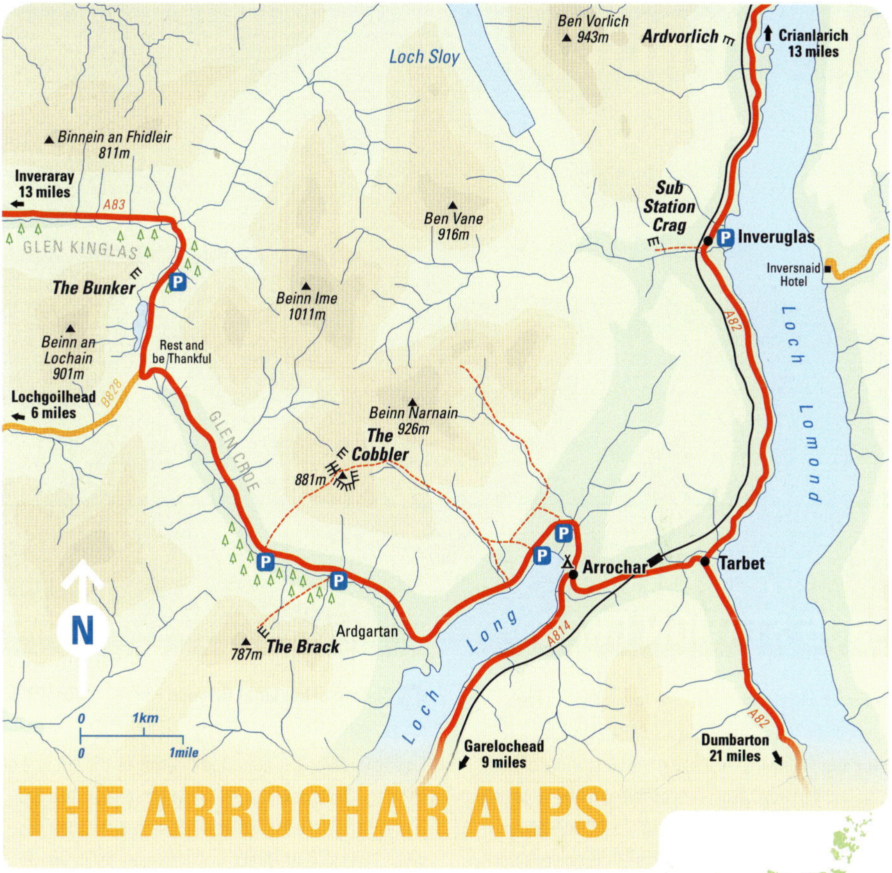

THE ARROCHAR ALPS

The village of Arrochar lies at the head of Loch Long. Five peaks to the north and west are collectively known as 'The Arrochar Alps', but it is the lowest peak, The Cobbler, that dominates, with the highest concentration of quality rock climbing of every grade. Like much of the Southern and Central Highlands, the rock is largely mica schist, folded and contorted, with many quartz intrusions. Many of the neighbouring hills also have good crags and cliffs.

Accommodation: Campsites at Glenloin Caravan Park & Campsite behind petrol station, Arrochar (NN 301 048; ☏ 01301 702239); Ardlui Holiday Park, Ardlui (☏ 01301 704243; www.ardlui.com); Beinglas Campsite (☏ 07957 626577; www.beinglascampsite.co.uk).

Youth Hostels at Inveraray (☏ 01499 302562; www.inverarayhostel.co.uk) or Crianlarich (☏ 01838 300260; www.hostellingscotland.org.uk). There are numerous hotels and B&Bs in Arrochar. Wild camping within Loch Lomond & The Trossachs National Park is subject to seasonal bylaws (www.lochlomond-trossachs.org).

Amenities: Slanj Restaurant / café / bar east of railway station on A83. Small convenience store, café, restaurants and chip shop all in Arrochar. Bar meals available in hotels.

BEN VORLICH
(MOUNTAIN OF THE BAYS)

NN 306 094 **Alt:** 200m 30min

SUB STATION CRAG

A fast drying slabby schist crag to the west of Loch Lomond on the southern slopes of Ben Vorlich.

Access: Follow the A82 along the west shore of Loch Lomond 4 miles/6.4km north of Tarbet to the large tourist car park (NN 3226 0987; 56.251697, -4.7089572) and viewpoint on the shore, just north of the power station. Travelling from the north this is 12.4miles/19.8km south of Crianlarich.

Approach: Walk south along pavement on west side of road for 0.5 mile/0.8km to a gate with lots of signs. Follow the tarmac road west up the hillside, taking the right fork just before the sub-station. Cut diagonally leftwards up to the crag.

Descent: Abseil from trees at the top of the crag.

1 The Pylon Effect ★   40m E3 5c
FA Ian Taylor & Colin Moody 3 November 1988

The fine left arête. Start at a short crack left of the great flake. Follow the crack until a leftward traverse leads to a flake on the arête. Climb the arête direct to the top. Belay well back.

2 Charge of the Light Brigade ★★★   30m E3 5c
FA Jim Divall & Ray Cluer 1985/6

Excellent reasonably protected climbing up the left side of the crag. Climb up to the left of the great flake, then the groove to a PR. Head diagonally right, then back left past several PRs to finish direct. Continue more easily above.

3 White Meter ★★   30m E4 6a
FA Colin Moody & Ian Taylor 3 November 1988

Gain the flake and from near its left end move up and step right, then up to a jutting quartz lump. Move up left to a bulge (F #1.5 in horizontal crack), and surmount it using a quartz hold, and up to trees. Quite bold and run-out on the easier lower section.

4 Power to the People ★   30m E2 5c
FA Jim Divall & Ray Cluer 1985/6

From the upper right end of the great flake climb direct past a PR to a horizontal break. Cross the small overlap and ascend the groove on the right. Step right to a ledge and continue easily up heather ledges to belay on the right.

5 Current Affair ★　　　　　　30m E2 5c

FA Jim Divall & Ray Cluer 1985/6

Ascends the wall left of the crack-line of 6.

1　**15m 5c** Climb up to gain two thin parallel cracks.
Up these, move left to a flake then right to PB.

2　**20m 4c** Move up left past a bulge to a grass ledge,
then up heather ledges to belay on the right.

6 Wired for Sound ★　　　　　　35m E1 5b

FA Jim Divall & Ray Cluer 1985/6

The crack-line left of a prominent rib at the right end.

1　**20m 5b** Up the crack and wall above
to a ledge on the right.

2　**15m 4b** Climb straight up to a grass ledge.

7 Live Coverage　　　　　　48m E1 5b

FA Jim Divall & Ray Cluer 1985/6

The rib at the right end of the crag.

1　**15m 4c** Climb the left side of the rib to
a break, then up the slab above.

2　**24m 5a** Up the slab on the right, move left
to a small corner and up crack above.

3　**9m 5b** Finish up the overhang above, or walk off
right below it reducing overall grade to HVS.

NN 3232 1218 **Alt:** 80m　**w**　15min

ARDVORLICH CRAG

A couple of fast drying, sunny, 'sportingly-bolted'
adjoining walls on a small knoll on the east side of
Ben Vorlich with grand views down Loch Lomond. It
should be noted that the routes are not complete
clip-ups and the second bolts on some look
alarmingly high (possible ground-fall
potential) – take care!

Access: From the south: Park in small lochside layby (NN 3267 1170; 56.268206 , -4.7035903), 1.3 miles/2.1km north of Inveruglas Visitor Centre, 400m south of farm entrance. **From the north:** Park in the same spot, which is 2.5 miles/4km south of Ardlui train station.

Approach: Walk north to gate at farm entrance, then follow right side of burn, just left of white cottage, cutting underneath railway line to pick up a vague path uphill. At a waterfall in the burn, traverse a short way right to the crag. **Some of the lower offs require extending with slings to reduce drag.**

1 Carnage — 5m F6c
FA John Watson 2006

The wee roof is butch. A bolted boulder problem, which can be extended to F7a by bouldering in along the break and boosting for the jug at first bolt, continue on to 7.

2 That Sinking Feeling ★ — 15m F6b
FA John Watson 2006

Excellent technical climbing up the left arête. Step right from the arête with crux move to layaways. Up jugs to clip, then more easily on good holds in the groove right of the wee roof.

3 The Groove ★★ — 15m F6a+
FA John Watson 2002

Good climbing up the juggy central groove and pocketed headwall. Veer left at top to a mantelshelf finish.

4 Drifting from Shore ★★ — 15m F6b+
FA Graham Harrison, et al. 2004

The bulge and headwall direct. The crux bulge can be bypassed on the right. Good headwall crimping.

5 Lake Lomond ★★ — 15m F6a
FA Graham Harrison, et al. 2004

Climb up behind the saplings to a crux, step left onto the wall. Follow the pale wall directly.

6 Dilemma ★★ — 15m F6a+
FA John Watson 2005

Pull on by a quartz crack (crux) and climb up to a thin section moving up and right, finishing direct.

7 Snake Eyes ★ — 15m F6a
FA John Watson 2005

Pull through central roof and move up and left to the meet 3. Step down and traverse left to bigger holds. Follow quartz straight up to a sapling, then easily right to LO.

8 Magic Carpet Ride ★ — 15m F6b
FA John Watson 2002

"You'll need your trad head for this one." The half-bolted, half-pegged route through the steepening overlap near the apex of the slab. Keep your nerve between bolts and pegs – the climbing is never desperate.

Jo George on the fingery Lake Lomond.
Photo John Watson, Stone Country Press

Centre Peak, The Cobbler. Paul McNally climbing.

THE COBBLER

NN 259 058 **Alt:** 800m 1hr 30min – 2hr

"That rock had beauty in it. Always before I had thought of rock as a dull mass. But this rock was the living rock, pale grey and clean as the air itself, with streaks of shiny mica and white crystals of quartzite. It was joy to handle such rock and to feel the coarse grain under the fingers."
– W.H.Murray, *Undiscovered Scotland*, 1951

With its distinctive craggy profile, the gnarled contorted mica schist, often studded with quartz, has attracted the attentions of generations of climbers from the Central Belt. Here are to be found as fine a cross section of routes as any mountain. The rock is particularly slippy when wet, (like standing on a wet fish!); though many of the routes dry quickly.

Access: From Glasgow and south: Follow the A82 to Tarbet then turn left and follow the A83 through Arrochar and up Glen Croe to park by the old bridge on the right at (NN 2427 0600; 56.214024, -4.8352632), just beyond the end of the forestry plantations on the east (right) side of the road, 4.7 miles/7.5km west from the large car park at the head of Loch Long. **From the west:** The parking spot is 1.3 miles/2.1km east from the Rest and be Thankful

(0.2 miles/0.3km beyond two large lay-bys on the north side of the road).

Approach: (A) Cross the burn by the old bridge and follow a path up the left side to reach a small dam after about 1.5 km. Follow a further path up the ridge (left side of the burn) on the right to reach the col between the North and Centre peaks. 1¼ hours.

(B) The popular alternative approach starts at the large car park (pay & display) by Succoth at the head of Loch Long (NN 292 050; 56.206845, -4.7552272), 0.6 mile/1km outside Arrochar. Follow a steep path directly up (through the trees to begin with) for about a kilometre (300m of ascent). The path then contours left round the hillside and follows the right bank of the Allt a' Bhalachain, past the Narnain boulders, before crossing the stream and heading up into the coire. 2 hours.

Accommodation: There are numerous small howffs in the coire, the best being high up, nearest the path (sleeps two). Wild camping anywhere in the hills.

SOUTH PEAK   NN 2405 0565 **Alt:** 770m

SOUTH FACE

The steep clean face overlooking Glen Croe.

Descent: From the large grassy terrace at the top of the wall scramble down rightwards and down the well-worn *South East Ridge*. For the single pitch routes on the lower slab, finish up either *Gladiator's* or *Ra*, or abseil from the ledge (sling sometimes in situ).

1 Glueless Groove ★★   45m E2 5b

FA Robin Smith June 1957

The shallow steep groove in the centre of the face, best done in one long pitch. Climb the cracked groove (most defined and leftmost) to a large flake and a ledge on the right. Continue quite boldly up the wall immediately above the flake, trending slightly left to some quartz blotches, then up to a ledge. Finish up the groove above.

2 Ithuriel's Wall ★★   45m E2 5b

FA Hamish MacInnes (some aid) August 1952;
FFA John Hutchison 1976

The prominent shallow left-facing corner high on the face. Start beneath broken grooves, down and left of the main corner pitch.

1 **10m 4a** Follow the grooves to a large ledge and block belay.

2 **25m 5b** Climb the corner through some bulges to gain the large ledge on 4. Belay in the corner on the right.

3 **10m 5a** Finish up the arête right of the groove.

3 Gladiator's Groove Direct ★★★   65m E1 5b

FA Bill Smith & Hamish MacInnes 1951;
Direct Start Bill Smith & Bob Hope June 1952

1 **35m 5a** A bold pitch, following a narrow slanting fault across the top of the slab. Start at a band of quartz at the left end of the slab. Up to a good hold below the quartz then on good holds to the right-slanting gangway. Follow this delicately to a large perched block at its end. Easily to a large ledge and belay.

2 **18m 5b** Move up and left to the base of a large right-facing corner, and up this with hard moves to gain a belay on the ledge on the left.

3 **12m 5a** Crack above to top.

4 Geb ★★   35m E4 6a

FA Gary Latter & Paul Thorburn 2 June 1995

A right-slanting diagonal line across the slab. Start up 3 to the quartz band. Traverse this past 2 PRs then up the vague crack-line as for 6 but continue rightwards to finish up the right side of the block.

5 Ra ★★   65m E4 5c

FA Gary Latter & Paul Thorburn 1 May 1995

A direct line up the left side of the slab. Quite bold and run out, despite the proliferation of pegs. Start down and right of 3.

1 **35m 5c** Up past PR onto a sloping ledge. Step left and up onto a small ledge above (skyhooks 0.5m above PR & out on right). Climb direct past 3 PRs to the rising traverse shelf on 3. Arrange protection (thin crack out left, or block on right) and climb directly up the wall above on improving holds to belay on a ledge.

2 **30m 5c** Follow 3, but where that route traverses left onto the ledges climb directly by a vague crack to finish up the last few metres of that route.

6 Osiris ★★★   35m E4 6a

FA Dave Griffiths 6 April 1988

Immaculate bold slabby climbing. Start at the lowest point of the crag, behind a large fallen boulder. Easily up broken ledges and left to PR below the centre of a

Rick Campbell on the second ascent of Ra.

roof. Step left and pull through the roof leftwards past a further PR to a thread. Step left and up a thin crack (PR at thread) trending slightly rightwards to finish easily up the left side of a huge perched block.

SOUTH-EAST FACE

Narrow slabby face overlooking Ardgartan and Loch Long.

1 Ardgartan Arête ★★   60m VS 4b

FA John Cunningham (solo) June 1948

1 **25m 4b** Start 5m right of arête. Up a slab over a bulge (poorly protected) and trend left to grooves in the arête. Up these to a large ledge and belay.

2 **10m 4a** Climb the wall and cracks to a finely situated ledge just right of the arête.

3 **25m 4a** Continue either up cracks in the wall above or the wall on the left overlooking the South Face to a large grass terrace. Move out right and scramble to the summit.

Paul McNally stepping out on the pleasant, well-protected, S Crack.

2 Ardgartan Wall ★   60m Very Difficult

FA Jock Nimlin, Jimmy Wynne, Wattie Neilson & Rab Goldie October 1937

Start about 10m right of the arête, beneath a shallow groove.

1 **15m** Climb the groove and wall.

2 **35m** Continue directly up the wall to belay on a grass ledge.

3 **10m** Traverse left along the ledge and finish up a steep little wall next to an overhang.

3 South East Ridge ★   100m Moderate

FA Gilbert Thomson & party October 1889

Follows a line near the right edge of the face. Also a useful descent. Follow well-worn crampon scratches and eroded paths. Continue to the summit, descend by *Original Route*. Avoid in the wet, as it can be very polished and slippery.

③ Nimlin's Direct Route ★   75m Very Difficult

FA Jock Nimlin & Andy Sanders April 1933

The edge formed by the North-East and North-West faces. Start about 30m up the face, where a ledge runs out leftwards to the edge. Traverse along the ledge, gain the edge, and follow this to finish at a conspicuous, semi-detached block near the summit. Fine open situations.

NORTH-WEST FACE

① Original Route ★   50m Moderate

FA William Naismith & Gilbert Thomson May 1894

Gain and follow a path along the main grass ledge then up and right over a short wall in a corner to the top. Very polished – avoid in wet conditions.

CENTRE PEAK

① The Arête ★★   60m Moderate

FA William Naismith & Gilbert Thomson July 1889

Above the col between the South and Centre peaks is a 10m hand crack up the centre of a slabby buttress. Follow this then the natural continuation up the edge to a perfect finish atop the summit block.

① Bow Crack ★   30m HS 4b

FA Bill Smith, W.Dobbie & Bob Hope June 1952

The crack in the steep wall left of the short prominent gully.

1 **16m 4b** Climb crack over a small bulge to a small ledge. Go left for 6m to belay in a small niche.

2 **14m** – Move up and right then over ledges to the ridge.

② S Crack ★★   40m VS 4c

FA John Cunningham & Bill Smith June 1948

A good sustained pitch up the left-facing cracked groove in the wall right of the short gully. Follow the hand-sized crack in the groove, finishing slightly leftwards to a grassy ledge. Well-protected with large nuts and Fs. Walk off left.

The classic airy scramble of The Arête, usually incorporated with the traverse of the South Peak.

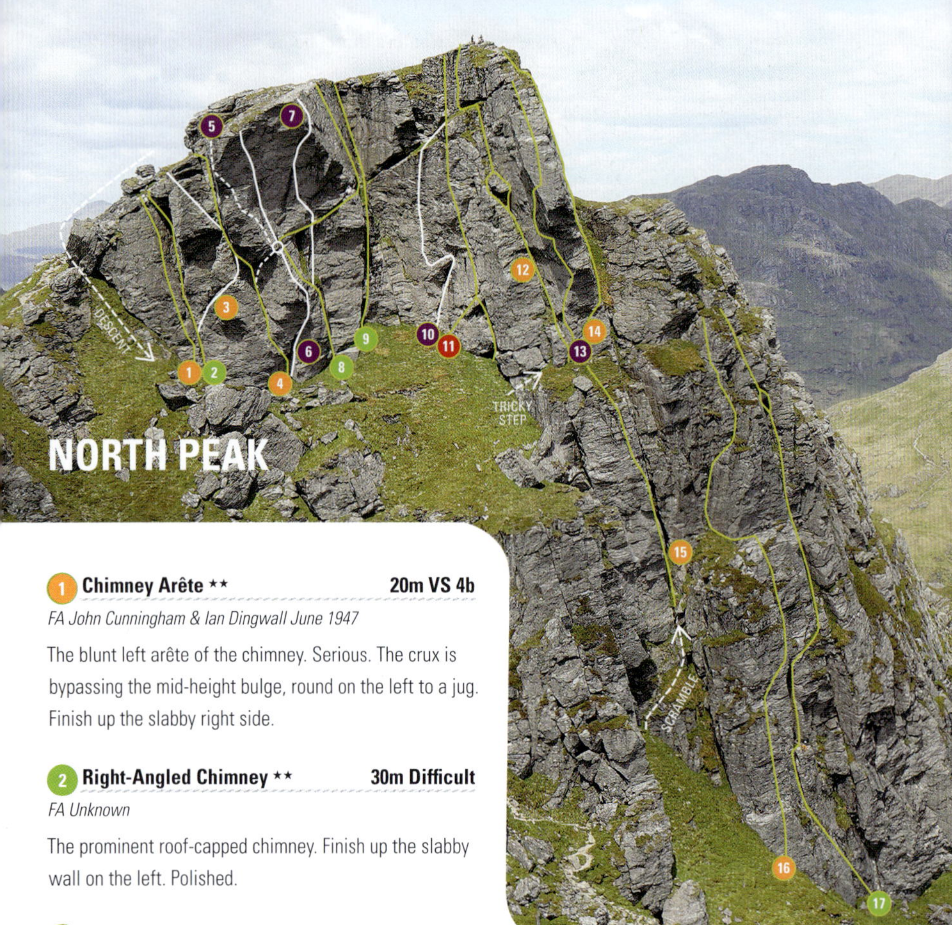

1 Chimney Arête ★★ — 20m VS 4b

FA John Cunningham & Ian Dingwall June 1947

The blunt left arête of the chimney. Serious. The crux is bypassing the mid-height bulge, round on the left to a jug. Finish up the slabby right side.

2 Right-Angled Chimney ★★ — 30m Difficult

FA Unknown

The prominent roof-capped chimney. Finish up the slabby wall on the left. Polished.

3 Cat Crawl ★ — 35m HS 4b

FA A.Lavery & A.N.Other 1936

Start below prominent fault trending right into larger groove.

1 **18m 4b** Direct to the start of the fault and traverse it for a few moves then direct up to a short left-facing corner. Up this to a good quartz spike belay above. (The original route traversed the entire fault then up the groove to the spike).

2 **17m 4b** Bridge up the exposed bulge above for a few moves, traverse left on good holds across the lip of the overhang then direct to the top.

4 Direct Direct ★★ — 30m HVS 5b

FA Upper Crack R. Muir & J.Wilson;
Lower Crack John Cunningham 1948

Start in the recess below a hanging left-facing groove. Climb the bulge (good protection in a crack above) to a good hold. Up the groove past the spike belay of 3 to finish directly up a steep crack above, or the wall on the left.

5 Wild at Heart ★★★ — 60m E6 6b

FA Pitch 1 Gary Latter & Brian Beer 3 September 1993;
Pitch 2 Gary Latter & Tom Keenan (redpointed) 27 September 1993

An outrageously exposed top pitch up the edge of all things. The main pitch attacks the stunning flying arête up the left edge of the scooped wall above the flake belay of 8.

1 **25m 5c** Start up 4 to the roof at 12m. Arrange protection, step down for 3m then ascend diagonally rightwards across the wall to better holds. Up to the roof and follow this right to belay on the *Punster's* flake.

2 **35m 6b** Go up the wall directly above to a crack then traverse left to quartz on the edge and up to a ledge and two poor PRs. Make committing moves leftwards off the ledge with a long reach to the first peg runner. Continue with difficult reachy moves slightly leftwards past a number of peg runners to good finishing holds near the arête.

6 Wild Country ★★★   50m E6 6b/c

FA Pitch 1 as The Nook Direct Start (1 pt. aid) Bill Smith & Mick Noon 1955; FFA with new top pitch Dave Cuthbertson & Rob Kerr 4–5 July 1979

A seminal route, ahead of its time and a pre-cursor of many hard mountain routes in the eighties. The pre-placed nut used on the first ascent has since disappeared. Start in the same recess as for 4.

1 **20m 5c** Climb a widening right-slanting crack past a large flake then easily leftwards up a slab to a block belay, as for *Punster's*.

2 **30m 6b/c** Up a slab and thin crack then move left to a large flat ledge. Follow a series of overhung ledges rightwards to a good hold at the base of a thin hanging crack. Up the crack (good Wallnut #4 on side), heading for a good small side pull out left (crux) to reach a recess then easily to the top.

7 Wide Country ★★★   40m E5 6b

FA Rick Campbell & Paul Thorburn (both led) 20 July 1994

The evil gaping slit up the right edge of the wildly overhanging face. Follow 8 heading direct for the widening crack. Thrutch up the offwidth crack to gain the niche and exit rightwards. A generous supply of tape, a high pain threshold and a Camalot #4 will all be found to be of use.

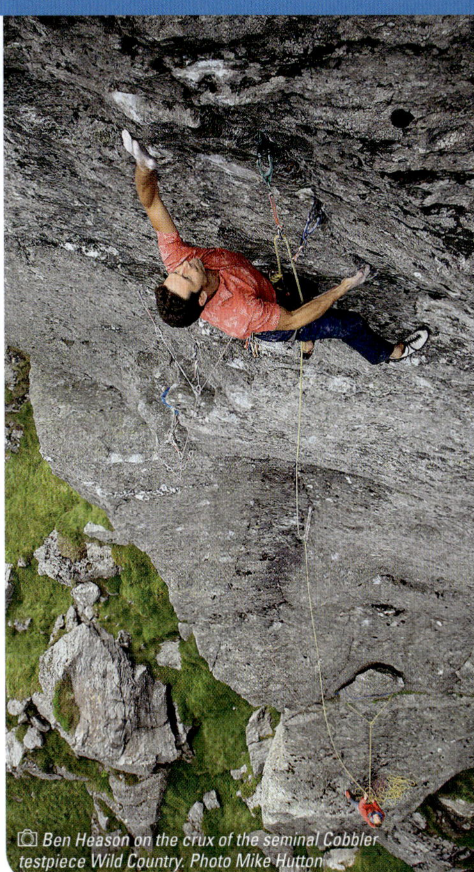

Ben Heason on the crux of the seminal Cobbler testpiece Wild Country. Photo Mike Hutton

8 Punster's Crack ★★★★   50m Severe 4b

FA John Cunningham & Bill Smith August 1949

One of the best routes of its grade in the country with superb positions. Start at a flat rock ledge directly below the prominent deep overhanging crack in the headwall.

1 **20m 4a** Up a short slabby stepped corner then trend left to a large block belay.

2 **15m 4b** Traverse up and right to a prominent wide slot (good large nut runner above for the leader). An awkward step across the slot leads to a shelf. Belay below the cracks at the right end of the ledge.

3 **15m 4a** Move diagonally left and follow cracks up the wonderfully situated steep slab above. Flake belay far back.

Andy Wren on the classic fifties testpiece Club Crack.

<table><tr><td>**9** **Right-Angled Gully** ★★</td><td>35m Very Difficult</td></tr></table>

FA William Naismith & J.MacGregor 1896

The prominent steep corner-gully at the back of the bay.

1 **17m** Up on polished holds to where it steepens.

2 **12m** Follow the ledge on the right to a thread belay at a large rock just before its highest point.

3 **6m** A final short steep corner on the wall on the left leads to the top.

Variation – *Direct Finish* ★ 10m Severe 4a. *FA Jock Nimlin (solo) 15 June 1930.* Follow the well protected crack up the back of the corner above the belay ledge at the end of the first pitch. Slightly dirty.

<table><tr><td>**10** **Rest and be Thankful** ★★</td><td>45m E5 6a</td></tr></table>

FA Dave Cuthbertson & Ken Johnstone May 1980

Superb meandering wall climbing, though quite slow to dry. Start at the foot of the right-facing groove in the centre of the wall.

1 **35m 6a** Follow groove to its top (often wet). Traverse left then step down to good holds. Traverse about 3m left then up to a good foothold. Continue up then right to a good horizontal break then up and left to reach a good ledge. Up the wall then easily right to a thread belay.

2 **10m 5a** The wall above at its highest point to the top.

11 Club Crack ★★★　　　　　　　50m E2 5b

FA Patsy Walsh & numerous Creagh Dhu members July 1957

Bold, sustained and strenuous climbing up the steep crack springing out of the cave at the right side of the wall. Start beneath the roof.

1 **40m 5b** Pull direct through the roof above the cave and follow the crack past an ancient and much fallen-on peg (rounded spike up on right). Continue in the same line, easing towards the top.

2 **10m 4c** Climb the highest point of the wall above the large ledge.

The following three routes start on a sloping grass terrace up and right of *Club Crack*. Thin people can reach the base by squirming through the back of the cave, salad dodgers by a short rock step on the right.

12 Right-Angled Groove ★　　　　48m VS 4c

FA Jock Nimlin (solo) 1934

1 **42m 4c** Follow the slabby open groove, with an excursion on to the left wall high up, to a thread belay on the long grassy ledge near the top. The corner crack can also be climbed throughout at VS 5a.

2 **6m** – Finish up the corner on the left.

13 Dalriada ★★★★　　　　40m E7 6b (F7b)

FA Gary Latter (redpointed) 20 September 1995

Spectacular and very sustained climbing up the wildly undercut prow directly under the summit of the North Peak. Start at the base of the arête at the same point as 12. Up the groove for 3m to a ledge then the flake crack above to 2 PRs. Move right round the arête and up to a superb thread. Straight up the thin finger crack and the arête past a poor PR to a good rest under the roof. Make hard moves left and up to reach the prominent diagonal crack and a line of incut jugs which lead past more PRs to the final capping wall. Continue with interest past two small finger pockets to pull out right to a ledge. Scramble up the ridge to belay just short of the summit.

14 Whither Wether ★★★★　　　　40m VS 4b

FA Hamish MacInnes & Bill Smith August 1952

A wildly exposed pitch, quite run-out in places, but the climbing is straightforward. Start from a belay just right of the base of 12. Up the centre of the steep wall, trending right. Pull round the edge and up a bold slabby wall to a short vertical crack then follow a leftward trending line to belay just short of the summit cairn.

15 Wether Wall ★★　　　　　　40m VS 4c

FA John Cunningham & Hamish MacInnes September 1951

A good pitch, providing a logical start to 14. Start by scrambling up to a grassy ledge about 40m up *Ramshead Gully*, above chokestones and below the deep chimney of the gully. Step left onto the wall and climb direct past a series of small ledges, heading for a small left-pointing flake. Pull over this and up a groove above to belay at the base of 12.

16 Incubator ★★　　　　　　　75m HS 4b

FA Above terrace John Cunningham & Ian Dingwall 1948; Below terrace John Cunningham, Bill Smith & Tommy Paul 1940's

A good direct line up the facet overlooking *Ramshead Gully*. Start just right of *Ramshead Gully*, (about 10m up left from the start of 17).

1 **25m 4a** Follow a series of broken walls direct to beneath a steeper wall.

2 **15m 4b** Up the crack in the bulge (above

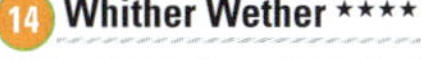
Kevin Howett on the eighties testpiece Rest and be Thankful.

Dave MacLeod on the second ascent of the spectactularly positioned (even if I say so myself!) Dalriada. Photo Dave Cuthbertson, Cubby Images.

and just right of a small juniper) and the fine corner above to the terrace.

3 **35m 4b** At the back of the terrace is a left-facing corner with twin cracks. Climb the right-hand crack, past a weird phallus at 10m. Move out right past grass ledges to finish by a short crack.

17 **Recess Route** ★★★   **90m Severe 4a**

FA Jock Nimlin, John Fox & Bob Ewing May 1935
Crux pitch (The Fold) climbed by Harold Raeburn and a large merry party, including two ladies 1904

A popular classic, following the prominent deep V-chimneys and grooves up the highest section of the face. Start by a boulder at the left side of a small amphitheatre just left of the lowest point of the crag.

1 **25m** Follow a left-slanting crack in the slab then the wide crack in the groove to a belay ledge.

2 **25m** Ascend the chimney past an overhang. Continue up the deeper chimney to belay on the Halfway Terrace.

3 **5m** Walk right along the terrace to a belay.

4 **10m 4a** Climb the steep left-slanting groove (The Fold) on its left wall to belay in a small cave.

5 **25m** Bypass the overhang above on either side then finish up the chimney.

CHOKESTONE GULLY   **Alt**: 750m 

There is a short steep 15m wall about 100m below the main face, just above the tourist path. 100m north of this is a gully wall containing a perfect hand and fist crack above a roof low down. Approach from above by cutting down diagonally left, then back right to the base.

1 **A Crack in the Clouds** ★★★   **20m E4 6a**

FA Paul Thorburn & Gary Latter 16 September 1995

Belay to a F #1.5 on the slab behind the route. Gain a niche left of a crack via a large initial roof on good holds. Pull into the thin finger crack with difficulty and follow it on widening jams. Belay on a boulder far back on the right.

 Cleaned 2018.

Bounding the right edge of the main face is the prominent vegetated scree-filled Great Gully, with a chokestone near its base.

18 **Spinal Rib** ★   **35m HS 4b**

FA Patsy Walsh September 1952

A good exposed pitch up the steep rib on the right wall of the gully. Scramble up under the chokestone of Great Gully. Pull up onto the rib and follow it over several bulges to a good rest beneath the final overhang. Climb this on the left (technical), or more strenuously on the right to join 21. Follow this to belay beneath a short wall. Finish up that route.

19 **North Rib Route** ★   **80m Very Difficult**

FA Jock Nimlin & John Fox 1935

Climbs the series of short steep walls on the right-bounding rib of Great Gully. Fine rock and good positions on the main pitch. Start at a groove at the lowest rocks, about 6m right of the gully. Start 2m up left from the right end of the gully wall. Move round right and up a short crack then pull round onto the right side of the rib and follow a crack to a large block belay on a ledge.

Paul 'Stork' Thorburn on the first ascent of the spectacular athletic jamming testpiece A Crack in the Clouds.

NN 246 034 **Alt:** 620m 1hr 30min

THE BRACK

Access: Follow the A83 0.5 mile/0.8km north of Ardgartan to the car park on the left (west) side of the road.
Approach: Cross the bridge over Croe Water and follow the forestry road west up the glen for about 1.8km to a stream just beyond the junction with a higher road then cut steeply up the hillside heading slightly rightwards, always keeping to the right side of the stream.
Descent: Down the left (east) side of the crag.

---

1 Edge of Extinction ★★★　　　　　　85m E6 6a

FA Pete Whillance & Pete Botterill 19 May 1980; Direct: Iain Small & Dave MacLeod July 2019

The big impressive arête in the centre of the main face left of *Mammoth*. The second pitch is serious.
 Cleaned 2019.

1 40m 6a Climb up the overhung recess (as for *Mammoth*) then directly up the crack above to a ledge at 12m. Move up onto the wall just right of the arête (2 PRs) and climb diagonally rightwards to a flake. Up the steep wall above on small holds to a ledge below a prominent hanging corner. Take the corner past a PR at start. Exit left at the top to ledges on the arête. Step down left to belay on smaller flake 2m below bigger loose/ expanding flake. 3 pegs replaced July 2019.

2 45m 6b Take the right side of the arête for 6m to a small ledge then up a ramp on the left to where it merges onto a steep wall (poor PR and nuts). Climb the steepest section of the arête from a sloping ledge, rather than traversing across the left wall and back right to the overlap (as for the original line).

2 Mammoth ★★　　　　　　　　　　88m E3 5c

FA Bill Skidmore, Bob Richardson & Jim Crawford (VS & A3) September 1967; FFA Dougie Mullin & Alan Pettit Summer 1978

The prominent hanging crack in the steep left wall of *Great Central Groove*. Start at a cave-like recess down and left of the crack underneath the arête.

1 18m 5a Climb up the overhung recess then swing round the right edge to gain a narrow ledge. Continue up the wide crack to belay beneath a prominent overhanging crack.

2 25m 5c Climb the wall and crack above the stance which leads to a recess. Continue through the overhang to a ledge (possible belay). Move up right to a grassy ledge then easily to a large ledge and thread belay.

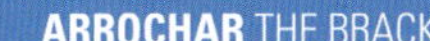

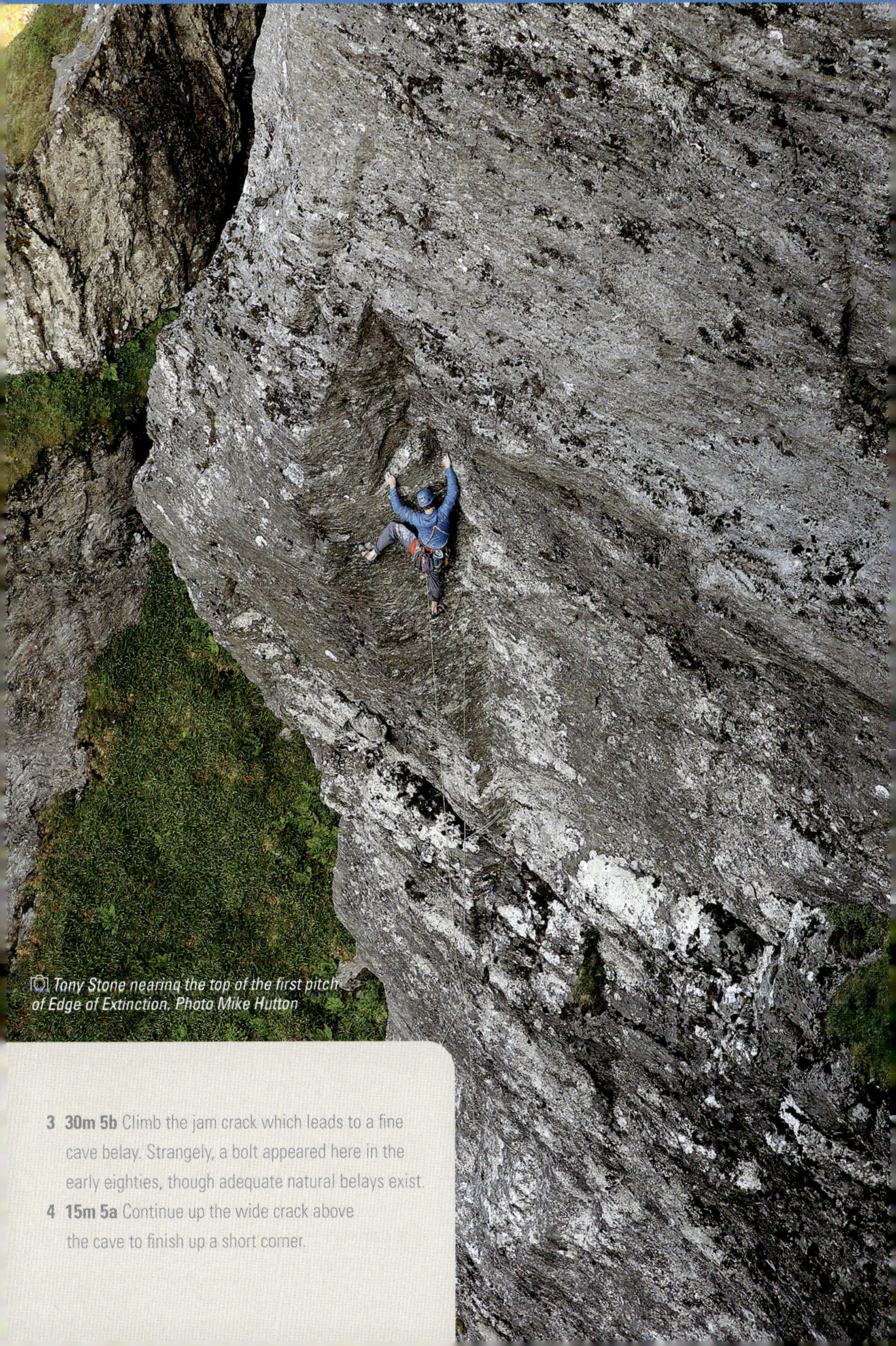

Tony Stone nearing the top of the first pitch of Edge of Extinction. Photo Mike Hutton

3 **30m 5b** Climb the jam crack which leads to a fine cave belay. Strangely, a bolt appeared here in the early eighties, though adequate natural belays exist.

4 **15m 5a** Continue up the wide crack above the cave to finish up a short corner.

THE BUNKER

NN 2290 0885 **Alt:** 420m **SE** 30min

A huge roof that overlooks the glen from the east face of Beinn an Lochain. Topped by a slab, from the roadside it resembles an old war bunker built into the hill. An hour from Glasgow, a massive spread of grades and plenty to climb even in unfavourable weather. With the large roof dominating the centre stage, this main spectacle holds several hard routes that tackle the steepness head-on. To the left, the steepness eases giving a few vertical and slabby walls that feel a little more familiar to the classic crimps and pocket schist climbs that other local crags are known for. Finally, to the right on the approach are several easy slabs, with short lengths (6-12m), friendly angles and good low-end grades.

Access: Park in large layby NN 2342 0881; 56.238924, -4.8508237 on the west side of the A83 0.9 miles/1.5km north of the Rest and Be Thankful summit/ B828 Lochgoilhead turn off.

Approach: Cross the stream and follow the initially very boggy path and its continuation up northeast ridge, to the left of the forestry plantation. Traverse left at a small cairn just above a steep rocky section.

ATOMIC WALL   **Alt:** 360m **SE** 20min

Vertical wall, up left of a smaller crag and large boulder, overlooking the approach path.

Approach: Leave the path after about 15 minutes and head left up the hillside.

1 Fallout ★★   16m F6a+
FA Oz Miller 23 July 2023

2 Meltdown ★   18m F6a+
FA George Beaton 10 September 2023

3 Uranium ★★   20m F6b+
FA Rory Watson 23 July 2023

4 **Oppenheimer** ★★ — 20m F6c
FA Willis Morris 23 July 2023

5 **Open Project** — 19m

6 **Open Project** — 19m

OVERLAP SLAB

NN 2295 0892

The first crag encountered on the approach, about 80m right of the other crags.

1 **Tiger I** — 11m F4b
FA Edvin Malinovskis 23 July 2023

2 **Danger Close** ★ — 12m F5a
FA Rory Watson 29 August 2021

3 **Friendly Fire** ★ — 12m F5c
FA Kasia Zadrozniak 12 June 2021

4 **Inglourious Basterds** ★★ — 13m F6a
FA Lanah May 29 April 2021

GRAVEYARD SLAB

1 **Tranquility** — 7m F4a
FA Eilidh Robinson 23 July 2023

2 **War and Death** ★★ — 6m F4c
FA Calum Brown 29 August 2021

3 **The Fallen Ghost** ★ — 6m F4b
FA Laura Jagowska 29 August 2021

PEA-SHOOTER SLAB

1 **Battle of the Bulge** ★ — 7m F5a
FA Jonathan Wilson 23 July 2023

2 **No White Flags** ★ — 7m F4c
FA Siobhan Young 13 June 2021

3 **I'll be a TANK One Day!** ★ — 7m F5a
FA Kasia Zadrozniak 15 May 2018

4 **Panzerfaust** ★ — 6m F6c
FA Willis Morris 23 July 2023

MAIN ROOF

Catches morning sun but shades over quickly due to the steepness of the roof. The roof protects the majority of the routes from the rain apart from in heavy downpours with a lot of wind, however several routes do have seepage so conditions are best found with one dry day beforehand. If climbing in the upper grades of the crag, there is almost always something dry to climb.

1 Flight Path ★   15m F7a+

FA Jamie Skelton & Morag Eagleson 9 August 2022

Steep route traversing through the roof.

2 Messerschmitt ★★★   20m F8a

FA Willis Morris 1 June 2022

Tackle the roof directly via the "crocodile jaw shaped" block before a tricky crux near the very end! Possibly harder, following loss of large block.

3 Dirty Bomb ★★   20m F7b+

FA Willis Morris 15 May 2018

Start between the two wet streaks. At the roof, step right across the wetness to gain jugs in the roof. Climb steeply via the obvious corner to a powerful crux at 5th clip which links into the easy top section. Finish dramatically on the horn-shaped spike.

4 Closed Project   20m F8c?

Bolted by Dave MacLeod 2020

5 Ricochet ★★   12m F7a

FA Willis Morris 20 April 2021

Steep juggy climbing through the roof of the arch underneath 6. Climb steeply with good holds to a large move to a quartz break. Step left to follow the roof proper to reach the lip of the slab. Small pockets take you across to the chains. Don't clip early, climb it all the way out to the lower-off!

6 Normandy ★★★   22m F6b+

FA Willis Morris 9 September 2017

The traversing slab gives the easiest path through the steepness. Gain the slab then traverse hard left to where the fun begins. With the roof confining you from above and the footholds growing ever closer to the drop off beneath you, you're trapped in a precarious position. Technical footwork and funky climbing allow a series of

committing moves on small undercuts to gain the corner beneath the final roof which is tackled head on by a difficult rock over. Extender draws are a must for alleviating rope drag on the traverse and best cleaned on second.

CANNON WALL

1 **Mortar** ★★   13m F6b
FA Kristina Ledingham 26 April 2021

2 **Flak Cannon** ★   13m F6b+
FA Matt Thomson & Purkle 26 April 2018

3 **Howitzer** ★★   13m F7a
FA John Milloy 23 April 2021

4 **Predator Missile** ★   12m F7a+
FA Nicholas Wylie 29 August 2021

5 **Ration Pack** ★   11m F6a
FA Scott Dearie 27 April 2021

5a **Empty Rations** ★   11m 6a+
FA Scott Dearie 27 April 2021

6 **Kill the Captain** ★   13m 6a+
FA Rory Watson 29 August 2021

THE RISING SUN WALL

1 **Hirohito** ★   15m F7b
FA Pete Roy 2021

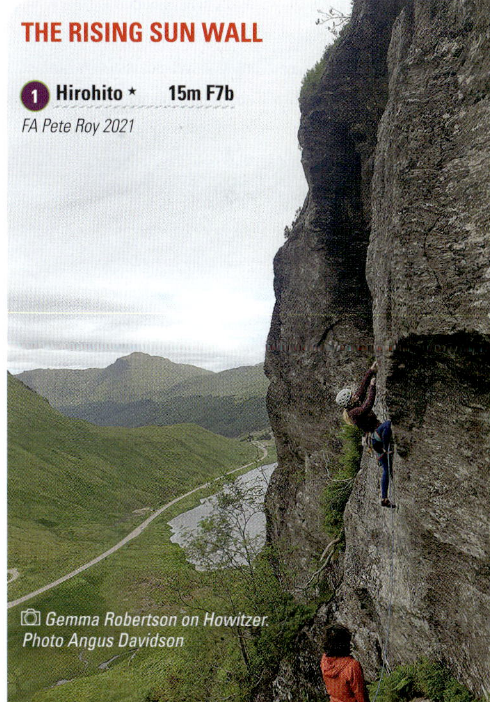

Gemma Robertson on Howitzer.
Photo Angus Davidson

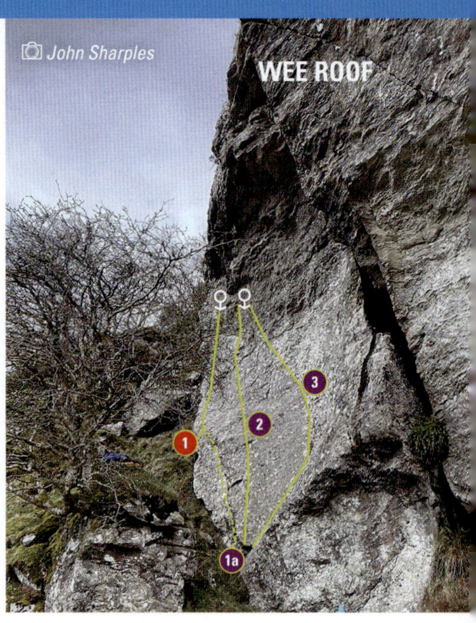

THE PADLOCK

NN 2061 0076 **Alt:** 190m **S** **W** 30min

Sitting up on the hillside, a huge padlock shape boulder can be seen above the village of Lochgoilhead. This boulder, others around and the crag behind, offer a pleasant climbing venue, with grand views across Loch Goil. The exposed position will catch the sun and any breeze, offering frequent protection from the summer midge. The crag still has scope for further development but already proffers some excellent climbing from F4 to F7b+.

Access: Park in the car park in Lochgoilhead village, opposite the Post Office.

Approach: Follow sign posts for the Cowal Way footpath towards Arrochar. A road, then track, past the Scouts' outdoor centre, leads directly up the hill. 300m from the car park, a forestry road is reached. Cross the forestry road, continue up hill to the left on the Cowal Way for a further 300m until another good path cuts back right through rhododendron and brambles. Follow this for 350m. At a hairpin under Scout Crag, cross the small stream to follow an occasionally vague, gently rising path until thick rhododendron is reached. Bypass the rhododendron by heading steeply uphill, from where the crag will be clearly seen 150m ahead.

THE PADLOCK BOULDER

NN 2057 0081 **Alt:** 170m **S**

The eponymous huge boulder, reached first on the approach.

1 Combination ★★ **14m F5c**

FA Steve Kennedy, Cynthia Grindley & Colin Moody
18 September 2022

2 Keyhole ★ **11m F5b**

FA Cynthia Grindley, Colin Moody & Steve Kennedy
18 September 2022

3 Gemmy Bar ★ **11m F5b**

FA Colin Moody, Steve Kennedy & Cynthia Grindley
18 September 2022

4 Paisley Boys ★ **9m F5c**

FA Steve Kennedy, Cynthia Grindley & Colin Moody
17 September 2022

5 Slim Jim ★ **7m F4a**

FA Steve Kennedy, Cynthia Grindley & Colin Moody
17 September 2022

WEE ROOF

1 Thoozins ★   9m F6c

FA John Sharples 23 April 2023

1a Hunners an' Thoozins ★   13m F6c+

FA Koon Morris 23 April 2023

2 Lost Track of Time ★★   13m F7a+

FA Koon Morris 20 May 2023

3 The Bonnie Arc Ae Goil ★★   15m F6c+

FA John Sharples 20 May 2023

BRAW SLAB   NN 2061 0076 **Alt:** 190m 

1 Haud Yer Wheesht ★★   15m F6a

FA Simon Smith 28 January 2023

An inviting hand crack up the vertical wall to join 1 on top of the plinth.

2 Awfy Braw ★★   13m F6a

FA John Sharples 3 July 2022

The striking corner feature on lovely holds.

3 Stushie ★★   13m F6a+

FA John Sharples 14 August 2022

Climb the pleasantly thought-provoking slab just right of 2.

4 Ice Breaker ★   12m F5c

FA Sven Jonuscheit 14 August 2022

Start as for 5, break off left just before the chimney and climb to the lower-off shared with 3. A steady introduction to the crag.

5 Lum-y-goil ★★   16m F6a

FA John Sharples 10 July 2022

Climb the West coast of the 'Scotland' shaped flake to gain the Grampian ledge under the wide chimney. Climb the chimney towards the chains. Sure to be become a well-loved classic.

Koon Morris on the first ascent of the fingery crag testpiece Bowser. Photo John Sharples

BOWSER WALL

6 Yogi ★   **14m F6c**

FA Koon Morris 3 July 2022

Climb the, initially slabby, then overhanging corner to clip the chains with either a long reach or good footwork. A fine test of power and technique repertoire.

7 Bowser ★★   **16m F7b**

FA Koon Morris 14 August 2022

The crag's wall climbing test piece. Weave up the bulging wall with constant interest, making the most of what will sometimes be found to be meagre pickings, when it comes to the hold selection offered.

8 Shoogly ★★   **15m F7a**

FA John Sharples 19 June 2022

Excellent sustained climbing. Mind the gap and establish yourself at the start of the route. Pre-clipping the first bolt should be considered to avoid risk of both climber and belayer disappearing into the 'bergschrund'. Climb the initial short techy slab and on towards a crack system through the overhanging bulge. Having dealt with the bulge, gain the ledge on the right for some respite before the final few moves up the wide crack to the lower-off. A nouveau classic.

9 Project ★   **15m F6c**

Potential line which has previously been top roped from a low start just left of 10.

10 Alfinn ★★   **15m F7a**

FA Koon Morris 19 June 2022

Glorious techy climbing up the deceptively steep wall, via a cruxy passage through the niche. Provides very satisfying sequences.

11 Goil Toil ★★   **12m F7b**

FA Jamie Skelton & Nicky Brierley 30 August 2022

The short line of glue-in bolts at the right end initially offers thin devious moves to a crux slab. Beyond, a more dynamic sequence gains the ledge and anchor.

FAR RIGHT

1 Size Isn't Everything ★   **15m E1 5b**

FA Simon Smith & Mina al-Murani 22 April 2023

The thin left crack.

2 Bigger is Better ★   **12m VS 4c**

FA Erin McLeod & Dalton Moore 28 February 2023

The right crack.

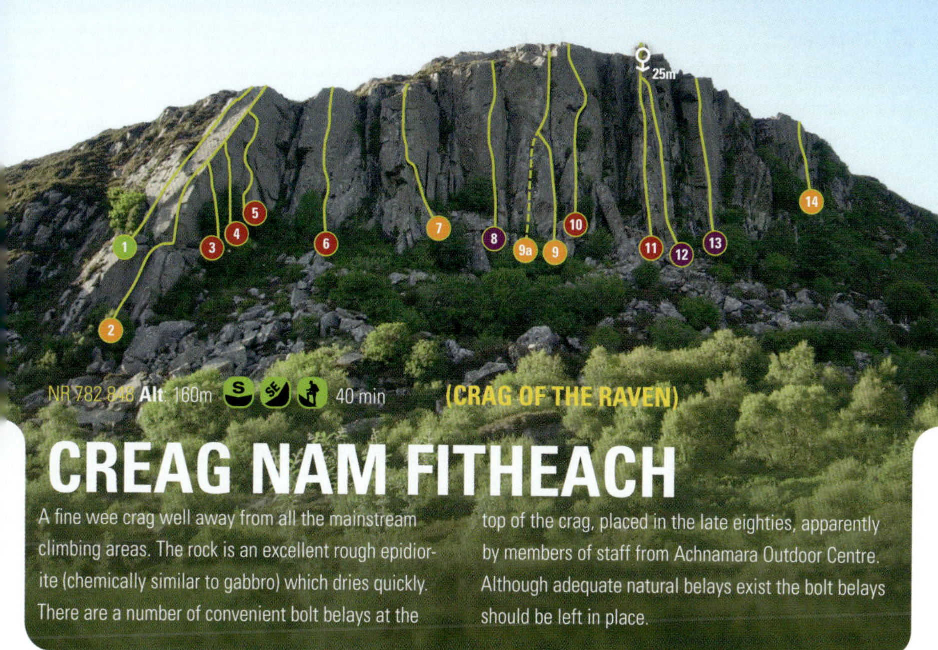

NR 782 848 **Alt**: 160m S SE 40 min **(CRAG OF THE RAVEN)**

CREAG NAM FITHEACH

A fine wee crag well away from all the mainstream climbing areas. The rock is an excellent rough epidiorite (chemically similar to gabbro) which dries quickly. There are a number of convenient bolt belays at the top of the crag, placed in the late eighties, apparently by members of staff from Achnamara Outdoor Centre. Although adequate natural belays exist the bolt belays should be left in place.

Access: From the south: turn left off the A816, 2 miles/3km north of Lochgilphead heading west along the B841 on the south bank of the Crinan Canal for 3 miles/4.5km to Bellanoch. From the north: turn right 1 mile/1.6km south of Kilmartin and follow the B8025 south-west across the Moine Mhor for 4.5 miles/7km to Bellanoch. At Bellanoch: head south down the B8025 for just over a mile/1.8 km then take the left fork down the single track C-class road through Achnamara to take a left turn, signposted Kilmichael of Inverlussa. Park on the grass in front of the church. (3.5 miles/5.5km).

Approach: Follow Land Rover track through the yard opposite the church which leads up round past a small hydro building. Continue more steeply up the track through a gate and up through the forest to its end at a small dammed loch. Gain the left end of the crag by a small path heading diagonally right from 100m before the dam (small cairn).

Descent: Down either end of the crag, or more conveniently by abseil. There is a good flake just back from the top of *America* which takes a thin sling.

1 Moby Dick ★★ — 40m Very Difficult
FA Alex Small & party 1940s

The pleasant slabby face forming the left side of the obvious arête. Start up the right side of some wedged blocks and climb a fine crack finishing slightly rightwards.

2 Captain Ahab ★ — 40m HVS 5a
FA Dave Hayter & J.Mallows early 80s

Below the slab is a prominent right-facing corner. Layback this to a ledge near the arête (possible belay). Continue up the obvious line just left of the arête (poorly protected) and finish more easily up the arête.

3 Maneater ★★ — 10m E3 5c
FA Dave Griffiths & Cameron Bell (1 PR) 21 February 1988; PR removed by Gary Latter prior to #2nd ascent 1988

The gaping off-width crack is not quite as ferocious as first appearances might suggest. Climb it mainly using a good finger crack for protection on its right wall. Stroll up the very much easier arête to finish, or lower off from a good spike a short way up this.

4 Temptation ★ 12m E3 5c

FA Dave Griffiths 20 July 1988

The slim left-facing groove just right of 3. Start behind a rowan. Climb the groove and flake and either finish up an easy arête or move left and lower off a good spike.

5 Crucifixion Crack ★★★ 20m E2 5c

FA Graham Little & Pete Linning 9 October 1982

Excellent well protected climbing. Start immediately right of a rowan tree. Up a ramp to make difficult moves to pull into a triangular niche. Pull over the overlap then hand traverse the diagonal flake-crack leftwards to gain the easy upper arête to finish.

6 Pocket Wall ★★ 20m E1 5b

FA Graham Little & Pete Linning 9 October 1982

The centre of the wall facing 5. Easily up on pockets to a ledge then continue with technical moves to stand on this. Either finish directly or escape out left (HVS).

7 The Razor's Edge ★★★ 20m HVS 5a

FA Graham Little 10 July 1981

Superb climbing up the prominent 'Africa-shaped' flake crack on the right flank of the third buttress from the left. Up a rib to a ledge to gain and layback/jam the fine flake crack and a short corner to finish.

8 The Changeling ★★ 20m E5 6a

FA Dave & Ian Griffiths (headpointed) 16 September 1995

The left arête of the pod-shaped groove (*Metamorphosis* E2 5c). Start just to the left of the groove at a wide crack. Up this to stand on top of a block. Now climb direct up the bulging arête (PR) to where the angle eases and swing into a hanging crack on the right. Follow this to where a move left to a ledge can be made. Up a ramp to the top.

9 America ★★ 25m VS 4c

FA Graham Little & Colin Ritchie 6 June 1982; Direct Start Gary & Karen Latter 5 November 2016

Start in the deep recessed groove 7m down, right of 8. Climb to the top of a large flake on the left. Go directly up

for 3m, move right into a crack and finish up this. 9a *Direct Start* gives a cleaner ★HVS 5a.

10 The Trial ★ 30m E1 5b

FA Graham Little & Colin Ritchie 6 June 1982

Start to the left of an enormous leaning block, which moves; beware! Climb a rib to a small bush. Cross an overhang to gain a roofed niche then direct up the wall above.

11 The Prow ★ 30m E1 5c

FA Graham Little & Colin Ritchie 5 June 1982

The prominent jutting rib. Using a flake on the left, gain and climb a finger crack parallel to the edge. Up the exposed arête and the short, difficult wall above.

12 Crystal Vision ★★★ 30m E5 6b/5.12a

FA Dave Griffiths 29 July 1988

The Fingerlicker of Scotland – superb sustained finger-jamming with a hard crux – very well protected. Start at the lowest point of the right wall of 11. Climb up and left to good slots then make hard moves rightward to gain the crack. Continue with difficulty to a good resting ledge. Continue up the groove above, exiting out left through a gap to finish up the crux wall of 11.

Cleaned in 2016.

13 Not Waving, But Drowning ★★ 30m E6 6b

FA Mark McGowan & Stuart Lampard (headpointed) August 1988

The striking square-cut arête. Climb up to below a roof at 6m. Reach round to a horizontal break (protection). Make moves up to a fingertip undercut (Roller #2) then make sequential slaps (crux) to a poor peg runner. Climb an arête, passing a F #2 placement on the left to a good ledge below a bulge (RP #4). An awkward move onto the ledge leads to a precarious move up the wall to a thin horizontal break (RP #2 & #3). Swing round arête to good holds and protection. Follow flake holds to top.

The 10m buttress bounding the far right end of the crag contains a trio of Severes, together with the fine diagonal jam crack of 14 *Chamonix Crack* ★VS 4c.

GALLANACH, ARDBHAN CRAIGS

NM 841 289 **Alt:** 10-30m (w) (NW) 2-10min

A superb selection of very accessible roadside crags on the outskirts of Oban, with splendid views over Kerrera and Mull. These conglomerate crags overlook most of the southern stretch of Gallanach Road, between the harbour and Kerrera Ferry Terminal. The rock quality is mixed, with better quality rock generally found in the lower regions, so although some crags are around 70m high, only the bottom half may have routes. Some sectors have a layer of solid limestone on top of the base rock with associated features like tufas and pockets which makes for interesting and varied climbing. On the whole the rock does not lend itself to trad climbing; most routes are sport climbs equipped with stainless steel or titanium glue-in bolts. With the exception of Oakley, the crags don't tend to seep, making climbing possible throughout most of the year. Take care as there is still a lot of loose rock on the newer routes; these will stabilise with further ascents. **Helmets advisable**. There is considerable scope for further development.
Access: From the main ferry terminal in Oban, continue south on Gallanach Road for 1.5 miles/2.4km to park on the verge just after the passing place (**NM 8376 2849; 56.399022 , -5.5056959**) just before ferry slip. If you drive past Rockies cottage on the left, you've gone too far. Alternatively, park a few hundred metres further on at the Kerrera Ferry Terminal and walk back up the road. At the height of the summer it can get busy; there is further parking on a wider section of the road 0.3 miles/0.5km further on, just before a meshed section of cliff. **Do not park in any of the passing places**.

THE ROCKIES

NM 8379 2844 **Alt:** 30m (w) 3min

The obvious clean open wall above Rockies cottage, bounded by a corner and fin on the right. Stakes at the top.
Approach: via a path leading up rightwards from **The Towers**.
Descent: down track on right to the ferry slip.

1 Beer Engineer ★★ 25m HVS 5a
FA Alex Thomson & Joe Turner September 2020

The hanging finger-crack splitting the headwall. Start at the right end of the lower slabs. Climb the easy slabs leftwards following the faint crack-line to below the overlaps (good but spaced gear). Step on to the slab on the right at the first overlap, then rejoin the crack through the second overlap and to the top. Good stake belay.

2 Kerreran Curry ★ 25m HS 4b
FA Alex Thomson, Solene Giraudeau-Potel & Joe Turner September 2020

The prominent corner to the right of 1. Quite traditional.

THE TOWERS

NM 8381 2851 **Alt:** 25m (W) 3min

Directly above the parking are towering crags with a large gully to the right.

Approach: Follow faint paths from the road up towards the huge leaning wall and corner.

1 The Braid ★★   35m E1 5b

FA Alex Thomson & Clár Nic Giollabháin April 2021

The corner systems and slab 5m left of 2. Follow the clean right corner for 5m. Step left below the chossy roof to below the second corner-crack. Climb this with the help of some surprising holds. Follow the crack above trending left and stepping left into another crack when needed. Wander up the easy slabs above on good but spaced gear aiming for good finishing cracks and the small tree at the top. Stake, sapling and boulder at top to belay. An excellent journey.

2 Vomitorium ★★★   30m F6b

FA Mike Tweedley May 2018

A very fine long pitch with nice moves and a great view. Probably the most climbed route in the area and very clean if you don't stray off route.

3 International Kidney ★★   30m F6b+

FA Dave MacLeod 13 August 2022

The right line on the smooth wall of 2, moving past the big detached flake at half height with a crux moving left on undercuts at the last bolt.

3a Kidney Harbour ★★★   30m F6b+

FA Dave MacLeod 2022

A really nice solid pitch on great rock throughout. Start up 3 until stood on the large detached flake, then step left into the last 3-4 bolts of 2.

4 Herbie Mhor ★   30m F7b

FA Dave MacLeod 2022

The great leaning wall. Start up the slab and tread lightly past the dubious rock pillar. The route beyond provides a superb lactic challenge on clean, sound rock.

5 Majestic ★★★   30m F6c

FA Dave MacLeod 13 August 2022

Excellent climbing and positions up the flying arête. Good pockets where needed.

6 Tabula Rasa ★★   30m F6b

FA Dave MacLeod 13 August 2022

The grooves and walls just right of the flying arête.

ROADSIDE

NM 8382 2857 **Alt:** 10m (W) 2min

The obvious buttress just metres from the road. The rock here is a very interesting mix of conglomerate with a layer of limestone under the overhang which extends to the ground on the right end.

1 DMZ ★★   15m F6a

FA Andy Corbe 28 May 2021

First route from the right side of wall up the pink coloured limestone tufa to the overhang. Step left to the LO.

1a Little Rocket Man ★★   40m F6a

FA Andy Corbe 1 June 2021

Start up 1, then traverse the entire base of the overhang to finish at the corner with the trees. Second pitch to be added, but for now there is an anchor installed on the tree. (60m rope gains base).

2 Bolt out of the Blue ★★   15m F6a

FA Andy Corbe 28 May 2021

Second route from the left through the obvious limestone pockets to finish at the same LO as 1.

3 My Button's Bigger than Yours ★★   18m F6b

FA Andy Corbe 6 June 2021

Line up the blank wall left of 2.

4 American Dotard ★★   18m F6a+

FA Andy Corbe June 2021

Starts left of 3. Up through the little roof and joins the obvious tufa by the main overhang.

5 Mission Creep ★★   35m F6a+

FA Dave MacLeod 25 August 2022

Start in a scoop near the left end. Climb up then left through a bulge on good holds and through the gap in the first overhang on good holds to the bottom edge of the slab above (rams horn here to redirect the rope when stripping the route). Balance leftwards on good foot ledges to a big hold in the left facing groove. Bridge up the groove to the LO.

6 Temple of Doom ★★   20m F6a+

FA Andy Corbe 5 June 2021

Start just left of the big tree growing against the wall. Take a line of black rock to the small corner, through the overhang and into the large hollow above to LO. Some of the best rock at this crag.

7 Nyx Got Zoomies ★★★   23m F6a

FA Andy Corbe 21 July 2022

Start 5m left of 6. Straight up the slab just right of the overhangs into the broken corner and then traverse left onto the superb slab above to finish. A fixed carabiner is in place to back clip on the descent.

8 Life of Pi ★★   20m F6c

FA Andy Corbe 5 September 2022

Start 10m left of 7, right behind the tree pressed up against the wall. Follow the leftward trending ramp and cracks to the overhang and then haul up rightwards through the overhang onto the beautiful slabby wall above. Shares the same LO as 7.

SIGMA BUTTRESS

NM 8396 2876 **Alt:** 10m (w) 4min

Very visible high buttress a few hundred metres north along the road, with a large obvious pit at the base. **Approach:** follow path skirting round the right side of the pit, starting from behind the passing place sign.

1 Kilbowie ★★   25m F6c

FA Alex Thomson July 2021

The left line, trending in towards the centre of the wall before heading back left. Potentially some loose rock if you head too far left in the first half.

2 Thin White Teuch ★   25m F7a

FA Alex Thomson 2021

The line to the right of 1. Bouldery moves through the bulge, followed by thin, teuchy moves, out right.

3 Arran Victory ★★★   30m F6b+

FA Dave MacLeod 2022

The leftmost line on the main buttress. Follow the crack in the slab then a left-facing groove. Where this peters out, move left (crux) to gain the upper cracked groove.

4 Pink Eye ★★   30m F6c+

FA Dave MacLeod 2022

Follow the crack in the slab as for 3 but break rightwards from the steep groove to reach an overlap. Pull through this to gain the upper steep left-facing groove.

5 Skerry Champion ★★   35m F6c+

FA Dave MacLeod 2022

Aims for the rightmost of the three left-facing crack/grooves at the top left of the buttress. Follow the pocketed slab and corner of 8 then continue up and left following a flake-crack to the spiky ledge. Crank between good holds up the overhanging wall to reach the crack.

6 Cream of the Crop ★★★   35m F7a+

FA Dave MacLeod 2022

Follow 5 to the spiky ledge. The smooth headwall directly above gives good sustained climbing, trending rightwards to join 7 at the last bolt. Excellent headwall.

7 Electra ★★★   35m F7b

FA Dave MacLeod 2022

Climb to a few moves along the right-slanting ramp of 8. Tackle the bulging overhang above with burly laybacking up the undercut flange.

8 Sigma ★★★   35m F6a

FA Dave MacLeod 2022

A great outing across wild terrain. The central snaking line of weakness, weaving its way through the steep wall. Climb the white pocketed wall and short corner above. Step slightly right and climb another white wall trending rightwards to gain a big right-slanting ramp. Move along this and then continue into the steep finishing groove (crux) of 10. A few long quickdraws are useful to avoid rope drag.

9 Irish Cobbler ★   35m F7a

FA Dave MacLeod 2022

Start up 10 but step back left on the pillar and follow a flaky crack to the right end of the ramp of 8. Attack the groove steep bulge above, just right of 7. The slab above is easy.

10 Miss Blush ★★   35m F6c

FA Dave MacLeod 2022

A steep start through a bulging right-facing corner, followed by a blunt prow trending rightwards to the top of the pinnacle. Step right into the overhanging groove and follow it to the top.

11 Melody Grooves ★   35m F6b

FA Dave MacLeod 2022

Climb the bulging right-facing corner and blunt prow but move rightwards onto grass ledges and a lower-off. Belay here if you want to climb it in two pitches, other wise use a few long quickdraws moving right onto the grass ledges. From the lower-off gain the open groove near the right edge of the crag and follow it to the top.

OAKLEY

NM 8407 2883 **Alt:** 20m  6min

This buttress features an obvious ledge around 15m up with an Oak tree growing out of it.

1 Fist Full of Green ★   15m Severe 4a

FA Andy Corbe 1 June 2021

Obvious left leaning offwidth crack with some nice fist jamming at the top. Currently finishes at tree with lower-off sling/ring, but with some more digging could be extended higher up the wall.

2 Gold Digga ★   15m F6a

FA Andy Corbe 1 June 2021

Start just right of the crack.

3 Blow the Oak ★   15m F6a

FA Andy Corbe 1 May 2022

Start 5m right of 2.

4 Blue Sky Mining ★   15m F6a+

FA Andy Corbe 29 April 2022

Starts a few metres right of 3. Line between the two diagonal crack lines.

5 Pocket Monster ★★   20m F6a

FA Andy Corbe 5 June 2022

Starts 5m right of 4, sustained climbing with occasional relief from giant pockets but beware the monsters that reside within.

OAKLEY LEFT   NM 8403 2879 **Alt:** 30m W

1 Cotyledon ★★   35m E1 5a
FA Alex Thomson & Andy Corbe 29 May 2021

The huge corner system about 60m up left from the main crag. Start up an easy slab to a ledge. Small wires and a cam in a pocket out left protect steep moves up and into the corner system. Follow the corner to a large ledge, then up to a roof. Rock on to the slab above from round to the left and continue up to atop the pinnacle. Step right and top out delicately through the undergrowth to a tree belay. Absorbing climbing, well protected with small cams and micro wires.

THE RICHES

NM 8443 2916 **Alt:** 30m W NW 10min

Crag is well hidden!

Approach: Head north back towards town for 0.6 miles/1km; 150m beyond some roadside houses is a passing place opposite the entrance to a house called Ardcuan. Head steeply up through the trees.

1 Ripper Ramp ★   12m F6c
FA Dave MacLeod 2022

A boulder move off the ground (stick clip useful) leads to easier climbing up the hanging ramp.

2 Ill Wind ★   12m F6c
FA Dave MacLeod 2022

An unusual route bridging and laybacking up the edge of the offwidth crack, with a tricky move rightwards to the anchor.

3 Coming from You ★★   15m F6c
FA Dave MacLeod 2022

Left side of wall.

4 Super Rich ★   15m F7b
FA Dave MacLeod 2022

Thin technical climbing leading to a good ledge, then easier headwall.

5 Mango Loco ★★   15m F7a+
FA Mike Tweedley 27 May 2021

Good climbing left of the blue streaks to a break, then easier above.

5a Mango Maxx ★   18m F7b+
FA Dave MacLeod 2022

Follow 5 to just past its crux, swing left to cross the headwall of 6 and continue left to the finishing holds of 3.

6 Ultra Rosa ★★   12m F6c+
FA Dave MacLeod 2022

Technical climbing leads rightwards to better holds up the right arête.

7 Glitterati ★★   12m F6a
FA Dave MacLeod 2022

Nice climbing on some of the best rock on the crags.

8 First News ★★   12m F5
FA Dave MacLeod 2022

The pleasant corner and slab above.

9 Potato Hack ★   8m F6c
FA Dave MacLeod 2022

Steep burly climbing up the left side of the prominent arête.

Prince of Thieves, Lower Tier, Erraid. Jules Lines soloing.

ISLE OF MULL

The second largest and one of the most accessible of the Inner Hebrides. Although it is one of the wettest islands of the Hebrides all the climbing described is situated around the west end of the low lying Ross of Mull peninsula, benefiting from a much lower rainfall than the hillier centre of the island.

Access: Caledonian MacBrayne operate vehicle and passenger ferries from Oban – Craignure (1 hour); Lochaline – Fishnish (15 minutes) and Kilchoan – Tobermory (35 minutes). (☎ 0800 066 5000; www.calmac.co.uk).

Accommodation and Amenities: Campsites: Shieling Holidays, Craignure (☎ 01680 812496; www.shielingholidays.co.uk); Pennygown Holiday Park, 4 miles north of Fishnish (07799 711125; www.pennygownholidaypark.com), Salen Bay Campsite, Salen (01680 300250; www.salenbaycampsite.co.uk). YH at Tobermory (☎ 01688 302481; www.hostellingscotland.org.uk); Craignure Bunkhouse (☎ 01680 812043; www. craignure-bunkhouse.co.uk); innumerable Hotels & B&Bs – **TIC** at The Pier, Craignure (☎ 01680 812377; www.visitscotland.com).

Wild camping is possible close to all the crags described, though there are few streams for drinking water. There is a wonderfully situated campsite (☎ 01681 700427; www.fiddenfarm.co.uk) at Fidden Farm, 1.3 miles/2km south from Fionnphort, overlooking a fine sandy beach at the north end of Erraid. There is a small store/post office in Bunessan, and pubs in Bunessan and Fionnphort.

Gary Latter starting up the initial finger crack on the first ascent of the immaculate and engrossing Dreamline, Paradise Wall (page 88). Photo Jules Lines

SCOOR

NM 410 188   15-20min

A fine assortment of schist crags on the south coast of the Ross of Mull. The area has a pleasant holiday atmosphere with a fine sandy beach in an idyllic setting. A herd of wild goats is in residence and there are often Golden Eagles around.

Access: Follow the A849 west across the island to Bunessan. Turn left (south) 200m beyond the school and follow the minor C-class road which becomes a rough though driveable track east past Loch Assapol, curving round right (south) to park overlooking ruined chapel (2.4 miles/3.8km from Bunessan).

Approach: Follow the track which heads out south-west (not the continuing road to Scoor House). At the third gate the beach is straight ahead. All are within 5 minutes from the beach. Garbh Eilean is the rough rocky 'island' straight ahead, while **Dune Wall**, **Beach Wall** and **The Slab** are all to the left (east), the latter hidden from view.

Tystie (page 77), The Slab. Gary Latter climbing. Photo Karen Martin.

GARBH EILEAN
(ROUGH ISLAND)

The 'island' on the left side of the beach. It becomes tidal only on spring tides or following a big storm when the sand is washed away.

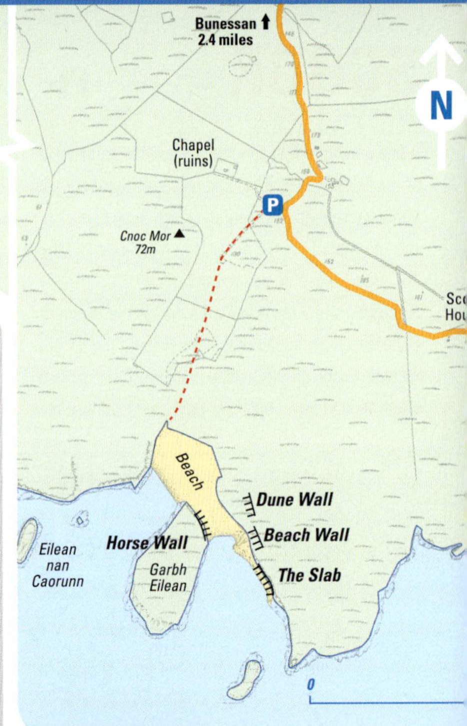

HORSE WALL   NM 411 185    15min

A clean gently overhanging wall on the tiny tidal island opposite **Beach Wall**.

1 Lauracheval ★★   13m E3 5c

FA Mark Garthwaite & George Szuca 5 September 1993

The left-slanting crack on the right side of the wall starting at the right edge of the dark stained wall.

2 See You in Hell Soldier Boy ★★★   14m E6/7 6b

FA Mark Garthwaite (headpointed) 15 October 1994

Start at good holds 1m left of the black-stained wall. Move up to a short sharp flake and move left to a very sharp finger jug, shake out and gear. Make very thin technical moves up the wall above to reach the diagonal crack. Pull out right to a finger hold, stand in the crack and reach over the top to finish. RPs and HBs #2 - #5 essential.

3 Eat My Shorts ★★   13m E5 6a

FA Mark Garthwaite & Colin Moody 25 September 1994

The layback crack at the left side of the wall. Flared and quite difficult to protect.

4 Wild Swans ★   11m E1 5b

FA Colin Moody & Mark Garthwaite 15 October 1994

The crack up on the left.

5 Photo Finish ★   8m E3 5c

FA Colin Moody & Louise Gordon Canning 28 June 1997

A bold start above a poor landing. Left of 4 is a small slanting overlap. Climb up right to the right end of the overlap, step left and climb the crack.

DUNE WALL NM 412 187 **Alt:** 15m (w) 15min

The first small wall left (east) from the approach. From a distance a coating of black lichen gives an impression of wetness but the wall dries quickly. If natural belays are found lacking there is a stake well up from *Wallcreeper*.

From the left are 1 *The Arête* HVS 5a; 2 *Flick-flake* HVS 5a – the groove and sharp layback crack immediately right & the thin crack of 3 *Tippidy Doodah* ★ E4/5 6b. The shallow corner is 4 *Red Shafted Flicker* E1 5b.

5 Wallcreeper ★   10m E2 6a

FA Colin Lambton, Colin Moody & Mike Rosser 1994

Climb the scoops in the centre of the wall. A steep start gains a left-sloping ramp. Mantel out right then make a thin move (crux) to good holds below the top.

6 The Crystal Ship ★   10m E4 6a

FA Tom Charles-Edwards 1996

The blind groove just right of 5.

7 Troglodyte ★★   10m HVS 5a

FA Colin Moody & Allan Petrie 27 September 1993

Climb the crack to the overhang, step left and layback to the top.

The crack to the right is 8 *Thorns* HVS 5a.

9 Dune Chough ★   10m VS 4c

FA Colin Moody & Allan Petrie 27 September 1993

The cracks on the right side of the wall. Climb up to a small overlap, step right and follow the flakes.

Tweekie Pie, Beach Wall. Gary Latter climbing. Photo Karen Martin.

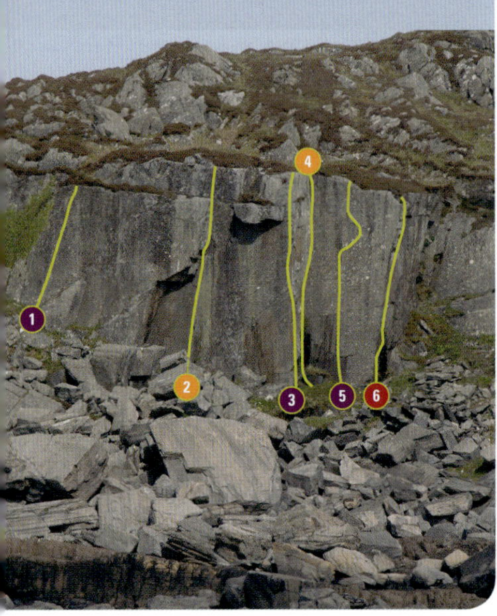

BEACH WALL   NM 413 186 **Alt:** 20m  15min

The short wall and the rocks beside it overlooking the boulders at the east (left) end of the beach. There is a large patch of ivy at the left end.

1 Better Than a Poke in the Eye ★   8m E4 6a
FA Mark Garthwaite 29 August 1993

The right-slanting crack at the left side of the crag just right of the ivy.

The crack and wide corner crack is 2 *Fungy* ★ HVS 5a.

3 The Reality Dysfunction ★★   15m E6 6a
FA Niall McNair May 2004

The prominent square-cut arête - low in grade but serious. Start on the arête and the 'drainpipe' feature before moving onto the arête proper at half height. Crux right at top.

4 Waves and White Water ★★   15m HVS 5a
FA Danny Brooks & James Marshall October 1991

The prominent square-cut corner, often with a wet spring at the base.

5 Tweekie Pie ★★★   15m E5 6c
FA Mark Garthwaite 5 September 1993

The thin cracks up the right wall of the big corner of 4. A very thin and technical crux is at the top, protected by RPs. Climb the thin crack to good edges in the horizontal break at two thirds height (small rocks). Move right then back leftwards up the shallow diagonal crack to a desperate thin section to better holds, finishing on a superb flake over the top. Belay stake in the bracken.

6 Sylvester ★   15m E2 5c
FA Mark Garthwaite & Colin Moody 29 August 1993

The w-i-d-e crack right of 5. Take some large cams!

7 Runout ★   15m VS 4b
FA James Marshall & Danny Brooks March 1992

Left of 9 is a recessed area, left again is another ridge. Climb this ridge to a bulge then the left-slanting crack on the left. Strenuous.

8 Bluebell Blues ★   15m E1 5b
FA Danny Brooks February 1993

The left wall of 9. Gain the base of the left-slanting ramp, up this and the corner and step right to finish.

9 Kilvickeon ★★   20m E2 5b
FA Danny Brooks March 1992

There is a boulder field west of The Slab and west again is a ridge. Start left of the base of the ridge. Climb a difficult overhanging layback then move right to gain a foot ledge. Move back left onto the arête and straight up to finish.

10 Mink ★   15m VS 4b
FA Danny Brooks 1992

The ridge high in the boulder field left of The Slab. Start up the right edge and finish up the left edge then traverse the neck to gain the boulder field. Although the crux is low down this is a bold route to lead and second – probably best soloed.

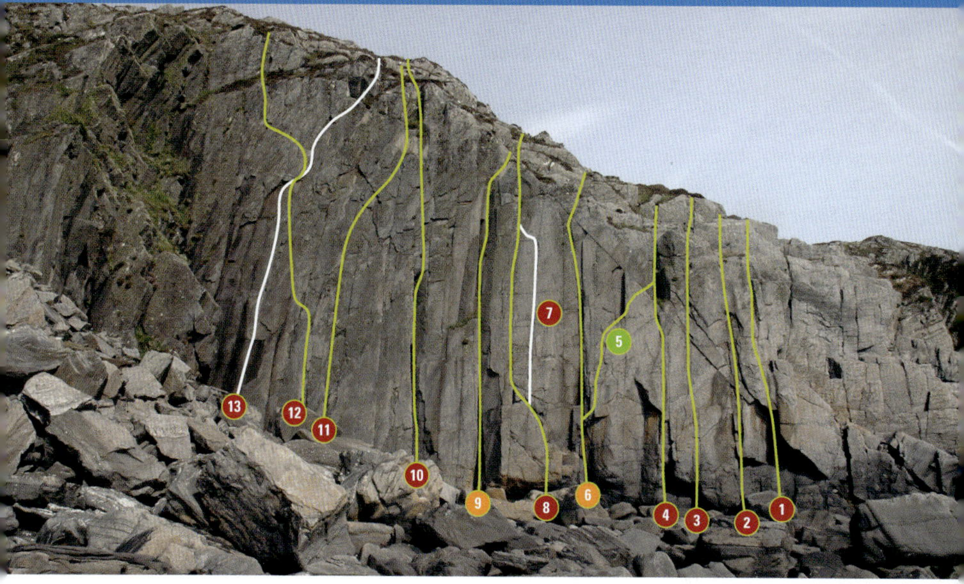

THE SLAB   NM 413 185      20min

Situated behind **Beach Wall** out of sight on the approach. The starts of the first three routes are affected by high tides. Protection is often behind flakes which might not take a big fall; therefore the adjectival grade is often higher than expected. Routes from right to left.

1 Head Butts ★   15m E2 5b

FA Danny Brooks & Colin Moody 18 October 1992

There is a small overhung niche at the right end of the crag. Gain this, climb the overhang to the ledge then continue up the crack in the shallow corner.

2 Tystie ★★   15m E3 6a

FA Colin Moody & Neil Horn 14 June 1993

The thin twin cracks just left of 1 with the crux near the top. Protection is *"a bit funky"*.

3 Bonxie ★   15m E2 6a

FA Gary Latter & Louise Gordon-Canning 7 August 1997

The prominent thin twin cracks in the wall left of 2. Make a hard bouldery start past a good friend slot at the start to better holds, finishing directly by good flakes in the upper wall.

4 Sawfish Crack ★   15m E1 5b

FA Colin Moody, James Marshall & Danny Brooks 28 September 1991

Start 6m left of the niche of 1 directly below a short left-facing corner. Climb to a triangular overhang, step right and follow the sharp crack leading to the finishing corner.

5 Pocket Razor ★   15m Severe 4a

FA Danny Brooks & James Marshall August 1991

Further left is a shallow corner crack. Start up this then move up rightwards on sharp flakes to finish up the final corner of 4.

6 Chough-less ★★   15m VS 4c

FA Colin Moody, Danny Brooks & James Marshall 28 September 1991

The shallow corner crack.

7 Splitting Hares ★   18m E3 6a

FA Michael Tweedley & Cynthia Grindley 2 June 2002

The thin crack right of 8. Wimp onto that route near the top.

8 Everything Flying By ★★   20m E1 5b

FA Colin Moody & Danny Brooks 19 August 1992

Left of the upper half of 6 are three thin vertical cracks. Move up left to gain and climb the left crack.

9 Seal Show ★★   20m HVS 5a

FA Colin Moody & Danny Brooks 20 October 1991

In the centre of the slab are two prominent faults starting as shallow chimneys. Follow the right fault, stepping right at the top. High in the grade.

11 Turnstone ★   25m E2 5b

FA Colin Moody & Mark Shaw 12 July 1994

Left of the left fault (10 *Greased Lightning* E1 5a) is a thin crack and left of this an easier crack, which is gained from the top of the highest boulder. Climb up passing a block at 7m then climb up to and over the bulge at two small overlaps to reach a flake. Follow the flake right to

a rock-scarred recess in the fault and finish directly to belay well back.

12 Where Eagles Dare ★★   25m E1 5b

FA Danny Brooks & Sean Morris October 1992

A fine intimidating wander. Start 7m left of the left fault just left of 10 and left of where the cliff base steepens. Climb a crack for 4m then move left 1m and follow another crack to the small overhangs. Step left and climb the open chimney for 2m. Hand traverse left across the wall to a sapling then direct to the top.

13 Hit and Run ★★   25m E2 5b

FA Colin Moody & Andrew Pedley 22 August 1993

A good route crossing 12. A bold start but low in the grade. Start 3m left of 12. Climb the crack direct to below the right side of the crescent-shaped overhang. Step right to join 12 up the open chimney and continue up and right to finish.

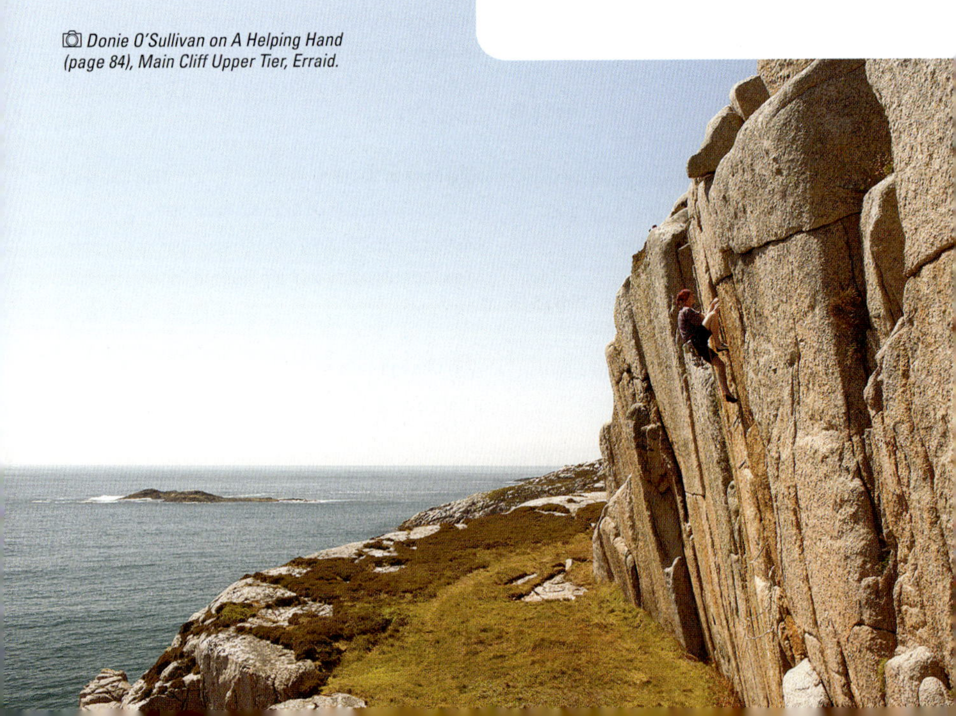

Donie O'Sullivan on A Helping Hand (page 84), Main Cliff Upper Tier, Erraid.

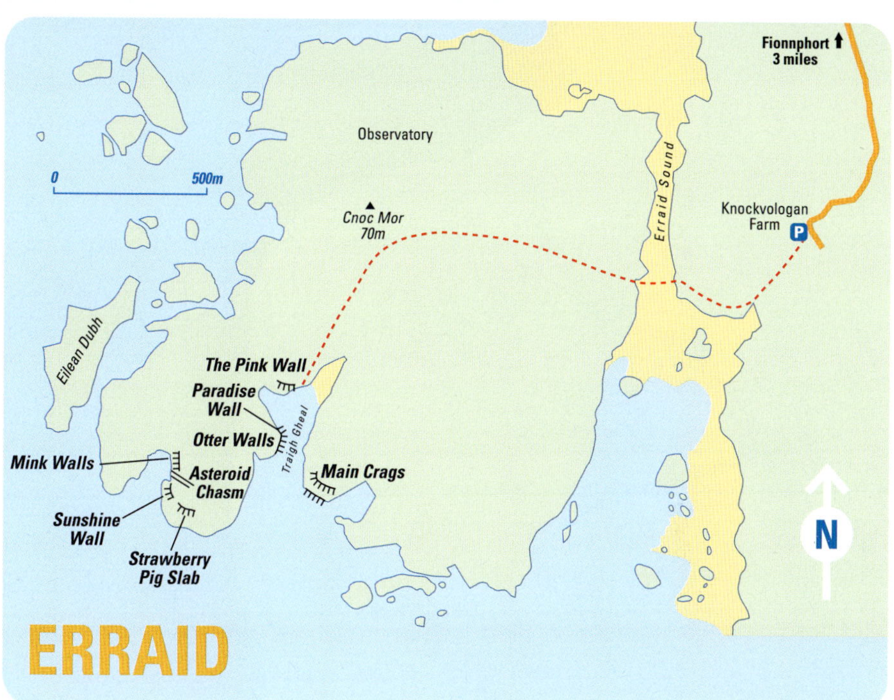

ERRAID

Erraid is a small tidal island situated just off the west-most point of the long peninsula of the Ross of Mull. It is accessible by foot across the sands most of the time, though the shifting level of the sand or spring tides may necessitate wading for approximately 2 hours either side of high water. The climbing is on a range of short but superb granite crags located close to the popular yacht anchorage Traigh Gheal (white sands) at the south-west corner of the island. The island featured in Robert Louis Stevenson's classic Scottish novel *Kidnapped*. Stevenson stayed for a short time at the north end of the island.

Accommodation and Amenities: The area around Traigh Gheal is a perfect haven, with beautiful white beaches and some fine wild camping. A small stream runs into Traigh Gheal, though during prolonged dry spells this dries out. Alternatively, the nearest fresh water supply would be the cottages at the north end, or the farm at Knockvologan. With lots of thick bracken, midges and ticks can be a problem.

Access: Follow the A849 west across the island to its end at Fionnphort. Turn left (signposted Fidden) at the car park and follow the narrow C-class road southwards for 3 miles/4.8km to park on the left, at Knockvologan Farm.

Approach: Follow the farm track south, bending round west down to the sands. Walk or wade across Erraid Sound and head west for 700m then south-west to arrive at the wonderful Traigh Gheal (35 min). For the **Main Crags**, scramble round the shore on the left to pick up a path cutting up then contouring right to arrive at the left end of the **Upper Tier** (10 min). The **Pink Wall** lies just a few minutes west of the beach, overlooking the small west-most beach. The **Paradise** and **Otter Walls** are about 5 minutes south from here. **Strawberry Pig Cliff** and **Sunshine Wall** lie about 10 minutes from here due south-west. **Asteroid Chasm** and **Mink Wall** are 15 minutes due west of the **Paradise Wall** overlooking a small bay.

Gary Latter on the first ascent of *The Wrecker* (page 83). Photo Karen Latter.

NM 293 195 **Alt:** 20m

THE PINK WALL 🟢 🟢 35min

The short crack-seamed wall overlooking the small west-most beach at Traigh Gheal.

1 Five Naked Women ★　　　　　7m HVS 5a

FA Colin Moody & Cynthia Grindley 5 October 2008

Steep crack at left end.

2 Panther　　　　　8m HS 4b

FA Colin Moody (on-sight solo) 2000

Leftmost slanting corner.

3 Pink One ★　　　　　9m HVS 5a

FA Jules Lines (on-sight solo) 17 July 2001

Twin cracks left of corner.

4 Floyd ★　　　　　9m HS 4b

FA Gary Latter (on-sight solo) 26 August 2000

Corner.

5 Elephants ★　　　　　9m HVS 4c

FA Jules Lines (on-sight solo) 17 July 2001

Cracks up left side of front wall.

6 Which One's Pink? ★★　　　　　9m HVS 5a

FA Jules Lines (on-sight solo) 17 July 2001

Next crack to right.

7 Pinky ★★　　　　　10m E2 5c

FA Gary Latter (on-sight solo) 17 July 2001

Central crack.

8 Sea Pink ★　　　　　8m E1 5b

FA Gary & Karen Latter 30 March 2012

The finger crack.

9 Perky ★　　　　　10m VS 5a

FA Gary Latter (on-sight solo) 17 July 2001

Crack up right side of wall.

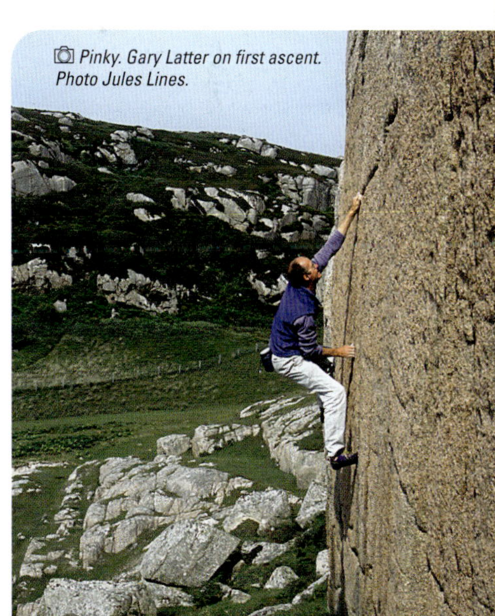

📷 *Pinky. Gary Latter on first ascent. Photo Jules Lines.*

UPPER TIER

NM 294 192 **Alt:** 30m    45min

Much steeper than the lower tier, with many short brutal pitches, often with hard bouldery starts above perfect flat grassy landings.

Descent: Down either side of the crag, the east descent by the easy rock ramp right of the highest section.

The first three routes are on the large block up and left of the main crag.

UPPER TIER – LEFT SIDE

1 Left-Hand Route   9m Severe 4a

FA Colin Moody 10 August 1997

Climb corner/chimney on left side of boulder then continue up the crack above.

2 The Gopher Hole ★★   11m VS 4c

FA Colin Moody & Derek Stuart 11 August 1997

Gain the shelf right of 1, step up then follow break right to finish up the scoop.

2a Gopher It ★   10m E2 5b

FA Michael Barnard & Ewan Lyons 10 April 2019

From the break continue directly up a line of flake edges.

3 One Dead Puffin ★★   9m HVS 5a

FA Michael Tweedley 11 August 1997

Climb the corner on the right to the break, move left and finish up the arête.

The short right-facing corner crack beneath the boulder is Severe 4a.

4 Ledge Route ★   12m VS 4b

FA Colin Moody 10 August 1997

Climb a jam crack then go directly up the slab above (bold), which faces the boulder.

5 Smelly Mussels ★   9m HVS 5a

FA Michael Tweedley & Louise Gordon Canning 14 August 1997

Climb the left side of the block left of 6. Traverse into the centre of the block and finish up the crack with an awkward finish.

The arête on the left is 5a VS 4c; the short chimney further left Severe 4a.

6 The Dead Pool   9m Severe 4a

FA Colin Moody 10 August 1997

Right again are two huge blocks. Climb the V-notch between them and finish up the ramp on the right.

7 Trapezium ★   8m E5 6a

FA Jules Lines & Gary Latter 16 July 2001

Bold thin climbing up the slab just left of the blunt arête.

7a Sea Fever ★★   8m E5 6a

FA Adam Russell (solo) 16 April 2016

The appealing slabby arête and wall. Climb initially on the slabby left side of the arête, with technical moves past the overlaps to a flat hold right on the arête at the base of

the hanging groove – gain a standing position on this to top out. Excellent climbing, no gear, low in the grade.

8 **Bacteria Soup** ★　　　　　　　**12m Severe 4a**
FA Colin Moody 10 August 1997

Right of 6 is a heather ramp with a short face on the right. Climb the flakes and jam crack in the centre of the short face. Step right and finish up the corner.

9 **Blood Orange** ★　　　　　　　**12m Severe 4a**
FA Derek Stuart, Colin Moody & Michael Tweedley 11 August 1997

The corner on the right to the shelf then move right and finish up a further corner.

9a **Blood Red** ★　　　　　　　**10m HVS 5b**
FA Michael Barnard & Ewan Lyons 10 April 2019

The obvious vertical crack.

10 **Rhythm and Stealth** ★★　　　　**12m E3 6a**
FA Adam Russell (solo) 16 April 2016

Excellent climbing up the blunt, initially blank looking arête. Start at a downwards pointing spike and climb direct to the big ledge via some obvious poor holds and well camouflaged good holds. Finish up 9.

11 **Tyke's Lead** ★　　　　　　**11m Very Difficult**
FA Louise Gordon Canning, Karen Martin & Colin Moody 10 August 1997

The chimney near the left side of the highest section of the crag. Above the chimney, step out left past a small spike then finish right, or up the arête at Severe 4a.

12 **Misunderstanding** ★★　　　　　**11m VS 4c**
FA Louise Gordon Canning & Michael Tweedley 14 August 1997

Start right of the chimney. Move up then left towards the chimney then finish up the steep ramp out right.

13 **Pharos** ★　　　　　　　　　**8m E2 5c**
FA Gary Latter (on-sight) 26 August 2000

The widening crack just right of 12.

14 **The Round House** ★　　　　　　**10m E1 5c**
FA Jules Lines (on-sight solo) 22 August 2002

Step off the right end of the boulder and make an awkward move to gain a grey foothold. Layback the flake to gain a grey jug and finish on jugs.

15 **Covenant** ★★★　　　　　　　**12m E4 6a**
FA Gary Latter & Colin Moody 10 August 1997

The prominent steep crack up the highest section of the crag. Sustained and well protected.

16 **Nite Lites** ★★　　　　　　　**10m E2 5c**
FA Colin Moody, Cynthia Grindley & Stephen Porteus 29 July 2000

The crack just right of 15, easing after a bouldery start.

17 **The Wrecker** ★　　　　　　　**10m E3 6b**
FA Gary Latter 31 March 2012

The hanging crack just right of 16. Move right from the first boss hold on that route into the crack, with difficult moves to get established in the finger crack. Continue more easily up the widening crack.

UPPER TIER – RIGHT SIDE

18 Access Route ★ 12m Difficult

Climb by either of the two wide cracks up the slabby section in the centre of the crag, where it changes direction.

19 Shorty 6m Severe 4c

The twin cracks.

20 Chickenhead ★ 10m E2 6a

FA Gary Latter (on-sight) 16 July 2001

The groove and short steep crack right of the slab where the crag changes direction.

21 I Hear of the Red Fox ★ 10m E1 5b/6b

FA Jules Lines (on-sight solo) 22 August 2002;
Direct: Gary Latter September 2012

Start just left of 22. Reach, jump, scratch! for a jug, go up to undercuts and stretch right into 22 for a move to gain a jug then go diagonally left to finish through a small crack in the headwall. Climbed direct it is E2 5c.

22 Walls without Balls ★ 12m E1 5c

FA Steve Kennedy & Dave Ritchie April 1991

The obvious deep crack in the wall just left of the ledge at ⅔ height. Awkward bouldery start.

23 Skerryvore ★★ 12m E3 6b

FA Gary Latter (on-sight) 10 August 1997

The steep crack leading to the left side of the ledge at ⅔ height. A difficult bouldery start leads to better holds in the niche at half height. Continue with interest to gain

the ledge. Finish easily above.

24 Stealth ★ 12m HVS 5b

FA Steve Kennedy & Dave Ritchie April 1991

Climbs the right crack to gain the upper ledge. Hard bouldery start.

24a Deviant Dreams ★ 12m E3 5c

FA Michael Barnard & Ewan Lyons 10 April 2019

Go up 24 to the first jug, then work out right aiming for an obvious hold in the crack system. Finish up this.

25 RLS ★★ 12m E3 5c

FA Gary Latter (on-sight) 10 June 2001

The rightmost of two crack systems just right of 24, starting up short easy right-facing groove.

26 A Helping Hand ★★ 12m E2 5b

FA Gary Latter, Louise Gordon Canning & Colin Moody 10 August 1997

The deep wide central crack. Large cams useful.

27 Catriona ★★ 12m E4 6b

FA Gary Latter 26 August 2000

The fading cracks in the wall just right of 26. Climb up to ledge from right then cracks to horizontal break and good Fs. Continue straight up past flange to a rounded finish.

28 Oliver ★ 13m HVS 5a

FA Steve Kennedy & Dave Ritchie April 1991

Start just left of the base of the descent ramp. Follow a steep flake up leftwards then direct by wide cracks.

 **29 Fagan** ★   **10m HVS 5a**

FA Dave Ritchie & Steve Kennedy April 1991

The wall and short crack above the base of the easy descent ramp.

 **30 The Silver Button** ★   **12m E2 5c**

FA Gary & Karen Latter 31 March 2012

The crack 4m right of 29. Climb the crack at the left end of the lower wall below the descent ramp onto the ramp, then move right and climb the crack above.

 **31 Stonecrop Groove** ★★   **6m E3 6a**

FA Jules Lines (on-sight solo) 16 July 2001

Left-facing groove up the left side of the short wall beneath the descent ramp.

 **32 Dubh Artach** ★★   **7m E5 6b**

FA Gary Latter 5 June 2011

The brutal twin cracks 2m right of 31.

 **33 Minor** ★   **9m HVS 5a**

FA Colin Moody & A.N.Other 1994

Left-facing flake, finishing up jam crack.

 **34 The Vagabond** ★   **10m HVS 5a**

FA Gary Latter (on-sight solo) 17 July 2001

Right-facing groove, stepping up left onto ledge to finish up jam crack.

35 The Islet ★   **10m E2 5c**

FA Gary Latter 31 March 2012

The hanging crack in the centre above the large roof.

Start either up the hand-crack up the left arête, or easier up the left side of the overhung recess, then left along the break. Move up to the horizontal break, then right and up the hand-crack.

36 The Merry Men ★   **8m HVS 5a**

FA Gary Latter (on-sight solo) 26 August 2000

Towards the right end of the crag is a prominent roofed recess. Climb the crack up the right wall.

37 Immersion ★   **10m E4 6a/b**

FA Adam Russell 17 April 2017

The arête and crack, with good climbing and good but awkward to place gear. Pull on and make a move up to a good hold on the left; gain a standing position on this using the crack and arête. Finish direct.

38 Dad of Yan ★   **10m E2 5c**

FA Tess Fryer, Cynthia Grindley & Colin Moody 5 May 2012

Climb the pillar using cracks on both sides, then through the block overhang.

39 The Dynamiter ★   **12m HVS 5a**

FA Gary Latter (on-sight solo) 17 July 2001

There is a short blunt arête at the far right end of the crag. Climb the cracks immediately to its right, scrambling to finish.

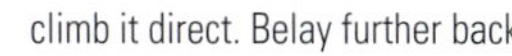 **40 Right Slab** ★   **50m Difficult/Very Difficult**

FA Colin Moody (on-sight solo) late 90s

At the far right end of the crag is a narrow easy-angled slab. Climb it, stepping out left to avoid a thin section, or climb it direct. Belay further back.

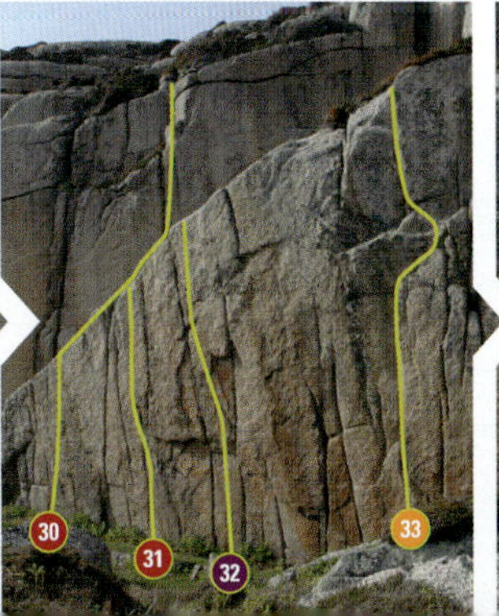

LOWER TIER NM 294 192     45min

A small enclosed bowl at sea level. Only a couple of the routes on the right side of the crag are tidal, though the crag could be affected by rough seas.

Descent: Scramble down the left (east) side of the slabby middle tier then negotiate a unique little narrow rock archway (fixed rope in situ) to gain the centre of the crag. There is also an easy (Mod) descent down the V-gully at the west end of the crag, behind the Sentinel. Routes on the right side of the crag can also be gained by an easy scramble in from the east at low tide.

The cracks in the gully wall at the back of the Sentinel are, from the left: VDiff, VS 4c and VS 4c.

1 Blurred Lines * 9m E6 6b
FA Gary Latter 18 April 2015

The thin diagonal crack emanating from sharp left arête. Start up the left side (pro), then swing round the arête and continue with difficulty to join and finish up the groove of 2.

2 Disingenuous * 7m E1 5b
FA Gary Latter 17 April 2015

The crack and shallow groove.

3 Kidnapped *** 12m E7 6c
FA Jules Lines (headpointed) 15 April 2015

The main arête. A very bold and technical lower section leads to an easier upper section passing a nice spiky jug.

4 Sentinel *** 12m HVS 5a
FA Steve Kennedy, Dave Ritchie & Colwyn Jones July 1990

The most prominent crack splitting the sea-ward face. Climb the cracks to a wide ledge before stepping back left to the main crack.

4a Sentinel Direct * 4m E1 5b

Climb the steep cracks just left of the start of the ordinary route which is joined just above the ledge.

5 Daylight Robbery ** 10m VS 4b
FA Dave Ritchie, Steve Kennedy & Colwyn Jones July 1990

The next crack right of 4, with a bouldery start.

6 Mullman * 8m HVS 5a
FA Steve Kennedy, Dave Ritchie & Colwyn Jones July 1990

The next crack to the right, starting up a short slab.

7 Erraid Shelter ** 8m E2 5b
FA Jules Lines & Gary Latter (on-sight) 18 July 2001

The short arête above the upper ledge, left of the arch.

8 Above * 6m HVS 5a
FA Gary & Karen Latter 4 June 2011

The hanging crack springing from the apex of the arch.

9 Beyond ** 7m VS 4b
FA Gary Latter (solo) 4 June 2011

The left-slanting fault-line has surprisingly good holds.

10 **The Long Way Around** ★   8m E1 5b

FA Gary & Karen Latter 21 April 2015

Twin cracks just left of 11.

11 **Weeping Corner** ★   7m HVS 5a

FA Dave Ritchie, Steve Kennedy & Colwyn Jones July 1990

The obvious black corner just right of the archway.

12 **Davie** ★★   8m E3 6b

FA Jules Lines & Gary Latter (on-sight) 18 July 2001

Surprisingly independent climbing up the right arête of 11. Place a low runner in the groove and move right to good jug on the arête (sling). Attain a standing position on this then continue pleasantly up the arête and crack.

> The shallow groove/arête on the right side starting up 13 was also climbed at E2 5c.

13 **Flood Warning** ★★   9m E2 5c

FA Colin Moody & Billy Gordon Canning 29 June 1997

Start a few metres right of 11. Climb an awkward bulge then the twin cracks.

14 **Tarzan** ★   10m HVS 5a

FA Steve Kennedy, Dave Ritchie & Colwyn Jones July 1990

Climbs the wall and crack starting at the right end of the undercut section.

15 **Jane**   10m VS 5a

FA Dave Ritchie, Steve Kennedy & Colwyn Jones July 1990

A difficult well protected start to gain the left side of the large incut ledge, finishing more easily up the deep corner.

16 **The Hole** ★★   8m HVS 5a

FA Steve Kennedy, Dave Ritchie & Colwyn Jones July 1990

The crack on the left wall of 17.

17 **Jungle Book** ★   8m VS 4b

FA Dave Ritchie, Steve Kennedy & Colwyn Jones July 1990

The obvious corner a metre right of 15.

18 **Spiral Arête** ★★   10m E1 5b

FA Steve Kennedy & Dave Ritchie April 1991

Start up a short corner and slab to the right of the arête. Spiral back up left and finish up a short corner.

19 **Prince of Thieves** ★   10m E1 5b

FA Steve Kennedy & Dave Ritchie April 1991

Sustained climbing up the striking crack just right of 18.

20 **Moody's Blues** ★   10m E2 5b

FA Steve Kennedy & Dave Ritchie April 1991

The next crack to the right. Gain a small incut ledge from the right. Climb a thin crack to a ledge, reach right and follow the deep crack.

> The two short cracks round to the right are Moderate and Difficult respectively.

MIDDLE TIER

This gives some fine easy-angled 18m routes on immaculate rock, generally climbing anywhere at Mod. – Difficult.

Karen Latter on The Hole.

NM 293 193

PARADISE WALL     40min

A fantastic deep-water soloing venue, on the west side of the bay, about 100m south of the small westerly beach.

Approaches: Routes 1 and 2 are approachable at all states of the tide, 3 at mid-low tide and the start of 4 is only reachable at low tide. For routes 1 – 3 descend a short chimney (Severe) on the east face, about 6m south of the wall then left along ledge to gain a flake line which allows a fine descending traverse down rightwards (5b) to the base of the routes. Route 4 is gained by scrambling down the easy ridge just north of the chasm forming the wall.

Nesting restrictions: Due to nesting shags on the easy-angled ridge overlooking the wall (used as approach for *Drowning in Adrenaline*), this crag should be avoided April-July.

1 The Brine Shrine ★★★ 12m F6b+ S0
FA Jules Lines (on-sight solo) 16 July 2001

The offset slanting off-width at the left side. A jammed chokestone low down is helpful, as are holds on the walls on either side. From a rest in the wide horizontal near the top, finish either directly up the wall, or hand traverse out left. High in the grade.

2 Please Rub Salt into my Wounds! ★★ 10m F6b+ S0
FA Andy Spink (on-sight solo) 1996

The flaky central crack, with the crux passing a small spike low down. Finish on a good jug and excellent jams. Low in the grade.

3 Dreamline ★★★ 12m F7a S1
FA Gary Latter (on-sight solo) 17 July 2001

The stunning hanging finger-crack (crux), widening to hands in its upper reaches. Finish by stepping into a recess out right.

4 Drowning in Adrenaline ★★★ 15m F7a S1/2
FA Jules Lines (on-sight solo) 18 July 2001

A fantastic rising traverse line. Start by bridging across the chasm (short people need not apply!) to left-facing flake a few metres right of the large schist intrusion. Move up left into the intrusion then gain the good flake out left and follow it into 3. Move up this to follow the next break out left to a reasonable rest in the next crack. Continue out left on the lower (cleaner) of two breaks, hand traversing the final wide slot of 1 to finish. Diving/jumping from the finishing ledge is obligatory.

Jules Lines cruising up Dreamline.

the top hanging groove. Difficult moves to enter this soon
lead to easier climbing.

 Ring of Bright Water ★★   50m F6b S0

*FA Jules Lines (on-sight solo) 21 August 2002; new finish Jules
Lines (solo) July 2012*

From the seaward end of the shelves, go down easy
slabs with grey nubbins and traverse along just above the
water line (best done at high tide) – technical in places.
Cross the flaky/juggy wall to the chokestone and then
rightwards across a square-cut bay, along jugs to where
the rock changes direction 90 degrees and becomes very
steep (6a above boulders, so no longer DWS). Climb
up the steep slab then descend the shelf to the corner.
Bridge out left to a foot hold on the arête and traverse
left and down to the ramp.

 Tarka ★   15m F6c S0

FA Jules Lines (on-sight solo) 21 August 2002

Start at the inland end of slabby shelves and climb down
on to a flaky/juggy wall. Move right for 4m to a large
granite chokestone. Squirm out above into a roof, slap for
a grey nubbin and climb a flake on the right to a big jug.
Ramble up a slab to finish.

3 Jules's Groove ★★   18m E4 6a

FA Gary Latter 7 September 2014

The undercut groove left of the corner. Climb the layback
flake steeply to a good flange and undercut spike beneath

Gary Latter on first ascent of
Jules's Groove. Photo Jules Lines.

Karen Latter on first ascent of The Shellfish Gene.

6 Black Eye Rib ★ 12m E1 5c F5+ S2

FA Jules Lines (on-sight solo) 20 August 2002

Start on the right side of an obvious rib, low tide necessary. Make crux moves on undercuts and barnacle-encrusted footholds to reach better holds (or start from further right, as per 8 and traverse at 5b). Climb the arête on the right then swing left around the arête to a spike and continue on its left.

7 Lobster Bisque ★★ 12m E3 6a F6c S3

FA Jules Lines (solo) July 2012

Move leftwards from the start of 8 into the central crack. Make a big move for the sloping top and mantel out on superb chicken heads.

8 The Otter's Breakfast Table ★★★ 15m E3 5c

FA Jules Lines (on-sight solo) 21 August 2002;
Direct Start: Gary Latter

Beautiful climbing on beautiful rock. Start 5m left of 9 on a large barnacle-encrusted boulder (the breakfast table). Climb a slanting groove/crack up the centre of the con-cave wall. Where it steepens, move rightwards on flakes to reach a large flake on the right edge. Continue up this to a jug and finish at twin cracks and a chokestone hold. 8a The prominent well-protected right-slanting groove can be climbed at no change in the grade.

9 The Longest Yet ★ 18m Severe 4a

FA Gary Latter (on-sight solo) 18 July 2001

The prominent open chimney.

10 Gourmet Crab Crack ★ 12m HS 4b

FA Jules Lines (on-sight solo) 20 August 2002

Cracks to the right of 9.

11 The Shellfish Gene ★★ 18m HVS 5a

FA Karen & Gary Latter 7 September 2014

At the right end of the ledges are twin right-slanting cracks. Jam the left crack and bridge up to pull over bulge on good holds. Continue slightly right on good flakes to pull onto sloping ledge by a good diagonal crack. Continue more easily above.

4 Amphibian ★★ 14m VS 5a

FA Jules Lines (on-sight solo) 20 August 2002

Climb the obvious square-cut groove. Low tide needed.

5 Amoeba ★ 12m E1 5b F4+ S2

FA Jules Lines (solo) July 2012

The centre of the wall, with the crux at the start.

NM 290 191 **Alt:** 20m  50min

STRAWBERRY PIG SLAB

A pleasant 8m slab on the peninsula to the west, with some obvious red intrusions, one of which looks like a pig.

All routes Jules Lines (on-sight solo) 19 August 2002, except Jules' Pistachios 21 August 2002.

1 Brambles   8m HS 4b

FA Karen & Gary Latter 8 September 2014

The broken cracks up the left edge.

2 Jo's Plums ★   8m VS 4c

An obvious straight crack at the left end.

3 Dari's Bananas ★   8m HVS 5a

Broken cracks.

4 Strawberry Pig ★★   8m E2 5b

Start at a short crack and make a hard move right to gain the pig then direct up the slab.

5 Jules' Pistachios   8m E5 6c

Start directly below the pig and climb up to a red blotch to the right of the pig. Scratch desperately to the top.

6 Raspberry Lips ★   10m E3 6a

Make hard moves to gain an undercut in an overlap. Go right and finish as for 7.

7 Ripe Mangoes ★   8m E2 5c

Go up past a red blotch and continue direct.

8 Rotten Pineapple ★   8m VS 4c

An obvious dog-leg at the right end.

SUNSHINE WALL

About 50m down and left (facing in) from **Strawberry Pig Slab**. A small crevasse forms the right side of the crag.

All routes by Jules Lines (on-sight solo) 21 August 2002.

1 Melanoma   10m Severe 4a

The initial corner.

2 Hydrogen   10m E2 5b

Climb the arête initially on the left via a flake then on the right.

3 Daniel's Dihedral ★   10m VS 4c

The off-width corner.

4 Vermelho Quente ★   10m E1 5b

The obvious arête where the crag changes direction, using the crack on its right side.

5 Spidery Cracks ★   10m VS 4c

The spidery cracks!

6 Black Square ★   10m HVS 5a

The crack with the black square.

7 Sun Spots ★   10m HVS 5a

A crack with a hanging corner and red blotch.

8 Grey Matter ★   10m E1 5b

A crack and roof with an aerated flake.

9 Where's Your Tan, Karen? ★   10m E1 5b

Go up to a short corner in an overlap. Pull through and up a hand crack.

10 Topping up your Tan ★   5/10m E1 5b

Either start from low down by a tongue of flake, or bridge across the void to gain grey footholds. Make delicate face moves using a small ear-type flake.

11 Anticyclonc ★★   10m E1 5b

Great climbing into the A-shaped roof and wall above.

12 Rays   8m VS 5a

A hand crack through a bulge.

13 Is this Scotland?? ★   8m E2 5c

The centre of the roof via a big hold then veer left along a diagonal.

14 Ellipse   10m HVS 5b

Start right of an arching flake. Move left to a bulge and climb twin cracks to the top past a red knob.

ASTEROID CHASM

NM 290 193 ⬛SW ⬛NE 🏄 ⛏ 🚶 45min

A stunning piece of rock architecture approx. 4m wide, caused by the erosion of a basalt dyke. The SW facing wall is just off vertical; the NE facing wall just overhanging. Also a huge block (the asteroid) wedged in the top. **Nesting restrictions:** Shags nest here and the crag should be avoided April – July.

SOUTH WEST FACING WALL

❶ Black Hole ★★　　　　　　　　26m E1 5c
FA Colin Moody & Cynthia Grindley 21 September 2002

Towards the right end of the wall is a right slanting ramp. Make thin moves to gain the ramp, follow it to the ledge then climb the steep corner crack. The start is slow to dry and was a bit damp.

❷ Shagged Out ★★　　　　　　　　25m E3 6a
FA Gary Latter & Jules Lines 9 April 2015

The narrow left-facing groove and flake 3m right of 3. Move rightwards along shelves to gain the line. Follow it, then the flake, finishing up crack in the headwall.

❸ Space Traveller ★★★　　　　　　28m E1 5b
FA Jules Lines (on-sight solo) 20 August 2002

The central curving twin cracked corner. Wonderful well protected climbing.

❹ Out of this World ★★★★　　　　　28m E7 6b
FA Jules Lines & Gary Latter 9 April 2015

The unique and superb line of flakes up the centre of the wall. Gain the middle of the three stepped ledges and step left off this onto a line of positive flake holds that go up at first before curving up and right to gain a diagonal break. Go right along the break a metre or so until it is possible to pull into the base of a bold left-facing layback flake. Continue up this to finish at a finger of rock. The bottom 6m usually has a veneer of moisture and is serious; most of the route is run out apart from the distinct crux that is adequately protected with small wires.

❺ Solar Collector ★　　　　　　　　15m E2 5b
FA Colin Moody & Cynthia Grindley 7 August 2005

Twin cracks up the corner across from 7. There is a large boulder at the start.

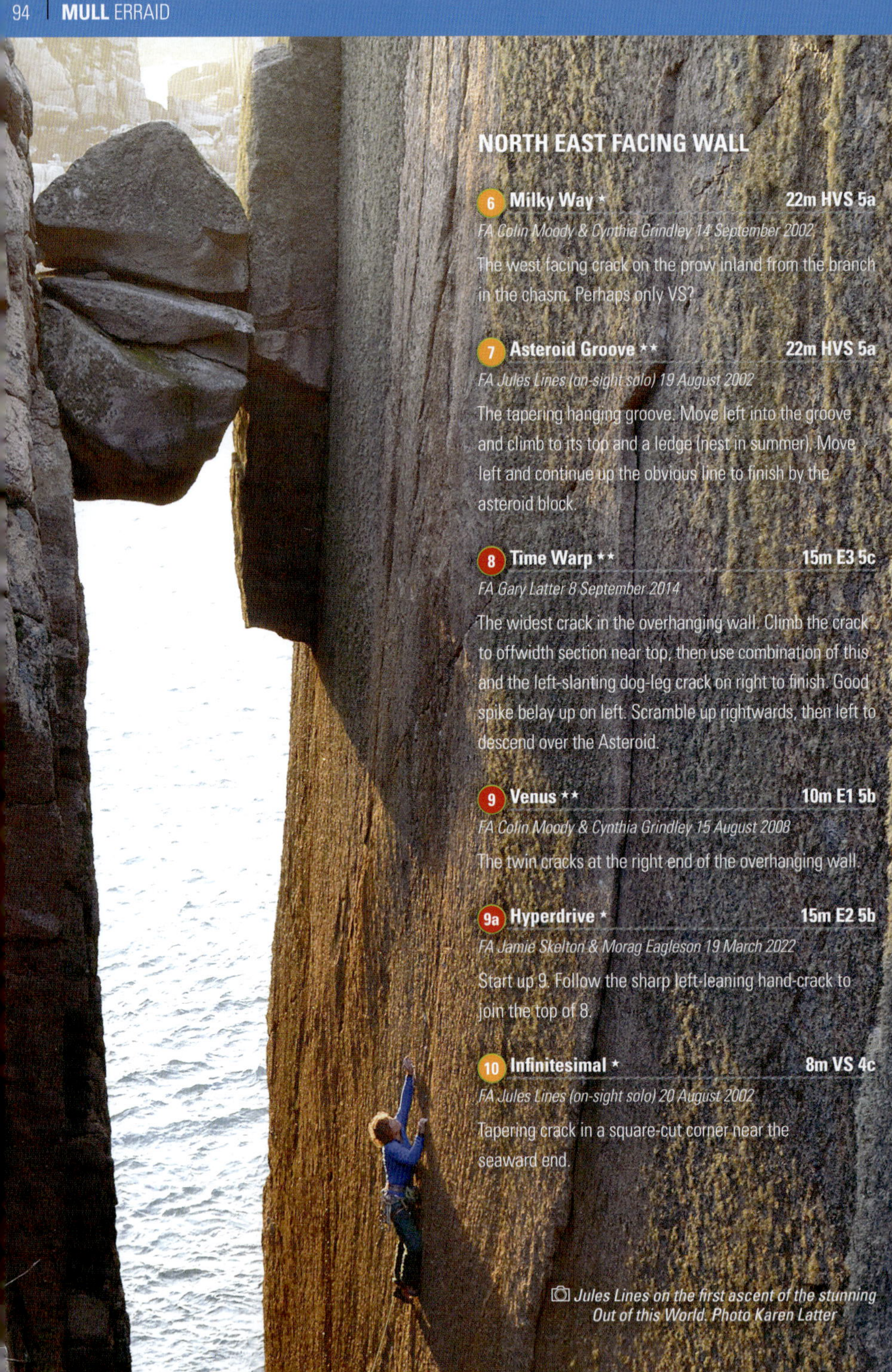

NORTH EAST FACING WALL

6 **Milky Way** ★   22m HVS 5a

FA Colin Moody & Cynthia Grindley 14 September 2002

The west facing crack on the prow inland from the branch in the chasm. Perhaps only VS?

7 **Asteroid Groove** ★★   22m HVS 5a

FA Jules Lines (on-sight solo) 19 August 2002

The tapering hanging groove. Move left into the groove and climb to its top and a ledge (nest in summer). Move left and continue up the obvious line to finish by the asteroid block.

8 **Time Warp** ★★   15m E3 5c

FA Gary Latter 8 September 2014

The widest crack in the overhanging wall. Climb the crack to offwidth section near top, then use combination of this and the left-slanting dog-leg crack on right to finish. Good spike belay up on left. Scramble up rightwards, then left to descend over the Asteroid.

9 **Venus** ★★   10m E1 5b

FA Colin Moody & Cynthia Grindley 15 August 2008

The twin cracks at the right end of the overhanging wall.

9a **Hyperdrive** ★   15m E2 5b

FA Jamie Skelton & Morag Eagleson 19 March 2022

Start up 9. Follow the sharp left-leaning hand-crack to join the top of 8.

10 **Infinitesimal** ★   8m VS 4c

FA Jules Lines (on-sight solo) 20 August 2002

Tapering crack in a square-cut corner near the seaward end.

Jules Lines on the first ascent of the stunning Out of this World. Photo Karen Latter

NM 289 193 **Alt:** 10m   45min

MINK WALLS

About 50m north from **Asteroid Chasm** lies a red slab with an arching overlap at the right end, with walls extending leftwards for 50m.

Descent: Either make a short abseil, easiest near the centre of the crag, or scramble down the gully (often wet) at the north end.

1 Tower Crack ★   **8m HS 4b**

FA Colin Moody & Cynthia Grindley 18 May 2008

At the left end is a small buttress higher up with a prominent south facing crack, the base gained by scrambling.

2 Aspen Grove ★   **22m HVS 5a**

FA Colin Moody & Cynthia Grindley 14 June 2008

Start right of the leftmost pool. Climb short flake crack, then left on large ledge. Follow twin cracks then a jam crack to a grass ledge. Continue up flake cracks to finish.

3 Guantanamo ★   **13m HS 4b**

FA Colin Moody & Cynthia Grindley 14 June 2008

Start just right of 2. Climb cracks to ledge, then move right and up short wide crack to another ledge and easy escape.

4 Iolaire ★★   **12m E2 5c**

FA Jules Lines (on-sight solo) September 2014

The clean wall and thin seams to the left of 5.

5 Pond Filler ★   **13m VS 4c**

FA Colin Moody & Cynthia Grindley 3 June 2006

Obvious crack at left end, where wall increases in height. Start at left side of pool and either climb direct or from the left. Final moves avoidable by a ledge.

6 Dirty Washing ★★   **11m E2 5c**

FA Jules Lines (on-sight solo) September 2014

Start by the pond and take the flake to the break, then the wafer thin flake in the arête above.

7 Pond Life   **8m E1 5a**

FA Colin Moody & Cynthia Grindley 23 July 2006

Off-width crack in left-facing corner at right side of pool.

8 Abby ★   **8m VS 5a**

FA Jules Lines (on-sight solo) 19 August 2002

Shallow corner just right of 7.

9 Emma ★   **8m E1 5a**

FA Jules Lines (on-sight solo) 19 August 2002

Just to the right are some flakes. Climb these and a shallow unprotected groove.

10 Iris ★   **7m VS 4b**

FA Jules Lines (on-sight solo) September 2014

The flake-line left of 11, taking the flake on the right at the branch.

11 Orbit ★   **7m E2 5c**

FA Colin Moody & Cynthia Grindley 23 July 2006

Right of 9 is a thin crack with a wider section 2m from the top.

12 Toad Hole   **7m Severe 4a**

FA Colin Moody & Cynthia Grindley 3 June 2006

Right of 11 is a black left-slanting seam. Climb a short right-facing corner to gain a ledge at the black seam then crack above slightly leftwards.

13 Toad Crack ★   **7m HVS 5a**

FA Colin Moody & Cynthia Grindley 3 June 2006

The fine crack just right.

14 Just Spitting ★   **7m HVS 4c**

FA Colin Moody & Cynthia Grindley 1 July 2006

Gain a flake left of 15 and continue up it. Protection can be placed before the top.

15 Jammer — 7m HVS 5a

FA Colin Moody & Cynthia Grindley 3 June 2006

The corner crack.

16 Caroline ★ — 8m VS 4c

FA Colin Moody & Cynthia Grindley 3 June 2006

Start just right of 15. Climb up to an undercut flake and then the flake.

17 Interrupted by Canoes ★ — 8m E1 5b

FA Colin Moody & Cynthia Grindley 3 June 2006

Good climbing up the cracks and flakes to the right.

18 Wrecked — 8m HVS 5a

FA Colin Moody & Cynthia Grindley 25 June 2006

The next line to the right, finishing up a right-facing flake.

19 Cave Dweller ★ — 8m E1 5b

FA Gary & Karen Latter 29 March 2012

Crack then direct above, just left of 20.

20 Neanderthal ★ — 8m E1 5b

FA Colin Moody & Cynthia Grindley 1 July 2006

Climb a jam crack and continue up a thinner crack.

21 Troglodyte ★ — 8m E1 5b

FA Gary & Karen Latter 29 March 2012

Thin crack 1m right of 20, then right and up flake.

22 Need an Inch ★★ — 8m E1 5c

FA Colin Moody & Cynthia Grindley 1 July 2006

The hairline crack.

23 Red ★ — 8m VS 4c

FA Colin Moody & Cynthia Grindley 25 June 2006

The cracks at the right end of the short wall.

24 Fussing ★ — 8m E1 5b

FA Colin Moody & Cynthia Grindley 18 May 2008

Shallow right-facing corner and continuation crack.

25 Access Route — 8m Very Difficult

FA Colin Moody & Cynthia Grindley 3 June 2006

The vegetated corner is useful.

26 Another Access Route ★ — 8m Severe 4a

FA Colin Moody & Cynthia Grindley 18 May 2008

The arête right of 25. Start on the left side and make an awkward initial move, then easily up and right. *Thrutchless Chimneys*, Very Difficult takes the chimneys further right.

27 The Sair Finger ★ — 12m HVS 5a

FA Cynthia Grindley & Colin Moody 18 May 2008

The right-facing corner. Move right before the roof, pull over the bulge and finish as for 28.

28 The Mink ★★ — 15m E2 5b

FA Jules Lines (on-sight solo) 3 June 2002

A fine route. Climb the right side of the slab, just right of a hairline crack, to gain the start of the overlap/arch on the right. Follow the overlap leftwards to finish up a corner-groove.

29 Helga — 15m VS 4c

FA Colin Moody & Cynthia Grindley 13 August 2006

The left-facing corner crack and the continuation crack.

30 Golden Strand ★ — 15m VS 4b

FA Colin Moody (solo) 24 July 2007

The cracks right of 29.

31 Antonia — 20m VS 4c

FA Colin Moody & Cynthia Grindley 24 April 2011

Climb the first corner left of 32 to the ledge, then the corner. Belay just below the top.

32 The Waverley ★ — 18m HVS 4c

FA Colin Moody & Cynthia Grindley 24 April 2011; Direct Start: Gary & Karen Latter 29 March 2012

Start up the chimney then move up the face and left to the ledge. Finish up the fine ramp. 32a *Direct Start* is E1 5b.

Karen Latter on Wrecked.

IONA

"We are now treading that illustrious Island, which was once the luminary of the Caledonian regions, whence savage clans and roving barbarians derived the benefits of knowledge and the blessing of religion"
– Dr Johnson

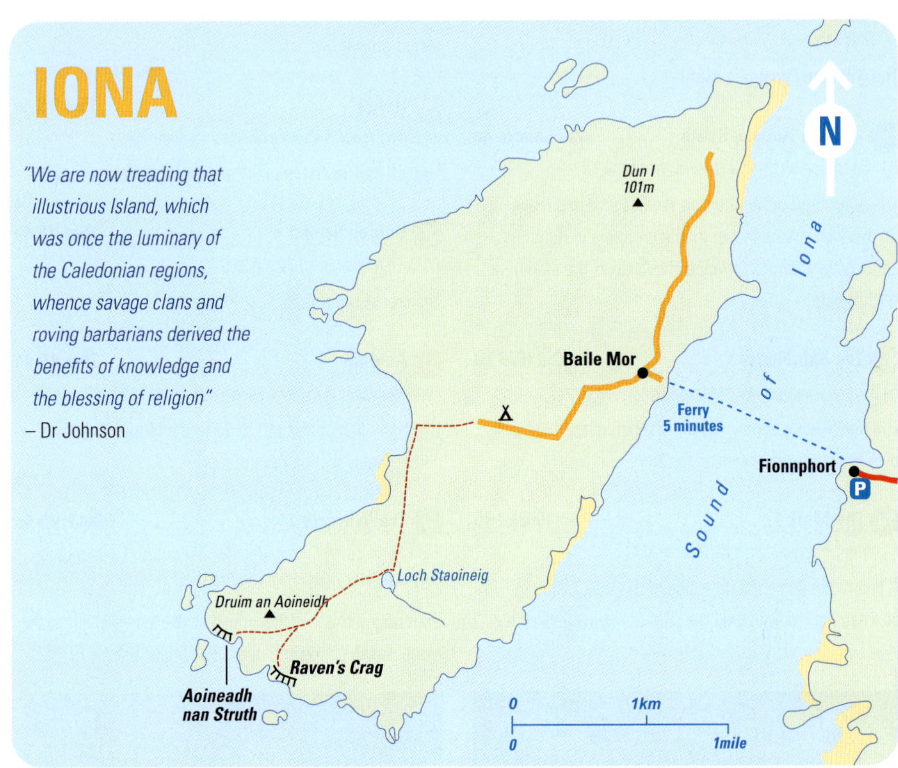

An incredibly popular (over 200,000 visitors a year) little island off the west coast of the Ross of Mull for those of a religious bent since an Irish murderer fled there in exile back in the 6th century. It is also the burial ground of 48 Scottish kings and the remains of an Iron Age fort are still evident. It is worth mentioning that the majority of visitors rarely stray from the paths and despite the commotion it is still possible to find solitude. The island has only been regarded as a climbing venue since the 90s.

Access: A regular passenger ferry (a 5 minute crossing) from Fionnphort at the west end of the A849 on the Ross of Mull to the jetty at St Ronan's Bay in the centre of the east side of the island. In summer (end of March – end of October) the first ferry crossing is at 08.35 with the last return 18.30 (18.15 on Sundays). More information at www.calmac.co.uk; ☎ 0800 066 5000.

Amenities: From the jetty, turn right for spiritual replenishment. Turn left for Martyr's Bay Restaurant for tea, coffee and meals. Iona Craft Shop (01681 700001; www.ionacraftshop.com) does bike hire, which may make a day trip a more reasonable option.

Accommodation: There is a campsite at Cnoc Oran, NM 276 237 (☎ 07747 721275; www.ionacampsite.co.uk). The heathery knolls at Loch Staoineig are good for wild camping. There are also many isolated beaches around the south west coast (carry drinking water with you). Numerous B&B establishments and a couple of hotels.

Note: Mountain bikes would make a day trip much more productive/ feasible and could be taken within about 30 minutes from the crags.

AOINEADH NAN STRUTH

This is the most south-westerly point on the island. There are three crags, described from south to north all within a short 200m stretch of the coast.

Appr: As for **Raven's Crag** to the summit cairn of Druim an Aoineidh then head due west, dropping down a short easy gully and across flat ground above the stony beach.

NM 253 223     1hr 15min

IRELAND WALL

The highest and most impressive of the crags on the headland directly opposite the north end of the long rocky island just offshore. Only the last two routes are tidal, though some of the others may be affected by rough seas.

Descent: Walk down an easy shelf descending northwards underneath the west face of the crag.

> 1 *The Shelf* is Difficult; leftward line, crack to finish up an easy corner is 2 *Sash Verte* VS 4c; 3 *Vatican City* Severe 4a follows crack then flake on right to corner crack.

 **Chinatown** ★   **25m E3 5b**

FA Steve Scott & John Adams 19 June 1993

Follow the crack up the black seam 10m right of the corner. Some loose rock.

5 The Good Book ★★   **30m E1 5b**

FA Colin Moody & Billy Hood 2 May 1993

The fine looking wide corner crack.

6 Heretic ★★★   **30m E3 5c**

FA John Adams & Steve Scott 20 June 1993

Excellent climbing on immaculate sea-washed rock, tackling the prominent crack up the centre of the south-facing wall at the base of the shelf. Follow an easy flake line leading rightwards to the crack. Cross the roof on good holds and move steadily up the wall above, finishing more easily up the upper crack. Belay well back.

NM 253 224
GULLY WALL   S W   1hr 15min

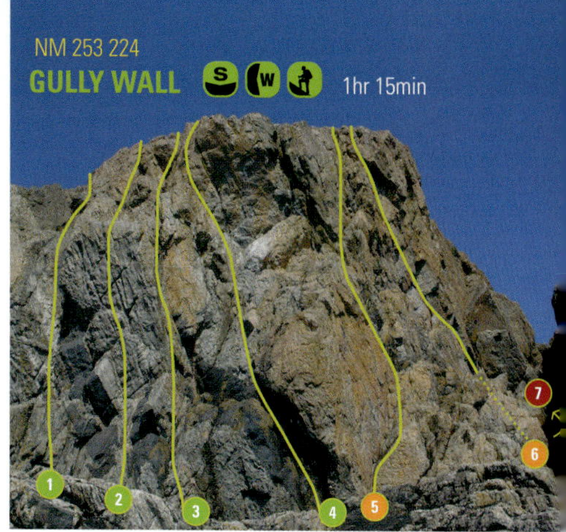

At its south end a deep narrow chasm cuts back into the hillside, containing a huge chokestone near its top.

Approach: From near the base of the shelf at **Labrador Wall** traverse south with some easy scrambling to gain a wide ledge running along the base of the crag.

1 The Black Streak ★   **12m Very Difficult**

FA unknown early 90's

The obvious line at the left side of the crag.

2 A Vicious Streak ★   **15m Severe 4a**

FA unknown early 90's

Follow the crack a few metres left of 3 then right to finish up a shallow corner.

3 The Quartz Crack ★   **15m Severe 4a**

FA unknown early 90's

Look for a prominent shallow quartz groove. Climb this and direct up the wall above.

4 Drunken Biker's Route ★   **15m Very Difficult**

FA Colin Moody, Billy Hood & John Ferrie 1993

Start 2m left of the arête, where the crag changes direction. Follow a shallow left-slanting groove to underneath the overhanging headwall. Step out left past a spike and finish easily.

5 The Man of Riou ★ — 15m VS 4c

FA Colin Moody & Billy Hood 1 May 1993

Start a short way right of the arête. Move up then step left onto a steep ramp then continue directly above.

6 L'Homme d'Iona ★ — 20m HVS 5a

FA Colin Moody & Billy Hood 1 May 1993

Climb up to the shallow overhang corner, up this swinging out left and finishing direct.

7 Il Uomo Di Roma ★ — 20m E2 5b

FA Colin Moody & Louise Gordon-Canning 12 July 1997

Start round right of 6. Climb an easy left-slanting ramp to a ledge and huge flake. Climb up right through an overhang right of the flake to a jug. Pull left into a crack then climb straight up to finish over a bulge.

8 Pontificating ★ — 20m E3 5c

FA Colin Moody & Louise Gordon-Canning 20 September 1997

Climb the easy left slanting ramp, as for 7. Step right above an overhang and traverse right below the overlap to follow the fault line to the top.

9 Pope on a Rope ★★ — 20m E2 5b

FA Colin Moody & Louise Gordon-Canning 12 July 1997

Start up another easy left-slanting ramp. Move right to gain and follow the line of corners.

10 Up Popes Another ★★ — 20m E1 5b

FA Danny Brooks, Colin Moody & Louise Gordon-Canning 22 September 1996

Scramble into the gully until under the huge chokestone. Follow the crack which slants out left below it.

11 Blood Donor ★ — 20m VS 4b

FA Colin Moody, Louise Gordon-Canning & Danny Brooks 22 September 1996

Start at a puddle in the gully gained by scrambling into the labyrinthine depths beyond 10. Follow a direct line.

12 Haemoglobin ★ — 25m VS 4c

FA Danny Brooks, Colin Moody & Louise Gordon-Canning 22 September 1996

Start at the puddle. Move farther into the gully, easily past the chokestone then direct up the wall.

Passage (page 104), Raven's Crag. Colin Moody climbing. Photo Cynthia Grindley.

Approach: Down an easy-angled wide ramp running under the base of the crag.

Descent: Down an easy gully down the right side of the **Orange Wall** or down the right side of the crag. The last four routes are on the small **Orange Wall** at the base of the shelf.

1 Force 5 ★   10m E1 5a

FA Barney Vaucher & Colin Moody 1993

Start in the centre of the scooped wall up right of 2. Climb the depression to a bulge, finishing up the steep flake out left.

The right facing corner crack is 2 *Quack* HVS 5a.

3 Infidels ★   12m E4 6a

FA Gary Latter 9 August 1997

Well protected climbing up the left side of the orange arête left of the square cut corner of 2. Climb the front face onto a ledge then a line just left of the arête to a good diagonal crack. Step right onto the arête, finishing on a good jug.

The recessed wall is 4 *Another Day* HVS 4c

5 Checking Out ★   12m VS 4b

FA Colin Moody (solo) May 1993

Start just right of the arête. Climb steeply up then step left and climb the rib.

6 Looking Around ★   12m Severe 4a

FA Colin Moody (solo) May 1993

Start just right of the descent gully, right of an arête. Pull onto a shelf right of the corner then up and left to finish up the corner.

ORANGE WALL

7 Fault Thing ★   8m Severe 4a

FA Colin Moody (solo) early 1993

The left-slanting fault bounding the right edge of the wall just left of the easy descent gully.

8 Crack Thing ★   10m E1 5b

FA Colin Moody, Billy Hood & John Ferrie 1993

The short crack on the right side of the wall, moving into 7 near the top.

9 Magnetic Wall ★   12m E1 5b

FA Colin Moody & Billy Hood May 1993

Start in the centre of the wall. Climb over a bulge on improving holds, heading leftwards to finish up the prominent flake crack.

10 *Chargold* VS 4c tackles the innocuous looking crack.

The Menhir

NM 255 220 **Alt:** 20m 1hr

MAOL NA CICHE
RAVEN'S CRAG

A long wall of generally gently overhanging gneiss on the coast at the south-west end of the island. The crag is set back about 50-100m from the sea. A few of the routes climb veins of darker (and softer) amphibolite.

Approach: Turn left at the ferry slip on the road heading south for 0.7km before turning right and following the road that runs west towards the golf course. At the road end follow the fence south (left) which meets a rough road heading up the hill to Loch Staoineig. Continue south over the track for a few hundred metres. The hill Druim an Aoneidh soon comes into view to the right. Head for the summit cairn then bear left (due south) to descend a shallow open U-shaped gully leading down to the left (north) end of the crag (overlooking the stony beach of Port nan Struth).

Nesting restrictions: A pair of ravens occasionally nests at the right end of the crag (route 21). If present (not there 2005) avoid the cliff from March-June.

Layout: The wall runs in a straight line with three rocks standing proud of the main face; The Menhir stands guard over the left end; roughly in the centre is The Pulpit, with the gap behind filled with large boulders. Right of The Pulpit is the impressive overhanging prow of The Altar. At the right end of the cliff the rock turns to face east and diminishes in height.

THE MENHIR

1 Welly Route　　　　　　　　**10m Difficult**

FRA Colin Moody (in his wellies!) 12 December 1992

The easiest way up the block. Either descend by the same line or attach a rope to the thread at the base and abseil down the west side.

MAIN WALL

The right-facing corner and short wall right of The Menhir is 2 *Eric the Red* ★ E2 5b.

3 Rod, Todd, This is God ★★　　　　　　**20m E4 6a**

FA Mark Garthwaite & Colin Lambton 24 September 1994

The thin crack in the red wall just right of the corner of 2, protected by RPs. Finish by a corner crack.

4 Allah Be Praised ★   25m E3 5c

FA Colin Moody, Michael Tweedley & Louise Gordon Canning 24 August 1997

Climbs the recess left of 5. Follow the left-slanting crack which gets better with height, step right when it ends. Go over a bulge then finish up the easy groove.

5 Yabbadabbadoo ★★   20m E5 6a

FA Mark Garthwaite & Dave Greig 28 August 1993

The left crack in the dark band of rock. A bold start leads to a small sloping ledge and gear. Move up right to the pocket on 6 (F #2.5). Move back left and climb the crack through roofs.

6 Mr Muscle ★★   20m E5 6b

FA Mark Garthwaite & Davie Greig 30 August 1993

The right crack in the dark rock band. Climb straight up to a pocket (F #2.5). Difficult moves lead to the roof (small friend in undercut slot). Move right to a jug on the lip and finish directly up the wall.

7 God is Dead ★★   20m E4 6a

FA Gary Latter & Colin Moody 8 August 1997

The crack up the black seam in the arête. Start on the arête. Up to the groove and up this steeply to good holds at its top. Continue more easily up the arête, moving left to finish past a hollow-sounding flake.

8 Jehad ★   20m E2 5b

FA Colin Moody & Stella Adams 19 June 1993

A fine line, though spoilt by some dubious black rock. The central corner system, trending slightly right. A nest was built by a big black bird on the route in 1994 but the grade remains the same.

9 Crusade ★   30m E2 5b

FA Colin Moody & Allan Petrie 26 September 1993

Right of the central corner of 8 is a small pinnacle half way up the cliff. This route climbs the crack on the left side of the pinnacle. Start below the crack. Climb a green slab that stands proud of the cliff base then continue up the crack, stepping left at the top of the pinnacle.

10 Solar Temple ★   25m E2 5b

FA Gary Latter & Colin Moody 8 August 1997

The shallow right-facing groove at the right side of the pinnacle. Climb a lichenous slab then over a bulge and up a groove to the pinnacle. Step right and up the wall on good incut holds.

11 Waco ★   20m E2 5b

FA Dave Greig, Colin Moody & Mark Garthwaite 28 August 1993

Start just left of the shallow cave. Climb the short right-facing corner and continue directly above.

12 Cul Dreimne ★   20m E2 5b

FA Dave Greig, Colin Moody & Mark Garthwaite 28 August 1993

Pleasant climbing but slow to dry. Start to the right of the shallow cave. Step left off the boulder and climb steeply to easier ground. Continue to a steepening then climb the slanting crack on the left to finish up the short wall.

THE PULPIT 

Many of the routes tend to be a bit unbalanced with steep starts and easier finishes.

13 Passage ★★   20m VS 5a

FA Colin Moody & Neil Horn 12 June 1993

Pleasant climbing up the prominent right-slanting ramp. Climb the crack and follow the ramp right until it runs out at a further crack which leads to the top.

14 Parable ★★   20m E2 5b

FA Colin Moody & Neil Horn 12 June 1993

A good route cutting through the ramp. Start 1m left of the left-slanting crack. Step up and right and follow the crack to the ramp. Continue in the same line over the bulge at a notch and direct to finish.

15 Fire and Brimstone ★   18m E3 5c

FA Colin Moody & Mark Shaw 20 August 1994

The red wall. A fine start but the upper half is a bit disappointing. Right of the crack of 14 is a short black crack at head height. Climb past the crack to reach the small ramp and follow this to its end. Pull out left and carry on up, trying to avoid 16.

16 Scripture ★   18m E1 5b

FA Colin Moody & Stella Adams 19 June 1993

Start up a bulging crack and follow it to the arête to a finish shared with 13.

17 Apocrypha ★   18m E1 5b

FA Colin Moody & Neil Horn 12 June 1993

Just right of the previous route are two seams of dark rock. Follow the right seam then the corner above on large flakes.

THE ALTAR 

18 The Incredible Dr Sex ★★★   12m E6 6b

FA Mark Garthwaite 4 September 1993

The stunning arête of the pinnacle. Start 1m left of the arête. Move up right to a jug on the arête and gear. Move up and slightly right to small flat holds then make some hard moves to gain and use an undercut on the right. Gain the next break and gear. More hard climbing leads direct to the top.

19 Blood Eagle ★★   12m E5 6a

FA Rick Waterton September 1997

Start just left of 20. Climb up leftwards past a R #1 placement on the right, finishing at an obvious slot.

20 Smoke Yourself Thin ★   11m E4 6a

FA Mark Garthwaite & Colin Moody 24 September 1994

Start near the right side of the pinnacle. Move up to then climb the left-leaning shallow groove. Continue up, stepping right at the top.

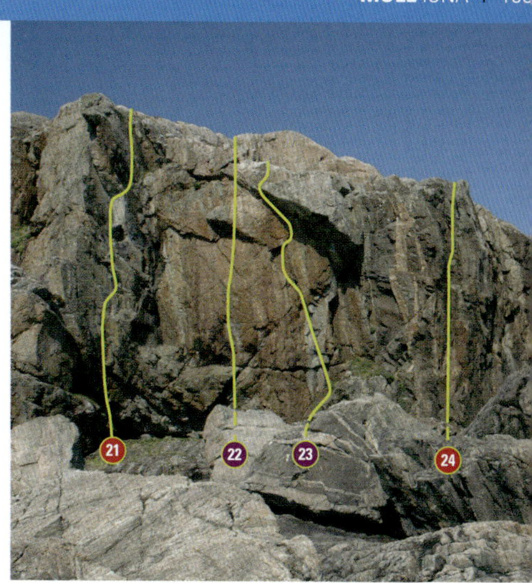

MAIN WALL (CONTINUATION)

21 The Bantry Boat ★   20m E3 5c

FA Colin Moody & Tom Charles-Edwards 22 July 1995

Below the nest are two corners. Climb the shallow left
corner and move right, brushing past the nest. A steep
wall then easier climbing leads to the top.

22 Smoking the Toad ★★★   20m E6 6b

FA Mark Garthwaite & Colin Lambton 24 September 1994

Start on the shelf at the prominent crack in the centre
of the red wall, right of the nest. Make a series of long
reaches on good holds to a PR. Hard moves above lead
to a good shake-out and gear under the roof. Pull through
the roof and finish direct. (The peg was placed in 1993).

23 Prodlgy ★★★   25m E4 6a

FA Dave Greig & Mark Garthwaite 30 August 1993

Climbs the right side of the red wall. Climb the grey crack,
step left and climb a ramp to a small overhang. Move
up right to main overhang and traverse left under the
overhang to its lip, then up and over.

EAST FACE

24 Mental Torment ★★   20m E1 5b

*FA Colin Moody, James Marshall, Danny Brooks
& Andrew Pedley 21 August 1993*

The stepped corner at the left side of the face. Worthwhile.

God is Dead (page 103).
Gary Latter on first ascent.
Photo Karen Martin

Buachaille Etive Mor.

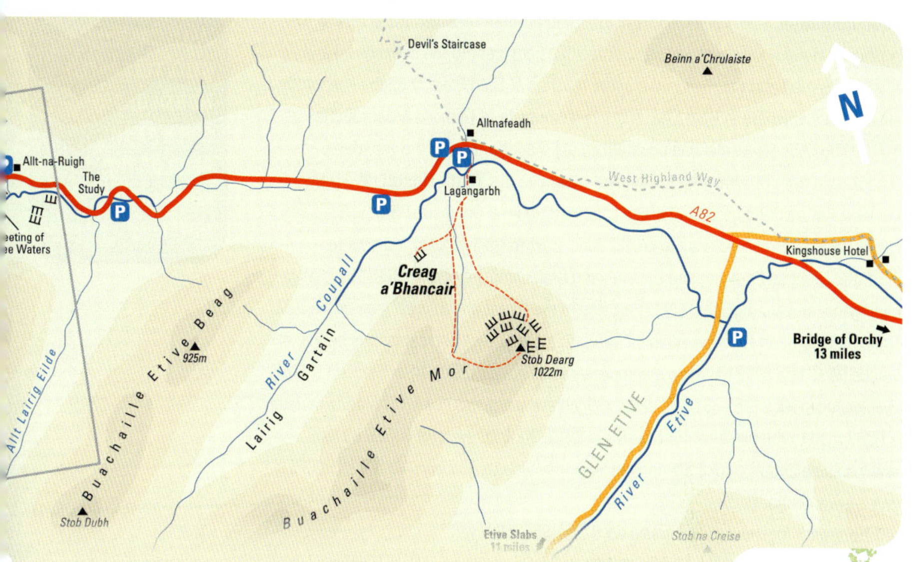

GLEN COE & GLEN ETIVE

COE & ETIVE

Glen Coe is probably the most famous of all highland glens, and certainly the most notorious, due to the tourist industry built around the infamous massacre back in 1692. Here are to be found some of the most accessible mountain routes around, rather like an altogether grander version of the Llanberis Pass. There are a superb range of climbs, from short roadside crags through to some of the most alpine routes in the country and everything in between, including the only mountain sport climbs in the UK. All the worthwhile rock climbing lies on the south side of glen.

The Glen runs west from the headwaters of the River Coe at the tiny Lochan na Fola, underneath Buachaille Etive Beag for 6 miles/10km to open out at the long sea loch of Loch Leven. The name (from Gleann Comhann) means narrow glen, popularly though wrongly supposed to mean the glen of weeping, due to the massacre. Although the distinctive conical Buachaille Etive Mor technically lies at the head of Glen Etive, by long custom Glen Coe stretches four miles further east to the head of Glen Etive. Glen Etive (Horrid Stormy Glen) is a fine picturesque glen with a more open pastoral nature than neighbouring Glen Coe. The River Etive has some splendid pools and a number of fine (though increasingly popular) roadside camping spots, while the backdrop of shapely mountains, from Buachaille Etive Mor and Beag at the head of the glen, Ben Starav across the loch and the distant Ben Cruachan afford a serene ambience lacking on most of the Glen Coe cliffs.

"Glencoe itself is perfectly terrible. The pass is an awful place. It is shut in on each side by enormous rocks from which great torrents come rushing down. In amongst these rocks on one side of the pass … there are scores of glens, high up, which form such haunts as you might imagine yourself wandering in the very height and madness of a fever …"
– Charles Dickens, 1841

Accommodation: Wild camping is still tolerated north of the river at the rear of the Kingshouse Hotel (with toilet and shower facilities adjacent to the bunkhouse) and between the road and the river in the upper reaches of Glen Etive, though is generally discouraged elsewhere in Glen Coe. There is a campsite at Glencoe Mountain Resort (☏ 01855 851226; www.glencoemountain.co.uk). About a third of the way down the back road from Clachaig Inn to Glencoe village lie the **Youth Hostel** (☏ 01855 811219; www.hostellingscotland.org.uk) and Red Squirrel Campsite (☏ 07538 763695; www.redsquirrelcampsite.co.uk). Also, next to the National Trust visitor centre, the Glencoe Camping and Caravanning Club campsite (☏ 01855 811397; www.campingandcaravanningclub.co.uk) is located in the forest on the west side of the A82, 1.1miles/1.8km SE of Glencoe village. There are also caravan and campsites nearby at Invercoe (☏ 01855 811210; www.invercoe.co.uk), 0.3miles/0.5km from Glencoe village; at Caolasnacon (☏ 01855 831279; www.kinlochlevencaravans.com) 3.5miles/5km down the B863 road to Kinlochleven; Cabins & campsite, Kinlochleven (☏ 01855 831902; www.macdonaldhotel.co.uk); Strathfillan Wigwam Village, Tyndrum (☏ 01838 400251; www.wigwamholidays.com).

Bunkhouses: Kingshouse Bunkhouse (☏ 01855 851259; www.kingshousehotel.co.uk); Blackwater Hostel & Campsite and West Highland Lodge (☏ 01855 831253; www.blackwaterhostel.co.uk) both Kinlochleven; Corran Bunkhouse, by Corran ferry (☏ 01855 821000; www.corranbunkhouse.co.uk); West Highland Way Sleeper, Bridge of Orchy (☏ 07778 746 600; www.westhighlandwaysleeper.com); By the Way Hostel & campsite, Tyndrum (☏ 01838 400333; www.tyndrumbytheway.com).

Mountaineering Club huts: Lagangarbh (SMC; NN 221 559); Blackrock Cottage, White Corries (LSCC; NN 268 530); Inbhirfhaolain, Glen Etive (Grampian Club; NN 158 507); Smiddy, Glen Etive (Forventure; NN 116 455); Waters Cottage, Kinlochleven (FRCC; NN 183 617); Alex MacIntyre Memorial Hut, North Ballachulish (Mountaineering Scotland/BMC; NN 044 611).

Amenities: The Clachaig Inn (☏ 01855 811252; www.clachaig.com) and the Kingshouse Hotel (☏ 01855 851259; www.kingshousehotel.co.uk), located at either end of the Glen, are the two most popular places for an après climb pint for most climbers. Both also offer bar meals. All the other hotels en route to Fort William also serve bar meals. For provisions, there is a large Co-op supermarket in Ballachulish, 1 mile/1.6km west of Glencoe village. There is also a small grocery shop/post office in Glencoe village together with a couple of cafes – try Crafts & Things at Tighphuirst. Petrol station in Glencoe village. For the occasional wet weather retreat (not that it rains much in Glen Coe!), the Ice Factor is an indoor climbing complex – Leven Road, Kinlochleven re-opening at the end of 2024. There is also a brewery next door!

Ric Hines on *The Pause* (page 114).
Photo Cassim Ladha

(NN 1116 4535; 56.562021, -5.0746762)

BEINN TRILLEACHAN
(HILL OF THE OYSTERCATCHER)
ETIVE SLABS

These granite slabs, unique to Scotland, lie on the slopes of Beinn Trilleachan above the head of Loch Etive. Exposed on the bedding plane at an angle of around 40°, they are generally smoother and with fewer cracks than granite which has cooled quicker nearer the surface. Several planes are separated by overlaps and walls, providing strenuous interludes to the more usual delicate 'padding'. Faith and friction are probably the two best attributes to endow a climber here, not to mention a cool head. With the exception of the corner lines on the left and *Spartan Slab* all the routes involve often quite long run-out sections of padding on bold open slabs, often including smooth holdless sections climbed on friction alone. It is recommended that new visitors choose a route lower than their usual leading standard as many find the climbing particularly bold on first acquaintance.

Access: Turn off south from the A82 about a mile/1.6km west of the Kings House Hotel down the single track road which runs down the entire length of Glen Etive. This culminates at a rough parking spot on the left, near the head of the long sea inlet of Loch Etive (13 miles/21km).

Approach: Continue down the road for 300m to a pier at the road end. Gain a path (very boggy at first) which heads diagonally across the hillside in the direction of the slabs, crossing a stream and negotiating a short rock step just beyond. A large flat boulder, the 'Coffin Stone' lies at the base at the right end of the main slabs, providing a convenient gearing up spot.

Descent: Carefully traverse right along a narrow path

across the top of the slabs and scramble with care down the prominent worn path bounding the right (north-east) edge of the slabs. Great care should be taken not to dislodge rocks onto parties on the slabs below.

① Jaywalk ★★★　　　　　　　　　　**210m E2 5c**

FA Jimmy Marshall, James 'Big Elly' Moriarty & Jimmy Stenhouse (4 PA) September 1960; FFA unknown

A very fine route, much less popular than the routes further right and not visible from the Coffin Stone. The crux sections gaining and leaving the upper slab are steep and well protected. Continue along the path beyond *Hammer*, up and along a ledge then by steep grass tufts to beneath the right edge of the slab 15m right of the vegetated left-bounding corner of *Sickle*.

1　30m 5b Ascend the two-tiered groove and its right-trending continuation to gain the upper slab. Step left then up to a belay.

2　45m 5a Move right and follow a series of grooves, then the left rib to a belay.

3　45m 5a Ascend the groove to a grass patch, move left a metre and continue by smooth slabs and a shallow scoop leading left into the large corner.

4　10m 4b Continue to a grass ledge and belay, remnants of small tree. A longer rope allows these pitches to be linked.

5　35m 5c From the thin grass ledge above surmount a short wall past a spike then climb the crack and groove above, moving right and laybacking into a grassy corner leading to a belay.

6　45m 4c Continue up the grassy corner past a short slabby section then up heather to a right traverse leading to the descent path.

2 Hammer ★★★   165m HVS 5b

FA Mick Noon & John Cunningham (1 tension) 7 April 1957

Excellent climbing up the big corner bounding the left side of the slabs. Scramble up to the base of the corner.

1 **15m 4a** Climb the corner to a tree belay.

2 **15m 4b** Ascend the cracked slab just right of the corner to belay near the top of a heather cone.

3 **25m 5a** Follow a shallow scoop leading obliquely up left (crux) into the main corner, which leads pleasantly to a good ledge and belay.

4 **40m 5b** Continue up the corner for 20m, traverse delicately right for 3m then follow cracks to a large overlap. Move right into a recess and pull over into the base of a corner.

5 **40m 4b** Move up the corner then undercut the overlap rightwards, then head diagonally up right to belay on large grass ledge with small trees.

6 **30m 4a** Continue directly over short steep wall to belay at descent path.

2a Hammer - Direct Start ★★★   35m HVS 5a

FA Gary & Karen Latter 24 May 2016

Climbs direct from the base of the corner. Scramble leftwards up grass from the top of the first pitch to nut belay at the base of the corner. Climb the main corner throughout, joining the original route above the scoop and continue to belay on the good ledge half way up the corner.

3 The Pinch Superdirect ★★★★   215m E3 5c

FA John Jackson & Rab Carrington 18 April 1968; Direct Start: Murray Hamilton, Brian Duff & Derek Jamieson 18 June 1978; Direct Finish Rab Anderson & Kenny Spence 31 March 1984

A superb combination taking the inset corner and slab overlooking *Agony*. Start as for *Agony*.

1 **37m 5c** Climb the first 3m of the *Agony* corner then move up onto the inset wall above. Continue up the corner past a narrow section to a PB.

2 **23m 5c** Continue up the corner to belay on a

Alastair MacLean at the top of the main corner on Hammer.

ledge. Running the first two pitches together on a 60m rope gives a stunning pitch.

3   20m 5c Move up right to a thin crack then climb up right past 'the pinch' to an obvious quartz pocket. Step right almost to the edge and climb a crack to belay just beneath a small overlap.

4   40m 5b Climb the crack direct, crossing a small overlap to a finger-slot 4m below another overlap. Either move right then up and left above the overlap or go up then left beneath the overlap. Both ways lead to a belay on *Hammer*.

5   40m 5a Climb up the slab above then corners onto the upper slab, traversing left below the final wall to a belay.

6   25m 5c Climb the steep chimney, move up to a ledge then up walls and across to a recess and tree.

7   30m 5a Climb the crack to the right of twin cracks to finish up slabs.

④ Agony ★★★   150m E2 5c

FA John Cunningham, Mick Noon & Bill Smith 6 April 1957; FFA Brian Duff & Willie Todd 4 June 1978

Excellent climbing up the great sweeping corner. Slow to dry, the route is in the sun until shortly after midday, after which it weeps when the corner comes into the shade.

1   35m 5c Up the corner to an old PR at 12m (just above good holds on the left wall). Move right across the slab heading for a small recess and protection in a flake just right again. Step back left and up the slab to a shallow left-facing flake and more easily up this to an overlap. Pull through this to belay.

2   35m 5c The sustained corner above to belay on a damp heather ledge at the base of the next corner.

3   35m 5c Often wet in the lower section, though still climbable in such conditions. The corner to the roof then round right to belay at an old ring peg 5m up and right of the top of the corner.

4   25m 5a Up the corner, traverse right and up the short corner and overlap to the capping wall. Traverse left to belay at a tree.

5   20m 4c Climb the wall behind the belay to easy ground leading diagonally right.

⑤ The Big Ride ★★   145m E3 5c

FA Dougal Haston & Robin Campbell (2 PA & 2 tensions) August 1964; FFA Rab Carrington & Ian Nicolson early 1970s

A direct line up the steep slabs right of *Agony*. Start below a crack in the slab just right of *Agony*.

1   30m 5b Climb the crack for 18m to a PR then traverse right into a grassy groove and climb it to a ledge.

2   30m 5c A bold pitch. Ascend the thin cracked fault above to a grass ledge in the corner.

3   40m 5c Traverse right 5m then climb up to a crack. Use a small pocket on the right to gain a weakness and follow this boldly to a ledge with old PRs. Continue to the next overlap and belay in the corner.

4+5   45m As for *Agony*.

⑥ Swastika ★★★   230m E3 6a

FA Mick Noon & Eric Taylor (some aid) 25 June 1957; FFA Willie Todd & Murray Hamilton 17 June 1977

Excellent climbing, though pitches 7 & 8 are much harder and a bit out of character with the rest of the route. Start below the rightmost of twin parallel cracks in a clean slab 40m left of the Coffin Stone.

1   35m 5b Take the right-slanting crack to a heather ledge.

2   25m 4b Move right across the ledge to a slab beneath a small block overlap.

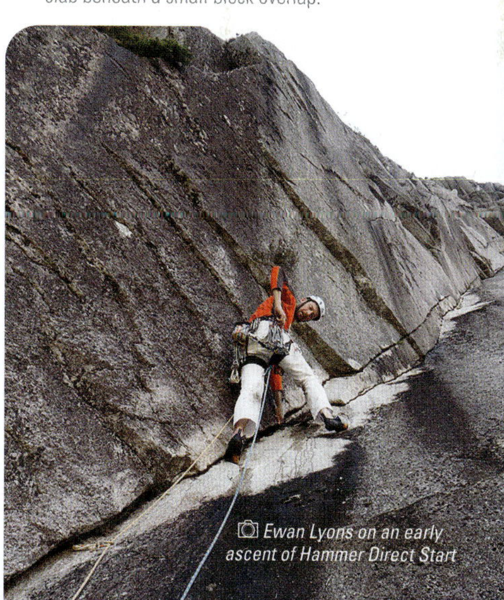

3 **10m 4c** Climb the slab crossing a small overlap to thread belay beneath the main overlap.

4 **25m 5a** Step up right and attain a standing position on the lip with interest. Traverse left along 'the moustache' to a grass ledge.

5 **35m 4c** Climb the quartz band above to a tree belay.

6 **30m 5a** Move up right to a ledge then climb the second quartz band above.

7 **20m 6a** Traverse left beneath the upper overlap for 2m then make hard move to pull over into a short corner and groove, moving up left to a tree belay beneath the final corner.

8 **30m 5c** Climb the steep awkward crack (very well protected) just left of the main crack, moving up and right to climb a corner to a ledge, finishing with a layback. Alternatively, the top pitch of *The Pause* provides a more reasonable finish.

9 **20m 4a** Continue up rightwards to gain the descent path.

7 The Pause ★★★★ 210m E1 5b
FA Jimmy & Ronnie Marshall, George Ritchie & Graham Tiso (4PA /1 tension) July 1960

Start at next groove to the left of *Spartan Slab*, 10m left of the Coffin Stone.

1 **25m 5a** Climb the groove, moving left high up to climb another groove to a large spike-hold. Swing out left and belay.

2 **20m 4c** Step right and ascend the layback crack on the lip of the groove then more easily to belay.

3 **40m 5b** Move up to the small block overlap, traversing around its right side to beneath the main overlap. Continue traversing delicately right (often wet) into the Crevasse.

4 **40m 5a** Gain the slab directly above then move into the groove on the left and climb this over a bulge to a ledge at 15m. Traverse right for 3m and climb a faint line of cracks to a small overlap. Move right to belay.

5 **40m 4c** Continue up right to the right end of the higher overlap then climb a thin crack to an easy groove leading to a grass ledge beside an overlap.

6 **30m 4c** Climb the overlaps on the left to the base of the final slab then traverse left to an undercut edge near the right side of the slab (5m right of *The Long Reach* sentry box). Go up the edge to gain grass ledge.

7 **15m 5a** Finish up the steep vertical corner on the left.

8 Spartan Slab ★★★ 210m VS 4c
FA Eric Langmuir, Mike O'Hara & John Mallinson (3 PA) 14 June 1954

The easiest worthwhile route on the slabs and consequently very popular. It follows an increasingly exposed rightward trending diagonal line close to the right edge of the main slabs. Start about 5m left of the Coffin Stone, at the most defined groove.

The Pause, Rob Kerr climbing.

1 **40m 4a** Climb the groove to its end then move rightwards across the slab to a ledge.

2 **25m 4b** Climb the slab and undercut flake heading up rightwards to a good ledge beneath the right end of the overlap.

3 **35m 4c** Move left and up to a recess beneath the roof. Make an au cheval move to get established over the lip. Continue up the gaping crack to a good horizontal crack. Traverse this right and continue round the rib to a belay near the edge of the slab.

4 **35m 4c** Climb the cracked groove above for 7m, step left into another crack and up to the overlap. Step down right with difficulty to gain a good flange. Cross the overlap and continue up good cracks to a belay.

5 **50m 4c** Continue with interest up the shallow right-facing, right-trending groove (sustained) which leads to an easier crack near the top. Belay at a rowan tree in a bay.

6 **25m** Climb cracked blocks directly above the tree to gain a ledge with some trees. Finish up the easy chimney fault on the left.

9 The Long Wait ★★   255m E2 5b

FA John Cunningham & Robin Smith September 1959

A long sustained route. Start behind the Coffin Stone.

1 **35m 5b** As for *The Long Reach*.

2 **40m 5a** Ascend twin grooves above to the large pocket on *The Long Reach*. Continue slightly right up a line of weakness in the slab to the overlap, traverse left and pull up into the Crevasse.

3 **15m 5a** Gain the slab directly above then move into the groove on the left and climb this over a bulge to a ledge.

4 **25m 5b** Descend diagonally left to an old PR just above the overlap then back up left to the right end of a grass ledge in a corner.

5 **25m 5b** Go up the corner to a P scar at 6m then break out left and move up left to follow the slab left of a grass column. Step right to belay.

6 **30m 5a** Follow the flange up right, continuing to belay at the top of another grass column.

7 **35m 5b** Move left to a diamond-shaped break in the overlap and cross this rightwards. Continue by a shallow groove and flange left until a traverse left can be made to belay at the right end of a grass ledge as for *Swastika*.

8 **35m 5b** Climb the slab to a small niche, move up left and ascend a corner to gain a tree-lined ledge. Traverse right to the base of *The Pause* corner.

9 **15m 5a** Finish up the vertical corner as for *The Pause*.

10 The Band of Hope ★★   215m E3 5c

FA John Newsome, Colin Stead, Ian Anderson & Ken Crocket 14 August 1971

A direct line between *The Pause* and *The Long Reach*.

1 **35m 5b** As for *The Long Reach*.

2 **40m 5c** Step left onto a spike and climb to the left end of the small snaking dyke on *The Long Reach*. Go up right for 6m to a hold (not visible from below) and continue up the slab slightly right then leftwards to reach the left end of the Crevasse. A serious pitch.

3 **15m 5a** Gain the slab directly above then move into the groove on the left and climb this over a bulge to a ledge.

4 **40m 5a** Continue straight up by bulges and small overlaps to belay in a black corner below a larger overlap.

5 **35m 5a** Climb straight up and belay below the short overhanging corner of *The Long Reach*.

6+7 **50m 5b, 5a** Finish as for *The Long Reach*.

11 The Long Reach ★★★★   220m E2 5b

FA John McLean & Bill Smith (1 PA) June 1963

An excellent direct line up the centre of the slabs. Start behind the Coffin Stone.

1 **35m 5b** Climb the centre of the slab to an overlap which is crossed to belay on the left as for *Spartan Slab*.

2 **40m 5b** Climb the twin grooves above onto a slab then step left along a small snaking dyke. Step boldly up left then trend diagonally up left across the slab crossing a small groove to belay by a small tree beneath the moustache.

3 **25m 5a** Cross the overlap above and left (as for

Swastika) on good holds then traverse the lip rightwards to ledges beneath a thin grassy break. Continue up and right for 5m to belay on small ledge and left-facing flake-crack on *The Pause*.

4 30m 5b Traverse left for 3m. then straight up past a quartz pocket to a small corner formed by a shallow right-facing groove and its continuation overlap. Make a *"long reach"* out left, then climb direct to belay on grass ledge near the left side of a small overlap.

5 40m 5b Cross the overlap and a bulge in the slab above, then direct to beneath the main overlap. Traverse right to belay beneath an overhung groove in the overlap.

6 35m 5b Move up left to below the groove. Gain the groove using a jug on the left wall (bold) and climb it to a ledge. Traverse left to gain a 'sentry box' and thin cracks leading to the tree-lined ledges.

7 15m 5a Finish up the final corner of *The Pause*.

The following three routes climb the shorter, less steep slabs on the right. Descend by traversing right on heather to gain the lower section of the path.

12 Vein Rouge ★★　　　　　　　　　**95m HVS 5a**

FA Rab & Chris Anderson 1 June 1991

Start left of three short cracks at the base of the brownish leftmost slabby tongue.

1 35m 5a Make a few moves up to gain first one pocket then another. Continue over some small steps then move up right into an obvious scooped depression in the wall, which leads to a grassy handrail.

2 30m 4c Ascend the crack above and continue to small ledges. A quartz band runs up the slab above. Move across and up right to belay at a crack.

3 30m 4c Move back down left to the start of the quartz band and follow this to the top then scramble to a belay in the corner.

13 Raspberry Ripple ★★　　　　　　**90m E1 5a**

FA Rab & Chris Anderson 1 June 1991

A very good second pitch climbing the rippled slab up the left side.

1 40m 5a Climb *Vein Rouge* to where it goes right into the scooped depression. Continue above, passing right of a heather ledge. Go up slightly left and continue to a small ledge at the base of the crack.

2 50m 4c The slab above is seamed with ripples. Follow the crack-line which leads to the rightmost ripple and climb this to a pocket midway up it (F #1.5). Continue up the ripple to holds and easier ground leading to a belay in the corner on rope stretch.

14 Seams Blanc ★★　　　　　　　　**95m E2 5b**

FA Rab & Chris Anderson 1 June 1991

An excellent second pitch. Start at the base of the lower rightmost narrow tongue of rock.

1 30m 5a Climb the tongue, heading for a prominent short corner in the wall. Climb this then move out left to a small ledge.

2 35m 5b Follow the seam directly above then the superb pegmatite vein directly above to share a belay with *Vein Rouge*. Steady but unprotected.

3 30m 4c Continue straight up the slab above then by grass steps to belay in a short corner.

(BIG HERDSMAN OF ETIVE)

BUACHAILLE ETIVE MOR

'The Buachaille', as it is affectionately known by generations of climbers, holds a special place in the history of Scottish climbing. It's distinctive conical shape, most striking when approaching over Rannoch Moor from the east, makes it easily the most iconic of all Scottish mountains. The mountain is actually a 7km long ridge with four tops, all the climbing concentrated on its most northerly top, **Stob Dearg** (red peak) the highest and most impressive.

Access: Park on the south side of the A82 at Altnafeadh (just west of a small stone barn), 1.8 miles/2.9 km west of the Glen Etive turn-off. Parking is also available in a long lay-by on the north side of the road 100m further west or down the rough track (4WD territory) nearer the river.

STOB DEARG NORTH-EAST FACE

Approach: Follow the path crossing the River Coupall by a footbridge then beside Lagangarbh Cottage. Follow the path cutting left after 200m, which ascends diagonally left across the hillside to the Waterslide.

Descent: Both *Curved Ridge* and *North Buttress* provide convenient scrambling descents. The lower section of *Curved Ridge* can be avoided (i.e. easier and much quicker) by going down **Easy Gully**, with a short step into the lower reaches of **Crowberry Gully**, picking up the approach path lower down just after crossing the stream. From the summit: follow the path south-west then west to flat bealach at the top of Coire na Tulaich. This is steep but straightforward – head diagonally left down a vague path then convenient scree slopes to gain the path above the west side of the burn. In the spring the large snow slope can be avoided either by descending the bergschrund on the east side or continuing north-west and down the ridge of Stob Coire na Tuileachan, skirting underneath the base of the west face of Creag na Tulaich.

LAGANGARBH BUTTRESS

The most westerly buttress on the north face with the top of the distinctive gaping slit of Lagangarbh Chimney visible from the car park. It has three faces bounded on either side by grassy gullies, its more broken west face overlooking Coire na Tulaich.

Approach: Leave the approach path at the first scree slope after 20 minutes and scramble up the rough slabby rocks to gain a right-slanting gully that runs up beneath the buttress. In the wet follow the heathery slopes left of the approach rocks.

Descent: Down the grassy gully on the right side.

1 Lagangarbh Chimney ★★   60m Very Difficult
FA P.Barclay & A.Ramsay September 1930

Excellent entertaining climbing up the huge gaping slit cleaving the centre of the face. Start by scrambling up from the gully to the base of the chimney.

  1 35m Climb the chimney over three strenuous choke-stones to belay on the left beneath the final chimney.

  2 25m Continue up the final more open chimney.

GREAT GULLY BUTTRESS

A good compact buttress overlooking the upper reaches of Great Gully.

Approach: Follow the path from Alltnafeadh, which crosses the great gash of Great Gully after about 30 minutes. The buttress on the right side gives a fine scrambling approach on immaculate rough water-worn rock. Routes from right to left.

Descent: Scramble down to the right.

1 Sundown Slab ★★   50m Severe 4a
FA Ian Clough & party 23 August 1967

Start at the leftmost of two grooves below and left of a prominent jutting flake. Climb the groove then go left by delicate slabs and open grooves to the top.

2 Ledgeway ★★   55m HS 4a
FA Bill Smith & Bob Hope 7 September 1952

Poorly protected. There is a 10m rib about 10m left of the right end of the face. Start at a groove immediately right of the rib.

  1 10m 4a Climb the groove to a grass ledge.

2 20m 4a Traverse right and up from the left end of the ledge to a shallow white-scarred fault. Follow this to pass a bulge on the left then climb up leftwards to the belay on *Direct Route*.

3 25m 4a Climb the flake then the crack to finish up the open groove above.

2a Direct Start ★   40m VS 4c

FA Jimmy Marshall 17 June 1956

Start just down and right of the normal start. Follow a fault to join the normal route at the shallow white-scarred fault.

3 Direct Route ★★   55m VS 4b

FA Sam Smith & Ian Dingwall October 1946

Similarly bold climbing to the adjacent *Ledgeway*.

1 10m 4a As for *Ledgeway*.

2 20m 4b From the left end of the ledge move up and slightly left for about 5m then head for a right-sloping shelf. Step up onto the shelf and continue up the fractured fault above to belay at a large pointed flake on grass ledge.

3 25m 4a Climb short steep wall left of the flake then finish up the slightly left-trending weakness.

4 June Crack ★★   60m VS 5a

FA Bill Smith & John Cunningham 12 June 1948

The prominent rightmost of triple cracks.

1 10m As for *Ledgeway/Direct Route*.

2 20m 4c From the left end of the ledge climb up and slightly left for 5m to gain and follow the prominent crack with an excursion onto the left wall to belay on a small shelf.

3 30m 5a Enter and climb the crack above either direct (hard but well protected) or by a short deviation on the right. Continue up the crack, finishing more easily.

5 *Playmate of the Month* E3 6a takes the prominent thin crack between 4 & 6.

6 July Crack ★★★   50m E1 5b

FA Robin Smith & Andrew Fraser June 1958

Fine sustained climbing up the thin central crack.

1 12m 4a Climb straight up to belay on a ledge below the crack.

2 38m 5b The crack.

7 August Crack ★★   50m HVS 5a

FA Bill Smith & John Cunningham 3 August 1955

The leftmost of parallel triple cracks.

1 12m 4a As for *July Crack*.

2 38m 5a Make a slightly descending traverse left for 5m to a thin crack. Follow this and the steep fault above to finish. This pitch can also be gained direct.

Alex 'Tam' Thomson on the superb Yamay (page 120).

GREAT GULLY UPPER BUTTRESS

NN 224 545 **Alt:** 820m   E)   1hr 15min

Approach: Either by a route on Great Gully Buttress
or by scrambling up the right side of that buttress.
Descent: Down the left side of the buttress.

1 Happy Valley ★   20m E1 5b

FA Ian Nicolson summer 1969

The steep wide crack 10m left of the central chimney-crack

2 Yam ★   30m E1 5b

FA Alec Fulton & John Cullen summer 1963

The chimney-crack cleaving the centre of the face with
the crux crossing the roof at one third height.

3 Yamay ★★★   40m E2 5b

FA Ian Nicolson & Kenny Spence 1 September 1968

The prominent crack and corner between 2 & 4. Climb to
a small roof, traverse the wall on the right and climb the
corner above. The 3a *Direct Start* is E2 5c.

3b Jamay ★★   35m E3 5c

FA Michael Barnard (2 PA) 10 July 2009;
FFA Ben Darvill & Gary Latter 11 July 2009

The obvious direct finish gives a fine pitch, steep and
sustained but well protected throughout. Instead of

traversing, continue straight up to an obvious slot in the
overlap. Pull over (crux) to gain crack above which is
followed with continued interest.

4 May Crack ★★★★   35m VS 5a

FA Bob Hope & Bill Smith 6 May 1952

Superb rock and protection make this one of the best
single pitches around. Start at a detached block right of 3.
Follow the thin crack which becomes wider and more
distinct in its upper half.

5 Façade ★★   45m HS 4b

FA Len Lovat 30 June 1957

Right of 4 is an open corner. Start up the wall to the
immediate right of the corner and take the line of least
resistance to a long horizontal fault below overhangs. Aim
right for a breach in the form of a vertical groove. Climb
the groove and trend left on the steep wall above to finish.

Unknown climber on the classic well-protected May Crack.

6 **World Class** ★   **45m HVS 5a**

FA Michael Barnard & John MacLeod 23 July 2011

A fine outing. To the right of the initial wall of 5 is a crack with an obvious leftward kink. Go up to climb the crack, following it up left before moving up to the ledge below the upper wall. Step left below the top groove of 5 and climb steep cracks to the top.

SLIME WALL
WEST FACE OF NORTH BUTTRESS

NN 223 545 **Alt:** 750m

A superb and atmospheric sheet of excellent rough compact bubbly rhyolite, which overlooks the tottering scree-filled depths of **Great Gully**. Despite modern protection and sticky boots many of the routes still have fearsome reputations; protection is not always abundant, the holds are sloping and exhilarating exposure makes the Slime Wall a memorable climb. It requires several days of dry weather to come into condition and a few of the routes almost always contain short seepage sections.

Approach: To the base of Great Gully Buttress then contour left, crossing the gully to reach the base.

Descent: From the top of the cliff scramble up and rightwards and traverse above the top of *Raven's Gully* to follow a diagonal shelf leading down into the tottering confines of **Great Gully**. Either tread warily down this (much loose scree) or more pleasantly, go down the crest of **Great Gully Buttress** heading back right into the gully near the base.

---

1 Bludger's Revelation ★★★   157m HVS 5a

FA Bludger's Pat Walsh, Hamish MacInnes & Tommy Lawrie 21 September 1952; Revelation Pat Walsh & Charlie Vigano June 1956; Link Pitch Jimmy Marshall, J.Griffin, George Adams & Ronnie Marshall July 1957

"I thought it was one of the most wonderful climbs on the hill, climbing up these long and beautiful silver-grey grooves leading up to the flake of Revelation." – Jimmy Marshall, *The Edge* 1994

Fantastic climbing threading a line up the left side of the face. Start beneath the left of two parallel grooves, 10m down from the base of *Raven's Gully*, where the face turns to face towards the road.

1 **18m** Climb up to belay at the base of the groove.

2 **24m 5a** Climb a detached flake to the right of the groove then step left into the groove and follow it on good holds to a ledge.

3 **30m 4c** The link pitch. Move left to the edge, climb up for 2m then traverse left into a vertical crack and follow this to a ledge and corner. Move round the edge onto the wall and climb excellent slabby rock to move left to belay.

4 **35m 4c** Step back right into the groove and up this rightwards following a line of good flat holds leading to the prominent wide flake crack. Follow this, turning the overhang on the left to belay at the right end of a narrow ledge where the flake starts to cut back right.

5 **50m 4a** Move right along the crest of the flake for 5m then directly to easier ground. Scramble off leftwards.

---

2 Lecher's Superstition ★★★   120m E2 5c

FA Dougal Haston & Jimmy Marshall (2 PA) June 1959; Superstition Davy Todd & Willie Gordon 1962; Lecher's Direct Kenny Spence & John Porteous August 1968

Another fine combination.

1 **18m** As for *Bludger's Revelation*.

2 **27m 5c** Climb detached flake behind to reach the foot of the groove. From top of flake step left onto steep wall and pull up onto the arête (crux). Return right to the groove and climb it

for 15m then move left 3m to belay just above the crux pitch of *Bludger's*. Slow to dry.

3 **35m 5b** Step right and ascend the prominent crack to belay at the top of the main pitch of *Doom Arête*.

4 **15m 5b** Climb the groove above for 10m then move up and right to an awkward belay in a corner (3m left of *Revelation* flake).

5 **25m 4c** Move up and left to finish up a prominent crack leading to easier ground.

③ Shibboleth & True Finish ★★★★   165m E2 5c

FA Robin Smith & Andrew Fraser 14 June 1958;
True Finish Robin Smith & John McLean June 1959

The finest route of its grade and length on the planet! Smith's finest achievement and one of the hardest leads of its day. Start 10m left of the base of *Raven's Gully*.

1 **25m 4b** Trend rightwards up grooves and a slabby ramp to twin blocks on a large ledge.

2 **20m 5c** Climb the crack for 5m then move left for 3m and move up and right into a left-facing groove (often wet) past an old PR (crux). Follow this to belay on a small ledge.

3 **20m 5a** The groove above then trend boldly leftwards across the wall to belay.

4 **25m 5b** Climb a delicate wall on the right (old PR and thread in a shallow groove on the left), heading for a small isolated overhang. Up the short corner directly above and continue to belay at old Ps below an overhanging wall.

5 **40m 5b** The shallow groove directly above then right along a ramp (the Original Finish, 35m 4b continues up the wall then right up grooves, finishing by a short overhanging corner). Traverse right to a good spike, move up and traverse right to a hanging groove. Easier up this to belay on a ledge.

6 **35m 5c** Traverse right above the void on good flat holds to the recess at the base of the crack springing from the roof of the Great Cave. Up the crack and continue more easily up the V-groove to finish.

Karen Latter on the superb final pitch of Shibboleth True Finish.

④ The New Testament ★★★★　　　　　**133m E4 6a**

FA Dave Cuthbertson & Joanna George 6 August 1995

A superb route *"one of the best in Glen Coe"*, taking a direct line up the centre of the cliff.

1 **25m 4b** As for *Shibboleth*.

2 **27m 6a** Climb corner above for 6m then follow a little stepped overlap going left to enter an obvious groove. (The slim hanging groove immediately right of the *Shibboleth* groove). Negotiate the 'slime factory' and enter the groove. At its top move right into a wet corner. Climb the corner and its right edge (there is another slim corner to the right which you enter towards its top). Climb the mossy thin cracks to belay on the left.

3 **27m 5c** Climb up and left from the two fingers of rock (forming the V) and climb a slim groove to a 3m tapering crack/groove. Trend left and follow a shallow groove/rib which becomes parallel and close to *Shibboleth's* 4th pitch. This leads to the right side of *Shibboleth's* isolated overhang. From a jug rail climb the wall above and enter a small left facing corner to reach a ledge and belay on *Apparition*.

4 **27m 5c** Step right and climb two tapering cracks to a ledge. Go up and right to a sloping shelf leading to the right edge of this steep section of cliff. Climb up and left to a square-cut hold then continue to a good side-pull beneath the bulge. Move left and join *Shibboleth True Finish* at the traverse into the hanging groove. Belay on a ledge above.

5 **27m 5c** From approximately half-way along the belay ledge climb a brown streak to gain the obvious stepped right-trending crack. Climb this in a fine position to easier ground.

⑤ Apparition ★★★　　　　　**145m E1 5b**

FA Jimmy Marshall & John McLean September 1959

Another superb climb often overlooked in favour of the more celebrated *Shibboleth*.

1 **30m 4c** Climb the right trending grooves (as for Shibboleth), then rightwards to the base of dark-stained corner. Traverse right 3m round awkward square-cut edge to a belay.

2 **45m 5b** Climb the prominent groove, stepping left into a parallel corner which leads to a square-cut roof. Pull over and climb a steep groove to a ledge then ascend the slab on the right to belay in a recess beneath a grass ledge.

3 **35m 5a** Climb up to then follow the diagonal crack in the steep wall leading to a slim ledge. Traverse left along this to join the slim groove and ramp of *Shibboleth* (pitch 5).

4 **35m 4b** The original *Shibboleth* finish. Continue up the wall trending right up grooves leading to a platform, finishing up a short overhanging corner.

⑥ Raven's Gully ★★　　　　　**135m HVS 5a**

FA Jock Nimlin, Barclay Braithwaite, Norman Millar, John McFarlane & Garry McArtney 13 June 1937

One of the hardest gullies in the country, setting a new standard in its day.

1-3　Three short pitches lead to a belay in a cave beneath a huge chokestone blocking the gully.

4 **5a** Climb the left wall until possible to pull over using a high hand hold above the slot. Belay well up the scree slope above.

5 **4c** Continue up narrow chimney past a jammed block then a deceptive groove.

6 **18m 4c** 'The Bicycle Pitch.' Ascend the left wall on small holds for the first 9m.

Continue up the gully. Above pitch 8, traverse left round the rib into grooves parallel to the gully and climb 45m of easier ground to a grass platform above the caves of the *Direct Finish*. Finish either by a narrow 3m chimney, or traverse left across slabs and ascend 12m chimney and fold.

⑥ₐ Direct Finish ★★★　　　　　**50m E1 5a**

FA John Cunningham, Bill Smith & Tommy Paul 30 May 1948

At the top of pitch 8 continue directly up the gully to chokestones and caves, climbing the first on the right then traversing onto the left wall to gain a dark cave. Strenuous bridging leads to rock shelves leading out left beneath huge chokestones to a grass ledge leading to the penultimate pitch of the ordinary route.

CUNEIFORM BUTTRESS  1hr

NN 223 545 **Alt:** 800m

The shorter upper continuation of the face, separated from Slime Wall by *Raven's Gully*, taking its name from the curious wedge-shaped markings evident at its base.

7 Raven's Edge ★   170m VS 4c

FA Dave Bathgate & Jim Brumfitt May 1964

The edge overlooking the gully, the highlight being the exposed final pitch.

1+2 60m Climb the left edge of Cuneiform Buttress to a point 12m above the fourth pitch of *Raven's Gully* at a large block belay below a vertical wall.

3 25m 4c Climb the wall then traverse left to the foot of a prominent corner. Climb the corner to a ledge and belay.

4 25m 4c Continue up the corner to belay on a platform.

5 30m Climb easily to a thread belay below the big roof on the extreme left edge of the buttress.

6 30m 4b Traverse left under the roof to emerge on the exposed right wall of *Raven's Gully* and finish up the deep crack which splits the wall.

7a Direct Finish ★   30m HVS 5a

FA John Porteous & M.MacDonald June 1970

From the thread belay at the top of pitch 4 climb the steep wall on the right. Finish up the chimney and twin cracks above.

NORTH BUTTRESS

1 North Buttress ★★   200m Difficult

FA William Brown, William Rose & William Tough July 1895

The broad easy-angled buttress lying immediately left of *Great Gully*, which is the second large scree funnel (stream) encountered on the approach path about 30 minutes from the car park. The base is distinguished by two prominent boulders low down. From the approach path scramble up heather and occasional rocky steps to the base of the buttress proper. Traverse left along a path immediately under the base (not obvious from below) to gain an open chimney in the centre of the buttress. Climb this for 20m then move diagonally right round the corner following the obvious (well-scratched) line to return back left. The route is well marked above, gradually easing leading directly to easy ground not far short of the summit cairn.

Cliff layout viewed from approach path at the Waterslide.

1hr

EAST FACE OF NORTH BUTTRESS

NG 226 546 **Alt:** 700m

The left side of the sprawling **North Buttress** is more open and has a sunnier aspect than the dark and gloomy west side (**Slime Wall**). Quick drying with a good spread of grades and much quieter than the adjacent Rannoch Wall.

Approach: Scramble up as for **North Buttress** and traverse up leftwards or continue along the path up past the **Waterslide** into Crowberry Basin then follow a path out rightwards and up to the base.

Descents: Traverse right and scramble down **North Buttress** or make a 40m abseil from sling & maillon on block at top of pitch 1 of 2 *Shackle Route*, gained by an easy scramble down a short open chimney from **Green Gully**. From the top of the **Upper Tier**, 40m abseils can be made from in situ anchors 10m up left from the top of 11 *Hangman's Crack*, or from block at top of 12 *Garrotte*.

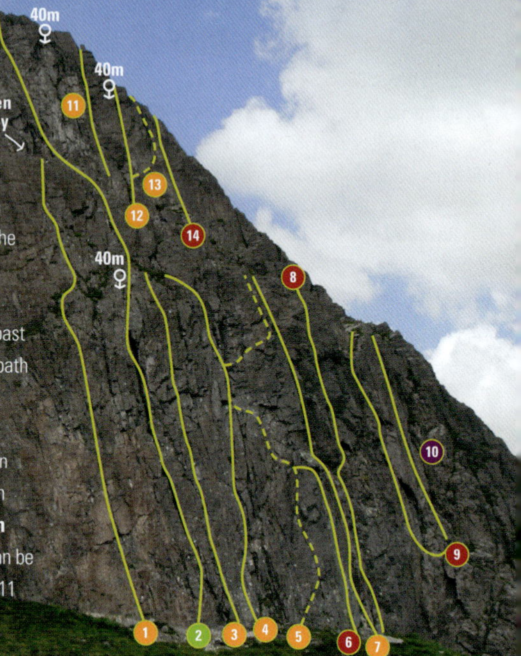

---

1 Brevity Crack ★   50m HVS 5a

FA Pat Walsh & Charlie Vigano summer 1954

A hanging crack starts about 5m up near the left end of the wall. Climb easily up to it then ascend the crack with more interest, easing higher up.

2 Shackle Route ★★   70m Severe 4a

FA Sidney Cross & Miss Alice Nelson June 1936

The wide crack splitting the left side of the face.

1 **40m 4a** Ascend the crack past a 'sentry box' (possible belay) on the left at 20m to easier ground near the top.

2 **30m 4a** A slender pinnacle/flake with a jammed block to its left lies directly above. Climb either the black groove right of the pinnacle or over the jammed block to gain a left-slanting groove. Finish over a steep wall leading to easier ground.

3 Shattered Crack ★★   40m VS 4c

FA John Cunningham & Pete McGonigle 23 June 1946

The long thin crack 3m right of *Shackle Route*. Ascend the wall on good holds past a large loose block heading towards the block overhang. Climb through this by a crack, finishing more easily.

4 Crow's Nest Crack ★★   85m VS 4c

FA John Cunningham & Pete McGonigle 24 June 1946

Good sustained climbing taking the crack above an overhung grassy recess. Start on a small flat ledge 2m left of the recess.

1 **45m 4c** Move up for 3m then trend slightly right for 7m to an awkward move at a corner. Traverse right into narrow crack and climb this, then make a long step left beneath a small roof onto a slab. Climb this to regain the crack where it is split by an overhung nose, following the leftmost crack to the terrace.

2 **40m 4a** Move up rightwards from the black groove of *Shackle Route* to follow a prominent crack.

5 Mainbrace Crack ★★★   45m HVS 5a

FA Pat Walsh & Bill Smith August 1955

Superb well protected climbing on impeccable rock. Start beneath the prominent vertical finger-crack 4m right of 4.

Climb the crack (crux) to better holds above, then the right-slanting groove. At its top, traverse left and go up shallow right-facing corner. At bulging wall move left and up short wide crack. Traverse back right to foot-ledge on arête, finish up fine groove above. Scramble up to belay at blocks.

6 White Wall Crack ★★   **50m E1 5b**

FA Bill Smith & Gordon McIntosh (2 PA) August 1955;
FFA Brian Robertson 1963

Round the edge is a noticeably white wall bounding the left side of *Bottleneck Chimney*.

1 **35m 5b** Climb the thin corner crack to pull onto a sloping ledge at 15m. Make a long step onto the rib on the right and climb to a ledge. Traverse left round a corner for 6m then climb up the open groove of *Mainbrace Crack* then traverse left to belay.

2 **15m 4c** Follow the wide crack of *Mainbrace Crack* and so to a small ledge on the arête. Leave this by an awkward step down, continue right for 5m to finish up a shallow rib.

7 Bottleneck Chimney ★★   **40m HS 4b**

FA Richard Donaldson & George McCarter summer 1941

Good well protected climbing up the obvious dark slot visible from the road. Climb the chimney, negotiating the bottleneck strenuously by good holds up on the right. Finish more easily.

8 Bluebell Grooves ★   **40m E3 6a**

FA John Cunningham & Frith Finlayson (5 PA) August 1958;
FFA Willie Todd & Dave Cuthbertson 3 June 1978

Worth doing as the quality of the rock improves noticeably on the upper section. Start immediately right of 7.

1 **20m 5c** Climb the sustained undercut groove on dubious rock to belay on a small grass ledge.

2 **20m 6a** Traverse right a few metres then ascend the overhung groove using holds on its left wall. Continue steeply to finish.

9 Gallows Route ★★   **25m E2 5c**

FA John Cunningham & Ian Dingwall June 1947

A serious and historic piece of climbing, the first extreme climbed in the Scottish mountains. A broad nose projects from the right end of the terrace split by a shallow chimney. Start by scrambling up to belay at the top of the chimney. Descend 3m and traverse left 3m to gain a steep scoop. Climb this, crossing an overhang on the left then a second scoop, turning an overhang on the right. Continue by a third scoop then traverse left to better holds and a belay. Scramble to finish.

10 Creag Dhon't Woll ★★   **20m E5 6b**

FA Mark McGowan & Cameron Bell September 1987

Bold climbing up centre of short compact wall right of 9. Climb up easily to belay as for 9. Ascend the white groove above to a good rest below a thin crack. Arrange protection in a hidden undercut on the right (F #1) then make awkward moves up the wall to a good slot and protection. Further hard moves up the crack lead to an easier finish.

UPPER TIER

Separated from the main lower tier by a long grassy upward curving terrace known as **Green Gully**. Approach via a route on the lower tier. Excellent compact rock.

11 Hangman's Crack ★★   **30m VS 4c**

FA Richard Donaldson & George McCarter summer 1941

The finely-situated square-cut corner. Scramble up to belay at the base. Move up and slightly right, make an awkward mantelshelf (crux) and move left into the corner which is followed throughout.

12 Garrotte ★★   **30m VS 5a**

FA John Cunningham & Mick Noon 4 August 1955

Superb well-protected climbing up the prominent crack in the wall right of 11. Climb the crack past a nook at 12m. Continue up past a grass ledge to finish on easy ledges.

13 Gibbet ★★   **30m HVS 5a**

FA Eric Taylor & Bill Smith 9 September 1956

Climb 12 to the first bulge then make a delicate traverse right into a clean-cut groove. Climb the groove.

The hanging groove just right is 14 *Gibberish* E2 5c.

D GULLY BUTTRESS   1hr

1 **D Gully Buttress** *   150m Difficult

FA William Newbigging & party 13 October 1903

On approach up the scree just beyond the **Waterslide**, the obvious slit of D Gully lies almost directly above. This route follows the buttress to the left of the gully. Start at the base of the gully and follow easy rocks leftwards. At about half-height a steep 10m wall is encountered. This is Hell's Wall, a poorly protected Severe 4a on sloping holds. Outflank this on the left and regain the buttress. Continue to the top and traverse right to gain the upper section of *Curved Ridge*.

NG 227 544 **Alt:** 580m    1hr 5min

CENTRAL BUTTRESS – NORTH FACE

Lies obliquely up left from the **Waterslide**, well seen in profile. Though very good in their own right the routes also provide logical approaches to the Rannoch Wall. A grassy gully leads up left to form Heather Ledge at just over ⅓ height.

Descent: Traverse easily right above the top of D-Gully to gain *Curved Ridge*.

1 **North Face Route** **   240m Severe 4a

FA Jim Bell & Sandy Harrison summer 1929

A good long route and a fine approach to the Rannoch Wall. Start by scrambling up easy slabs to a recess beneath a bollard at 3m at the lower north-east edge of the buttress.

1 **30m** Follow a series of shallow grooves to a ledge.

2 **25m** Move up leftwards and follow the wide left-most crack on huge holds to belay at its top.

3 **40m** Scramble easily up to belay at the top right end of Heather Ledge.

4 **25m** Move up a short way then traverse right round the edge on good holds and up a shallow cracked groove to underneath the overhung recess. Traverse right and drop down to belay on ledge beneath a steep wall.

5 **25m 4a** Climb the wall (crux – quite well protected) to gain a right-slanting shelf at 3m.

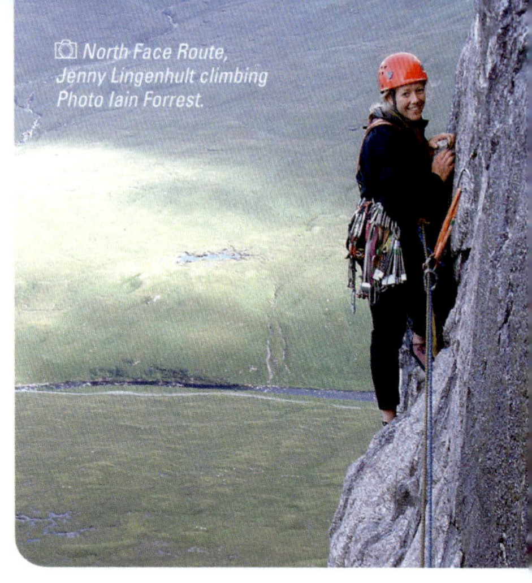

Go easily along this then up rightwards to belay at the base of a deep overhung chimney slot.

6 **25m** Go up the chimney, stepping left and up the wall at 10m then step back right into the chimney above the overhang. Continue more easily to belay on the right.

7 **40m 4a** Move left down the rock-strewn ledge to near its end then make difficult moves up to climb a crack near the edge then the edge above.

8 **30m** Continue up the edge above on immaculate rock. Scramble up easily then traverse rightwards above the top of D-Gully to gain *Curved Ridge*.

UPPER TIER

2 **Hiccup** **   70m VS 4c

FA Brian Robertson & Jimmy Houston 24 September 1962

Climbs the steep wall just left of the edge. Start below a prominent crack on Heather Ledge 3m left of the large belay at the start of the traverse on the *North Face Route*.

1 **20m 4c** Climb direct up the crack just left of the arête to belay on small ledge.

2 **50m 4c** Continue up the initially poorly-protected arête above which soon eases, then continue fairly directly in the same line above.

NN 2261 5427 **Alt:** 600m 🟢 🟢 1hr 20min

CENTRAL BUTTRESS – SOUTH FACE

An excellent quick drying buttress with the obvious exception of Waterslide Gully.

Approach: Scramble round left from the base of the North Face.

Descent: Easiest by two abseils, first from a large block at the top of *Pontoon* then from the flake on *Pegleg* (slings usually in situ). Alternatively traverse right to gain *Curved Ridge*.

1 Gravity and Grace ★★★　　　　　　　**55m E2 5c**

FA Willie Todd & Ian Rae June 1986

Another bold pitch giving good climbing on rough rock up the wall left of *Pontoon*. From the foot of the easy gully just left of *Pontoon* scramble up to a belay on a heather ledge at the foot of a shallow, left-facing corner. Climb the corner and crack above. Move left below a smooth wall and cross this back right, passing a tiny spike to the foot of a steep crack. Climb the crack to gain a left-slanting ramp/groove. Follow this to a break below the final overhang. Cross the overhang and belay in a small corner on the left.

2 Pontoon ★★★　　　　　　　　　　　**80m E1 5b**

FA Jimmy & Ronnie Marshall & James 'Big Elly' Moriarty April 1959

Another excellent route with two contrasting pitches. Bold and fingery on the first; strenuous and well protected on the second. Start about 12m up left from the base of *Pegleg*.

1　**40m 5a** Ascend the grey wall on the left side of the grey rib. Move up left to reach a thin crack and climb this to step left to below the main crack.

2　**40m 5b** Avoid the initial grassy section by a deviation on the left then climb the steep crack for 25m. Continue up a thin crack in the smooth slab (crux) leading to easier ground and the terrace.

3 Pegleg ★★★　　　　　　　　　　　**90m E1 5a**

FA Jimmy Marshall & George Ritchie September 1957

"Lovely climbing on immaculate rock." Good sparsely protected climbing up the centre of the wall. So named as there may be much standing around on one 'shaky' leg at the crux! Start 7m left of *Waterslide Corner* at a reddish, clean-cut left-slanting rake.

1　**20m 4a** Climb the left-trending weakness to belay on the second large ledge.

2　**35m 5a** Go up slightly right on steep, water-worn rock then traverse hard left (crux) above the belay to gain a ledge. Ascend the steep groove above then traverse left to a crack. Climb it to reach the top of a large flake.

3　**35m 4a** Continue more easily to finish up a steep crack.

4 Poker ★★　　　　　　　　　　　**85m HVS 4c**

FA Michael Barnard & John MacLeod 24 July 2017

Excellent bold climbing, a fine companion to *Pegleg*. Start as for that route.

1　**45m 4c** Go up the left-trending weakness, then move right to below a slight rib in the wall above (6m right of the *Pegleg* belay). Start up the rib and climb directly up the wall to eventually reach a left-trending line of flakes; follow these more easily and continue up to belay at a small corner which is level with the obvious undercut roof on the right (3" & 3.5" cams).

2　**40m 4c** Step right to gain the crack just left of the undercut roof. Follow the crack until possible to traverse into the white groove on the right. Climb the groove and continue directly up the corner and crack above.

5 Waterslide Corner ★★　　　　　　　**70m E1 5b**

FA Dave Jenkins & Ian Fulton August 1970

"Great climbing off the belay on second pitch. Slightly bold but still E1." Start beneath the watercourse.

1　**20m 4a** Climb by steep red slabs up to the prominent shallow groove.

2　**30m 5b** Climb straight up the steepening groove. Step right, climb a smaller corner on the right for 3m then step right onto a steep wall. Climb this delicately to finish with a tricky move onto a sloping ledge.

3　**20m** Finish more easily up a groove above.

Curved Ridge. Climbers Malcolm & Karren Smith.

CLIMBS ABOVE CROWBERRY BASIN
1hr 15min

The large basin below the base of Crowberry Ridge where **Easy Gully** and **Crowberry Gully** diverge. **Approach:** Just beyond the **Waterslide** the path zig-zags up the scree before leading out right to the **Crowberry Basin**. From the **Waterslide** climb steeply up on a variety of well worn paths over scree, veering slightly right but keeping to the left side of the stream that runs down over the **Waterslide**. Follow the path to cross the watercourse by a line of good holds rightwards up a slab above an exposed drop into the gully. Further scrambling over one short rock step soon leads to the base of *Curved Ridge*.

CURVED RIDGE NN 2263 5445 **Alt:** 655m

1 Curved Ridge ★★★   **240m Moderate**
FA George Bennett Gibbs July 1896

A great way up the mountain with grand views of the Rannoch Wall. The base of the ridge proper starts at about 680m. *Curved Ridge* lies almost directly above the Crowberry Basin, separated from the large dome-shaped Rannoch Wall by the mainly scree-filled *Easy Gully* just to its right. The line up the ridge is open to much variation but the best line sticks fairly close to the crest, keeping out of the side wall of the gully which is a bit loose. If in doubt look for the crampon scratches! About two thirds of the way up the ridge levels out, affording excellent views of the Rannoch Wall and a popular luncheon spot. Above, trend slightly left, a short slabby left-facing corner close to the top marking the last of the difficulties. A large cairn marks the top of the ridge from where loose scree leads up then rightwards into Crowberry Gap, behind the tower. The summit lies less than 10 minutes from here.

CROWBERRY RIDGE

This lies between **Easy Gully** on the left and the deep recesses of **Crowberry Gully** on the right. I reckon it should actually be called Blaeberry Ridge – *"The writer's chief difficulty so far (in reaching the foot of the ridge, has been to get his companion past the clumps of ripe crowberries growing everywhere, and this circumstance has suggested a name for the climb."* – The botanically challenged William Naismith, who made the first ascent of *Naismith's Route* with William Douglas in 1896. Anyone who has tasted crowberries (*Empetrum nigrum*) will concur that he's actually describing blaeberries (*Vaccinium myrtillus*), also known as whortleberry or bilberry in Scotland, an altogether more enticing approach appetiser.

NG 226 544 **Alt:** 760m  1hr 15min

SOUTH EAST FACE – RANNOCH WALL

The distinctive mass of the Rannoch Wall sweeps round from the broad broken **Crowberry Ridge**, overlooking the upper reaches of **Easy Gully** and *Curved Ridge*. Quick drying and with many superb exposed routes it is perhaps justifiably the most popular cliff in the glen. Apart from the two easiest classics, *Agag's Groove* and *January Jigsaw*, the majority of the routes are fairly bold. Catches the sun until mid afternoon.

Approach: Via scrambling up the lower section of *Curved Ridge* then dropping down into **Easy Gully** to gain the base. The upper section of *Curved Ridge* can also be accessed via a route on **Central Buttress** or **D Gully Buttress**.

Descents: Traverse easy slabs left then continue diagonally leftwards to gain an exposed path traversing left immediately above the top of the cliff with a couple of short steep steps (Moderate) to gain the top of *Curved Ridge* (cairn). Descend this back to the base. Faster alternatives include abseiling down from a block belay at the top of *January Jigsaw* to the large block belay on *Agag's Groove*, retrieving the gear on a subsequent ascent; or by scrambling right and down *Crowberry Ridge*. Alternatively, a 60m abseil from sling and maillon on large flake at top of *Whortleberry Wall*.

1 Grooved Arête ★★★ 80m VS 4b

FA John Cunningham & Bill Smith October 1946

Superb poorly protected climbing up the right edge of the wall. Start at a square-cut groove on the right side of the arête just left of the prominent polished 6m chimney leading up to the First Platform on the North-East Face.

1 **45m 4b** Climb the groove direct on small holds, stepping left and up the arête overlooking *Agag's Groove* at about 20m (possible belay on *Agag's* on left). Traverse back right into the groove (above a more difficult bulging section) and follow this with better protection, trending left to belay beneath a groove.

2 **35m 4b** Climb the exposed groove, soon leading to easier steps slightly rightwards up the broad ridge on immaculate rock.

1a Grooved Arête - Right Finish ★★★ 45m HVS 4c

A superb single pitch on impeccable rough rock. Instead of moving left (to the arête overlooking *Agag's Groove*), continue up the groove, trending rightwards to the abseil block on *Fracture Route*.

2 Agag's Groove ★★★★ 105m Very Difficult

FA Hamish Hamilton, Alex Anderson & Alex Small August 1936

A well-trodden classic giving spectacular exposed climbing on good holds. It follows the left-curving groove starting from the right edge of the wall, clearly visible from the road. Start at a large rectangular detached block at the extreme right edge of the face.

1 **30m** Climb the crack above the block leading into the groove. Follow this to a huge block belay at the base of a wide crack in the groove.

2 **25m** Continue up the easy ramp above to a large block belay or N & PB on a ledge just above.

3 **30m** Continue in the same line to a sloping shelf leading out left onto the exposed nose. Continue steeply up the cracked groove above on superb incut holds. Move left to a block belay. A stunning pitch.

4 **20m** Move left along the ledge and follow fine cracks on good holds to an abrupt finish.

3 January Jigsaw ★★★★ 75m Severe 4a

FA Iain Ogilvy & Miss Esme Speakman 10 January 1940

Superb exposed climbing through improbable-looking ground at the grade, following a line cutting through *Agag's Groove*. Start from the pinnacle on the ledge midway between *Agag's Groove* and a large semi-detached flake in **Easy Gully**.

1 **20m 4a** Climb a short right-facing square-cut groove then move up and left on large steps to

Zoe Anderson nearing the top of the finely situated *Grooved Arête - Right Finish*.

a large flake. Traverse hard right along a ledge
to a flake belay directly above the start.

2 **20m** Move right and follow a flake and wall
leading to the block belay on *Agag's Groove*.

3 **15m 4a** Step off the top of the block and
traverse right (not easy to protect) round the
corner into a slanting groove. Follow this into
a triangular niche (The Haven) then traverse
left and move up to belay next to a metal spike
below the overhanging crack of *Satan's Slit*.

4 **20m 4a** Traverse up and right round an edge
into a groove. Climb it for a couple of metres
then move left and up a steep wall. Either finish
direct or by an easier groove on the right.

4 Satan's Slit ★★ 85m VS 4c

FA Iain Ogilvy & Miss Esme Speakman 5 September 1939

A good counter-diagonal line to *Agag's Groove* crossing
that route at mid-height. Start at the base of the easy
chimney up the left side of the large semi-detached flake
at the base of the wall.

1 **30m 4a** Follow the chimney and steep easy
ground to some prominent flakes. Traverse
left 6m then move back rightwards to
belay immediately above the start.

2 **18m 4b** Climb on small holds for 6m, moving
slightly left then traverse delicately right
(quite bold) to belay in *Agag's Groove*.

3 **15m 4a** Follow *Agag's Groove* for 5m then break
right up a shallow scoop and climb direct to a
small stance beneath an overhanging crack.

4 **18m 4c** Follow the crack, soon easing after the
initial 4m, continuing more easily to finish.

5 Line Up ★★ 75m HVS 5a

FA Con Higgins & Ian Nicolson 1 June 1969

Excellent climbing aiming for the prominent slim corner
high on the face. Start below a prominent narrow overlap
in the centre of the reddish slab.

1 **25m 4c** Climb up to the overlap then straight
up to the left side of a small hanging slab
belay. Move up and right to belay ledge.

2 **25m 4c** Climb corner above for 5m, step left and
continue to belay at the base of the upper corner.

3 **25m 5a** Climb corner and roof direct.

6 Whortleberry Wall ★★ 105m HVS 4c

*FA Pitch 1 John Cunningham & Bill Smith October 1946; John
Cunningham & Bill Smith 16 September 1956*

A bold meandering route. Start at the left side of the slab
a few metres left of *Line Up*.

1 **30m 4c** Move up then trend right, heading for the
left side of the small hanging slab (common to
Line Up). Move up and right to belay on ledge.

2 **20m 4c** Traverse horizontally left for 5m then
gradually upwards to a shallow groove
which leads to a small juniper ledge.

3 **25m 4c** Traverse horizontally right for a metre
then make an upward traverse to the groove
of *Line Up*. Step right round the edge and cross
the face rightwards to a belay on small ledge.

4 **30m 4b** Climb the crack directly above,
finishing up easier ground.

*Karen Latter running the first 2 pitches of
January Jigsaw together*

CROWBERRY RIDGE, NORTH-EAST FACE

All the routes start from the First Platform, a terrace reached by a short polished chimney (Difficult) leading from **Easy Gully** to its left end.

Descent: For the first 5 routes, descend as for the **Rannoch Wall**. Alternatively abseil back down the wall (25m from in situ wire strop & maillon on spike at top of first pitch of *Fracture Route*).

7 Symposium ★★   70m E2 5c

FA Andy Tibbs, Dave Hainsworth & Alan Winton 4 May 1985

A good sustained pitch up the thin cracks just left of the more prominent *Engineer's Crack*.

1 **30m 5c** Climb cracks directly, moving left at the top just above a small overlap. Continue up easier ground to belay on the right.

2 **40m** Finish up *Fracture Route*.

8 Engineer's Crack ★★★   65m E1 5b

FA Hamish MacInnes, Charlie Vigano & Bob Hope September 1951; FFA Kenny Spence 1960s

A superb well-protected pitch; low in the grade.

1 **25m 5b** Climb the crack to a small ledge, continuing up the thin crack above. Where the crack peters out, follow a good line of holds rightwards to finish up the final easy section of *Fracture Route*.

2 **40m** Finish as for *Fracture Route*.

9 Fracture Route ★★   65m VS 4c

FA Kenny Copland & Bill Smith October 1946

Start at a pillar lying against the wall.

1 **25m 4c** Climb easily up to the base of the crack, then climb this, moving left to small block belay at its top.

2 **40m** Traverse left round the edge and follow a crack left of the nose.

10 Dingle ★★   30m HVS 5a

FA Dougal Haston & James Stenhouse September 1958

Climb easy lower wall to base of corner, then the fine corner. At the top move out left to the small block belay, as for 9.

11 **The Orphan** ★　　　　　　　　**35m VS 4c**

FA Colin Read, Colwyn Jones & Geoff Swainbank 3 Sept 2000

The prominent left-facing corner right of *Dingle*. Climb the corner crack until it abuts a small overhang, pull out left and go up until a step right to a ledge. Climb crack to broken ground, then traverse left to abseil point.

12 **Direct Route** ★★　　　　　**237m Severe 4a**

FA George & Ashley Abraham, Jim Puttrell and Ernest Baker May 1900

"Forty feet of boulder climbing – not justifiable by the ordinary run of climbers" – SMC members after a failed attempt in 1905. *"George Abraham led up in brilliant style. Anxiously we watched him quit the platform, stepping out upon a tiny ledge with his left toe, and, moving cautiously leftward and gradually up, disappear from our sight. The rope went out by inches, and we waited in dead silence for a shout to say that the distant platform was won. The shout was a long time coming"* – Ernest Baker, *The Highlands with Rope and Rucksack*, 1923
A remarkable lead at the time. The crucial second pitch is much harder – out of character with the rest of the climb. Start about 10m from the right end of the First Platform.

1 **20m** Up left-facing grooves and a crack in the wall above which leads to Abraham's Ledge.

2a **12m 4a** The crux. Move left round the rib and up past two polished foot-holds to enter a smooth scoop. Move up this slightly rightwards to better holds and continue to belay on the Upper Ledge.

2b **30m** There are numerous variations out right avoiding the main challenge: **Greig's Ledge**, Difficult, climbs up from the right end of Abraham's Ledge for a short way to traverse right into an open corner. Make an awkward move traversing right round the right edge to gain the ledge. Make a rising traverse left from the far end of the ledge to gain the Upper Ledge.

3 **30m** From left side of ledge climb steeply to belay on a shelf where the angle eases.

4 **25m** Move left to the crest and up to a belay at the top of the right side of the Rannoch Wall.

Continue much more easily up broken ground then by a narrow ridge to the base of Crowberry Tower (110m). Climb this easily by an obvious line on its north side (40m). The easiest descent is by a spiralling line down ledges on its west flank leading into Tower Gap. The summit lies less than 10 minutes up the slope from here.

SOUTH-EAST FACE OF STOB DEARG 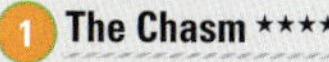

20min

Much quieter (and sunnier) than the more frequented cliffs on the north-east side of the mountain.

Access: Drive down the single track road down Glen Etive and park on the left after 0.9 miles/1.5km, or at a long lay-by on the left (south-east) side of the road, 0.5 miles/0.8km further on, directly beneath *The Chasm*.

Approach: Direct up the hillside from the road to the lower reaches of *The Chasm*.

Descent: From the top of *The Chasm* traverse left and descend slopes south-east into Glen Etive, easiest further left. For other routes either go over the summit and down the path via Coire na Tulaich or traverse right and descend *Curved Ridge* and the lower reaches of **Easy Gully**, heading south from just above the Waterslide across the moor back to Glen Etive.

1 **The Chasm** ★★★★　　　　　**450m VS 4c**

FA R.Stobart, Noel & Mrs Odell 13 April 1920 (partial winter ascent); Direct Finish Ian Jack & James Robinson 30 August 1931

"A grand day of strenuous rock work in magnificent surroundings." – Jim Bell

The prominent stepped gully cleaving the south-east face of Stob Dearg some 1.5km down Glen Etive beyond Coupall Bridge. Flanked by shorter, easier and less distinct gullies on either side it is one of the longest and best gullies (and routes!) in the country, mainly on good clean rock. It retains its hardest pitches to the end; the exit out of the Devil's Cauldron. The first climbing starts about 20 minutes above the road at NG 227 536, about 370m. The route follows in general the line of the watercourse with obvious route finding for most of the way. A selection of the most interesting pitches includes:

5 25m The Red Slab. Start up a shallow scoop on the left wall close to the watercourse. Climb the slab (loose and poorly protected) trending out right at the top. Belay further back.

6 A huge chokestone (dislodged in July 1976!) blocks the way. Bypass this by the corner on the right wall.

8 30m The 100 foot pitch. A huge dyke now crosses the gully. Climb the steep clean wall 6m right of the main watercourse on good holds and immaculate rock. An excellent pitch.

9 15m 4a The Piano Pitch. Step down and cross the watercourse to the left wall. Move up for few metres then make a right traverse *"above a small but beckoning pool"*, ending by a difficult move onto a sloping chokestone.

10 25m 4c The Converging Walls. The gully now narrows to less than 2m. Scramble up easy steps right of the watercourse to a large ledge beneath the clean smooth walls. Follow a line of holds up and left on the left wall to stand on top of large chokestone then bridge both walls towards the back of the gully. Large boulder belay further back. An exhilarating pitch.

11 20m Above, another huge chokestone bars the way. Bypass this on the right by a line of good holds starting close to the back.

12-14 A number of short easier pitches lead obviously to the main fare:

15 18m The Devil's Cauldron. Climb directly up the watercourse to belay in a cave.

16 22m 4c Chimney well out from the cave heading for a prominent foothold on the right wall. Gain and pass two small chokestones about 3m apart where back and foot work or straddling leads to a gradual easing in the difficulties. Approximately 300m of scrambling leads to the summit ridge and a short stroll to the summit itself.

1a The South Wall of the Chasm ★★   30m VS 4b

FA Jim Bell, Colin Allan & Miss Violet Roy June 1934

The easiest (i.e. driest!) finish out of The Devil's Cauldron. The technical grade is Severe 4a but the finish is normally damp and graded accordingly. Start on the left wall at a chimney. Climb this to a runner at 6m then make very awkward moves to gain a ledge on the right wall. Traverse right round the edge onto a broad ledge then follow the easiest line to finish.

2 The Chasm to Crowberry Traverse ★   1000m Moderate

FA George Tertius Glover & Collinson 10 April 1898

A natural slanting line sweeping across the South-East Face. The line starts beneath the first steep wall on the north (right) side of *The Chasm* where a well marked path shows the way. This leads past scree slopes to pass underneath an impressive undercut wall on the left about halfway along. Move out right round the corner at a steepening and cross the top of the left fork of Lady's Gully, coming up from the right. Continue up smooth, often wet rock then go up a rounded broken rib to the right of a wet groove. Continue slightly right up a shallow recess then up scree slopes fairly directly to gain the large cairn at the top of *Curved Ridge*.

NN NN 2157 5508 **Alt:** 430m

CREAG A' BHANCAIR

NW · 30min

" … **VERTICAL AND BULGING IN GREAT PORPHORYTIC WALLS OF APPARENT IMPREGNABILITY.** "
– Jimmy Marshall,
Hard Rock, 1974

An extensive low lying cliff with a collection of extremes. It is the most accessible mountain cliff in Scotland (aside from the practically roadside crags above the Bealach na Ba in Applecross). All the worthwhile routes are in the extremes – the cliff contains the highest concentration of hard climbing in the glen. The Tunnel Wall usually stays dry most of the summer, though the upper section of the main cliff can weep for a few days after prolonged rain.

Approach: Follow the path from Alltnafeadh crossing the River Coupall by a footbridge then by the side of Lagangarbh Cottage and up into Coire na Tulaich. After about 1km break off right from the path (by some large boulders) and cut across the wide rocky (usually dry) stream bed to pick up a path which cuts underneath the base of a number of small outcrops. Continue up this to arrive at the right end of The Tunnel Wall.

Descent: From the top of the cliff traverse right (south-west) on rough slabs then continue to gain some shallow vague gullies. Descend these and traverse back east to the base of the cliff.

1 Carnivore ★★★ · 170m E2 5c

FA John Cunningham & Mick Noon (some aid) 9 August 1958;
Direct (Villains) Finish Don Whillans & Derek Walker 13 June 1962

The first route to breach the main wall, and then only
after several attempts by many of the leading climbers
of the day. It follows a rising rightward traverse line
amidst impressive surroundings. Start about 10m down
right from the left end of the cliff, at a bulge beneath a
right-trending weakness.

1 **40m 5b** Pull over the bulge and move up rightwards
   on good holds to a well-hidden thread runner.
   Continue up by a vague groove to a small ledge and
   PR at the start of the traverse. Climb down rightwards
   to gain a line of better holds and follow the fault more
   easily rightwards to a shelf which leads to a point
   beneath a long ledge. Climb up steeply onto the ledge
   then walk right to a PB at the end of the ledge. It is
   wise to protect the second by back-roping from the
   PR at the start of the traverse (karabiner often in situ).

2 **20m 4b** Climb the slabby green scoop on the
   right to a ledge.

3 **20m 5a** Climb over ledges and an obvious line of
   weakness to a right slanting crack. Climb this to belay
   on small shelf beneath overhanging black recess.

4 **40m 5c** Climb up into the recess, up this and
   the layback crack above which leads to the
   base of a slim groove. Stand in this with
   difficulty then continue easily to a ledge.

5 **50m** Continue easily past some ledges to the top.

2 Twilight Zone ★★ · · · · · · · · · · · · · · · · · · · 40m E5 6a

FA Graeme Livingston & Alasdair Ross 20 June 1987

A bold single pitch climbing a weakness in the wall left
of 3 to finish up an obvious cleaned white streak. Start
beneath a small bulge below a gap in the overhang.
Climb the groove boldly to a serious section through the
bulge (R #2) then join 1. Climb the steep wall above to
gain the top of the *Le Monde* ramp. Climb directly up from
here until it is possible to hand traverse right on spikes
to the bottom of the white streak. Ascend the wall above
directly with some technical moves at the top then climb
the slab above to an in situ PB. Abseil descent.

3 Le Monde ★ · 50m E4 5c

FA Nick Colton & Willie Todd (on-sight) 4 June 1976

Of historical interest. A very bold on-sight effort. Start
20m right of 1 at a groove beside a small pedestal.

1 **25m 5c** Climb groove until it fades, traverse right
   to shallow scoop, climb the scoop and bulge,
   step left then go straight up to the PBs on 1.

2 **25m 5a** Follow the leftward-slanting ramp/
   fault to a PB below roof. Abseil descent.

4 The End of Innocence ★★★★ · · · · · · · · 95m E7 6c

FA Iain Small & Blair Fyffe 28 June 2014

One of the best hard routes in Scotland, high in the
grade. It takes the superb steep grey wall to the left
of *Romantic Reality*. Named so as it was climbed 100
years to the day after the Arch Duke Franz Ferdinand was
assassinated.

1 **25m 6a** Climb *Celtic Dawn.*

2 **35m 6c** Climb the grey wall above the belay (small
   wires), trending leftwards to a good jug, good
   sky-hook on another jug 1m up and left of this.
   Step right and make a dynamic move up to a good
   hold and move right to a crimpy break (small cam),
   then up to a long thin roof. Make a strenuous and
   technical traverse right on undercuts to gain a
   flake-line. Climb this for a few metres, before more
   hard moves up and right to gain another small roof
   (moderate to large cams). Make more hard moves
   left and up to gain better holds and a slight easing
   of the angle. Continue up and left to a grassy ledge.

3 **35m 5b** Continue up easier ground to the top.

4a Celtic Dawn ★★ · · · · · · · · · · · · · · · · · · · 25m E5 6a

FA Graeme Livingston, Alasdair Ross & Dougie Dinwoodie 24 May 1987

A sustained and serious pitch. Start up the rightmost and
most prominent of twin grey overhanging grooves, about
5m left of *Romantic Reality*. Move left at the top of the
initial groove to gain some good holds in niche at about
8m, then up to an obvious hole. Exit rightwards through
the bulge onto a green slab and peg. Move right to a small
overlap, then direct up the wall above on good holds.

5 **Romantic Reality** ★★★　　　　　**100m E7 6b**

FA Pitch 1 Pete Whillance & Derek Jamieson 31 July 1980;
Pitches 2 & 3 Dave Cuthbertson & Kev Howett May 1984

A stunning route, the main pitch breaching the pale wall
dominating the upper left side of the cliff. Start 15m right
of *Le Monde* and just left of The Tunnel Wall.

1 **25m 6a** Up a short steep wall to a ledge at 5m.
Continue up to a little block and move left to a
ramp then up an overhang barrier, pulling over into
a prominent rightwards-slanting crack. Follow the
crack and where it ends climb steeply over a bulge
and up to the grass ledge and PB on *Carnivore*.

2 **25m 6b** Up a short wall to a slab beneath the
overhang. Pull over this on sloping holds to a
short diagonal crack and protection. Move left
and up a groove to an overlap. Pull out right and
ascend blindly to a 'thank God' hold. Up over a
bulge to belay at the left end of a cleaned ledge.

3 **50m 5b** From the right end of the ledge pull rightwards
onto the wall and ascend this to easier ground.
Finish up a right trending ledge and mossy rock.

6 **The Constant Gardener** ★★★　　　**100m E6/7 6b**

FA Iain Small & Blair Fyffe 19 July 2014

A fine sustained bold route. The second pitch, gained by
the first (E4 6a) of *Romantic Reality* gives a superb well
protected E6 pitch, one of the best of its grade in the Glen.

1 **25m 6b** Start below a small diagonal overlap just to
the left of the sports wall. Pull up to the overlap,
and from its left end climb directly up into a slight
scoop below a bulge (critical Number 1 wire). Make
hard moves through the bulge to the right end of
a horizontal break (gear). Step right and climb a
faint diagonal crack to below a bulge (large cam
in pocket). Pull over this to the belay ledge.

2 **30m 6b** Climb up rightwards to a ramp line which
steepens to become a vertical corner. From the
top of the corner make hard moves right. Continue
up the gradually easing wall above to a belay.

3 **45m 5a** Continue up the wall over various
bulges to reach easier ground.

7 **Symbiosis** ★★★　　　　　　　**35m E8 6b**

FA Paul Thorburn & Dave Cuthbertson (both led)11 September 1995

The shallow groove system right of *Uncertain Emotions*.
Very serious in its lower half, sustained with difficult
but sound protection above – a comprehensive selection
of micro wires is required. Start below an undercut
left-trending flake-line above a rocky ramp. Follow the
undercuts then move up past a poor skyhook. Move up
left then back right to gain better protection in a flake.
Traverse left to near *Uncertain Emotions* then follow the
faint crack-line to the bulge guarding entrance to the
scoop. Gain this then exit right and move up to finish.

8 **Up With the Sun** ★★★　　　　　**100m E7 6b**

FA Grant Farquhar & Gary Latter 29 July 1991

A direct uncompromising line filling the obvious gap
cutting through *The Risk Business*. Start 6m down left
from the tree at a prominent triangular foot hold.

1 **50m 6b** Go leftwards and up on cleaned edges to a
thin crack. Make difficult moves through bulge rightwards to better holds and up rightwards to the PR
on *TRB*. Along this to just before the belay and up
a shallow groove in the arête past overlaps to pull
onto the capping wall with difficulty. Continue directly to join the belay below *Carnivore Direct Finish*.
Note: the lower section of this pitch was climbed
after top-rope practise then led using three ropes.

2 **20m 6b** Move out leftwards above the belay to a
prominent undercling and climb directly through
this with difficulty to gain a mysterious old in
situ thread. Step left and directly to a roof and
protection. Traverse diagonally leftwards to share
the belay at the top of the third pitch of *TRB*.

3 **30m 4b** Easily to the top.

9 **The Risk Business** ★★★★　　　　**100m E5 6a**

FA Pitches 1 & 2 Pete Whillance & Ray Parker 28 & 29 May 1980;
Pitches 3 & 4 Pete Whillance & Pete Botterill 24 August 1980

A grand route with three varied pitches starting up the
prominent leftward-slanting fault bounding the right
margin of The Tunnel Wall. Start at a large rowan tree
beneath the line.

1 25m 6a Move up left to a small ledge and poor protruding PR. Move up to a good F #2 placement under the overlap. From the PR traverse left 3m then make difficult moves up and left to reach better holds. Continue up on widely spaced holds until a short traverse left leads to a ledge and PB at the base of a prominent groove.

2 15m 5c Climb the groove boldly to grass ledges then easily up right to the PB below *Carnivore Direct*.

3 30m 6a Walk 6m left along the ledge and up a short wall to a ledge below a small overhang. Take the overhang on its left and climb a short groove to where it ends at another small overhang. Traverse left for 2 metres to an obvious small spike. Arrange some dubious protection, then climb straight up the groove to good holds, leading to a belay on grass ledge.

4 30m 4c Either head up leftwards for 10m to abseil anchor, or continue directly to finish.

10 Walk With Destiny ★★　　　　**100m E2 5b**

FA Dave Cuthbertson & Dougie Mullin 28 May 1978

Good sustained climbing up the central crack and wall above. Start at the base of a slim groove.

1 40m 5b Follow the groove until it is possible to gain a crack system by a left traverse. Follow this with a slight deviation on the left and continue to a ledge and belay. Walk left a little to the right end of the *Carnivore* traverse.

2 18m 5b Climb the wall and bulge to a horizontal break. Cross the bulge above and continue to a ledge and belay.

3 42m 5a Climb the shallow groove above then continue up the wall above to a ledge and belay. Easily to the top.

THE TUNNEL WALL

In the centre of the crag is an extensive dome-shaped overhanging pink wall so called due to its supposed resemblance to the side walls of a railway tunnel. Here are to be found the only bolted sport routes on any mountain cliff in the UK. All are long sustained tests of stamina; all except the routes at either end require a 60m rope and around a dozen quickdraws.

1 The Third Eye ★★　　　　**25m F7c**

FA Dave MacLeod (redpointed) May 2004

Superb technical climbing up the pale wall and bulge up the left side. Start just right of a small cave. Difficult climbing leads to a boss and rest. A thin sustained section up the pale wall gains a line of good holds below the bulge, which is crossed on good holds to a good rest. The final black wall is breached using small crimps leading to a LO at the big horizontal.

2 Axiom ★★★　　　　**30m F8a**

FA Dave MacLeod (redpointed) May 2004

A brilliant parallel line left of 3. Start up 1 to the second bolt then break off rightward with increasing difficulty to a hard section on sidepulls to gain good edges in a vague niche (technical crux). More sustained climbing leads through the bulge to a decent rest. Move up then rightward to another difficult section, finishing slightly easier.

3 Admission ★★★　　　　**30m F8a**

FA Graeme Livingston 20 June 1987; retro-bolted 1998, rebolted 2023

A parallel line just left of 4 giving more powerful climbing. Start under a small bulge about 6m left of 4. Climb up to a small ledge then rightwards through the bulge to gain a stopping place. Climb the wall above to reach the left end of a small overhang. Move rightwards underneath it and surmount it on large undercuts. Attain a standing position on a small ledge then using a prominent undercut move rightwards then back left to a LO.

4 Fated Path ★★★★　　　　**30m F7c+**

FA Graeme Livingston July 1986 ; retro-bolted 1998, rebolted 2023

An outstanding pitch direct up the centre of the wall. Start in the middle of the wall at the left side of a block in the initial roof. Move left from the block and then directly up until it is possible to swing right into a shallow groove. Traverse rightwards to a small overlap with good holds above. Climb the wall above past three obvious horizontal

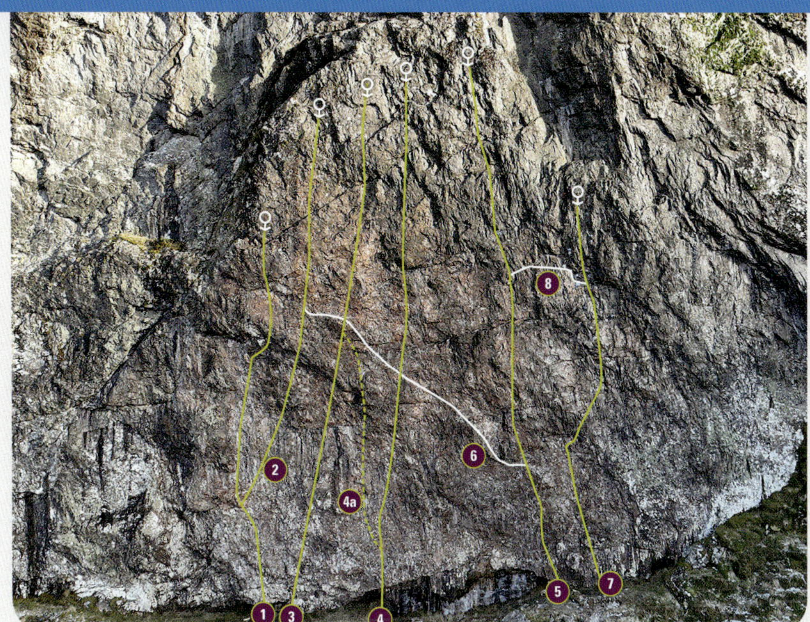

breaks which cross the wall at about 7m intervals with hard moves past the third break. Continue more easily up the final less overhanging headwall to a LO.

4a Fated Mission ★★★   30m F7c+

FA Dave Birkett 2002

A logical combination with a distinct crux. Follow *Fated Path* to the 4th bolt and make some difficult moves before moving hard left to join *Admission* above its crux bulge.

5 The Tribeswoman ★★★★   30m F7c+

FA Paul Laughlan June 1990; retro-bolted 2001; Direct Start Dave Cuthbertson 1995

Essentially a long direct start to 8, making it independent of 7. Start beneath the prominent right-trending hand-rail. Climb up on undercuts to the left end of the hand-rail and follow a direct line to finish up the crux wall of 8.

6 Vector Space ★★   38m F8a

FA Dave MacLeod July 2011

The obvious diagonal overlap across the wall, to finish up *Axiom*. The hardest line on the wall, despite missing out all the cruxes of the routes it crosses.

7 Uncertain Emotions ★★★   25m F7b

FA Dave Cuthbertson June 1986; retro-bolted 1995, rebolted 2017

Follows a line of scoops bounding the right edge of wall. Start at a small flat boulder at the far right end of the wall just before the cliff starts to slope up to the right. Pull over the initial block overhang and climb steadily up the shallow groove to an overlap. Move slightly right to gain a small but positive hold. Continue up and rightwards on improving holds to a second overlap. Pull over this using a good ramp hold and continue to a good shake-out beneath the top bulge. Climb up and rightwards into the middle of the wall above the bulge to finish abruptly at a LO.

8 The Railway Children ★★★   30m F7c

FA Dave Cuthbertson 3 July 1987; retro-bolted 1995, rebolted 2017.

Excellent climbing up the right edge of the upper wall with a short technical section. Gain the shake-out in the top recess on 7 just before that route breaks out right. Hand traverse left on a slightly descending break then climb the hard wall on side-pulls and pinches to gain better holds. Easier climbing soon leads to a LO

Paul O'Reilly starting up the delectable Lady Jane on the Lower Walls, Aonach Dubh (page 153). Photo Dominic Oughton

THE THREE SISTERS AREA

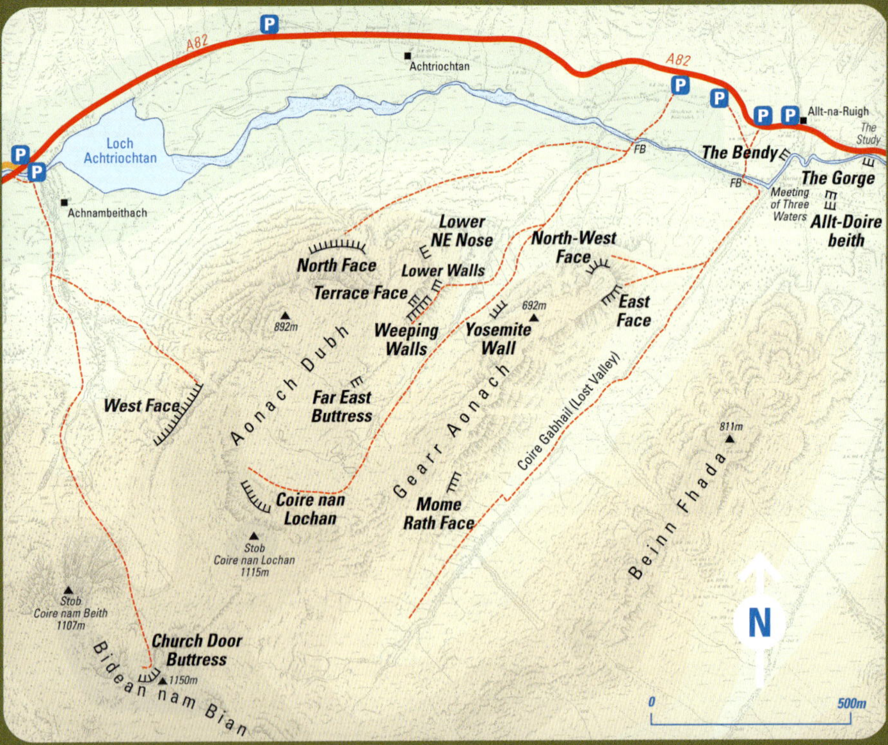

ROADSIDE CRAGS

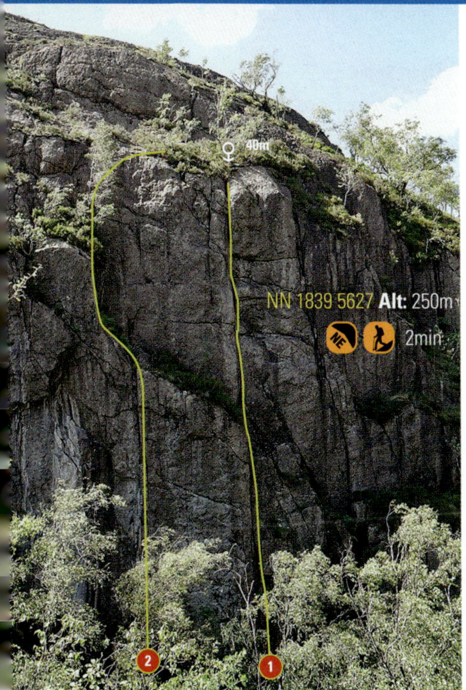

GLEN COE GORGE

About 100m downstream from the waterfall that flows from the Allt Lairig Eilde into the River Coe near the top of the glen. The climbing is on the steep crag facing the road, due south of The Study.

Access: Park in the large lay-by 160m upstream from the gorge.

Approach: Scramble down easy rocks and jump across the river about 50m downstream from the prominent crack of *Chariots of Fire*.

Descent: Abseil from trees at the top of the crag.

1 Chariots of Fire ★　　　　　　　**40m E1 5a**

FA Martin Lawrence & Rab Anderson 26 June 1981

The widening crack. Near the top climb the wall on the left. Tree belay far back.

2 Delusions of Grandeur ★　　　　　**35m E2 5c**

FA Andy Tibbs & G.Jones 8 May 1988

Left of 1 is a thinner steeper crack. Follow this to exit right to a wide ledge at 20m (possible belay). Finish up the corner crack and easier ground above.

THE BENDY

The steep wall overlooking the pool where the River Coe bends shortly before The Meeting of Three Waters. It is very sheltered, fast drying, and must be one of the most accessible crags in the Highlands. After prolonged wet spells the bottom of the routes are often saturated with spray from the waterfall but the routes remain climbable.

Access: There is limited parking for a couple of cars on the north side of the A82 at Allt-na-ruigh (NN 175 566), the white cottage on the north side of the A82 opposite Beinn Fhada – the eastmost of the Three Sisters.

Approach: Descend the west bank of the Allt-na-ruigh for 130m, then abseil to base from tree.

 All routes cleaned 2023.

1 The Roaring Silence ★★ 20m E4 6a

FA Kevin Howett & Mark Charlton 7 April 1987; Direct Start: Andy Nelson 7 Jul 2020

Reasonably well protected climbing up a series of thin incipient cracks running up the left edge of the crag. From the big boulder move up to the left end of the lowest ledge. Climb bouldery moves passing a small crack (micro cam) to mantel onto the ledge at 6m. Step right and direct up wall past a bulge, then over a second bulge rightwards to a good stopping place and finish up a short technical wall.

2 Sportline ★★ 18m 7b

FA Andy Nelson & Brian Bathurst 22 Jul 2020

A direct route up the central slender prow.

3 In Seine ★★ 25m E3 5c

FA Mark Charlton & Kev Howett 8 April 1987

Hard for the grade. Strenuous and well protected on good holds. Start at the base of the central vegetated ramp. Up this for a few feet then break out left and up a thin crack-line to gain the right side of the long ledge system. Step right and up the wall on good holds to reach a triangular niche. Move diagonally right to the wall to beneath an arching overlap. Pull over this slightly leftwards to finish past a good spike just below the top.

ALLT DOIRE-BHEITH (STREAM OF THE BIRCH GROVE) 🅂 Ⓝ🆆 🧗 10min

NN 179 562 **Alt:** 250m

The hidden gorge on the lower slopes of Beinn Fhada above The Meeting of Three Waters. The main crag is a long slightly slabby wall facing away from the road. With the exception of *Squirrel's Crack*, an extra rope is required for a satisfactory belay to the boulders 50m further back.

Access: Park where the old road leaves The Study at NN 179 565 or further east as for Glen Coe Gorge.

Approach: Drop down the hillside to cross the River Coe then head due south for 300m to the top of the crag.

Descent: Down either side of the crag.

1 Sweltering ★ 20m E2 5c

FA Gary Latter & Paul Farrell 19 July 1987

A direct line through the roof at its widest point left of the central crack. Climb an easy stepped corner and up to the roof. Undercut this leftwards to good holds. Step right round the bulge and make a long reach to the next break. Climb directly to the top.

2 Squirrel's Crack ★ 20m HVS 5a

FA probably The Squirrels 1960s

The central crack. Belay close to the edge at the top.

3 The Smouldering ★ 20m E4 6a

FA Gary Latter 16 June 1987

Start to the right of the central crack just left of a steep smooth slab. Up to good footholds and undercuts, pull right onto a slab with difficulty and continue boldly to twin breaks. Step right at the break and up to a diagonal line of holds and a good nut crack. Pull out slightly right and up to finish.

4 Neeh ★ 20m E1 5a

FA Gary Latter, Paul Thorburn & Paul McNally 9 June 1995

At the right end of the crag are a cleaned slab and a very shallow left-facing groove. Up the slab or the easier groove then the wall direct to a good break. Direct above past another break to finish easily.

A small leaning wall low down on the south side of the gorge catches the evening sun and contains four micro-routes from E1–2. *Sin Nombre* ★ E1 5b takes the well protected groove and thin crack on the right.

5 Inertia ★ 10m HVS 5a

FA Davy Gunn & Mark Tennant May 1997

Further up the gorge is a steep wall high up in the trees with a deep crack hidden by a birch tree. Well protected climbing up the crack leads to a birch tree and the top.

GEARR AONACH

The Bidean nam Bian massif throws out three long blunt ridges to the north, popularly dubbed 'the three sisters of Glen Coe'. Beinn Fhada, Gearr Aonach and Aonach Dubh, become progressively steeper, rockier and more impressive from east to west.
The varied cliffs of both Gearr Aonach and Aonach Dubh contain the most accessible easy to mid grade mountain routes in Scotland (excepting the handful of routes on the practically roadside crags above the Bealach na Ba in Applecross).

Access: Park at the upper (eastmost) of the two main car parks on the south side of the road.

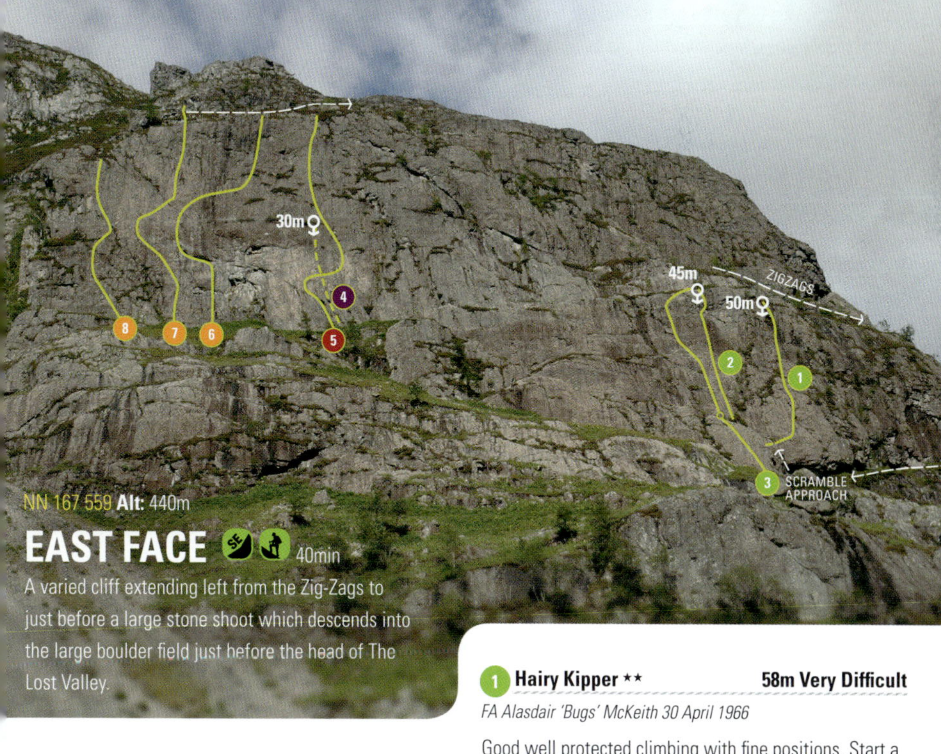

NN 167 559 **Alt:** 440m

EAST FACE 40min

A varied cliff extending left from the Zig-Zags to just before a large stone shoot which descends into the large boulder field just before the head of The Lost Valley.

Approach: As for the North Face to pick up the path leading up to the base of the Zig-Zags then contour left to gain the right end of the cliff.

Descent: Either abseil from trees or walk off right and down the Zig-Zags, the prominent well-worn path containing a few easy scrambling sections about halfway down.

1 Hairy Kipper ★★   58m Very Difficult

FA Alasdair 'Bugs' McKeith 30 April 1966

Good well protected climbing with fine positions. Start a few metres along the first zig just left of a roof. Scramble up a shelf leading left to a crack on the right wall at 6m.

1 **35m** Climb the crack, traverse right and slightly up for 10m on good ledges. Move back leftwards up cracks to a ledge below a block recess.

2 **23m** Move up rightwards then climb the open fault to belay at a rowan in a small recess.

2 Harebell Wall * — 45m Severe 4a

FA Gary Latter & Margaret King 28 August 1999

Another fine well protected pitch. Start up the short vertical crack 6m left of *Hairy Kipper*. Climb the crack then move leftwards and up on good holds, veering slightly right to prominent crack in headwall. Climb this which leads into a short left-facing groove. Exit either directly or steeply rightwards then direct to belay on birch tree.

3 High Flying * — 75m Very Difficult

FA Alasdair 'Bugs' McKeith 30 April 1966

Start at the lowest buttress a few metres left of the start of the Zig-Zags, separated from the main face by a wide shelf.

1 **30m** Climb the small buttress, walk back to belay in a short corner. Alternatively scramble up the shelf as for *Hairy Kipper*.

2 **45m** Climb a cracked wall to a ledge beneath a prominent cube of rock (possible belay). Continue up the steep wall just right of the cube on good holds to a tree belay on the terrace.

4 Eyes of Mica ** — 30m E4 6a

FA Andy Nelson & Kev Howett 17 June 1988

Start up the wall just right of the diagonal fault of *Marshall's Wall* about 50m up left from the start of the Zig-Zags. Climb the wall to a horizontal crack. Pull over a bulge and then go left into the pink scoop of *Marshall's Wall* (PR). Climb the thin crack and wall above to the right end of the final steep wall with a long reach to better holds. Abseil descent from in situ ring peg.

5 Marshall's Wall ** — 80m E2 5b

FA Robin Smith & George Ritchie May 1960

A good though wandering route following a natural weakness up the crag. Start just right of an overhung recess.

1 **20m 5b** Gain and follow thin cracks leading up left to steeper rock. Move out right above the bulge. Step down to a cramped P & NB in a small recess.

2 **20m 5b** Traverse right 3m then up onto a ledge. Traverse right (quite bold) for 3m to a small ledge. Move up then back left to P & NB on grass ledge.

3 **40m 4c** Move left and climb to ledges, finishing by a short steep corner.

6 Via Dolorosa ** — 100m HVS 5a

FA Brian Robertson & Alasdair 'Bugs' McKeith 23 May 1964

Start at the base of the wide crack, which lies just left of a yellowish recess.

1 **50m 5a** Ascend the crack to overhangs (possible belay) then traverse left 6m. Continue directly for about 15m then move back right to belay.

2 **10m 5a** Move right to a belay.

3 **40m 4b** Ascend the wall above, finish up a short crack.

7 The Mappie * — 85m VS 4c

FA James Moriarty & Jimmy Marshall April 1959

Start beneath a clean thin crack 5m left of a wide crack.

1 **45m 4c** Climb up to just beneath the left end of the roof then traverse diagonally left for 12m to a small ledge. Go directly up thin cracks then back rightwards to a stance and spike belay above the start. It is more difficult to leave the traverse line sooner.

2 **40m** Continue more easily directly up the wall. Belay far back.

8 Herbal Mixture * — 70m HS 4b

FA Jimmy & Ronnie Marshall & George Ritchie September 1957

Good climbing despite the name. Start beneath the left end of the overhangs about 10m left of a wide diagonal crack.

1 **30m 4b** Climb heather grooves up and left to a ledge at 15m. Continue up and left to a good ledge in a corner.

2 **40m 4b** Move up slightly right to a ledge at 12m then up and left across the final steep wall (poorly protected) finishing on easier ground.

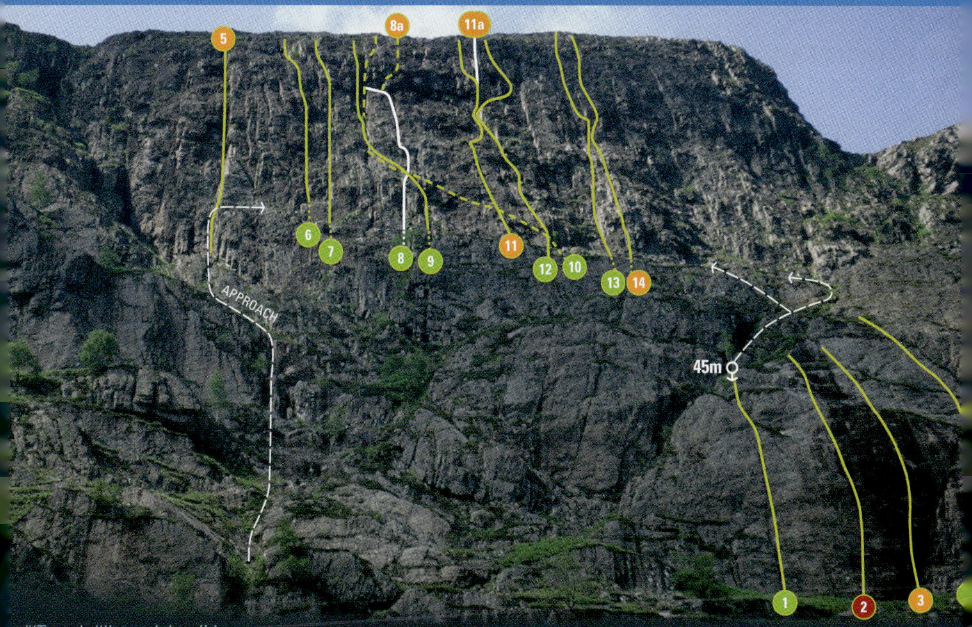

> "'Twas brillig, and the slithy toves
> Did gyre and gimble in the wabe:
> All mimsy were the borogroves,
> And the mome raths outgrabe.
> "Beware the Jabberwock, my son! ..."
> from *Jabberwocky* – Lewis Carroll

MOME RATH FACE

A large impressive face in a delightful situation overlooking the upper reaches of Coire Gabhail (Corrie of the Booty, but more commonly known as The Lost Valley). The face takes a lot of drainage but is well worth a visit after a period of dry weather. The face can also be used to provide a useful approach to the routes on Stob Coire nan Lochan (35 minutes) or Far Eastern Buttress on Aonach Dubh (about 20 minutes).

Approach: Cross the River Coe by the eastmost of the two bridges and follow a well worn (very popular with the touroids) path up into Coire Gabhail. Head across the flat grass and screes to pick up the path again and follow this for 100m then head diagonally left up the hillside then by easy scrambling up the lower reaches of the left-slanting Lost Leeper Gully to gain the left end of the lower cliff. The main face can be gained by continuing scrambling up the gully at about Moderate.

Descent: Scramble back from the cliff to reach easy ground then walk a long way left to reach easy angled slopes beyond a large open gully. Descend to join the well-worn path close to the west bank of the Allt Coire Gabhail.

LOWER CRAG

NN 1599 5512 **Alt:** 500m

Right of the open Lost Leeper Gully. From the top of the routes scramble up the right-slanting fault for 70m to the grassy terrace at the base of the main face. All the routes lead to a belay on a small rowan towards the left end of the grass ledge at the top.

1 Slimcrack ★ — 45m Severe 4b

FA Ian Clough & C.Slesser 17 October 1965

The prominent diagonal crack near the centre of the wall. Follow the crack with difficult moves on the right wall at mid-height (overhead protection at this point) leading to tree belay on the left

2 The Burning ★ — 50m E1 5a

FA Andy Tibbs & Stuart Cameron 12 May 1990

Start midway between 1 and the left edge of the large brown streak. Climb the wall directly (possible belay at 25m), continuing more easily up the upper wall.

3 Batura Wall ★ — 50m HVS 5a

FA Andy Tibbs & Helen Shannon 2 July 1989

Start immediately left of the large brown streak. Climb up then left along an obvious break to a small ledge beneath a bulge. Traverse left above the bulge then up a crack then the easier wall trending left to finish.

4 Flake Groove ★ — 65m Very Difficult

FA Ian Clough, George Brown, J.Donnison & R.Logan 12 June 1967

The wide right-slanting fault towards the right end of the face is slow to dry.

1 **25m** Climb the groove, passing a large flake at 20m with care to belay on ledge a short distance above.
2 **10m** Make an airy traverse left then move up to a small thread belay.
3 **30m** Continue leftwards up walls and slabs.

MAIN FACE

5 Rainmaker ★★★ — 70m VS 4b

FA Ronnie Marshall & James 'Big Elly' Moriarty 13 Sept 1958

A frabjous route on perfect rock up the large left-facing recessed corner at the left end of the cliff – one of the best routes of its grade in the Coe.

1 **20m 4b** Climb the corner (bold at first) to better holds and protection, to belay on a large ledge.
2 **15m 4b** Continue up a short wide crack to pull out of a recess on huge holds.
3 **35m 4b** Continue up to beneath a formidable-looking bulge. Layback round this in a fine position then more easily above to belay on grass ledge at top. Scramble right and up to finish.

6 Slithy ★★ — 120m Severe 4a

FA Ian Clough, P.Macleod & R.Morgan 18 August 1967

Good climbing up the left side of the wall. Start at the left end of the alp beneath a rib with grooves on either side.

1 **20m** Move up left and follow the groove round left of the rib awkwardly to a ledge.
2 **25m** Continue up the cracks above.
3 **30m** Continue up the fault to a bulge then move left to a prominent chimney.
4 **45m** Ascend the chimney, exiting left to a grassy bay. Finish up a big flake leading to easier ground.

7 Toves ★★ — 110m Very Difficult

FA Gary & Karen Latter & Colin Whiston 3 September 2000

Good climbing with a superb final pitch. Start beneath the next groove right of *Slithy*.

1 **20m** Ascend the groove passing a steep bulge on the left, continuing more easily to a good ledge and block belay.
2 **40m** Continue directly up the cracks above to belay on a ledge.
3 **50m** Continue up the fault on excellent holds and protection to level with a rowan in a bay out left. Step up right into the fine hanging groove and finish up this, passing some blocks near the top. Scramble to finish.

8 Brillig ★★   110m Very Difficult

FA Dr Patricia Littlechild, Cliff Ogle & Gary Latter 21 July 2000

A good direct line cutting through the easy shelf of *Mome Rath Route* finishing up that route's final pitch. Start beneath an open groove 3m left of *Outgrabe Route*.

1 **40m** Move up into the V-groove and climb steps in the slabby left wall then direct by the crack-line to block belay on the wide shelf.

2 **30m** Move out right and follow fine slim vertical groove to a good ledge on the right at the top.

3 **40m** Traverse leftwards and finish up *Mome Rath Route*.

8a *Direct Finish* ★ VS 4b climbs out left round the first roof, traversing back right to climb slightly rightwards through the upper bulge leading to easy ground.

Andrew Meldrum on the excellent first pitch of Jabberwock.

9 Outgrabe Route ★   115m Very Difficult

FA Ian Clough & party 6 June 1966; pitch 3 J.Brockway & Stuart Orr 17 May 1954

Start beneath two prominent crack-lines 10m down and left of *Mome Rath Route*.

1 **35m** Climb to a recess then move left up a short slab corner to the leftmost crack, which leads to a large block belay.

2 **30m** Continue directly to the large grass ledge of *Mome Rath Route*.

3 **50m** Follow a direct line just left of the chimney.

10 Mome Rath Route ★★   120m Very Difficult

FA Stanley, Mrs M & Miss C Stewart 16 May 1954

Start at the base of an easy angled grey shelf slanting left. Climb easily up the shelf to belay on a good grass ledge at the base of an open chimney. Continue directly, pulling steeply out rightwards at the top. Scramble to finish.

11 Jabberwock ★★   100m VS 4c

FA Ian Clough & C.Kynaston 27 August 1966

Start beneath the vertical crack up the wall just right of *Mome Rath Route*.

1 **50m 4c** A fantastic pitch on wonderful rough pocketed rock. Climb the crack to belay on large overhung ledge on the left (past possible belay on ledge at 30m).

2 **20m** Climb easily up rightwards across slabs and up an easy-angled left-facing groove to belay at the base of a steep crack.

3 **30m 4c** Climb the steep crack directly above on good holds until it is possible to swing onto large holds on the left edge. Continue more easily up the wall to the top.

The obvious direct continuation of the crack is 11a *Borogroves* ★ HVS 5a.

12 The Wabe ★★★   95m Very Difficult

FA J.Brockway, Stuart Orr & D.Parlane 15 May 1954

Excellent and sustained climbing on very good rock.

Start beneath the wide flake crack 8m right of the grey shelf of *Mome Rath Route*.

1 **50m** Climb the flake and continue trending up left then fairly directly, crossing an overhang on the right on good holds. Trend left to spike belay beneath an easy left-facing slabby corner. Other belays possible.

2 **45m** Climb the slab and corner (as for *Jabberwock*) then traverse easily right along ledge to the right side of a large open recess. Climb up then move diagonally left along a prominent hanging slab above the roofs. Continue direct to belay in a recess. Scramble to finish.

13 Whimsy ★★   95m Severe 4a

FA Ian & Niki Clough 28 August 1966

Very good climbing up the shallow left-facing groove near the right side of the face. Approach the base of the route by scrambling along a grass terrace from the right.

1 **45m 4a** Climb the groove and continue up the easier upper groove to belay beneath prominent twin cracks.

2 **50m 4a** Climb the right crack for 6m then traverse left steeply on good holds and up the left-hand crack which soon eases. Continue in the same line to belay up on the left. Scrambling remains.

14 Gyre ★★   125m HVS 5a

FA Gary Latter, Tom Cameron & Scott McQueen 26 July 2000

A good direct line up the cracks just right of *Whimsy*. Start 3m right of *Whimsy* beneath a hanging groove.

1 **50m 5a** Climb the crack which leads to the right side of a bulge. Step left and cross the bulge on good holds past two grass clumps then continue up the easy groove. Climb the fine rib on the right edge to belay beneath the right-most crack as for *Whimsy*.

2 **25m 5a** Traverse rightwards to a hidden hanging groove. Move up this for a couple of metres then traverse back left to cracks and climb direct on excellent pockets leading to a belay beneath a large flake on the right.

3 **50m –** Continue direct on easy excellent rock passing triple blocks to belay at the very top of the cliff.

Slimcrack (page 148), Harry McCaffery on this fine approach pitch to the main cliff.

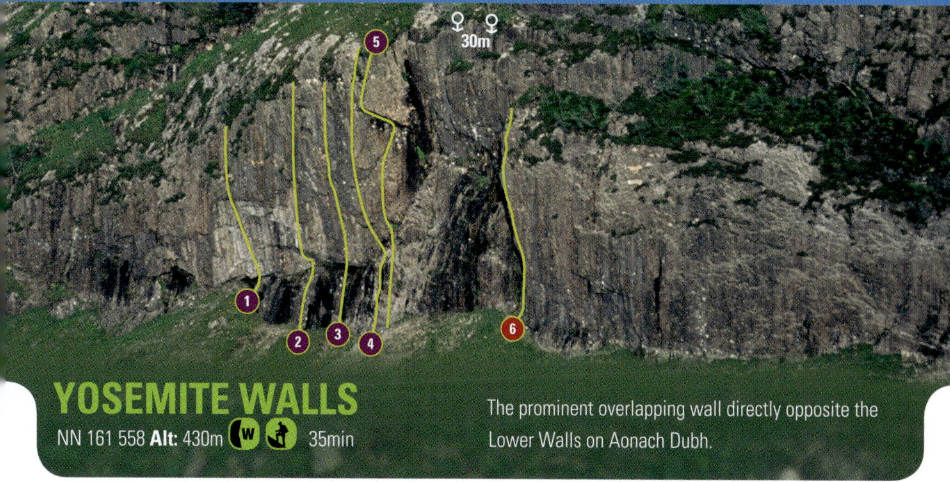

YOSEMITE WALLS

NN 161 558 **Alt:** 430m w 35min

The prominent overlapping wall directly opposite the Lower Walls on Aonach Dubh.

Approach: From the lower of the two main car parks on the south side of the road (NN 168 569) follow the path down the hill to cross the River Coe by the lower bridge. Continue up the path then steeply directly up the hillside to the base of the cliff.

Descents: Traverse right crossing the gully bounding the right edge of the crag to some large boulders then contour back to the foot of the crag. Alternatively abseil from one of the trees overhanging the top of the square-cut recess in the centre of the crag.

1 Rock Lord ★★★   **30m E7 6b**

FA Dave Cuthbertson & Rab Anderson 26 June 1995

In the centre of the overlapping wall an obvious right-facing corner provides the first weakness through the lower steeply overhanging barrier. Start about 4m to the left of the corner at an undercut cave. Difficult initial moves lead to a PR (pre-clipped on FA). Better holds lead up and rightwards to a break level with the top of the aforementioned corner. Move left and up to good holds at a break protected by an assortment of cams and small wires. Continue up the overhanging wall on undercuts to the final bulge. Pull over this (crux) and trend slightly left to a break. Now move up and right on mossy rock to a hollow flake. Step left and up to tree.

2 A Sweet Disregard for the Truth ★★★   **35m E6 6b**

FA Paul Thorburn & Gary Latter (both redpointed) 13 June 1995

Follows a direct line through a series of roofs in the centre. Start beneath a short right-facing groove at the left end of the long roof. Up an easy slab and groove to a break. Step left and pull up leftwards to a good slot

(R #9). Pull right and up to a good hold then direct to good undercuts under the first roof. Undercut rightwards then up to good jams. Step left to good holds at a large perched block then either direct or leftwards to a good small ledge then on good incut holds to an incut ledge. Step right and more easily up the right side of a crack to a nut and tree belay on a cleaned ledge.

3 Boiling Point ★★   30m E4 6a

FA Rab & Chris Anderson & Dave Cuthbertson 25 June 1995

The slim groove in the wall which runs into a crack-line just left of the arête. Direct entry to the groove is possible but prevented by very wet slabs beneath the bulge. Climb up to the right side of the arête, swing around left and traverse to the base of the groove. Climb the groove and the ensuing crack to reach the top of the crag.

4 The Mystery Trend ★   25m E4 6a

FA Paul Thorburn & Gary Latter 9 August 1995

The right arête. Scramble up easy slab to belay underneath the arête. Move up a groove to gain and follow a left-slanting crack through a low blocky overhang. A thin crack leads to the right side of the mid-height roof. Traverse left under this and make difficult moves round the arête. Continue up then right to easy ground.

5 Magnitude ★★   30m E5 6a

FA Dave Cuthbertson, Chris & Rab Anderson 25 June 1995

A steep sustained pitch up the wall right of the arête. Just right of the start of 4 is a second shorter shallow left-facing groove. Climb this (spike runner on the left wall) which is only adequately protected, to a rest beneath the right side of the overlap. Pull leftwards round this to a more comfortable position and good protection. Follow the thin crack above with increasing difficulty, culminating in a bold move out left beneath the large capping roof. Finish more easily up a groove.

The steep cracked arête right of the central alcove is 6 *Three Tarp Shugs* ★ E2 5b.

Rick Campbell crucifying himself on The Mystery Trend.

AONACH DUBH

THE LOWER WALLS

A long low line of walls (often water-streaked) below the Weeping Walls and the Lower North-East Nose.

Approach: From the lower of the two main car parks on the south side of the road (NN 168 569) follow the path downhill to cross the River Coe by the lower bridge. Continue up the path, leaving the main path 100m beyond where it crosses an open scree gully and a small stream just before a steep section. Follow the less obvious old path to a large boulder just above the prominent waterfall. Cross the stream and continue up a vague path steeply leftward.

Descent: Down either end of the crag or by abseil.

1   Thing of Beauty ★★   40m E3 5c

FA Gary Latter & Dave Greig 17 August 1993

A direct line up the left edge of the face just right of the wet crack. Start up 2 and climb direct up the diagonal crack past a series of horizontal breaks to a bold run-out section. From the ledge step left and up a blocky dyke to belay in short twin corners.

2   Lady Jane ★★★   25m E2 5b

FA Dave Cuthbertson & Derek Jamieson July 1977

Bold steady climbing on immaculate rock. Start 5m right of the wide wet crack. Climb direct by a faint depression to a prominent diagonal crack. Follow this rightwards to a wide horizontal break then direct (left of a wet streak) to small tree. Abseil descent (sling in situ).

3   Charlotte Anne ★★   25m E4 6a

FA lower half – (as Mr Bates) Rab Anderson 12 June 1988; Direct Finish Paul Thorburn & Gary Latter 12 June 1995

A good direct line cutting through 2. Start just right of that route at a short leftward sloping ramp. Move up and make difficult moves over the bulge then direct up the wall heading for a shallow pale groove (3m left of a tree). Gain the groove with a hard move then boldly on improving holds to finish.

4   Angeline ★★   25m E3 5c

FA Gary Latter 11 July 2014

Good sustained climbing, taking a direct line to the finish of 2. Start at the base of the short easy ramp, just right of 3. Climb up then right to good horizontal slot, place protection in short diagonal crack above, then move left and up to good horizontal break. Continue directly up tiny right-facing groove, then direct, moving right to good ledge. Climb direct from the left end of this to gain the good horizontal break on 2. Finish up this.

5 Sir Chancealot ★★　　　　　　**25m E1 5b**

FA Brian Duff & Kenny McCluskey 27 May 1978

A good pitch up the centre of the wall, heading for a tiny tree on the ledge. Start 10m right of 2. Climb direct to a short right-trending ramp then the wall above, trending slightly right near the top to a thread belay in the cave 2m left of the tiny tree. The top pitch (through the roof above the belay – E2 5c) is probably not worth doing. Descend by abseil.

6 Name Unknown ★★　　　　　　**30m VS 4c**

FA the ancients, sometime in the dark ages

The most prominent feature of the wall is a stepped left-trending groove. Climb this. The continuation above is often wet. Climb the rib either side to a thread belay 2m left of a tiny sapling. Abseil descent (sling in situ).

7 Double Exposure ★　　　　　　**70m VS 4b**

FA Ken Crocket & Chris Gilmore May 1978

Some 6m right of the lowest rocks is a corner with small, multi-coloured slabs. Start below the corner.

1　**25m** Climb the corner for 15m, break out right and go up to belay under a roof.

2　**45m 4b** Move up and left for a few metres, step onto a steep wall, move right up to a block then follow a groove on the left. Easier rocks lead to the top.

8 Lament ★★　　　　　　**55m Severe 4a**

FA John Cullen & Charlie Vigano April 1951

Start 20m up right from the lowest rocks beneath a left-slanting corner with a tiny rowan at the base.

1　**35m 4a** Climb the corner to pull out right at the top. Follow a direct line for 20m, move right and up a shallow groove leading to a belay in a recess.

2　**20m** Climb more easily directly to finish.

9 The Challenge ★★　　　　　　**20m E3 6a**

FA Dave Cuthbertson, Rab Anderson & Dougie Mullin 27 May 1978

50m right of the wall containing *Lament* is a short undercut buttress just right of a scrappy chimney capped by a large roof. Start 2m right of the chimney. A fierce bouldery start leads past a good nut slot to good holds at 5m. Step left and continue directly up a faint crack-line to a large ledge level with the rowan on the left. Move out rightwards to belay on a ledge above. Scramble off right along a grassy ledge to abseil from a small rowan.

LOWER NORTH-EAST NOSE

A good gently overhanging cliff with a varied collection of steep extremes, many staying dry during the rain.

Approach: As for The Lower Walls approach to cross the stream above the waterfall, drop down again and cross the top of a further wooded stream to gain and cross a grass rake.

Descent: Either make a 50m abseil from a small tree at the top of *Spacewalk* or down broken ground at the left side of the cliff.

1 Turnspit *   60m VS 4c

FA Dougal Haston & Robin Smith 2 October 1961

Climbs a line close to the right edge of the slender buttress at the left extremity of the cliff. Start a few metres up from the right edge in the grassy gully.

1 **30m 4b** Climb the steep wall then move right and up the groove in the edge which leads to a poor belay.

2 **30m 4c** Continue up the groove then out onto the left wall. Climb past a creaking flake and finish up the wall above.

2 Stormtrooper *   60m E3 6a

FA Kenny McCluskey & Colin McLean 11 June 1978

The steep wall and overhangs at the left side of the face between *Boomerang* and *Little Boomerang*. Start below the right crack.

1 **40m 6a** Climb the crack to the first overlap and go left a metre into a broken crack. Climb this trending rightwards into the centre of the wall. Continue up to a small ledge under the roof. Climb up the left side of the roof to good holds then traverse rightwards through overhangs on a horizontal crack and up to the belay ledge on *Boomerang*.

2 **20m 4a** Finish up *Boomerang*.

3 Boomerang *   90m HVS 5b

FA John Cunningham & Mick Noon 2 August 1955

A good main pitch following the left curving corner at the left side of the main face.

1 **30m 5b** Climb the wall and crack to a ledge and over a steep section above (crux). At the top make a long step left to a large ledge.

2 **30m 4b** Continue up the crack to reach a small cave and chokestone belay.

3 **30m** Finish up the grassy crack.

4 Freak-Out ***   65m E4 6a

FA Dave Bathgate & Alasdair 'Bugs' McKeith (A3) 1 July 1967;
Aid reduced to 1 PA by Dave Cuthbertson April 1977;
FFA Dougie Mullin & Jim Melrose May 1979

The crag classic, taking the central vertical crack leading to the prominent Λ shaped roof at half height. Start by scrambling up to a ledge and tree at the base of the crack.

1 **20m 5c** Climb the wall to gain the crack, which leads to a short traverse and a small ledge and belay on the right.

2 **30m 6a** Regain the crack and follow it to a rest of sorts in the niche. Pull over the roof on good holds then continue up the crack to an inverted flake poised beneath the final roof. Pull out on undercut jams leading rightwards through the roof to easy ground.

3 **15m** Finish easily up the unpleasant wall above.

5 Crocodile ★★★　　　　　　　　　　**50m E3 6a**

FA Pitches 1 & 2 Willie Todd & Rab Anderson 4 June 1977;
Pitch 3 Dave Cuthbertson & Murray Hamilton 5 June 1977

Start at a left-curving line of overhangs 5m right of
Freak-Out.

1　**20m 5c** Climb steeply to the overhangs then pull out
right to gain a small ledge. Continue up the shallow
groove above to P and nut belay on small ledge.

2　**10m 5b** Follow the groove on the right for
6m then traverse right round the edge of the
wall to a ledge below a prominent groove.

3　**20m 6a** Climb the groove with difficulty to a
ledge at the top. Hand traverse this easily left
then scramble up leftwards to the tree belay
on *Spacewalk*. Abseil descent.

6 Revengeance ★★　　　　　　　　　　**50m E6 6b**

FA Pitch 1 Dave Cuthbertson, Rab Anderson & Martin Lawrence
26 June 1981; Pitch 2 Dave Cuthbertson 27 June 1981

A slightly hybrid direct line up the wall right of *Crocodile*,
starting as for that route. A bold main pitch.

1　**25m 6b** Climb the crack in the bulge rightwards
to join *Spacewalk* and up this to its junction
with *Crocodile*. Step right and climb the wall to
a tiny overlap (least prominent of several – good
stopper #2 at right end in thin diagonal & RP #4
in slot to right). Step right again and continue
with difficulty up the wall, easing with height, to
belay at the end of *Crocodile's* second pitch.

2　**25m 5c** From the top of cracked blocks climb
the wall and very shallow groove parallel and
right of *Crocodile* then overhang direct to top.

7 Spacewalk ★★★　　　　　　　　　　**50m E5 6b**

FA Ken Johnstone & Pete Ogden (1 PA) 17 April 1978;
FFA Murray Hamilton & Dave Cuthbertson 6 June 1980

A good steep line up the wall right of *Freak-Out* with a
short well protected crux section. Start directly beneath
steep crack in steep wall about 6m below and to the right
of the start of *Crocodile*.

1　**25m 5c** Up the crack then up and
slightly right to roofs. Pull through the
roofs on the left to join *Crocodile* at old PR.
Continue to belay common to *Freak-Out*.

2　**25m 6b** Up a groove on the right past an old PR to
underneath a roof. Step left, up a short groove
and continue up then right to a tree belay.

Mark Glaister on the first pitch of Crocodile.
Photo Dave Cuthbertson, Cubby Images.

WEEPING WALLS

A fairly extensive off-vertical wall with distinctive drainage weeps down its left side. Away from the weeps some of the routes dry reasonably quickly. There is a large flat-topped boulder at the base of the centre of the cliff. Routes from right to left.

Approach: As for The Lower Walls then continue by a well-worn path that leads diagonally up left underneath the cliff.

Descent: Follow a vague path then descending shelves down rightwards (facing out). Go up slightly across a slab just beyond *The Lower Bow* then down a final short step (see diagram). For routes from *Quietude* leftwards scramble left along grass ledge and drop down to a lower ledge. Either abseil (25m) from tree or scramble down from far end of ledge.

1 Drainpipe Corner ★★　　　　　**50m Severe 4a**

FA Bill Smith & Charlie Vigano summer 1951

A good sustained route up the right-bounding corner, though often a watercourse.

  1 **23m 4a** Climb the corner, steeply at first to belay at a good ledge where the angle eases.

  2 **27m 4a** Continue up the corner to belay on a birch at the top. Descend by abseil or scramble.

2 Dangle By-Pass ★　　　　　**65m Very Difficult**

FA Hamish & Catherine MacInnes Summer 1965

The highlight is a fine exposed rising traverse. Start at a short stepped groove just left of the right edge of the wall and just right of where the path gains the base of the crag.

  1 **20m** Climb slightly leftwards then trend back right to nut and thread belay just right of the right end of a long narrow grass ledge.

  2 **15m** Move up a few metres then follow obvious traverse line on large holds rightwards.

  3 **30m** Continue more easily by cracks and grooves up the slabby left side of the arête to tree belay on large heather ledge at top. Either abseil off (50m) or scramble up the blunt edge and easy slabs leftwards above the large rowan at the left end of the ledge.

3 Curving Crack ★★　　　　　**85m Severe 4a**

FA Len Lovat 5 October 1952

Start 5m right of the large block in the centre of the face.

  1 **20m** Climb up directly to belay at a small rowan at the base of the crack.

  2 **30m 4a** Climb the crack and the wall above to belay on a long ledge.

  3 **35m** Continue more easily up broken rocks above.

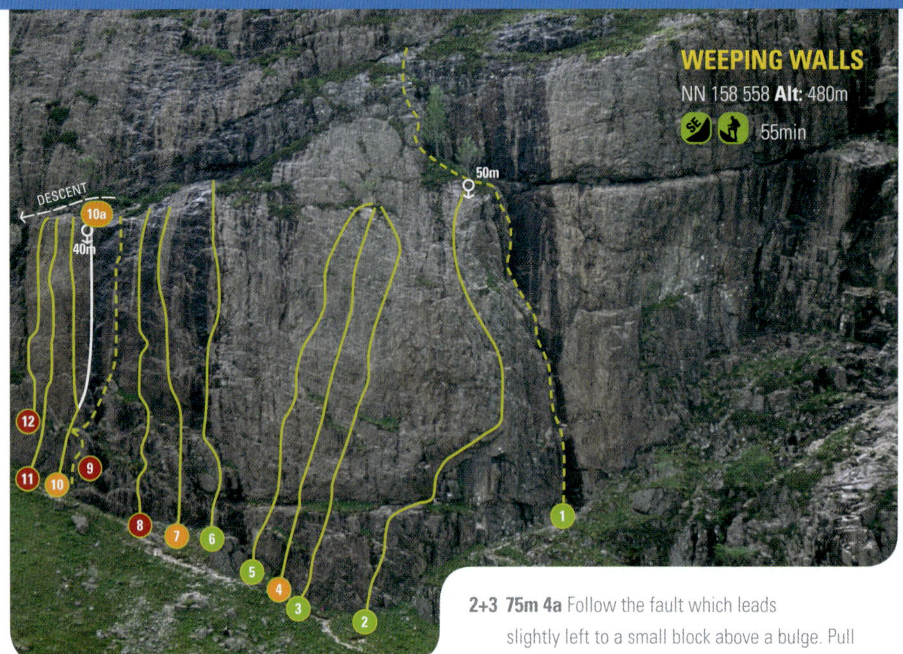

2+3 75m 4a Follow the fault which leads slightly left to a small block above a bulge. Pull over the block and follow parallel grooves then more easily directly to the Terrace. Finish more easily up the slabby wall above.

4 *The Straight Climb* ★ VS 4b takes a direct line between the two cracks.

---

5 **The Long Crack** ★★ 90m Severe 4a

FA Len Lovat & J.Johnstone 6 June 1953

Follows a diagonal crack leading to the rowan in the centre of the top of the main section of the wall. Start just behind the right side of the large block boulder at the base.

1 22m 4a Climb fairly directly to belay down left of the start of the crack.

2 28m 4a Move out right and follow the crack mainly on its right side then direct to the rowan.

3 40m Continue trending leftwards to belay far back.

6 **Weeping Wall Route** ★★ 90m Severe 4a

FA Douglas Scott & J.Henderson August 1947

Excellent sparsely protected climbing, though slow to dry, following a slightly left-trending fault up the right edge of the weeps. Start just left of the block boulder.

1 15m – Climb direct to belay in a recess beneath the fault.

The following 7 routes all end at a long narrow grassy terrace. Either finish more easily by about 25–30m of easy angled rock or traverse left to abseil (25m) from a tree at the far left end of the terrace.

7 **Quietude** ★ 50m HVS 5a

FA Martin Hind & S.McCabe 11 June 1977

Start beneath a prominent black overhang, 15m right of the prominent cracks of *Spider*.

1 12m Climb the wall to a ledge below the overhang.

2 38m 5a Cross the overhang by a left-slanting crack to a ledge. Continue leftwards up the wall to a shallow groove and up this to the terrace.

8 **Short But Sweet** ★★ 50m E2 5c

FA Dave Cuthbertson & Roy Williamson Summer 1981

Bold climbing taking a line midway between 7 & 9. Go up the lower wall to the right end of the grass ledge then continue up the wall, starting past two prominent finger pockets, moving up and right to a slight overlap. Step left and climb the steepening wall with final tricky moves about 2m left of the final corner on *Quietude*.

9 Solitude ★★★   48m E3 5b

FA Dave Cuthbertson, Rab Anderson & Willie Todd 3 June 1977

Fine bold wall climbing on immaculate rock. Low in the grade.

1 **8m** Climb steep wall to the ledge below the crack-line of *Spider Right-Hand*.

2 **40m 5b** From the right end of the ledge climb straight up to a prominent block at 6m then directly up the wall above to the terrace.

10a Spider Right-Hand ★★★   43m VS 4c

Excellent reasonably-protected climbing tackling the rightmost of the twin cracks bounding the left margin of the prominent dark seepage stain. Slow to dry.

1 **8m** Scramble up past a heathery ledge at 6m to the base of the crack.

2 **35m 4c** Climb the crack leading into a shallow groove in its upper reaches and follow this to belay on terrace.

10 Spider ★★★   43m HVS 5a

FA Jimmy Marshall, Ronnie Marshall & Archie Hendry April 1957

Excellent sustained sparsely protected climbing.

1 **8m** Scramble up past a heathery ledge at 6m to belay below the leftmost of the twin cracks.

2 **35m 5a** Move up and climb a shallow right-facing groove to its top then move up right to good holds and protection. Move up to a small foot-ledge then a vague hidden crack to finish directly.

11 The Fly ★   42m E3 5c

FA Dave Cuthbertson & Willie Todd 9 July 1977

A serious pitch up the wall left of *Spider*. Start at a prominent crack 10m left of *Spider*.

1 **12m 4c** Climb the crack to a ledge on the right.

2 **30m 5c** Climb directly up the wall above then move up and left into a scoop. Move slightly left then up the wall above to belay on the terrace.

12 Bivvy Wall ★★   42m E3 5b

FA Dave Cuthbertson & Paul Moores August 1994

More attractive than *The Fly* but equally serious. Start just right of an obvious crack. Climb the wall (without using the crack). Now make an awkward move up the arête then trend slightly right to a scoop (poor wires and spike). Climb the wall above to gain a line of holds trending slightly right. Easier climbing leads to the terrace.

◁ *Drainpipe Corner (page 157), Weeping Walls. Chris Durn climbing.*

NN 157 557 **Alt:** 520m

BARN WALL  1hr

The extensive easier angled continuous mass of rock bordering the left margin of the Weeping Walls.

1 Lower Bow *   65m Moderate

FA Donald McIntyre & Bill Murray (both solo) May 1947

This forms a prominent easy-angled open chimney fault higher up. It starts as a shallow right-slanting groove, 2m right of a small rowan and at the base of a slanting shelf.

1 **25m** Climb the groove which soon leads to an easy-angled slab.

2 **40m** Continue more easily up the open fault.

NN 157 558 **Alt:** 600m

TERRACE FACE  1hr 10min

This lies above The Terrace with the fine barrel-shaped buttress at the left side harbouring some ultra-classic easier routes.

Approach: Either by a route on the **Weeping Walls** or the right side of the **Barn Wall** or by scrambling easily up the descent route, which starts as for *The Bowstring*.

Descent: Gain a small worn path heading right (north) then scramble down a narrow gully formed by an eroded basalt dyke (often wet). At the base of this a narrow scree cone leads back down to The Haven. Alternatively, make a 45m abseil from in situ sling on block just below the top of *Arrow Wall*, which can be gained by a traverse from the top of the adjacent routes.

1 Quiver Rib ****   55m Difficult

FA Donald McIntyre & Bill Murray May 1947

Superb steep well-protected climbing with improbable situations for the grade. Start at the left side of the buttress just right of a wide chimney.

1 **25m** Climb the rib then move up right to belay beneath a steep wall.

2 The Bowstring *   115m Difficult

FA Donald McIntyre & Bill Murray May 1947

Steady climbing on good clean rock. Start at the same point as the descent route, 20m left of the *Lower Bow* on a shelf with small rowans at its right end.

1 **20m** Climb straight up then slightly right up a glacis to belay in a short left-facing corner.

2 **45m** Climb the corner and a short steep crack then more easily, veering left into an open chimney for a few moves before returning to climb the right side to belay on a large ledge. Either walk right to the base of The Terrace Face or:

3 **50m** Continue up the left side of the chimney.

3 Barn Wall Route *   110m Moderate

FA Donald McIntyre, Trevor Ransley & Bill Murray May 1947

Follow the 'line of least resistance', starting from the highest section of the base of the wall.

2 **30m** Continue in a fine position up the left-trending shallow groove above the belay.

2 Arrow Wall ***   60m Very Difficult

FA Len Lovat & J.Johnstone 7 June 1953

Another excellent route through some unlikely territory for the grade. Start midway between *Quiver Rib* and *Archer Ridge*.

1 **30m** Follow the shallow groove avoiding an overlap at 12m on the right. Continue in the same slightly wandering line to belay on the ledge as for *Quiver Rib*.

2 **30m** Climb the steep black groove directly above the belay on good holds and continue in the same line to the top.

3 Wounded Knee **   60m Severe 4a

FA Ken Crocket & Brian Dullea 14 April 1991

Good though fairly sparsely protected climbing up the wall to the left of *Archer Ridge*. Start midway between *Arrow Wall* and the arête.

1 25m 4a Climb a steepening groove to belay on a ledge on the left.

2 35m 4a Step left onto a steep wall, climb up and left for a few moves then traverse across the bulging wall rightwards to the edge. Continue on good holds to the top.

4 Archer Ridge ★★★   60m Very Difficult

FA Bill Murray & Donald McIntyre May 1947;
Direct Len Lovat, Ian McNicol & A.Way May 1954

The blunt right-bounding rib completes the trilogy of classics. Start at the left side of the rounded ridge.

1 35m Follow the crest (possible belay on the right at 25m) then out slightly rightwards beneath a steeper wall. Continue back up and left onto the crest which leads to a belay in a short corner recess.

2 25m Move out right a short way then follow a crack and the short final steep wall which leads to a belay a short way back. The *Direct* (Severe 4a) follows the crest to the left, moving steeply up slightly left from the belay. With careful rope management, possible in one superb 50m pitch, moving up left to gain the block belay just below the top of *Arrow Wall*.

High up in the centre of the face is a distinctive recessed crag above a large heather and rowan tree infested bay known as the Basin. Gain the base by easy scrambling from the left.

5 De Vreemde Stap ★   30m E1 5b

FA Martin Hind (Harpic) & H. Van Ryswick July 1978

A useful direct entry to the Basin. Start about 10m right of a chimney high up on the left side. Climb a shallow groove until a wide step right and up to gain the continuation groove. Climb this then by a left-trending crack leading into The Basin.

The following two routes are situated above the Basin. High up there is an area of distinctive curving grooves.

6 Hesitation ★★   50m HVS 5a

FA John Cunningham (1PA) 3 July 1966;
FFA Ken Crocket & Ian Fulton summer 1972

This route takes a line left of the distinctive pink slab. Good climbing with a steep finale. Start above and left of the trees beneath an obvious shallow groove left of a slabby scoop.

1 35m 4c Climb the groove then head up and right to a ledge at 25m. Climb diagonally right to a small stance in the large groove capped by a huge roof.

2 15m 5a Climb up to the roof, traverse left 5m then follow an overhanging groove out right (crux) to finish on a large ledge.

7 Terrace Arête ★   35m VS 4b

FA Patsy Walsh & John Cullen May 1954

Bold climbing taking the stepped arête delineating the right edge of the crag. Climb the lower arête on good holds to a large ledge. Step out left with some tricky moves to become established on the arête proper and continue on improving holds to protection. Continue more easily up a shallow groove in the arête.

Archer Ridge. Graham Bremner climbing.

NN 154 555 **Alt:** 740m SE 1hr 20min

FAR EASTERN BUTTRESS

An excellent crag of rough rock, possibly some of the best in the glen. Well worth the extra walk to escape the crowds.

Approach: Continue up the path on the left side of the stream, crossing the stream where it starts to level out above the waterfalls then head directly up to the base of the crag. Alternatively, from the other cliffs lower down on the East Face continue up the hillside underneath the cliffs for a further 25 minutes beyond Barn Wall.

Descent: The easiest descent is by moving up and left and down the large wide grassy gangway sloping down underneath the small crags bounding the left side of the crag. Alternatively, from the top of the narrow gully bounding the right side of the crag containing a chokestone, traverse a small rock shelf round the corner and down easy slabs then skirt round to the base of the crag.

1 Farewell Arête ★   70m Very Difficult

FA Ken Crocket & Alastair Walker 19 May 1990

At the right edge of the buttress is a deep gully. This route follows the arête forming the right wall. Climb the arête with a step left at 6m then continue to a ledge and belay. Continue more easily in a further two pitches (or one long pitch) in the same line.

The following two routes start by scrambling up 15m to a belay at the base of the wall.

 Shibumi ★★   45m VS 4b

FA Ken Crocket & G.Jefferies 12 May 1990

A good pitch on excellent rock up the rightmost of two parallel right-slanting cracks. Move right to good foot-holds before the corner then follow the crack leading to easier ground at the top of the corner. Finish up the edge.

 Satori ★★   45m VS 4b

FA Ken Crocket & G.Jefferies 12 May 1990

The left crack. Start beneath the left arête of the wall. Climb up then move right into the crack and follow this and the direct continuation to belay far back.

 Yen ★   75m Very Difficult

FA C.Kynaston & J.Garster 29 August 1966

Start 8m right of *Nirvana Wall* near the right edge of the slab.

1 **25m** Climb steep slabs to beneath a steep cracked wall.

2 **50m** Move right and climb the crack with difficulty to a grassy ledge. Finish up the groove.

5 **Nirvana Wall** ★★★   60m HS 4b

FA Ian Clough, C.Kynaston & J.Garster 26 August 1966

The highlight is a well positioned thin crack and groove up the highest, central section of the crag. Start by some wonderful quartz streaks towards the left end of an immaculate smooth slab.

1 **35m 4a** Climb the thin crack then continue up easier ground to belay beneath the steep upper crack.

2 **25m 4b** The thin crack, finishing up the right-facing roof skirting the right end of the roof. An excellent well protected pitch. Thread belay 15m further back.

6 **Eastern Promise** ★★   65m E1 5b

FA Rab & Chris Anderson 6 July 1991

A good little route following the arête and thin crack-line left of *Nirvana Wall*. Start beneath a grassy groove.

1 **40m 5a** Go up the groove a short way then into a thin crack-line on its right wall. Climb this and the arête to a block belay beneath the upper wall.

2 **25m 5b** Step down right then follow the thin crack up the centre of the wall past the right side of the wall to the roof. Pull rightwards through this then move back left above it and up to the top.

7 **Rough Slab** ★★   50m Severe 4a

FA Bob Richardson, Peter MacKenzie & Bill Skidmore 2 September 1962

A fine aptly named route up the left side of the face. Start beneath a groove in the clean wall overlooking the broken gully on the left. Follow the groove past a large block to a good ledge beneath small overlaps. Traverse slightly right and climb a groove through the overlaps. Move round left above to finish up a fine steep clean slab.

Karen Latter on Nirvana Wall.

STOB COIRE NAN LOCHAN
(PEAK OF THE CORRIE OF THE LITTLE LOCHS)

This north-east facing corrie formed by the long ridges of Aonach Dubh and Gearr Aonach harbours some of the finest situated routes in the glen. Care should be taken with perched blocks, particularly near the top.

Approach: From the lower of the two main car parks on the south side of the road (NN 168 569) follow the path down the hill to cross the River Coe by the lower bridge. Continue steeply up the path on the left (east) side of the stream up into the corrie.

Descent: Traverse right and down the well worn path and easy-angled ground beyond the right (north) end of the cliffs.

SOUTH BUTTRESS   NN 148 552 **Alt:** 900m

The leftmost clean buttress.    2hr

① Unicorn ★★★   100m E1 5b
FA Jimmy Marshall & Robin Campbell 18 June 1967

The classic line of the corrie taking the eye-catching left-facing corner. Start by scrambling up leftwards to the base of the corner.

1 **35m 5b** Climb the corner for 8m then move right and up the right rib for about 10m to step back left into the corner. Follow this to belay at a small foot ledge.

2 **35m 5a** Continue up the wide crack in the corner. At the top carefully negotiate some tottering blocks to belay in a chimney formed by two large (solid) blocks.

3 **30m 5a** Move round right and up a short steep chimney to a large flake. Move out right and up good flake cracks which lead abruptly to easy ground. Belay further back. Scramble up the easy ridge above to finish.

From the base of the final pitch, a 60m abseil descent is possible (slings usually in situ) then climbing *Scansor* (climbing the final chimney common to both only once).

2 Shadhavar ★★★ 100m E3 6a

FA Ian Taylor & Tess Fryer 21 July 2013

The finely positioned crack on the right wall of the upper section of the *Unicorn* corner. More amenable than appearances would suggest.

1 **35m 5b** Climb *Unicorn* until level with the triangular roof on the right arête.

2 **35m 6a** Traverse out right to the thin crack and follow it directly with excellent protection. Near the top swing round the right arête for a few moves, then go hard left (to avoid some loose blocks) into *Unicorn*, which is followed to the ledge.

3 **30m 5a** Follow *Unicorn* to the top.

3 Scansor ★★ 120m E2 5b

FA Paul Braithwaite & Geoff Cohen 2 September 1972

A well situated route up the pillar right of *Unicorn*. Care should be taken with loose flakes on generally easy ground. Start at the same point as *Unicorn*.

1 **45m 5b** Follow the obvious groove right of *Unicorn* to a small ledge on the arête. Move up left and climb the steep crack to level with a small roof on the left. Traverse right and continue by a steep arête to belay at a block.

2 **25m 5b** Make an awkward move to an upper ledge on left. Go up the wall then traverse right to a small foot-hold in the centre of a pillar. Now make several strenuous moves up and right to a ledge and belay.

3 **20m 4c** Follow the groove on the right to ledges then up and left across a minefield of loose blocks to belay in a chimney formed by two large (solid) blocks.

4 **30m 5a** Finish up the top pitch of *Unicorn*.

4 Gecko Wall ★★★ 60m E6 6b

FA Iain Small, Tony Stone & Blair Fyffe 25 August 2013

Start in a small gully below and right of the *Scansor* pillar.

1 **40m 6b** Climb up and into a pod on the left wall of the corner. Make tricky move rightwards into corner. Follow this towards a roof until an escape can be made to easier ledgy terrain on the left. Climb this until a tricky move right gains a crack. Follow this, pulling strenuously over the left end of the top roof to directly gain the *Scansor* belay.

2 **20m 6a** Climb directly up the thin crack above to the *Scansor* belay. Abseil off or finish as for *Scansor*.

CENTRAL BUTTRESS

NN 148 552 **Alt:** 900m   2hr

5 Central Grooves ★★ 115m VS 4c

FA Ken Bryan & R.Robb (1 PA) 3 July 1960

A popular route up the prominent groove-line cleaving the centre of the buttress. Start at the base of the groove.

1 **25m 4c** Climb the groove to belay on small ledges.

2 **20m 4c** Continue up the grooves to belay at the top of a pedestal on the left.

3 **50m 4c** Continue in the same line to a block belay on a broad terrace.

4 **20m** Finish easily up the crest.

Andy Stewart & Paul Riley nearing the top of the second pitch of Unicorn. Photo Dominic Oughton

NORTH FACE OF AONACH DUBH

The great gaping slit of Ossian's Cave dominates this broody atmospheric face. The base of the cliff rises immediately above Sloping Shelf, a diagonal fault that runs up right directly beneath the base of the cliff. This forms the boundary between the scrappy lower andesite band and altogether more appealing rhyolite layer above.

Approach: From the lower of the two main car parks on the south side of the road (NN 168 569) follow the path down the hill to cross the River Coe by the lower bridge. Cross the burn and head diagonally right up the hill side, skirting round either side of a couple of small crags low down. A ledge system leads up right to the base of the Sloping Shelf, which in turn is followed by crossing a deep stream-filled gully (path high up). Just beyond this a short awkward greasy slab regains the path leading up under the cliff. There is a stream running down the shelf directly underneath the base of the crag.

Descent: Routes end on Pleasant Terrace (a bit of a misnomer). Follow this carefully (much loose rock) to make a 35m abseil from an in situ PB near its right end.

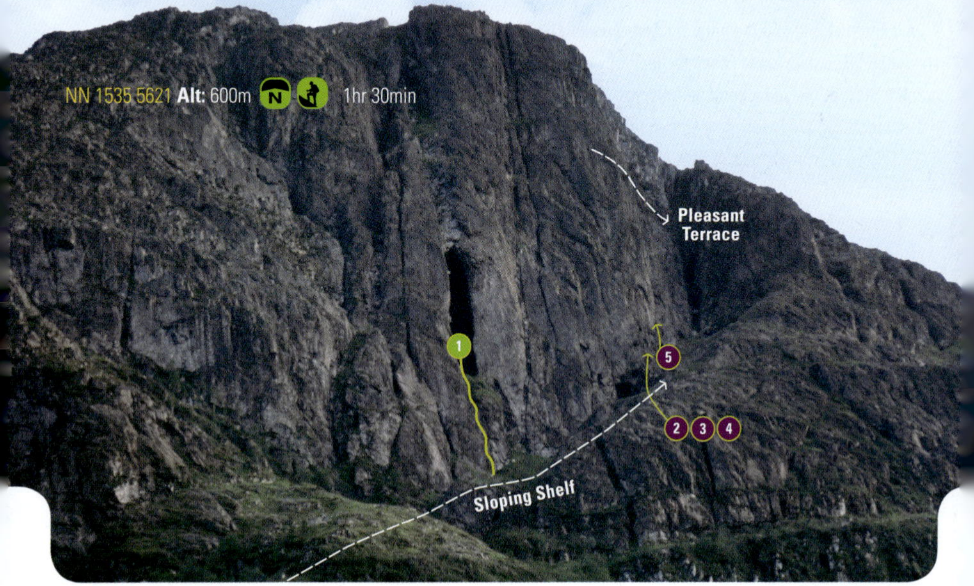
NN 1535 5621 **Alt:** 600m
1hr 30min
Pleasant Terrace
1
5
2 3 4
Sloping Shelf

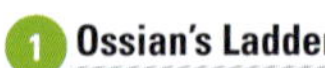

1 Ossian's Ladder — 45m Hard Very Vegetated

FA Neil Marquiss 1868

The first recorded 'climb' in the glen – of historical interest only. Recommended if you're insane or into botany in a big way. The route consists mainly of copious quantities of lush vegetation. The cave itself has been formed by a huge block falling out of a dyke, leaving behind a broken 45 degree sloping base.

3 **40m 5c** Follow the groove just right of the arête (the second most obvious groove left of 4) boldly until it is possible to traverse right to the poor belay on 4.

4 **25m 5a** Continue up the rib to Pleasant Terrace.

3 Repossessed ★★★ 40m E5 6a

FA Martin Crocker 30 June 1995

A sustained and superb pitch following a direct line above the roof where *Eldorado* steps left. Follow the main pitch of *Eldorado* for 10m to the roof. Step right to an undercut and up a wide crack to better holds. Continue more easily to the overhanging wall above and up this to awkward sloping jugs (crux). A long reach gains better holds and easier ground leading to the long shelf.

4 The Clearances ★★★ 105m E4 6a

FA Ed Grindley, Cynthia Grindley & John Main August 1976
(climbed over 2 days & some aid used);
FFA Murray Hamilton & Willie Todd 1977

The hanging crack-line in the wall left of *Yo-Yo*. A sustained classic with a bold lower section.
Start 6m left of *Yo-Yo*.

1 **45m 5c** Gain the left-slanting shelf and follow this, then move diagonally left on the slab above. Move up right to a small ledge (PR) then boldly up the wall to step right into the crack and follow it more easily to a shelf.

2 **40m 6a** Move up the shelf and follow the shallow left-facing groove 5m left of the corner to an old nut under the roof. Step right under the roof (crux), pull over on good holds and continue to a poor belay on small ledge.

3 **20m 5a** Continue more easily to the top.

5 Eragon ★★★ 40m E6 6b

FA Guy Robertson & Blair Fyffe June 2009

A stunning and at times bold pitch up the big wall between *The Clearances* and *Yo-Yo*; protection is there when needed most. The route aims for and passes through the obvious niche, finishing up the slim groove in the upper part of the wall. Start as for *Yo-Yo*, but continue left to the end of the initial break (protection). Continue diagonally up and left until good flat holds lead more

2 Eldorado ★★★ 125m E5 6b

FA Ken Johnstone & Mark Worsley (3 PA) 22 June 1977;
FFA Dougie Mullin & Murray Hamilton 17 May 1980;
FA Top Pitch Dave Cuthbertson & Ken Johnstone 18 May 1980

Excellent sustained climbing up a series of cracks through the bulges in the wall left of *The Clearances*. Start just right of obvious crack.

1 **18m 5c** Move up and left into the crack and climb this to belay on the left.

2 **40m 6b** Follow the groove to an overhang, step right to climb a very short corner. Step left above a roof onto a steep wall and a good resting spot. Follow the overhanging crack over a second smaller overhang with difficulty then go up and left to a large ledge. Continue up beside a crack to a terrace.

directly up to a rest a few metres below the niche. Go up through the niche (crux), step left, then trend back right for a few moves before heading back up leftwards to a better rest below the final groove. A precarious wobble up the groove leads to the big ledge and junction with the other lines hereabouts.

6 Yo-Yo ★★★★   **90m E1 5b**

FA Robin Smith & David Hughes May 1959

The great corner of the crag. The bottom 15m often seeps. On the first ascent Smith spent 4 hours drying the rock with a towel (hence the name). Start directly beneath the corner.

1 35m 5b Move left round an overhang and up the left-facing groove. Pull right 6m above the ground to climb black seepage streaks leading slightly rightwards to the corner proper. Up this to belay on a large shelf.

2 25m 5b Continue up the corner/chimney to belay above the main overhang.

3 30m 5b Continue in the same line with a short excursion to the left near the top.

7 Bungee ★★★   **100m E3 6a**

FA Blair Fyffe & Guy Robertson (on-sight) June 2009

A superb exposed and well protected route, starting up *Yo-Yo* then breaking out round the right arête onto the wall to its right. A prominent feature is the striking crack high on the headwall.

1 30m 5b *Yo-Yo* pitch 1.

2 30m 5c Step delicately down right and traverse under an obvious overlap to a precarious perch near the edge. Move up into a hanging groove and climb this then a crack on the left with difficulty to a sloping ledge. Traverse right along this to climb the first obvious groove with an awkward exit onto a ledge.

3 25m 6a Above is a clean-cut triangular niche with a finger crack above its right end. Climb steeply up to gain and follow this crack, then pull left into another crack which is followed with increasing interest to another good ledge.

4 15m 5b Climb the steep wall above to the terrace.

Repossessed, Paul Thorburn on second ascent.

WEST FACE OF AONACH DUBH

The main bulk of the West Face is split vertically into distinct buttresses by six prominent vertical gullies. There are also two horizontal terraces across the face, effectively dividing the buttresses into three tiers. The lower narrow ledge is called Middle Ledge and lies at the base of all the routes. Above is a broader sloping terrace known as The Rake. The lower tier is composed of loose vegetated andesite of no interest to the rock climber. In contrast the steeper middle tier of rough pink rhyolite contains the bulk of the climbing. The one other area of interest is the miniature corrie-like feature of The Amphitheatre, situated between the upper tiers of E and F Buttresses.

Approach: Follow a well constructed path leaving the A82 at the west side (right) of the bridge over the outflow of Loch Achtriochtan directly opposite the Clachaig turn-off. This leads steeply up the hillside, higher up keeping close to the right (west) bank of the Allt Coire nam Beithach.

Cross this stream below the lowest waterfall and head steeply up to the base of **Dinner-time Buttress**, then cross **No. 2 Gully** on the right, low down and climb its right flank (below **B Buttress**) to gain the left end of **Middle Ledge**. Follow this right to the buttress of your choice.

E BUTTRESS

NN 1438 5557 **Alt:** 630m    1hr 30min

This is the impressive steep buttress left of No. 4 Gully with its finest feature, the south-west face overlooking the gully.

Descent: Traverse right and descend a sloping shelf into **No. 4 Gully**, then scramble down this with care.

1 The Big Top ★★★★   **160m E1 5b**

FA Robin Smith & Jimmy Gardner August 1961

A classic. The highlight is a spectacular wildly exposed pitch up the left arête of the huge leaning left wall of the *Trapeze* corner. Start at a block belay below the arête.

1 35m 4c Go up left for 15m, climb a slabby corner and then up and right to a belay on top of a large flake.

2 35m 5a Climb the arête to the right and a bulge above. Continue by a crack on the edge of the arête to an easing of the angle. Belay at the base of a large flake.

3 45m 5a Move right into a diagonal line of slabby grooves and climb these until it is possible to move left into a 3m crack. Climb the crack to a large ledge and belay.

4 45m 5b Now climb a huge flake on the right to reach a wall. Move left up the wall and then right and traverse right across a groove to a slab. Climb the slab and a wall and finish up a broken groove to the top.

2 Trapeze ★★★ 130m E1 5b

FA Jimmy Marshall & Derek Leaver summer 1958

Start down and right of the large corner near the left edge of some vegetated rakes.

1 15m Scramble up the rakes to the base of the corner.

2 20m 5b Climb the strenuous corner to beneath an overhang.

3 40m 4c Turn the overhang on the left and follow the now easy corner to slabs, continuing to a mossy bay.

4 5m Traverse right to a well defined platform.

5 40m 5a Quit the platform on the left and trend left across a steep wall to a ledge and corner above. Turn the corner on the right and make an ascending right traverse to a groove and crack. Climb these to a slab below an overhang and traverse right to a rock bay and belay.

6 10m Now follow a groove rightwards and finish by a short crack.

Trapeze, Dave Griffiths on the main (crux) corner pitch.

The Big Top, Scott Muir on the wildly exposed second pitch. Photo Dave Cuthbertson, Cubby Images.

BIDEAN NAM BIAN
(PEAK OF THE MOUNTAINS)

The highest mountain in Argyll at 1150m, the massif projects the long ridges of Beinn Fhada, Gearr Aonach and Aonach Dubh to the north. These ridges mostly obscure views of the summit from the road except from Loch Achtriochtan where the twin buttresses of Diamond and Church Door are visible beneath the summit screes. The former is broken and vegetated, unlike its superlative neighbour.

CHURCH DOOR BUTTRESS

"This is just wonderful, totally brilliant – there's even some sheep" – Paul Thorburn, below base of cliff, 1995
One of the finest cliffs in Glen Coe, it also happens to be the furthest from the road. With the base of the routes lying at an altitude of 950m/3200 feet, the summit lies just a few minutes away from the top of the cliff. The routes are relatively slow to dry and it is unlikely those on the West Face will dry out before June due to the large quantity of vegetation on the slopes above.

Access: Park on the south side of the A82 next to the outflow of Loch Achtriochtan opposite the Clachaig turn-off.

Approach: Follow a well constructed path leaving the A82 at the west (right) side of the bridge over the outflow of Loch Achtriochtan directly opposite the Clachaig turn-off. This leads steeply up into the coire beneath the crags of Stob Coire nam Beith (1 hour +) and continues up to the flat coire floor beneath the broken Diamond Buttress. From here scramble up screes to the base of the cliff.

Descent: From the top of the routes go to the summit then down to the col on the right. Descend steeply down the scree slope running back underneath the base of the cliff. This slope holds much snow well into June and in such conditions the buttress crest further left (facing out) should be followed. In favourable conditions a shorter descent can be made by traversing right (west) across the screes about 50m short of the summit, heading for just above a large pink boulder then diagonally left to gain the scree slope at the base of the cliff. Alternatively, a 50m abseil from a flake at the top of *Temple of Doom* gains the base on rope stretch.

EAST FACE

Go up the scree-filled gully between **Diamond** and **Church Door Buttresses** to a prominent chimney opposite the neck of Collie's Pinnacle on the left.

1 Flake Route ★★　　　　　　**130m Very Difficult**
FA Harold Raeburn, John Bell & Graham Napier July 1898

"I went forward to the end of the ledge to try and field him should a slip occur, while Napier jammed himself in a hole in the ledge, worked Raeburn's rope over the small hitch and anchored me. This time Raeburn was successful, and wild cheers broke out … " – Bell.

The original route up the buttress. A huge flake is separated from the buttress by a crack. Ascend the crack to the crest of the buttress. Make an awkward step up and right then direct until a traverse left leads to an arch formed by two enormous boulders. Cross the arch and ascend the shallow chimney (crux). Finish steeply to easy ground.

2 Crypt Route ★★★   135m Very Difficult

FA R.Morley, M.Wood, John Wilding, & Fred Piggott 15 September 1920; Gallery Variation: Ian Norris, P.Barker & A.N. Other 15 July 1952

A unique subterranean excursion into the very bowels of the cliff, of interest to speleologists, troglodytes and other such perverts. The only route in the country where a torch is de rigueur. Ascend the chimney for 20m to a corridor cleaving into the cliff. Go to the rear of the corridor where there is a choice of routes, though the honeycomb of subterranean passages are all rather similar!

The Tunnel Route: through a narrow passage in the left wall to a chamber then another tunnel to a further chamber from which a long narrow tunnel leads upwards to a 0.5m diameter hole in the cliff face 6m beneath the right end of the arch. Ascend easy slabs for 6m to another hole in the lower right end of the arch.

The Through Route: through the cave-like end of the corridor to a smaller cave, exiting this with interest using the top of the chokestone. Continue past a grass ledge, a jumble of boulders and a cracked block to the arch.

The Gallery Variation: From the second cave of *The Through Route* enter another chimney in the fault to gain a smaller third cave. Enter the 'gallery' (2m x 1.3m x 6m) above, descend 1.3m from the 'gallery' floor, and facing out, traverse rightwards 15m to climb up to the arch. All three variations finish by *Flake Route.*

> **Top tips: (1)** take long stick and big hook to recover head torches and assorted gear from deep pit **(2)** light breakfast and no lunch recommended **(3)** head torches no use – better off with mouth-held hand torch **(4)** leave rucksacks at base!

WEST FACE

1 Fundamentalists ★★★   100m E4 6a

FA Gary Latter & Pete Craig 15 September 2002

The crack-line at the left side of the face just left of *Lost Arrow.* Start at the block belay as for that route.

1 25m 5b Move up leftwards onto a ledge then up left over blocks. Step right and climb a wall above the block to belay on the right as for *Lost Arrow.*

2 25m 6a Climb the slab on the left into the base of the grey corner. Step right into the crack with difficulty and follow it to belay in a small recess just right of the prominent square-cut roof.  Currently very overgrown and probably unclimbable without cleaning – gain upper section by the excellent pitch 2 of *Lost Arrow.*

3 20m 6a Undercut left and pull spectacularly round the roof, or go straight up and move left to the same point. Climb the crack above to a good no hands rest on top of the huge block forming the roof. Continue up a difficult crack to belay on a ledge. A superb pitch on impeccable rock.

4 30m 5c Move diagonally up left and climb a short arête on its front face. Continue trending leftwards to a deep corner crack. Step left and climb a short finger crack with difficulty to belay on a ledge above. Scrambling remains.

2 Lost Arrow ★★★   100m E3 6a

FA Gary Latter & Paul Thorburn 10 August 1995

The crack and corner system up the left side of the clean face left of *Kingpin.* Start at a block belay at left end of grassy ledge.

1 25m 4c Up a groove and wide crack to belay on a slab below a small roof.

2 35m 6a Pull through a crack in the lower roof to a slabby ledge beneath a crack in the right side of the roof. Pull through this and up a crack (crux) past an old PR on the left. Continue up the crack to a long sloping ledge on the right wall where the crack narrows and bends. Pull out right to the edge of groove and up midway between

both to the easier groove. Thread and nut belay beneath main corner. A superb well-protected pitch on immaculate and very rough rock.

3 **40m 5c** Climb the rib 3m right of the corner (good nut high in the corner) and move back into the corner. Easily up this and traverse right under the first roof and up a flake to the large capping roof. Undercut this right with a hard move pulling round the right edge of the flake to belay. Scramble up then left to summit screes.

3 **Wall of the Evening Light ★★** 90m E6 6b

FA Iain Small & Blair Fyffe 10 July 2014

A fine hard route between *Lost Arrow* and *Kingpin*. Start as for *Kingpin*.

1 **40m 6a** Climb *Kingpin* to the groove junction. Climb the left groove and wall above to step right to the belay at the top of the second pitch of *Kingpin*.

2 **50m 6b** Climb the groove on the left to step back right

to a sloping ledge (possible belay). Ascend a wall to another ledge, then take thin cracks up the centre of the wall to reach the black streak. Climb this boldly to the roof and pull left and through it to gain a diagonal crack in the headwall that leads to a niche. Climb the crack out of this to belay on ledge above.

4 **Kingpin ★★★** 105m E3 6a

FA Wall Thomson & John Hardie (8 PA), 17 August 1968;
FFA Pitch 1 Murray Hamilton & Dave Cuthbertson 1977;
FFA Complete Dave Cuthbertson & Dougie Mullin 1978;
Direct Finish Murray Hamilton & Rab Anderson summer 1982

A brilliant sustained route, following a direct line up the left side of the prominent pillar up the highest central section of the face. Start just right of a block belay directly beneath a prominent large arrowhead-shaped recess. Pitches 1 and 2 may be better run together. Linking the first pitch of *Lost Arrow* into pitches 2 & 3 gives a fine E2.

Paul Thorburn nearing the top of the first pitch of Temple of Doom.

1 20m 6a Climb crack to a slab then left and up to a steep shallow groove and up this with difficulty to swing right to a poor belay at a groove junction.

2 20m 5c Move up right and follow a black groove (often wet) to an awkward mantelshelf onto a small ledge beneath the short hanging chimney. Up this and step left at its top to belay on a ledge.

3 30m 5b Move back right and up the fine sloping ramp to a small hooded recess. Exit slightly left then up to a ledge. Continue to belay at the foot of the prominent corner.

4 35m 5c Follow the corner to the roof, step left and up the continuation corner above. Swing right at its top to finish up a prominent short flared chimney.

5 The Lost Ark ★★★   90m E5 6a
FA Pete Whillance & Ray Parker 26 July 1983

Excellent sustained climbing. Just right of the pillar of *Kingpin* is a large open white-speckled groove.

1 45m 6a Up the left side of the groove then the wall to the roof on the arête. Step left and climb boldly up the left side of arête on rounded side pulls for 5m (crux) to better holds. Move up right and traverse right across the top of the groove in a sensational position, then move straight up to good ledge and belay at base of corner.

2 45m 5b Climb the corner above to a good ledge then past a large keyed in block. Avoid the dirty section by moving out onto the right wall then back into the corner. Continue more easily leftwards to finish up the final chimney of *Kingpin*.

6 The Holy Grail ★★   35m E5 6b
FA Paul Thorburn & Gary Latter 8 August 1995

A fine sustained pitch with good protection between *The Lost Ark* and *Temple of Doom*. Climb the prominent easy lower V-groove and the shallow white groove above to a roof. Pull out left to a good rest then make hard moves up rightwards into the stepped upper groove and up this to pull to belay as for *Temple of Doom*.

7 Temple of Doom ★★★★   75m E3 6a
FA Murray Hamilton, Rab Anderson & Graeme Livingston 21 July 1984

The prominent V-groove and hanging stepped corner gives superb, well-protected climbing.

1 30m 6a Climb the easy lower groove to a large flake. Up the smooth groove above past bombproof runner placements and swing left at the top to a large hanging flake. Pull over this and belay beneath deep V-groove.

2 45m 5c Move right and easily up crack for 6m then step back left. Pull over the roof above and continue in a superb position up the hanging stepped corner system. At the top of this follow the continuation crack for a short way until it is possible to step right into the corner on the right. Up this to the top. A brilliant pitch – the situations are strictly space-walking.

7a Temple of Dumb ★   45m E4 5c
FA Ian Taylor & Tess Fryer 29 July 2011

Follow the second pitch of *Temple of Doom* for 6m and where it moves left, move right until below a large corner. Gain the base of the corner by a circular move out right to avoid a loose looking flake, then continue up the sustained corner, pulling left onto a ledge at the top. Finish easily.

8 The Last Crusade ★★   50m E3 5c
FA Rab Anderson & Johnny May 30 May 1992

The prominent V-groove (another one!) and wall right of *Temple of Doom*. Start just below a huge block at the base of the crag beneath a wide crack in a shallow corner.

1 10m 5a Climb the corner crack, pull out left to ledges and move up left to the foot of an open corner.

2 20m 5c Move right then up to the foot of the V-groove. Climb the groove to its top, swing out left then move across to climb steeply up the left side of a small roof (F #4 useful, though not essential) and pull onto a small ledge.

3 20m 5c Climb the cracks in the wall above.

STAC AN EICH (STAC OF THE HORSE) NN 031 593 **Alt:** 100m    5min

A steep little granite crag in the woods of Leitir Mhor (Lettermore) overlooking Loch Linnhe. Many of the routes stay dry all summer, though it is very sheltered and midges can be a problem late in the season.

Access: From the roundabout south of the Ballachulish bridge, head west down the A832 Oban road for about 1.3 miles/ 2km. Turn off left about a kilometre beyond (south) of the old ruined Ballachulish pier up a forestry track (signposted 'The Monument') just before a telephone box. Drive through a gate and up the forestry track for a couple of hundred metres to park at the bend.

Approach: Continue up the track then steeply directly up the slope at the next bend.

Descent: By abseil from trees or fight through the undergrowth and down the steep vegetated descent gully at the right end of the crag.

The crag is dominated by an imposing wall in the centre, split by three short groove lines. Right of this is the central corner line of *Marathon* then a further overhanging wall split by a couple of cracks bounded on the right by an easier angled area of rock.

---

1 Let Sleeping Dogs Lie ★★ 21m E4 6b

FA Murray Hamilton & Rab Anderson 19 May 1985

The deceptive central groove with a perplexing crux. Climb the groove to a good flat hold, make a long reach for a side pull on the right then regain the groove (crux) which leads to easier ground. Move up onto the slab and belay.

2 Seal of Approval ★★ 24m E4 6a

FA Gary Latter & Ian Campbell 6 July 1985

A spectacular line across the centre of the crag. Start beneath twin cracks in the centre of the overhanging wall just right of the previous route. Up twin cracks to gain a scoop. Up this and a flaky groove to good incut holds (nut placement in undercut). Swing right from here into a shallow groove system. Pull leftwards from the top of this to a tree belay. Either scramble up broken ground or abseil off.

3 Bill's Digger ★★★   25m E5 6b

FA Dave Cuthbertson June 1984

Excellent well protected climbing with a very reachy, contorted crux. It takes the shallow, stepped groove in the arête delineating the centre of the crag. Pull over the initial bulge leftwards on improving holds and move up to a thread runner in a recess. Continue up the steepening groove with a powerful reachy move to gain good holds in a horizontal break. Step right and continue up an easier crack.

4 Marathon ★★   30m E1 5b

FA Ed Grindley, Mike Hall & Fiona Gunn 14 November 1981

The central corner line. Well protected, with some weird moves. At the top traverse left onto the slab to belay on the rib. Either scramble rightwards to finish or abseil off.

5 Gunslinger ★★   25m E3 5c

FA Ed Grindley & Davy Gunn 22 October 1981

Well protected, strenuous climbing on big holds up the deceptively steep cracks in the highest part of the wall right of 4. Start at an undercut flake. Up this and the crack above, passing a small resting ledge at mid height to finish on good blocks. Tree belay further back.

6 The Monument ★   10m E3 5c

FA Ed Grindley 26 March 1982

The crack and capping roof at the right end of the frontal face. Enter the ramp and up a crack to pull over the roof on good holds. Trend easily rightwards to a belay.

The crag continues round right to form a shorter, easier angled west-facing wall.

7 Appin Groove ★★   9m HS 4b

FA Ed Grindley 1981

The left-facing layback groove at the highest point of the wall.

Let Sleeping Dog's Lie. Rick Campbell climbing.

Ewan Lyons starting up the long, sustained The Pincer (page 188), Garbh Bheinn.

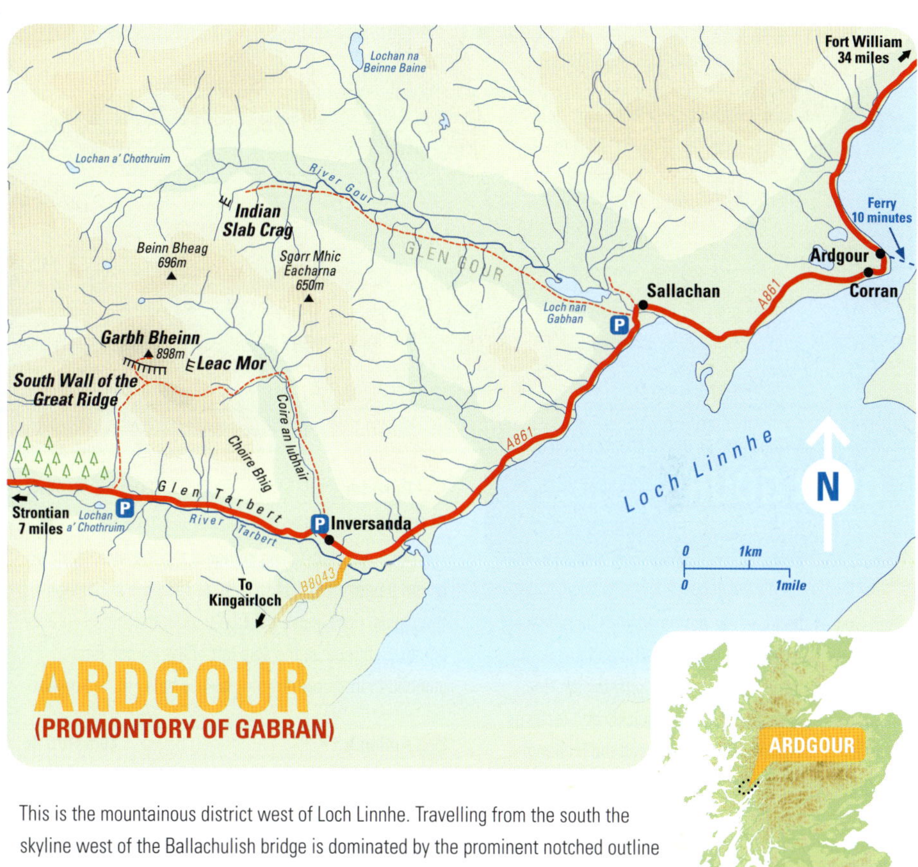

ARDGOUR
(PROMONTORY OF GABRAN)

This is the mountainous district west of Loch Linnhe. Travelling from the south the skyline west of the Ballachulish bridge is dominated by the prominent notched outline of Garbh Bheinn. Being further west and lower lying the district often benefits from fine clear weather when Glen Coe is enshrouded in cloud. It is also considerably quieter.

Accommodation: Wild camping anywhere in the hills, though permission from the estate may have to be sought during the stalking season – (www.ardgourestate.co.uk). A better option (sea breeze to dissuade midges) may be to camp on one of the many beaches (plenty of driftwood) thus avoiding the attentions of the landowners. **Caravan & campsites:** Sunart Camping, Strontian (☎ 01967 402080; www.sunartcamping.com); Ardnamurchan Campsite, Kilchoan (☎ 01972 510766; www.ardnamurchancampsite.com). **Bunkhouses:** Ariundle Centre, Strontian (☎ 01967 402279; www.ariundlecentre.co.uk); Highland Base Camp, Lochaline (07824 541901; www.highlandbasecamp.com). Numerous B&Bs locally. Chalets and self-catering accommodation further west in Strontian and Salen. **Amenities:** Small supermarket & filling station at Strontian. The Inn at Ardgour is the nearest pub.

"On the one hand ranged a vast array of the mainland mountains … on the other the Atlantic Ocean and the small isles of the west. This truly is the combination to which the Scottish hills owe all worthiness – rock, water, and the subtle colours of the seaward atmosphere."

– W H Murray, Undiscovered Scotland. (J M Dent & Sons,1951)

N ↟ 1hr 45min–2hr 🚲 1hr 30min
NM 918 645 **Alt:** 250m

GLEN GOUR
INDIAN SLAB CRAG

Glen Gour is the large flat-bottomed glen running west from Sallachan, north of Garbh Bheinn and just south of the Corran ferry. Despite being north-facing, this fine gneiss crag dries remarkably quickly and receives a fair bit of sun due to its open aspect and low angle. The routes give excellent climbing with fairly spaced protection, generally on superb rock.

Access: From the Corran ferry, follow the A861 road west for 2.4 miles/3.8km to a small loop road running close to the north bank of the river flowing into Camas Shallachain. Park on the south side of the bridge.

Approach: Follow the track west close to the south side of the river and Loch nan Gabhar then along the south side of the glen until it peters out after about 5.5km. Head diagonally left up the hillside to the base of the cliff. Although rough going at first, **mountain bikes** can be taken 3.5km along the glen as far as the sheepfold (free on the ferry).

Descent: From the large heather terrace at the top, traverse left and across the stream bed then down the slope just to its right (east), re-crossing the stream lower down and slanting down left to regain the base.

① **Outrider** ★★　　　　　　　　**80m VS 4b,4b,4b**

FA Les & P Brown Easter 1972

The prominent left-facing corner at the left side. Walk up left to a path crossing the crag. Ascend the initial slab then move right above the tree to gain the initially grassy corner. Continue up the slab left of the corner. Best climbed in three pitches (no obvious belay at mid-height).

② **Ambush** ★★　　　　　　　　**40m HVS 4c**

FA Mike Pescod, Rose McKie & Donald King 15 May 2003

The left edge of the hanging slab above *Outrider*. Climb *Outrider* to above the tree and step right to belay in the higher corner.

③ **Time Lord** ★★　　　　　　　**205m VS 4b**

FA Colin Moody & Cynthia Grindley 1 July 2000

Slightly better protected than *Indian Slab*. Start right of a vertical grassy crack up and left of *Indian Slab*.

1　**50m 4b** Climb up using a flake then direct to belay beneath the path. Walk left to belay at the base of the rib.

2　**40m 4a** Climb the rib and continue to belay beneath a black bulge.

3　**50m 4b** Move left round the bulge to follow the left edge of the obvious slab.

4　**50m 4a** Straight up.

5　**15m 4a** Finish up left on ripples.

④ Indian Slab ★★★   220m VS 4b

FA Ian Davidson & Les Brown Easter 1972

Good climbing, particularly on the first and third pitches. Start at the base of the black-streaked slabs, down and left of a steep section.

1 **50m 4b** Follow the slab to a grass ledge beneath a steepening.

2 **50m** Continue up to cross the path then follow ribs and heather leading up left to beneath the prominent slab.

3 **50m 4b** A superb sustained pitch. Climb the slab crossing the overlap at 30m.

4+5 **70m 4a** Trend up right to finish.

⑤ Mullennium Direct ★★★   200m Severe 4a

FA Colin Moody & Cynthia Grindley 1 July 2000; pitch 1 Gary Latter & Jeremy Birkbeck 16 August 2002

Four excellent full length pitches towards the right side of the slab. Start at the very toe of the crag, down right of the shelf at the base of *Indian Slab*.

1 **50m 4a** Move up right over initially broken ground to gain the superb smooth slab and follow this, taking the cleanest line trending slightly right to gain the base of the original route.

2 **50m 4a** Ascend the pale slab to belay a few metres left of a group of small rowans.

3 **50m 4a** Traverse left to the edge and ascend the slab overlooking *Indian Slab*.

4 **50m 4a** Continue directly to finish.

PALE FACE

Obliquely up and left of the main crag, on the opposite side of the open gully is an obvious pale slab.

Descent: Down the right side of the crag.

⑥ Paleface ★   30m HVS 4c

FA Gary Latter (on-sight solo) 16 August 2002

The prominent thin crack. Climb the crack, finishing by an easier ridge at the top.

⑦ Sun Dance ★   30m E2 5a

FA Donald King, Rose McKie & Mike Pescod 15 May 2003

The front face of the slab.

Indian Slab, Jeremy Birkbeck on pitch 3.

GARBH BHEINN
(ROUGH HILL)

The most southerly significant outcropping of gneiss in the country, Garbh Bheinn is an excellent mountain with as varied a range of routes as anywhere, from one of the best ridges on the mainland to a clutch of excellent extremes, and everything in between. The fact that the summit lies just below the three thousand foot mark makes it all the better for that, being devoid of all those boring Munro-baggers. The panorama from the summit is stunning – choose a fine clear day and savour.

Access: Take the ferry over Loch Linnhe at the Corran narrows to Ardgour (7.00-21.00 in summer; 10 minutes crossing time) and follow the A861 south-west (left) for 7 miles/11 km, turning off right 0.3miles/0.5km beyond the Kingairloch turn off, onto the old road which loops round. Park just before the old bridge.

Approach: (A) Follow a stalkers path on the right side of the stream (Abhainn Coire an Iubhair) up the very boggy strath of Coire an Iubhair (Corrie of the Yew Tree) to cross the stream after about 4.5km (1 hour). Continue steeply south-west up the right side of the burn emanating from the Garbh Choire Mor to reach the base of the cliffs.

(B) For routes on the **South Wall of the Great Ridge**, a faster, drier and steeper approach up the coire at the back can be made. Drive a further 2 miles/3km west along Glen Tarbert to park on the old road on the left overlooking a tiny lochan (100m before the road crosses the Allt a' Chothruim). Follow a vague path steeply up the right (east) bank of the burn flowing down Coire a' Chothruim (Corrie of the Balance, unnamed on 1:50,000 map, but immediately south of the summit) then slog up hillside rightwards (north-east) to the bealach between Sron a' Gharbh Choire Bhig and Garbh Bheinn. From the bealach continue left up the ridge by a path then cut across easy-angled slabs to gain the upper left end of the **South Wall of the Great Ridge.** 1hr 30min.

(C) An alternative approach to the bealach, much drier underfoot than the first described and not as tortuous as the second, is to cut up left from the old bridge and follow the ridge which eventually drops down into the bealach 2hr 15min.

Descent: Head left (west) along a well worn path along the summit ridge and down to the left end of the cliffs, or continue further to a path into the coire from the bealach.

1 NM 906 623 **Alt:** 600m 1hr 45min -2hr

 *The Great Ridge. Rob Kerr climbing.*

LOWER CLIFF

The cliff left of the obvious *Great Gully*.

1 The Great Ridge Direct Start ★★

165m Severe 4a

FA Dan Stewart & Donald Mill 12 April 1952

A good sustained approach to the upper ridge. Start down and right of the prominent right-slanting ramp on the right side of the crag.

1 **20m** Up the shallow steep ramp, starting on huge pockets then slightly left to belay at base of huge right-sloping ramp.

2 **50m 4a** Up the ramp (belay possible at 25m, at block just right of old PR) and continue in the same line to move up a steep flake then a short slab to belay on a long grassy ledge.

3 **20m** Move left round the edge and up easy slabs to belay on grassy ledge below a prominent flake chimney.

4 **30m 4a** The awkward chimney then leftwards over jumbled blocks then up a slab by a wide crack. Flake belay on the left at the back of a grass slope above.

5 **45m** Scramble up right then follow a rib, traversing right then up grass to a block belay up to the left of the base of *The Great Ridge*.

2 The Great Ridge ★★★

250m Difficult

FA John Bell & Willie Brown April 1897

A fine long mountaineering expedition, especially when combined with the *Direct Start*, with a stunning somewhat abrupt finish right on the summit of the hill. The ridge projects south-east from the summit, with the distinctive *Great Gully* cutting deeply into the face just to its right. The climbing becomes very much easier just below half-height. Continue up left from the base of the *Direct Start* to just before the base of a steep 50m cliff just beneath the bealach. Ascend a short step then traverse diagonally right along a grassy rake which leads almost into *Great Gully*. Climb the right edge of a slabby buttress overlooking a shallow open gully then walk up right to a block belay just up left from the base of the ridge proper (45m). Climb over sharp flakes left of the edge and move right to the crest. Continue up this to belay at the base of a short steep V-groove (35m). Climb the groove on good holds then a short ramp on the left (15m). Continue easily up grass to beneath a steep wall. Traverse left along the ledge then climb diagonally right to regain the crest of ridge. Continue up this to belay on the ledge above (40m). Continue up the crest, outflanking a steeper section on the left to a large grassy ledge. Climb a short right slanting grassy gully on the left, with a short rock step at its top then more easily up the obvious line to easier rocks leading to the summit cairn.

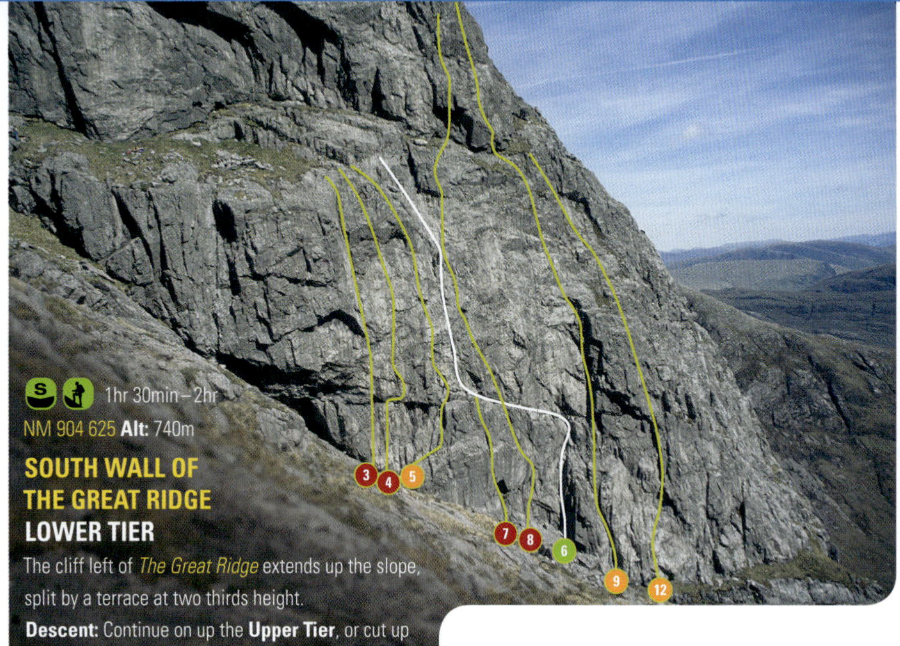

S **🚶** 1hr 30min–2hr

NM 904 625 **Alt:** 740m

SOUTH WALL OF THE GREAT RIDGE LOWER TIER

The cliff left of *The Great Ridge* extends up the slope, split by a terrace at two thirds height.

Descent: Continue on up the **Upper Tier**, or cut up the easy-angled slabs leftwards beneath the base of the **Upper Tier** then down the path to the bealach.

---

3 Gralloch ★★　　　　　　　　**45m E2 5b**

FA Mark Diggins & Allen Fyffe June 1981

Follow cracks and ramps up left into the big open right-facing corner. Ascend this more easily to the terrace.

4 The Gay Blade ★★　　　　　　**40m E3 6a**

FA Gary Latter & Paul Thorburn 24 June 1996

The prominent thin crack-line up the wall left of *Scimitar*. Start 3m left of the broken rising ledge system of *Scimitar*. Climb the initial cracked wall with difficulty (crux) to move right at the prominent horizontal break to easier ground. Continue up the crack, moving right on side-pulls into the steep finger crack which soon relents. Continue more easily in the same line, past a short steep wall near the top.

5 Scimitar ★★　　　　　　　　**105m VS 4c**

FA Dan Stewart & Donald Mill 13 April 1952

From the base of the broken rising ledge system, about

30m up left from the huge boulder at the base of *Butterknife*.

1 **30m 4c** Follow the ledge up right until it becomes horizontal. Go up a steep crack above the left end to an overhang, move right spectacularly to good holds on an edge. Move up and left to a ledge and belay at the base of a corner.

2 **25m 4a** Continue directly up slab and open chimney leading to the terrace.

3 **50m 4c** Climb a smooth vertical groove then move right to a flake. Continue either up the left-facing corner or the slabs on its left leading to easier angled slabs to finish on the crest of the ridge.

6 Razor Slash ★　　　　　　　　**75m Severe 4a**

FA Jimmy Marshall, Len Lovat & Archie Hendry 1 April 1956

Start at the huge boulder set against the face about 30m down from *Scimitar*.

1 **25m 4a** Climb the boulder and step into a dièdre which is climbed to a platform.

2 **20m 4a** Traverse horizontally left along the ledge for 8m to the base of an obvious layback slab edge. Climb this ledge, occasionally laybacking,

and at the top move out delicately over a nose (crux) then right and back left to belay.

3   30m 4a Follow the left diagonal fracture cutting across the prominent chimney of *Scimitar* to finish on the terrace.

7 Leviathan *   20m E3 6a

FA Paul Thorburn, Neil Craig, Rick Campbell & Gary Latter 28 June 1997

The wide overhanging crack left of *The Golden Lance*, so named because it is *"nasty, brutish and short."* Gain the crack and follow it forcefully to end on a broken ledge. Scramble off left and down easy ground.

8 The Golden Lance *   100m E3 6a

FA Rab Anderson & Alan Russell 30 June 1984

The prominent thin crack-line left of *Butterknife*. Start on top of the large boulder.

1   20m 6a Climb the thin crack then up left and back right to belay on the traverse ledge of *Razor Slash*.

2   40m 5c Step right and climb the thin crack-line then over a short leaning wall. Continue in the same line to the terrace.

3   40m 5b Above is a short corner terminating at a small roof. Climb the corner and pull over rightwards to reach easier ground. Move up and leftwards into the centre of the wall then climb up to a short leaning wall. Pull over this and finish up easier ground.

9 Butterknife ****   105m HS 4b

FA Jimmy Marshall, Archie Hendry, George Ritchie & Ian Haig 15 September 1956

Stunning climbing, with a particularly fine second pitch. Start directly beneath the main corner, 10m right of the large slanting boulder leaning against the base of the crag.

1   25m 4a Up the groove which slants left to belay on a block-strewn ledge below the corner crack.

2   25m 4b Up the superb corner on excellent holds to belay at its top. Well protected – large hexes/Fs useful.

3   25m 4a Easily up the slab above to belay below a roof at the right end of the terrace.

4   35m 4b Cross a small roof low down and follow a direct line to the top.

4a   VS 4c *Direct Finish* – A fine 45m pitch takes the prominent thin vertical crack midway between the short left-facing corner of *The Golden Lance* and the original finish. Cross the initial overhang on good holds and follow the crack over a steepening to a short diagonal left-slanting crack. Up this and trend slightly rightwards on easier ground to join the crest of *The Great Ridge*. Well protected.

10 Bodkin *   75m E1 5a

FA Ken Crocket & Stuart Smith 10 June 1979

The right arête of *Butterknife*. Start 3m right of that route.

1   25m 5a Climb to a steepening at 15m, step left (crux) and continue up and left to belay on the edge below some bulges.

2   25m 5a Move right then up to an overhang. Move left to the edge and continue more easily up this.

3   25m 4a As for *Butterknife* to the terrace.

11 Poniard **   60m HVS 5a

FA Gary Latter & Dave Greig 28 June 1997

Surprisingly reasonable climbing up the wall between *Butterknife* and *Mournblade*. Start 10m right of *Butterknife*, beneath a prominent shallow pale groove. Climb the groove then move rightward and climb to an undercut flange beneath a small overlap. Cross this and the main overhang above on good holds. Move up a short way to another overlap, step left and finish up the groove, easing towards the top.

12 Mournblade **   65m VS 4b

FA Ken Crocket, Colin Grant & John Hutchinson 31 July 1976

The corner parallel to and 12m right of *Butterknife*. Start at a rough flake 6m right of *Butterknife*.

1   30m 4b Climb up then right to the base of the corner. Follow the corner, step right into a groove and climb it to a good stance at a pinnacle.

2   35m – Climb the bulge directly above on good holds and continue more easily up the wall on the left.

UPPER TIER

This is the uppermost cliff on the mountain, with many of the routes ending just a few metres short of the summit.

Approaches: Head steeply up the back of the coire to the bealach then by a path towards the summit, cutting down easy-angled slabs along the base of the crag. Alternatively the right end of the grass terrace at the base can be gained by following *The Great Ridge Direct Start*. Another possibility is by scrambling up a rocky step at the base of Bealach Gully Buttress (at the foot of the prominent deep gully) then cutting rightwards and up by grass and wood rush. Avoid in wet conditions.

Descent: Follow the well worn path left down the summit ridge to the bealach, or make 35m abseil from sling & maillon on block at top of *The Peeler*.

13 Sgian Dubh ★★ 50m Severe 4a

FA Jimmy Marshall & Len Lovat 1 April 1956

Fine steep climbing up the left side of the wall. Start beneath the prominent deep chimney, just left of the very impressive smooth overhanging wall.

1 **20m 4a** Climb the open chimney, making use of a fine hand crack in the back, to a ledge. Walk along the ledge to belay in its centre.

2 **30m 4a** Climb diagonally leftwards following a line of flake cracks to beneath a bulge. Step left and pull up on good holds to gain the base of an easy-angled right-slanting ramp. Go up this then steeply on good flakes to a thread and nut belay on the ledge above. Scramble up to finish.

14 Sala ★ 30m E1 5b

FA Alan Taylor & Rab Anderson 20 June 1982

Climb a flange/crack just left of *Menghini* to where a diagonal crack comes in from the right and continue up to a ledge. Move right and finish up the wall above.

15 Menghini ★★★ 30m E1 5b

FA Alan Taylor & Rab Anderson 20 June 1982

The prominent crack just left of *The Peeler*.

16 The Peeler ★ 47m HVS 5b

FA Robin Smith & James Moriarty June 1961

The hard climbing is well protected and concentrated in the first 9m. Start on the outer edge of the *Sgian Dubh* flake.

1 **12m 4b** Climb the crest of the flake to belay on the platform.

2 **35m 5b** Climb the groove on the right to pull up and leftwards round a roof. Continue up a short steep crack which soon falls back into a groove leading to the top.

17 Cutlass ★ 60m E3 6a

FA Rab Anderson 6 June 1992

The leftmost crack-line on the leaning wall, immediately right of *The Peeler*. Good climbing, though a bit close to

The Peeler at times.

1 **40m 6b** Climb the awkward short groove in the arête just left of *Sgian Dubh* to gain ledges. Climb to beneath the roof, pull round left then move up and right to follow the prominent left-slanting crack to a junction with *The Peeler*. Either belay here or move up and right to a grass ledge.

2 **20m** Easy ground to top.

18 Sabre ★★   60m E5 6b

FA Rab & Chris Anderson 6 June 1992

Excellent climbing up the thin crack near the left side of the leaning wall.

1 **40m 6b** Climb the initial corner of *Sgian Dubh* to the ledge. Place the high runners on *Cutlass* and extend them then hand traverse out right for 3m to attain a standing position below the crack. Climb the crack to beneath a small roof, step up left and pull up to easier ground. Move up and climb the cracks in the wall to the right of *Cutlass* to reach a grass ledge.

2 **20m** Easy ground to top.

19 The Epeeist ★★★   50m E5 6b

FA Paul Thorburn & Gary Latter 23 June 1996

Excellent varied climbing up the central blocky crack in the leaning wall, directly above the very prominent black seep. From the top of the initial chimney of *Sgian Dubh*, first hand then foot traverse the shelf out right with increasing difficulty to the base of the crack. Pull the ropes and move the belayer to below (or drop a 3rd rope). Up the crack with a hard move low down to follow excellent holds which lead out left near the top. Finish up the easier wall above on excellent rock to spike belay. Scramble off.

20 Kelpie ★★★★   45m E6 6b

FA Murray Hamilton & Rab Anderson 21 June 1986

A stunning line, giving one of the best routes of the grade anywhere. Start 5m right of the hanging flake, at the right side of the leaning wall.

1 **25m 6b** Pull over the roof and up to good nuts in the crack. Traverse left into the steep hanging flake and up this with hard moves up slightly rightwards to good holds at the base of the crack. Powerful sustained climbing up the crack leads to a belay at the top of the leaning wall.

2 **20m 5b** Continue up the crack leftwards until a groove can be followed to easier ground leading to the top.

Neil Craig getting to grips with the amazing crux upper crack of Kelpie.

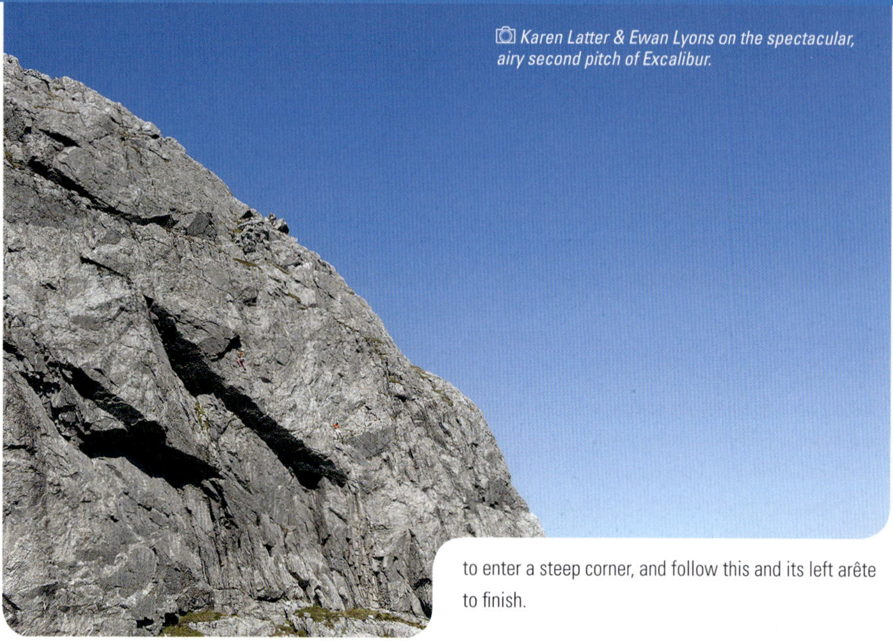

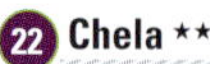

Karen Latter & Ewan Lyons on the spectacular, airy second pitch of Excalibur.

21 Tru-Cut * 50m E4 5c

FA Murray Hamilton & Rab Anderson 13 June 1982

Start as for *Kelpie*. Gain the groove above the initial overlap from the left. Follow this to pull out left and continue up to a move right to reach a ramp/groove. Follow this over the initial bulge to reach a belay, or continue to the top.

22 Chela ** 45m E4 6a

FA Murray Hamilton & Al Murray May 1981

"Hard moves back to back with so so gear in the groove; great and sustained E2 with brilliant gear after initial groove." The prominent smooth left-facing groove. Bold & sustained.

1 **25m 6a** Climb the groove to a nut belay.
2 **20m 5b** Continue more easily directly above.

23 The Pincer *** 45m E2 5b

FA Dougie Dinwoodie & Bob Smith August 1978

Fine open climbing up the right side of the right arête of the *Chela* groove. Follow the arête, passing the left side of a bulge to reach a small overhang. Turn this on the left to enter a steep corner, and follow this and its left arête to finish.

24 The Contender *** 50m E3 5c

FA Rab & Chris Anderson 11 August 1994

Brilliant sustained climbing following the thin hanging crack up the left side of the white wall. Many small wires (R #1 – 5) required. Start in the centre of the wall, behind a prominent projecting block embedded in the ground. Climb directly to the right end of a short left-slanting crack at 6m and follow this to gain a jug up on the left. Move up to a large rounded pocket, step left and follow a thin crack over bulges into a groove. Cross a bulge at the top of the groove and follow a ramp a short way then swing out right and up a short crack to ledges. Ascend a niche then a rib on the left finishing up a slab to a thin grass ledge just below the top.

25 White Hope *** 50m E5 6a

FA Pete Whillance, Murray Hamilton & Rab Anderson 5 May 1984

Excellent sustained climbing, following a direct line up the centre of the clean white wall. Start immediately behind a large embedded flake, below a thin vertical quartz seam in the centre of the wall. Climb this to a right-slanting flake at 12m. Make hard moves directly up from its right end to gain jugs beneath a small isolated

roof. Pull over its right side and climb direct, moving slightly left then back right to below the final leaning wall. Move up leftwards to a short crack and pull over to a belay. Scramble to the top.

26 The Clasp ★★   60m E1 5a

FA Jimmy Marshall April 1960

A left-trending line below the leftmost end of the lower of the two large roofs. Start beneath the right end of the roof.

1 **15m 4c** Climb the steep wall and trend left to a belay.

2 **45m 5a** Continue up under the roof, traverse left to a shallow groove then up to a chimney trending left to the top.

27 The Foil ★   80m E2 5c

FA Paul Moores & Mick Tighe (1 PA) 29 May 1978

Sparsely protected climbing across the steep wall sandwiched between the two roofs. Start at a short wall beneath the right end of the upper roof.

1 **40m 5c** Move up to the roof, move left directly below it and follow it left with difficulty to where it fades. Exit left onto steep slabs.

2 **40m** Finish up the cracks in the slab.

28 Excalibur ★★★   65m HVS 5a

FA Ken Crocket & Colin Stead 10 June 1972

Impressive situations, with two fine contrasting pitches. It gains and traverses the lip of the smaller second roof system. Start at the pale open groove beneath the right end of the long roof.

1 **20m 5a** Climb the central and deepest of three faint groove-lines on good holds to pull out right to a good spike on the rib. Continue steeply up this with hard moves to gain a good ledge level with the lip of the roof. Belay here.

2 **20m 4c** Traverse left above the lip of the roof by a line of good hand holds, past a prominent thin diagonal crack-line then up to the base of the steep bottom-less corner. Up this to a good spike at its top. Step round the edge and down to belay on a good ledge.

3 **25m 4a** Move left 3m and easily up slabs to finish.

29 Guenevere ★★   35m HVS 5a

FA Rab Anderson & Mark Garthwaite 3 September 2000

Fine airy climbing. Climb steeply up the arête/rib immediately right of *Excalibur* to where that route swings round from the left. Move up slightly right then back left to climb a short, smooth leaning wall (good gear in the horizontal breaks). Bold climbing directly above (passing just to the left of a large detached flake/block) on good holds gains vague cracks in the wall/tower feature directly above. Continue up the cracks, moving up left just below the top. Easy ground gains the top of the hill.

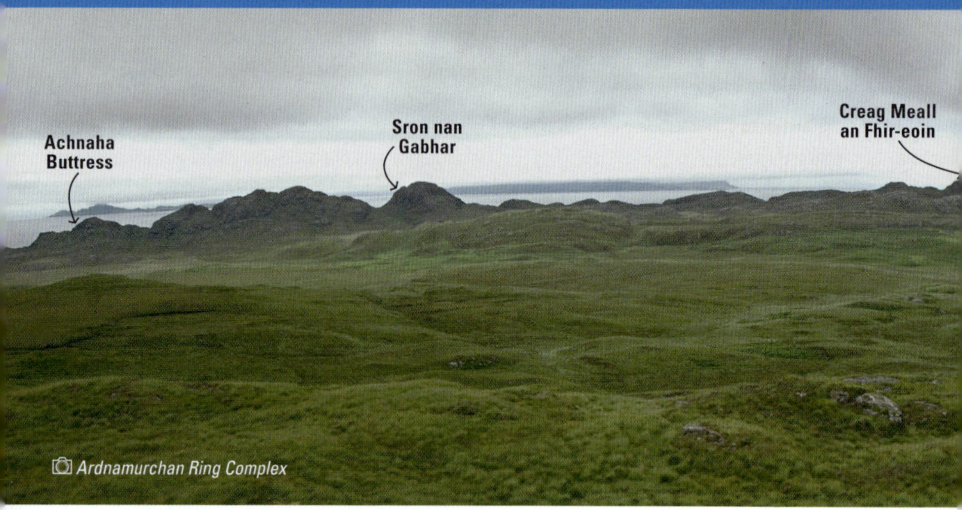

Ardnamurchan Ring Complex

ARDNAMURCHAN
(PROMONTORY OF THE SEA OTTER / SEA-VILLAINY)

This wonderful peninsula presents the most westerly point on mainland Scotland, offering a variety of climbing in a truly idyllic setting well away from the crowds. The outcrops on the north side of the road just before the road end at Sanna Bay offer a number of short routes on immaculate glaciated gabbro. The views towards the Inner Hebridean islands of Canna, Rum, Eigg and Skye would alone make the journey there worthwhile. The immaculate quality of the climbing makes it a wonderfully peaceful place to climb, well away from mainstream climbing venues. Worth combining with a trip over to climb on Garbh Bheinn.

All the inland crags are within what is called the Great Euchrite, more commonly termed the Ardnamurchan Ring Complex, a very distinctive tertiary volcanic intrusion, easily identified from the road. Though there are innumerable smaller crags throughout the area, only the biggest and best crags are fully described.

Access: The quickest approach is via the Corran ferry over Loch Linnhe then west along the A861 for 24 miles/38.4km to turn west onto the B8007 at Salen. Follow this for 19.6 miles/31.4km, turning right 0.75 miles/1.2km beyond Kilchoan towards Sanna Bay. The parking spot for the first three crags is reached about 1.9 miles/3km after the turn-off, about 200m north of the bridge over the Allt Uimha na Muice, in a small quarry on the right, by the start of a Land Rover track (NM 469 978). For the majority of the remaining crags, continue for a further 1.6 miles/2.5km to park on the left side of the road (NM 458 687) about 0.3 miles/ 0.5km beyond where the Allt Sanna (burn) runs beneath the road, 47 miles/75km from Corran – 1½ hours.

From Fort William, take the A830 road towards Mallaig, turning south onto the A861 at Lochailort for 22.2 miles/35.5km to Salen then as above. 70 miles/112km – 2 hours.

CRAGS OF THE RING COMPLEX

The crags are grouped within three distinct areas, on the three main rocky knolls of Meall Meadhoin, Meall an Fhir-eoin and Meall Clach an Daraich forming the eastern and northern perimeter of the ring. They are described anti-clockwise, as encountered on the approach.

 1hr

NM 496 687 **Alt** 290m

MEALL MEADHOIN
(MIDDLE LUMP)

THE APRON SLABS

Meall Meadhoin is the highest (388m) and most easterly point visible from the road in the ring complex, containing a 90m high slabby crag down and left of the summit.

Approach: Follow a good path which heads north about 200m north of the bridge over the Allt Uimha na Muice (NM 469 677; 2.9km beyond turn-off) for about 1.7 km until just beyond a ruined croft at Glendrian. Leave the path and head north-east across rough ground in the direction of the slabs.

Descent: Down either side, best on the left.

Nesting restrictions: The crag is very close to a very important nest site and should not be climbed between March and the end of July.

1 Dance on a Volcano ★   **70m E1 5b**

FA Rab & Chris Anderson 17 August 1997

The blunt edge formed between the frontal face and the side wall on the left.

1 **45m 5b** Climb directly up the edge just left of *Ne'er Day Corner* and continue up easier ground to a belay.

2 **25m 4a** Scramble up slabs to the top.

2 Ne'er Day Corner ★★   **75m VS 4c**

FA p.1 Colin Stead & John Newsome 1 January 1971; p2 August 1972

The obvious corner and slanting crack leading up and around the left side of the roof.

1 **45m 4c** Move up the corner and follow the crack which slants up left towards the edge of the buttress. At the level of the roof step right and climb a wide crack then the slabby rib to a small stance.

2 **30m 4c** Continue up slabby ribs rightwards to finish on the summit block.

3 Gift of the Gabbro ★★   **75m E2 5b**

FA Rab & Chris Anderson 17 August 1997

The centre of the slab midway between the two corners.

1 **45m 5b** Climb the initial wall by a flange then the slab to a small ledge. Move up right to a thin crack in a shallow scoop and place the last wire for some distance. Pull out left at the top of the scoop then go up left to the base of a blind flange. Move right to a thin crack and follow this to a flat-topped spike (sling). Step left and up to the overlap and pull through this by the obvious flange. Continue for some way up and left to belay on the slabby rib as for *Dance on a Volcano*.

2 **30m 4c** Step back right into the groove and climb up right to follow a parallel line to *Dance on a Volcano* up short walls and slabs. Climb the final short wall centrally by an obvious 'ear'.

4 Toulouse Booze Cruise Blues ★   **70m E1 5b**

FA M.Harris & D.Balance 30 May 1998

The thin bottomless crack and slanting crack through the bulge. Start at the base of the right-facing corner.

1 **45m** Move left round left arête into a thin crack and ascend for 10m. Move right and follow a right-facing flake and the crack over the bulge leading to a large ledge. Traverse right 5m to belay beside a large block.

2 **25m** Continue directly up slabs.

The slabs on the right side of the crag are generally climbable anywhere at about Severe, on immaculate rock, though not easily protected away from cracks.

5 Leac Glas ★★   **80m HS 4a**

FA John Newsome & Colin Stead August 1972

Fine, steady climbing with spaced protection. Start just right of the diagonal grassy fault.

1 **40m 4a** Follow the blunt grey rib past a superb bucket hold to a ledge below a crack.

2 **40m 4a** Finish up the crack and flakes above.

6 Solas ★★★ — 80m HS 4b

FA Colin Stead 1970s; Variation Colin Stead 1980s

The best and hardest of the slab routes, better protected than appearances suggest. Follow a line up the slabs starting a few metres right of *Leac Glas* keeping right of a curving crack, heading for the prominent straight crack on the skyline. A variation climbs the two curving cracks on the left.

7 Gall ★★ — 80m Severe 4a

FA Colin Stead 1970s

Excellent climbing, quite bold after the initial crack. Start at the foot of black seepage marks (small flake).

1 **25m 4a** Follow the right-slanting crack for 8m (crux, well protected). Trend right and go up to a narrow ledge then direct to a belay.

2 **38m 4a** Continue up good rock on the right side of the fault to belay on a large terrace.

3 **17m** Finish up the same line on easier ground to the top.

8 Leac Louise ★★ — 70m Severe 4a

FA Gary Latter & Mike Pescod (both solo) 30 August 1999

The prominent triptych of discontinuous cracks on the right side of the crag. Start 5m right of *Gall*.

1 **45m 4a** Follow the crack, stepping left onto clean rock beneath some vegetation. Continue up the second crack to a good ledge.

2 **25m 4a** The fine crack above, finishing either by easier slabs or escape out right.

CREAG MEALL AN FHIR-EOIN
(CRAGGY LUMP OF THE EAGLE)

A good slabby crag of clean glaciated gabbro, clearly visible on the hillside due north-east of the farm at Achnaha, 1.2 miles/2km before the road end at Sanna. The best and most varied of all the crags.

Approach: Follow a good track which heads north from the parking spot for about 1.7km until just beyond a ruined croft at Glendrian. Head north-east across rough ground in the direction of the crags.

Descent: Down either side of the crag, or make a 45m abseil from sling & maillon on large block at the top of the main pitch of *Yir*. This is easily gained from other routes by scrambling down from the top.

NM 482 698 **Alt:** 120m  50min

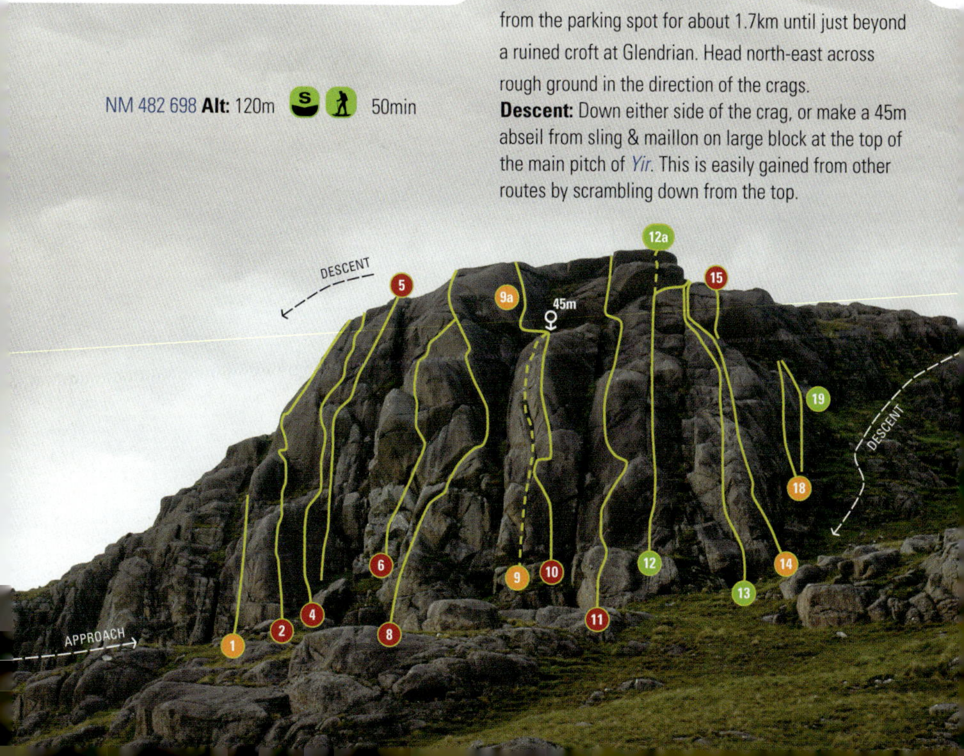

1 Crater Comforts ★ 60m VS 4c

FA Rab & Chris Anderson 3 August 1997

The crack-line running up the left side of the crag, immediately right of a thin grassy gully, to finish up the obvious prominent crack in the upper right side wall. Start at the lowest rocks beneath the crack next to a pointed flake just left of the edge which turns into a slab.

1 **25m 4c** Follow the crack and its continuation up a whaleback to where it thins to form a hollow flange. Step down and around the flange then step up right to belay just left of the thin crack in the left side of the smooth central wall.

2 **35m 4c** Step down into the grassy gully and after a few moves pull back onto the rock and climb the side wall to reach the obvious left-slanting diagonal crack and finish up this.

2 Volcane ★★ 50m E1 5b

FA Joanna George & Dave Cuthbertson 3 August 1997

The prominent crack below the smooth central wall. Start just right of the toe of the buttress beneath a crack on the slabby left wall of the grassy sloping recess.

1 **30m 4c** Climb the crack to the ledge beneath the left side of a smooth wall, and continue up cracks on the left to the shoulder to the left.

2 **20m 5b** Follow a short steep ramp to a break above, make a delicate step up and move right to a crack leading to a ledge. Continue up the crack and shallow corner to a large ledge below the top.

3 Trauma Crack ★ 45m E3 5c

FA Dave Cuthbertson & Joanna George 3 August 1997

The prominent curving crack on the right side of the smooth central wall. Good climbing but slightly contrived. Start as for 2 and belay down on the left of the crack.

1 **20m 5c** Climb the tapering groove and crack to a point very close to 4 (protection). Avoid possible escape on to that route and follow the crack which now bends back to the left (crux) to the ledge.

2 **25m 5a** Climb the continuation cracks and corners with a tricky move above a ledge to the large ledge below the top.

4 Magma Force ★★ 50m E1 5c

FA Rab & Chris Anderson 3 August 1997

Start up the sloping grassy recess just right of 3 at the base of a heathery crack. Step up then move out left to climb the centre of the slab. Climb the short left-facing corner and continue up the left side of the rib to below a wide crack. An awkward move up a groove gains the crack directly. Climb the crack to a ledge, step right and climb via short steps and slabs to a short corner leading onto the flat ledge at the top of the crag.

5 Vulcanised ★ 50m E2 5c

FA Rab & Chris Anderson 3 August 1997

A parallel line just right of 4. Scramble to the top of the grassy recess then go up left to climb the crack and short right-facing corner. Continue up the slabby rib to a heathery ledge and climb a thin crack to beneath a steepening with a thin crack. Climb the steepening, continue above then step right to finish up the right edge of the crag.

6 Star Wars ★★★ 50m E3 5c

FA Rab Anderson & Dave Cuthbertson 4 August 1997

Takes the crack up the right side of the steep buttress. Start at the base of the crag well below the crack.

1 **30m 5c** Scramble up grass and rock to climb a short crack leading to the leaning cracked wall. Climb this past a niche into a diagonal break running up right then step up left and climb a steep crack to a ledge and belay just above.

2 **20m 5b** Cross the heathery garden above then climb the centre of the wall to gain and follow a slanting flake-line-come-crack up right. Finish up a slabby rib.

7 Night Falls ★★ 45m E2 5c

FA Gary Latter & Charlie Prowse 8 July 1997

The prominent right-slanting diagonal crack. Start beneath the centre of the buttress.

1 **25m 5c** Move up and step left past a small juniper bush to a short crack leading up into a niche. Pull out right and over the roof on a superb jug. Continue more easily up the flake crack then by a fine thin crack up the slab to belay on large ledge.

2 **20m 5a** Climb the blunt rib which soon eases.

 **Return of the Jedi** ★★   45m E1 5b

FA Rab & Chris Anderson 3 August 1997

Climbs the right side of the steep buttress. From the right side scramble up a heathery slab to a steep stepped groove/crack. Climb this then go up right along a horizontal break to the edge of the buttress. Step right around the edge and make some bold moves to gain the diagonal break of 7. Climb the crack just right of the edge and continue to the top.

 **Yir** ★★★   55m VS 4c

FA Charlie Prowse, Rob Kerr & Gary Latter 8 July 1997

Just right of the central grass-filled fault is a prominent crack leading up into a slabby groove.

1 **45m 4c** Follow these to a horizontal break at the top of the groove. Move right to the easy-angled crest and continue up this past a further break to belay on large ledge just below the top.

2 **10m 4a** Step left and climb the wide curving crack, or thinner crack out left at the same grade.

> 9a Variation 15m 4b Move left across the grassy groove. Pull over bulge via a wide crack, move left and follow a thin curving crack.

10 **Minky** ★★   45m E2 5b

FA Dave Cuthbertson & Joanna George 3 August 1997

A direct line up the short slabby rib right of 9.

1 **45m 5b** Pull over a small overlap and follow a break slanting up to the left to a junction of cracks with 9. Go up a metre, step right and take a direct line up the whaleback by some thin poorly protected climbing to belay where the angle eases.

11 **Up Pompei** ★★   60m E1 5b

FA Rab & Chris Anderson 4 August 1997

A direct line up the front of the second slabby rib. Start at the lowest rocks.

1 **10m 4a** Gain the large ledge by easy unprotected climbing up the front. The undercut crack to the left can be gained with difficulty (E2 5b), or walk in from either side.

2 **45m 5b** Follow a short crack in the left side of the rib and continue up into the centre. Step up left and follow the left side of the rib to easier ground. Continue to the headwall (possible belay), swing left and climb the short crack to a spacious ledge.

3 **5m** Climb the short wall above via the obvious step.

12 **An Deireadh** ★★   55m Very Difficult

FA unknown 29 June 1997, or before?

'The End'. The top of the right side of the buttress is split by a prominent short steep crack. Start directly beneath this, from a ledge at the base of the crack.

1 **46m** Follow the crack to belay on a small sloping ledge beneath the final steep crack.

2 **9m** Climb up to the crack and escape out right on good holds.

12a **Direct Finish** ★   16m Severe 4b

FA Rob Kerr, Charlie Prowse & Gary Latter 8 July 1997

The crack above the belay.

13 **An Toiseach** ★   50m Very Difficult

FA Margaret Riley & John Stevenson 22 July 1997

'The Beginning'. The crack up the third slabby rib, just left of the arête and right of the heathery groove-line between the ribs. Large gear.

1 **30m** Go up the corner to the crack and follow this to a very large ledge.

2 **20m** Continue up the slab behind and left of the belay to the large crack which comes up from the heathery groove-line between the ribs then climb directly up the slabs.

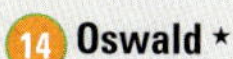 **Oswald** ★   55m HS 4b

FA Charlie Prowse, A.Simpson & T.Harper 17 July 1997

Start just right of the blunt rib at the right end of the crag. Follow the left of twin cracks for 5m then move out left to follow another crack and easier ground to finish.

15 Tremor Crack ★ — 10m E3 6a

FA Dave Cuthbertson & Joanna George 4 August 1997

Attack the steep and strenuous overhanging crack in the headwall to an awkward exit. Gained easily by scrambling in from the right, or via 13 or 14.

16 Bloody Crack ★ — 10m E4 6a

FA Mike Pescod & Nick Carter July 2002

The short 'elephant's arse' crack right of 14. Strenuous.

Around the edge, right of the third slabby rib and up the slope is a short slabby wall with two right-slanting cracks.

17 Oisean Bheag ★ — 20m Severe 4a

FA John Stevenson & Margaret Riley 22 July 1997

Pleasant climbing up the 'little corner' left of the cracks.

18 An Rathad Ard — 15m HS 4b

FA John Stevenson & Margaret Riley 22 July 1997

The leftmost crack, starting below a triangular niche, leading to a belay below an overhanging wall.

19 An Rathad Losal ★ — 15m Severe 4a

FA Margaret Riley & John Stevenson 24 July 1997

The lower, rightmost of the two cracks. Follow this to below the heather ledge then step left to finish as for 18.

DOME BUTTRESS

50min

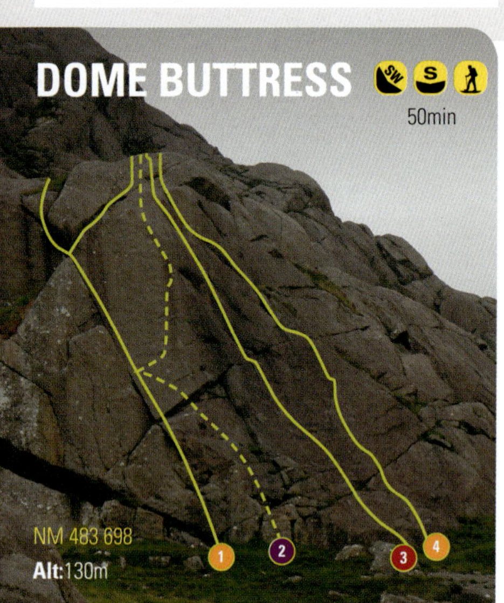

The smaller buttress with a slabby right face 100m right of the main crag.

1 Claude ★★ — 20m VS 5a

FA T.Harper & Charlie Prowse 17 July 1997

Start at a large detached flake. Climb rightwards up the flake for 3m to gain the fine crack and up this to an overlap. Traverse left below the overlap to a steep groove, pulling through on a mega-jug to finish. Finishing out right along the break at the overlap is the same grade.

2 Subduction Zone ★ — 25m E3/4 5c

FA Rab & Chris Anderson 17 May 1998

Bold climbing up the wall right of 1. Climb broken ground up the left edge of the slab, pull over a small roof and move up left to place a runner in the crack of 1. Step back right and climb the wall to reach a thin crack on the right edge. Move onto a foothold on the edge then make a couple of thin moves to reach the top.

3 Lava Lout ★ — 25m E1 5b

FA M.Harris & D.Ballance 29 May 1998

A line close to the left edge of the slab. Start at the same point as 4, but climb the slab directly above to a small left-facing corner near the edge. Climb the corner to a small roof then step right into another left-facing corner which is climbed to the top.

4 Greta Gabbro ★★ — 20m VS 4c

FA John Stevenson & Margaret Riley 24 July 1997

Well protected, the difficulties increasing with height. Start at the toe of the slab and head up right to a flake then go up a left-facing corner to the top.

The main fault up the right side of the slab is 5 *Canna Do It* Diff, the narrow crack just right 6 *Rum Do* S 4b.

NM 484 697 **Alt:** 160m SW 50min

MEALL AN FHIR-EOIN
(LUMP OF THE EAGLE)

The area of slabs, walls and ribs lying on the main summit knoll of Meall an Fhir-eoin.

WEDGE BUTTRESS

Between **Dome Buttress** and **Hooded Wall**, at a slightly higher level.

 **Fairy Ring** ★   **50m Very Difficult**

FA Steve Kennedy, David Hood & Cynthia Grindley 28 June 1998

The central crack system. Start at the lowest point and follow the crack to a prominent protruding flake at 10m, step right and follow a parallel crack to the top.

2 **Fox** ★   **50m Severe 4a**

FA Stuart Campbell & Chris Cartwright 22 April 2000

Start 2m right of 1. Follow a left-slanting crack system to ledge then a curving crack and groove slightly left then cut back right to an obvious water-worn pink patch. Finish up the crack above, trending left then right.

 **Hounded** ★   **30m Very Difficult**

FA Stuart Campbell & Chris Cartwright 22 April 2000

Start in a prominent black recess. Climb a wide overhanging crack to a spike then the continuation crack, crossing 2. Finish straight up slabs when the crack runs out, to belay as for 1.

4 **Foxed** ★   **20m Severe 4a**

FA Stuart Campbell & Chris Cartwright 22 April 2000

Start 5m up right of 3 at an inverted teardrop flake. Up the left edge of the flake then a crack until it peters out, continuing up flakes to belay as for 1.

HOODED WALL

A small slabby buttress 50m right of and slightly higher than **Dome Buttress**, overlooked by a beak of rock.
Descent: Scramble down right before the slabby ribs.

1 **Vesuvius** ★   **20m E1 5b**

FA Steve Kennedy & Cynthia Grindley 16 May 1998

The slim corner close to the left edge. Start by a pointed undercut (often damp). Steep moves over the initial bulge lead up rightwards to the corner/crack. Continue up the slab left of the upper corner to finish just left of the beak.

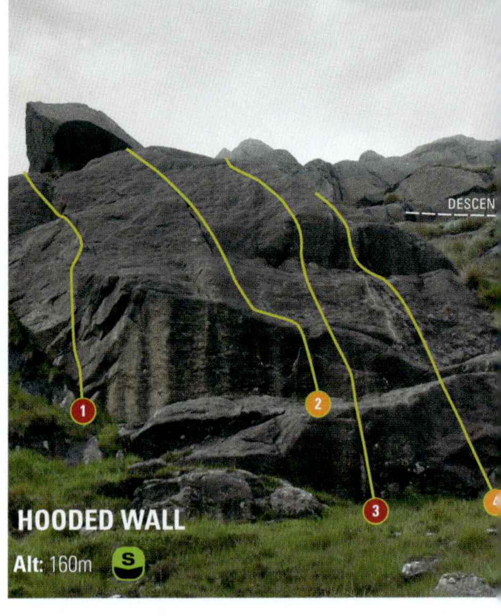

2 Krakatoa ★★ — 20m HVS 5b

FA Steve Kennedy & Cynthia Grindley 16 May 1998

The next crack line to the right. Surmount a bulge directly below the crack then climb the crack directly.

3 Etna ★★ — 20m E1 5b

FA Steve Kennedy & Cynthia Grindley 16 May 1998

The crack right of 2. Initial moves up a steep slab lead directly to a steep wall at mid-height which forms the crux. Hard for the short.

4 Stromboli ★ — 20m HVS 5b

FA Steve Kennedy & Cynthia Grindley 16 May 1998

The rightmost crack-line. Climb the lower slab directly to a short steep wall at mid-height. Surmount the wall and finish up a corner.

LOWER TIER

Some 50m right of **Hooded Wall** are two prominent whalebacks. The first three routes described climb the slabs just left of the left whaleback, by a grassy terrace.

Descents: A 25m abseil from sling & maillon on block at top of 8 *Mirka*. From the summit easy slabs on the right lead to the rake between the tiers. Either go left down this, or right round the crest to descend more directly.

5 Vulcan ★ — 65m VS 4c

FA Steve Kennedy, David Hood & Cynthia Grindley 28 June 1998

Start at a black streak on the left of the slab.

1 **30m 4b** Gain and climb a prominent crack running up the left side of the slab to the grassy terrace.

2 **35m 4c** Climb the slab and flake above for a short distance then move left to reach the obvious crack-line running up the left side of the upper slabs. Climb the crack then slabs up rightwards to a break in the upper wall. Pull through the break by a groove.

6 Dead Ringer ★ — 65m HVS 5b

FA Steve Kennedy, David Hood & Cynthia Grindley 28 June 1998

A more direct line up the slab. Start at a short undercut wall below a crack 2m right of 5.

1 **30m 5b** Steep moves lead into the crack which is followed to a slab. Climb the slab near the right edge to the terrace (joining 5).

2 **35m 5a** Climb the slab directly above the belay, but instead of moving left (on 5), continue to a break. Climb directly up the slab near the right edge (poorly protected) to a break in the upper wall. Finish up 5.

7 Western Front ★   55m HVS 5b

FA Rab & Chris Anderson 17 May 1998

The right crack, sharing the same start as 6.

1 **45m 5a** After a bouldery start climb the crack then follow a slabby rib to heather ledges. Step right to climb a short buttress then go left round a rib and climb a groove up its left side (or the rib itself) to belay beneath a leaning barrier wall.

2 **10m 5b** The short, thin Y-shaped crack provides a fine sporting move to gain the slab leading to the upper terrace. The **Upper Tier** lies just above, providing a logical continuation.

8 Mirka ★★   30m E2 5c

FA Dave Cuthbertson & Joanna George 17 May 1998

The left whaleback. Ascend the slab to a steepening left-slanting crack leading to triangular block. Navigate around the block onto its top and an easing in the angle. Step right and climb a crack in the whaleback to a ledge. Continue by easy grooves in the crest.

9 Beth's Route ★   120m Very Difficult

FA Davy Virdee, L.Curtis & E.Vokurka 9 May 1998

A series of short pitches, starting up the broad chimney between the whalebacks.

1 **25m** Ascend the chimney then cracks and the slabby right wall of the gully.

2 **20m** Follow a crack up the rib up on the left to gain broad terrace.

3 **20m** Climb the V-cleft splitting the slab on the left then direct to the terrace.

4 **25m** Traverse across a right-slanting slab to an undercut layback. Use this to ascend the edge of the slab to a corner leading to easy ground.

5 **30m** Scramble up slabs.

10 Ringmaster ★★   40m VS 5a

FA Steve Kennedy & Cynthia Grindley 16 May 1998

The prominent corner immediately right of the right whaleback. A bouldery undercut crack leads to the corner, which is followed throughout. Very well protected.

11 Ring of Fire ★★   40m HVS 5a

FA Steve Kennedy & Cynthia Grindley 5 September 1998

A spectacular route up the rightmost whaleback. Climb the corner of 10 for 7m then hand traverse the break out left onto the crest of the whaleback. Climb the crest to finish up a crack in the centre of the short finishing wall.

UPPER TIER

Above lies a grassy bay overlooked by a steep wall and a prominent roof. This bay is gained easily by scrambling in from either end, or via any of the previous routes.

12 The Great Eucrite ★   30m E4 5c

FA Steve Kennedy & Cynthia Grindley 13 September 1998

Bold sustained climbing up the steep slabby wall just right of the prominent vegetated corner on the left. Start at a black streak near the centre. Climb to a small knob and leftwards-slanting crack (protection) at 4m. Make bold moves up and right to reach the right end of a ledge above. Finish more easily up steep left-slanting cracks.

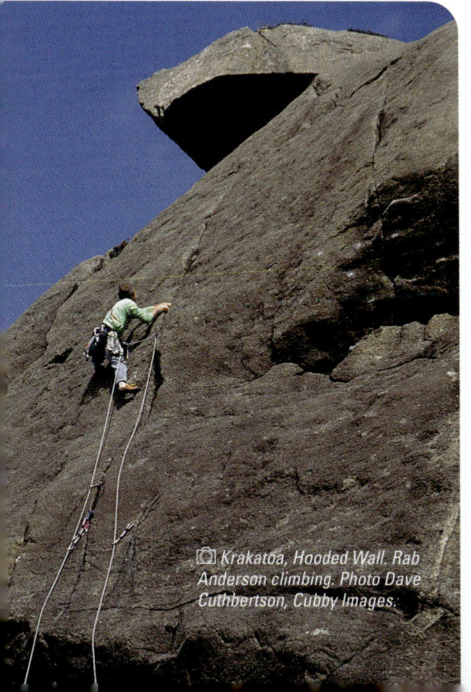

Krakatoa, Hooded Wall. Rab Anderson climbing. Photo Dave Cuthbertson, Cubby Images.

13 Xenolith ★ 30m E1 5b

FA Steve Kennedy & Cynthia Grindley 16 May 1998

Start by the large detached flake at the right edge of the steep initial wall of 12. An easier slabby wall lies just to the right, leading up to the left end of the prominent roof. Climb the slab right of the flake to the ledge below the steep upper wall. Climb the left-slanting cracks in the wall to finish.

14 Barbarella ★★ 20m E4 5c

FA Julian Lines (on-sight solo) 4 June 2001

The beautiful square-cut gritstone style arête. Start from the left to gain the arête, which is climbed to a break. Pull through the capping roof and continue easily.

📷 *An Deireadh, Creag Meall an Fhir-eoin. Charlie Prowse climbing.*

15 Pyroclast ★★ 45m Severe 4b

FA Steve Kennedy & Cynthia Grindley 16 May 1998

The far right end is defined by an easy angled corner with an overhung left wall. Climb an easy slab then step up left across a steepening into the corner proper, which is followed directly to the top.

15a Variation Finish ★★ VS 4c

FA Rab & Chris Anderson 17 May 1998

From the ledge in the corner above the steepening, step left to the base of a fine snaking crack up the immaculate slab. Ascend the crack, finish directly up the slab.

16 Fear of Flying ★★ 35m VS 4b

FA Steve Kennedy, Bob Hamilton & Pete Harrop May 2002

A good route in a wildly exposed position for the grade, following parallel cracks running horizontally left above the overhanging wall just left of 15. Climb the initial corner of 15 before pulling out left into the cracks. Traverse these horizontally left to the edge, continue left a short distance then finish directly.

MEALL CLACH AN DARAICH
(HULL-SHAPED ROCKY LUMP)

The rocky ridge forming the closest section of the ring complex on the north side of the road.

Access: Park on left side of the road at NM 458 687, 0.5 miles/0.8km beyond the farm at Achnaha.

ACHNAHA BUTTRESS   15min

NM 461 696 **Alt:** 40m

A steep wall on the first rocky dome from the road.

Approach: Head in the direction of the crag across boggy ground and the burn after 200m to the crags.

1 Plocaig Rock ★ 25m HS 4b

FA John Stevenson & Margaret Riley 21 July 1997

Start just left of the arête, on the west (sea-facing) end of the crag. Climb an obvious crack over a bulge (crux), continuing to a right-slanting crack. Keep left of the arête to finish up the slabby wall.

2 Wheesht! ★★ 20m E2 5b

FA Joanna George & Dave Cuthbertson 16 May 1998

Scramble up onto the grassy terrace. Climb the first obvious crack up the left end of the crag with strenuous climbing to the niche.

The crack just right of 2 is 3 *Acrophobia* ★ E2 5c.

4 Pooper Scooper ★ 10m Difficult

FA Simon Richardson & Chris Cartwright 10 October 1998

The obvious barrel-shaped scoop cutting through centre of the **Upper Tier**.

5 Soul Mining ★ 10m E4 5c

FA Dave Cuthbertson & Joanna George 16 May 1998

Poorly protected climbing up the orange wall just right of 4 then black streaks above the terrace. Take a more or less direct line up the fingery wall midway between the black streak and the crack of 6.

6 Coal Mining ★ 10m VS 5a

FA Joanna George & Dave Cuthbertson 16 May 1998

The obvious crack following the right line to the top.

7 Nicht Thochts ★ 10m VS 4c

FA Joanna George & Dave Cuthbertson 16 May 1998

The rightward slanting seam right of 6. Join the horizontal crack for about a metre then break up left to finish up the wall above. A *Direct Start* up the wall just right is 5a.

8 Uisge ★ 15m E2 5c

FA Dave Cuthbertson & Joanna George 16 May 1998

Right of the grassy gully splitting the crag is an orange scoop. Climb the wall starting at a pale streak to gain the scoop. Leave it following the rightmost crack, moving leftwards to the top.

A prominent scooped recess at the right side of the lower tier contains three cracks, taken by 9 *Achaye*, 10 *Achrobat* (both E1 5a) and 11 *Achtung* ★, E2 6a (HVS 4c started further right).

12 Plocaig Walk 22m Very Difficult

FA Margaret Riley & John Stevenson 21 July 1997

There is a dark slab below a roof near the right end of the main face. Climb the slab then easier ground to a terrace with a small wall. Go up this past a leaning block to the top.

13 Bondi Beach ★★ 15m E2 5b

FA Chris Cartwright & Simon Richardson 10 October 1998

The blunt scooped arête left of 14. Start up a small left-facing corner left of arête. Climb past a white streak, moving right and up to a blunt spike. Finish up the wall above.

14 Shark Attack ★ 15m E3 6b

FA Dave Cuthbertson & Joanna George 16 May 1998

A painful, powerful boulder problem up the short steep crack on the steep east-facing side wall of the crag. Climb the leftmost crack in the wall, finishing up the easier wall above.

SRON NAN GABHAR (NOSE OF THE SHE-GOAT)   30min

NM 469 698 **Alt:** 80m

A prominent whaleback ridge on the right (eastmost) dome of the hill due north of Achnaha.

Approach: Head north-east skirting the right edge of the rocky ridge.

Descent: Down either end of the crags.

The broad slabby buttress on the right has a fine slab at its foot split by three thin crack-lines, the leftmost of which has a prominent triangular block set into it.

3 Locksport ★ 45m E1 5c

FA Danny Carden & C.Hutchinson 15 March 2015

Gain the undercut crack left of the prominent crack. Gain thin cracks to reach a horizontal break. Surmount the bulge above, then trend slightly left up easier ground towards a small triangular block beneath the large flat belay ledge.

1 Mjollnir ★★ 50m HS 4b

FA Brian Davison & A.Richardson 29 May 1999

The left groove on the whaleback. The steep start is the crux.

2 Thor ★★ 50m Severe 4b

FA Charlie Prowse, Rob Kerr & Gary Latter 8 July 1997

The prominent left-facing groove and crack up the centre. Scramble in from the left to belay beneath a short steep groove. Climb the groove and pull steeply out left onto the left-facing groove above. Up this and continue in the same line to finish more easily up the slabby ridge above.

4 Ozone Layer ★ 40m HVS 5a

FA Steve Kennedy, Cynthia Grindley & Mark Shaw 11 July 1998

The leftmost flake crack. An awkward undercut start leads into the crack which is followed to a ledge. Continue up the crack then directly on easier ground to finish.

5 High Plains Drifter ★★ 40m HVS 5b

FA Rab & Chris Anderson 16 May 1998

The central crack then a short groove, continuing up excellent compact rock to the top.

LOCHAILORT CRAGS

There are a number of recently developed crags flanking the A830 Fort William to Mallaig road. They benefit from much better weather than Nevis.

BOATHOUSE CRAG

NM 8008 8308 **Altitude:** 110m **S** 15min

A compact wall of lovely clean rock overlooking the west end of Loch Eilt. There is a short vertical crack which takes a 3" cam (yellow Camalot or Dragon) for belays at the top of the leftmost 3 routes.

Access: Park (NM 8007 8269; 56.883282, -5.6115611) at the derelict boat house (cairns on opposite side of road) 7.2 miles/11.5km west of Glenfinnan Station.

Approach: Walk west down road for 200m, then direct to the crag.

📷 *Gary Latter on the superb Very Gneiss Wall. Photo Karen Latter.*

1 Hypertension ★★　　　　　18m VS 4c

FA Steve Kennedy & Cynthia Grindley 18 August 1999

The short corner leading past a bulge into the prominent vertical crack, climbed on very positive holds throughout.

2 Priapism ★　　　　　18m E2 5b

FA Ali Rose & Matt Rowbottom 22 May 2018

The blank looking wall and rib between. Very bold in its lower half. Climb the wall and rib above to the horizontal break. Continue more easily directly above.

3 Very Gneiss Wall ★★★　　　　　18m HVS 5a

FA Jules Lines & Paul Higginson 22 August 1997

"Possibly the best single pitch HVS in Lochaber." Climb the obvious clean line up the centre of the slabby wall. The rock and the climbing are immaculate - it's a pity it's so short. Protection is good but spaced.

4 Sedative ★　　　　　18m VS 4b

FA Steve Kennedy & Cynthia Grindley July 1998

The vague runnel at the right side of the wall. Climb to the horizontal break, step right then direct.

A south-west facing slab with a fine outlook south to the hills and west down to Lochailort. All routes end in lower offs, except 5 *Brown Crack*.

Access: Head west from Fort William on the A830 towards Mallaig. Pass Glenfinnan and Loch Eilt to the small Lochan Dubh with a pine tree covered island. There is a house and railway yard on the right, at a left bend. Park at layby (NN 7833 8330) on the left (south side of road) 200m west of the bend. 8.1 miles/13 km west of Glenfinnan Station; 1.2 miles/1.9 km east of Lochailort. The crag is directly above the parking.

Approach: A direct approach involves crossing the railway, so walk 200m back east to the railway yard at the bend. Follow the footpath under the railway, cross the burn then head diagonally up leftwards, crossing a couple of smaller burns en route.

1 Leftover — 12m F5c

FA Colin Moody & Cynthia Grindley 12 April 2021

2 Alarm ★ — 14m F6a

FA Colin Moody & Steve Kennedy 6 March 2021

3 The Hole ★ — 16m F6a+

FA Steve Kennedy, Cynthia Grindley & Colin Moody 11 April 2021

Start at a small hole and climb up leftwards to LO of 2.

4 Route One ★ — 7m F6a+

FA Colin Moody, Cynthia Grindley & Stan Pearson 22 June 2021

Start at the hole and go up right.

5 Brown Crack ★ — 8m VS 4c

FA Colin Moody, Steve Kennedy & Cynthia Grindley 21 April 2021

The short right-curving crack. Belay up right on a large spike. To descend, go right then down.

6 Premolar — 10m F6b

FA Steve Kennedy, Cynthia Grindley & Colin Moody 15 July 2022

Line just right of 5.

7 Memory Lane ★★ — 25m F6b

FA Steve Kennedy & Colin Moody 26 May 2021

The obvious girdle, following the main central fault rightwards. Start up 5 then foot traverse right into 8, step up then hand traverse the fault all the way to the LO of 16. A couple of small to medium cams are useful to supplement the bolts.

8 Molar ★★   **10m F6b**

FA Colin Moody, Steve Kennedy & Cynthia Grindley 11 April 2021

Climb through the tooth mark.

9 The Stripper ★★   **10m F6b+**

FA Colin Moody & Steve Kennedy 16 May 2021

10 Hollow Shield ★★   **12m F6a+**

FA Colin Moody, Steve Kennedy & Cynthia Grindley 19 March 2021

Climb through the shield.

11 Big Crack ★★★   **12m E1 5b**

FA Colin Moody, Steve Kennedy & Cynthia Grindley 20 April 2021

The hanging crack. Clip first bolt of 12 to start. Large blue/ grey cam useful.

12 Great Wall ★★   **12m F6c**

FA Steve Kennedy, Cynthia Grindley & Colin Moody 19 March 2021

Start below 11 and climb up to the horizontal break. Hand traverse rightwards to a flat foothold (occasional damp streak) then up the wall between the two cracks.

13 Fly Wall ★★   **12m F6c+**

FA Morag Eagleson & Jamie Skelton 29 August 2021

Takes the slightly left trending line up the middle of the slab, joining 12 just before its crux. A long reach is very useful for getting off the ground. Hard move to gain the final breaks.

14 Cry Wolf ★★   **12m E1 5b**

FA Colin Moody & Steve Kennedy 1 May 2021; Direct Start: Steve Kennedy 18 May 2021

Climb up to the scoop of 15, move left and climb the fine left-slanting crack. Can also be climbed at the same grade with a quick pull on the first bolt of 15. 14a *Direct Start* is an awkward E1 5c ★★ with overhead protection.

15 April ★★   **14m F6a**

FA Steve Kennedy, Cynthia Grindley, Colin Moody & Billy Hood 22 April 2021

Pull up right into scoop to start then leftwards up the rib.

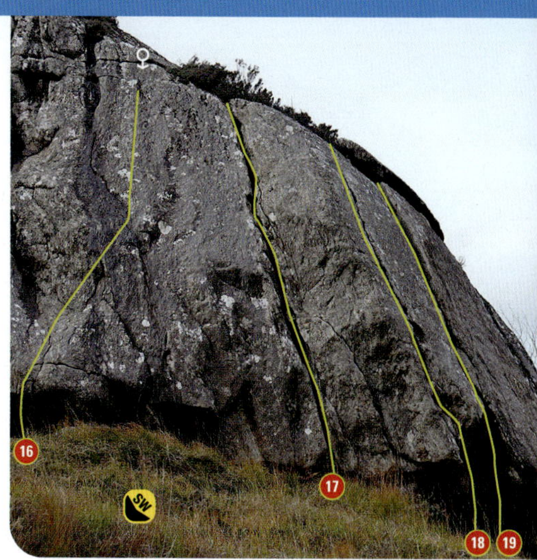

16 Fumble   **10m F6a**

FA Colin Moody, Cynthia Grindley & Steve Kennedy 20 May 2021

Start up 15 then move right.

17 River ★   **10m E2 5c**

FA Gary Latter 29 October 2022

The short wide crack, easing after the brutal start.

18 Kashmir ★   **10m E3 6a**

FA Gary Latter 29 October 2022

The thin flake, then direct past a bolt.

19 Piglet ★   **10m E1 5c**

FA Gary Latter 29 October 2022

The wide flake, soon easing.

SOUTH EAST FACE

20 Illegitimate Groove ★   **10m F6a**

FA Colin Moody & Steve Kennedy 9 July 2021

Interesting moves up the shallow groove and wall above.

CLACH A' PHRIONSA (PRINCE'S STON

Two sport crags situated in a fine setting close to the north shore of Loch nan Uamh. The differing aspects of the crags often result in it being a sun trap in good weather. The routes are short but varied and generally of good quality.

Access: Park in the long layby for The Prince's Cairn (NM 7189 8447; 56.895289, -5.7471071), 3.7 miles/6 km west of Lochailort.

Approach: Walk west along pavement for a few hundred metres until at the sweeping bend. A small path leads from the bend down onto a shingle beach. Thereafter, head right (west), cross a burn, a further shingle beach, then walk through a deep rocky gully on the left (scramble over rocks to the side at high tides). Continue along the rocky shore then scramble up to a left-slanting shelf just before the crag.

WEST WALL

1   HRH ★   15m F6c+

FA Jamie Skelton & Rory Brown 10 April 2021

2   Jugobite ★★   14m F6c

FA Morag Eagleson & Jamie Skelton 19 March 2021

3   The Last Stand ★★   15m F6b

FA Steve Kennedy & Cynthia Grindley 2 May 2019

4   Jacobite Rebel ★★   15m F6b

FA Steve Kennedy, Cynthia Grindley & Colin Moody 20 April 2019

5   Picnic Pleaser ★   16m F7a

FA Jamie Skelton & Morag Eagleson 19 March 2021

6   The Bonnie Traverse ★★   30m F6a

FA Nathan Adam & David Wood 21 April 2021

PROW EAST FACE   E

7   Aided by Flora ★★   13m F6a

FA Steve Kennedy & Chris Docherty 26 May 2018

8   Highland Laddie ★   12m HVS 5a

FA Steve Kennedy, Cynthia Grindley & Colin Moody 20 April 2019

A mixture of trad and bolted climbing; the lower section is well protected with medium sized cams.

SOUTH WALL 

9 Culloden ★ — 13m F6a
FA Steve Kennedy & Colin Moody 10 August 2021

10 Flight to Safety ★ — 12m F6a
FA Cynthia Grindley & Steve Kennedy 17 April 2019

11 Heaven's Darling ★★ — 11m F5+
FA Steve Kennedy & Cynthia Grindley 19 April 2019

12 The Highland Charge ★★ — 11m F6a
FA Steve Kennedy & Cynthia Grindley 24 May 2018

13 Jacobite Trail ★ — 11m F6a
FA Steve Kennedy & Cynthia Grindley 11 November 2018

13a Honnold of the North ★ — 11m F6b+
FA Jamie Skelton & Rory Brown 10 April 2021

14 Usurper ★ — 10m F6a
FA Steve Kennedy & Cynthia Grindley 24 May 2018

15 Martinet ★ — 12m F6a
FA Jamie Skelton & Morag Eagleson 12 August 2022

EAST FACE

16 Young Pretender ★★ — 12m F6a
FA Steve Kennedy & Cynthia Grindley 24 May 2018

17 Speed Bonnie Boat ★★ — 11m F6b
FA Steve Kennedy & Cynthia Grindley 13 April 2019

18 Raising the Standard ★★ — 12m F6a+
FA Steve Kennedy & Cynthia Grindley 8 May 2019

JACOBITE GOLD AREA — Altitude: 10m

NM 7075 8437

Located behind and left of the main crag. Approach on the right, then go left around the back of the knoll before short steps lead down towards the shingle beach.

1 All the Way Over There? ★★ — 18m F6c+
FA Jamie Skelton & Morag Eagleson 21 April 2021

2 Pass the Snacks — 12m F6b+
FA Morag Eagleson & Jamie Skelton 21 April 2021

3 Jacobite Gold ★★ — 12m F6b+
FA Steve Kennedy & Cynthia Grindley 27 June 2019

4 Jugzilla ★ — 11m F6c
FA Jamie Skelton & Morag Eagleson 21 April 2021

UPPER WALL — Altitude: 10m

5 Squeezing It Out ★ — 6m F7a
FA Jamie Skelton & Morag Eagleson 21 April 2021

6 Dogs on the Run ★ — 8m F6c
FA Morag Eagleson & Jamie Skelton 21 April 2021

The Gutter, Pine Wall Crag, Polldubh.

GLEN NEVIS

This splendid glen, celebrated as one of the most beautiful in Scotland, runs from the north end of Fort William south-east for 4 miles/6km, before turning east for a further 6 miles/10km. In a compact 4km central section over 70 crags flank the lower slopes of the mighty Ben Nevis massif and Sgurr a' Mhaim on either side of the glen.

The sheer variety and range of routes on offer, the sheltered quick drying southerly aspect of most of the crags, the quality schist and the short approaches (5–40 minutes) would alone make it an attractive venue. Combine these factors with the magnificent surroundings; stunning mountains, a spectacular waterfall, a unique gorge, and it all adds up to the best outcrop climbing venue in Scotland.

Many trees line the lower slopes and indeed sprout from many of the crags, including native oak, birch, ash, rowan and Scots pine, creating a relaxed friendly atmosphere and an often convenient means of retreat. In recent years local climbers, with the support of SNH and landowners, have cleared much of the encroaching tree cover around the popular lower crags at Polldubh, resulting in generally cleaner, faster drying and less-midgey venues. The higher crags at Polldubh and above the Gorge are more open, and are often blessed with a breeze. On some of the harder routes, the nature of the rock necessitates a good selection of RPs and micro-wires, and on some tied down skyhooks. In contrast, many of the popular routes are well protected, making it an ideal introductory venue.

Access: Turn off the A82 (signposted Glen Nevis) at the roundabout at Nevis Bridge at the north end of Fort William. Follow the road for 4.9 miles/7.8km to cross the River Nevis by a bridge over the lower falls at Achriabhach, from where the Polldubh crags are clearly visible. The road now continues as single track with passing places. For **Polldubh**, park at a long lay-by on the right 0.5 miles/0.8km beyond the bridge, or by the roadside near the lay-by. For **Whale Rock**, continue for a further 0.4 miles/0.7km just round the bend to park on the left, just before a footbridge down on the right. For the remaining crags in the **Car Park Area**, **The Nevis Gorge** and **Steall Meadows**, continue for a further 0.8 miles/1.3km to a large car park at the road end, which can become very busy (full) with touroids in the summer.

"In its farthermost reach, Glen Nevis is a desolate moor. In its central section of 5 miles (from the great bend under Sgurr a Bhuic to Polldubh), it is one of the most beautiful glens in Scotland. The lower stretch of 4 miles to Bridge of Nevis gives a pastoral scene of green fields flanked on one side by forest and on the other by the huge bare slopes of Ben Nevis."

– W H Murray, Highland Landscape (The Aberdeen University Press/ National Trust for Scotland, 1962)

FORT WILLIAM

Since around the turn of the century, the self-proclaimed "Outdoor Capital of the UK" (aye, right!) has expanded to serve and exploit the constant stream of mainly foreign tourists that seem to invade throughout the summer months. In many ways it can be likened to a highland version of Chamonix, with a high street mall littered with loads of tartan knick-knack and woolly jumper shops in place of the somewhat more appealing street-side cafes.

Amenities: Many cafés, restaurants and pubs in and around the pedestrian precinct on the High Street. Morrisons supermarket at An Aird, midway between the rail and bus stations should provide all provisions. Also, Lidl at Camanachd Crescent, behind Morrisons. Late opening Spar store in Claggan (first right off A82 Inverness road after Glen Nevis roundabout). **Climbing Walls:** 3 Wise Monkeys Climbing (☎ 01397 600200; www.threewisemonkeysclimbing.com) has bouldering and leading walls, café and shop, and is within walking distance of the town centre. Outdoor shops: Nevisport (☎ 01397 780011; www.nevisport.com); Ellis Brigham (☎ 01397 706220; www.ellis-brigham.com); Cotswold Outdoor (☎ 01397 719118; www.cotswoldoutdoor.com).

Accommodation: Masses of hotels, guest house and bed and breakfast accommodation, with the mile long southern approach road dubbed locally 'the golden mile', such is the profusion of B&B and guest house accommodation available. Bookings and information from **TIC** in the High Street (☎ 01397 701801; www.visitscotland.com). The Glen Nevis Visitor Centre is situated 1.5 miles up the glen. **Glen Nevis Youth Hostel** (☎ 01397 702336; www.hostellingscotland.org.uk) at the start of the tourist path up the Ben, 1.8 miles/3km along the Glen Nevis road. **Bunkhouses:** Fort William Backpackers, Alma Road (☎ 01397 700711; www.fortwilliambackpackers.com); Bank Street Lodge (☎ 01397 700070; www.bankstreetlodge.co.uk); Chase the Wild Goose Hostel, Banavie (☎ 01397 748044; www.chasethewildgoosehostel.co.uk); Glenfinnan Sleeping Car (www.glenfinnansleepingcar.com); Grey Corrie Lodge, Roy Bridge (☎ 01397 712241; www.greycorrielodgebunkhouse.co.uk); Aite Cruinnichidh, Roy Bridge (☎ 01397 712315; www.highland-hostel.co.uk); Station Lodge, Tulloch (☎ 01397 732333; www.stationlodge.co.uk).

Club Huts: Alex MacIntyre Memorial Hut, North Ballachulish (Mountaineering Scotland/BMC; NN 044 611); CIC Hut, Ben Nevis (SMC; NN 167 722); Steall Hut, Glen Nevis (JMCS; NN 178 683); Raisg, Roy Bridge (Climbers' Club; NN 271 812).

Campsites: Glen Nevis Caravan & Camping Park (☎ 01397 702191; www.glen-nevis.co.uk) is a large official and very busy campsite, 1.5 miles/2.4km down the Glen Nevis road at NN 125 722 overlooking the tourist path up Ben Nevis. Wild camping discreetly further up the glen, with popular riverside sites below Polldubh (not permitted in the lower glen). Ben Nevis Park, Camaghael (☎ 01397 703446; www.highlandholidays.com); Linnhe Lochside Holidays, Corpach (☎ 01397 772376; www.linnhe-lochside-holidays.co.uk); Gairlochy Holiday Park (☎ 01397 712711; www.theghp.co.uk); Stronaba Farm Caravan & Camping Site (☎ 01397 712259; www.scottishcamping.com) both Spean Bridge; Bunroy Park, Roy Bridge (☎ 01397 712332; www.bunroypark.co.uk).

NN 145 694 **Alt:**130m (W) 30min

CREAG AN FHITHICH BEAG
(LITTLE CRAG OF THE RAVEN)

This is the first crag to come into view when driving up the glen and one of the most impressive, characterized by an overhanging front face broken by several overhangs on its right side.

Approach: The quickest approach is by crossing the river at a ford (NN 139 691) at a point about a kilometre downstream from the lower falls at Achriabhach, where the river runs close to the road, just beyond a large flat area of ground on the left (north) side of the road. (4 miles/6.4km from the roundabout). From the other side of the river (north), gain a path on the left and follow this to a stream (Allt na Dubh-ghlaic) just beyond a gate. Cross the stream and cut directly up through the trees by an ill-defined ridge to gain the left end of the crag. If the river is too high, or for those with hydrophobia, approach can also be made by starting along the track at Achriabhach Bridge (NN 145 685).

Descent: Down either side of the crag, easier on the left.

1 **Hollow Wall** ★★  9m VS 4c

FA A. Wallace 1975

On the short steep lower tier, directly underneath the left section of the crag. The crack near the left end, stepping left near the top.

2 **Virtual Reality** ★  25m E5 6b

FA Rab Anderson, Duncan McCallum & Johnny May (red pointed) 23 May 1992

The thin crack and hanging groove in the left wall of the crag. Reasonably well protected, though difficult to place.

The following two routes share a common start, a quartz intrusion on the left arête.

3 Liminality ★★★ 25m E7 6c

FA Dave Cuthbertson (red pointed) 22 May 1987

A subliminal excursion up the edge of all things – or just a neat little arête? The prominent left-bounding edge of the crag. Up on quartz holds then left to a ledge on the left side of the arête. Follow the thin crack on the right side (good cam and RP #5), then pull left and go up the arête to the top.

4 Caterpillar ★ 27m E3 5c

FA Dave Cuthbertson & Ed Grindley 10 May 1984

The diagonal quartz vein then easier up a ramp to finish up a short steep wall.

5 The Dream of the Butterfly ★★ 30m E6 6b

FA Gary Latter (red pointed) 14 September 1993

The horizontal fault emanating from the initial crack of 6 provides very safe strenuous climbing. Follow the fault past some PRs with a difficult section stepping down to better holds leading to the quartz recess on 4. Finish up that route.

6 The Handren Effect ★★★★ 25m E6 6b

FA Dave Cuthbertson April 1983

One of the best pieces of wall climbing in the glen. Bold and sustained at a high standard. Start at an obvious inverted L-shaped crack. Up this past a pair of PRs with difficulty to reach a good flake hold (RPs behind this; situ nut above). Up a shallow runnel with hard moves to reach a superb bucket hold at the top of the wall. Stroll easily up the short corner above.

7 Juggernaut ★★★ 30m E7 6b

FA Dave MacLeod 18 August 2002

Outstanding climbing up the wall in the centre of the crag, featuring fairly well protected but sustained strenuous climbing. Start up the left-facing scoop and step left to the PRs. Launch directly up the wall to reach a jug (good wire behind). Move up and left to the deceptive break (RP) and dyno to a good hold directly above (crux). Move left and mantel onto the sloping ledge giving access to easy finishing rib on the left.

8 The Monster ★★★ 30m E5 6a

*FA Andrew Wielochowski & Pete Webster (A3) 1975;
FFA Dave Cuthbertson & Alan Moist 16 June 1985*

A strenuous well protected pitch. Start beneath a shallow left-facing scoop, as for 9. Climb directly, initially on good holds then continue up the wall above to an overhang. Go left to undercuts (F on left) and good footholds, and along an awkward jam crack with a hard move onto a ledge. Easier climbing up and rightwards to finish.

9 Steerpike ★★ 45m E2 5c

FA Noel Williams & Andrew Wielochowski (1 PA & 1 tension traverse) 1975; FFA Dave Cuthbertson & Gary Latter 1982

Start just right of the inverted L-shaped crack, beneath a shallow left-facing scoop.

1 **20m 5c** Up this on good holds past some old pegs to arrange a runner underneath the roof. Cross the gangway on the right (crux) to easy ground, and belay at the foot of the quartz groove.

2 **15m 5b** Up the quartz groove directly above and pull rightwards onto a ledge and belay at an oak tree.

3 **10m** Climb the tree and move right into a chimney, or abseil off.

10 Spring Fever ★★ 25m E3 5c

FA Dave Cuthbertson, Kev Howett & Callum Henderson 30 May 1985

A fine route when combined with 9. Start from the hanging belay at the top of pitch one of that route. Climb thin crack and quartz overhang, as for that route, to a slab right of a hanging corner. Step down and swing wildly left for a jug then the obvious line leftwards.

11 Exocet ★★★ 40m E6 6b

FA Andrew Wielochowski & Noel Williams (A3) 1975; FFA Dave Cuthbertson 18 May 1982; pitch 1 freed the following month

A direct uncompromising line through the overhanging prow up the steepest section of the crag. Start at the foot of a prominent diagonal crack-line up and left of an overgrown left-facing corner crack.

1 **15m 6a** Swing right into the crack and up this to a hanging belay at the foot of a quartz groove (common to 9).

2 **25m 6b** Undercut the roof leftwards for about 3m (very awkward) to a large flat hold above. Attain a standing position on this with difficulty and follow a line of holds trending leftwards to a resting ledge. Continue more easily above.

Note: The first pitch, although slow to dry, provides a very fine pitch in its own right at E3 6a – an excellent combination with *Spring Fever*. The section above the roof can also be gained from *The Monster*, providing a long exposed pitch at no change in the overall grade.

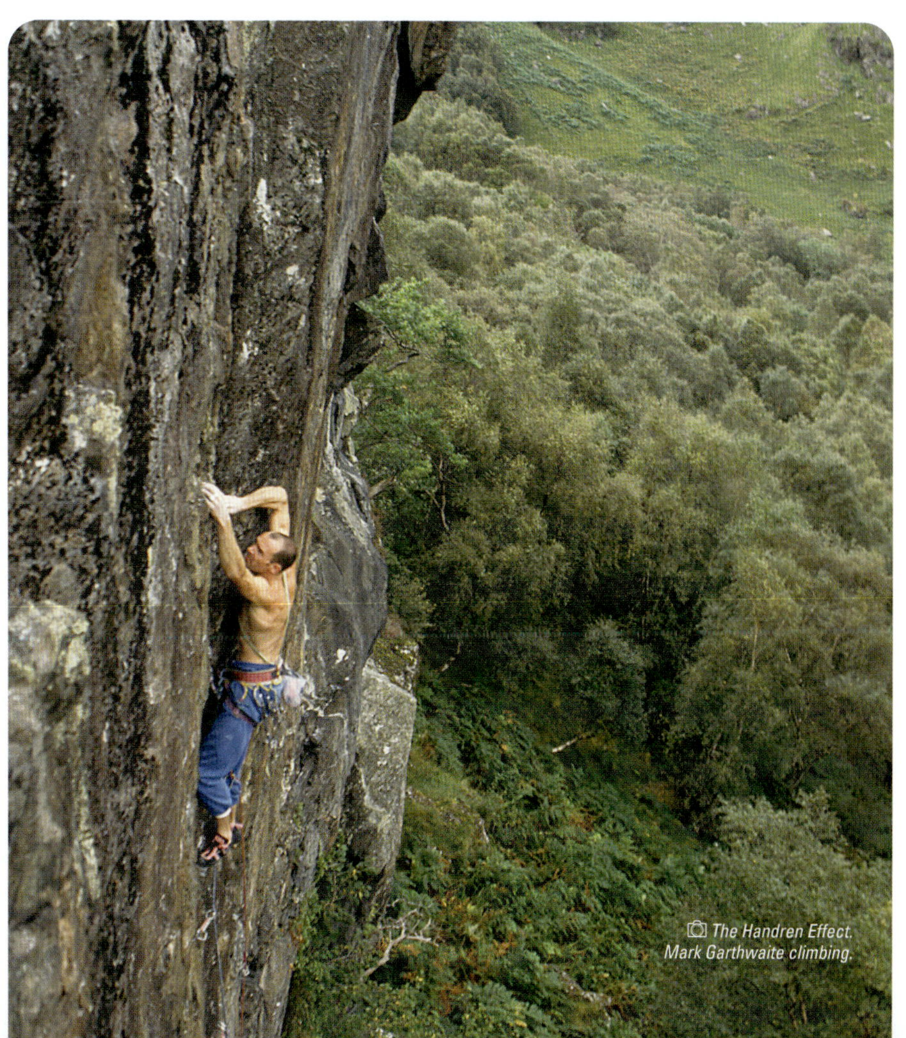

The Handren Effect. Mark Garthwaite climbing.

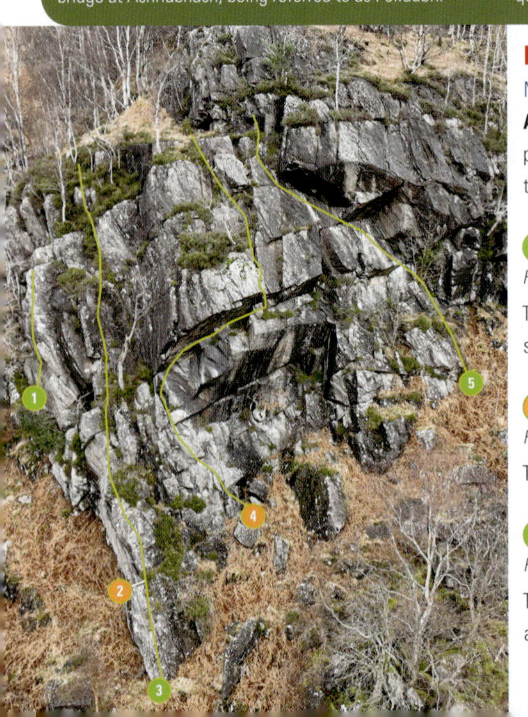

POLLDUBH

The name Polldubh (literally the black pool) correctly refers to a pool in the River Nevis, just upstream from the lay-by at NN 153 685 below **Pinnacle Ridge**. Common usage has resulted in the frontage of crags on the hillside, extending for over a kilometre from the bridge at Achriabhach, being referred to as Polldubh.

The crags are many and varied, numbering around 40 in all, ranging from mere overgrown boulders a few metres high to fine 90m plus crags with two and sometimes three pitch routes. The obvious advantage of linking numerous routes within separate tiers is quite apparent, giving up to 300m of ascent.

HANGOVER BUTTRESS

NN 1475 6864 **Alt:** 110m   (W) (SE)   15min

Approach: Head fairly directly up the hillside on a vague path from one of the passing places, midway between the lower falls and the main Polldubh layby.

1 **Hangover Crack** *   **8m Difficult**

FA Geoff Hewitt July 2014

The short steep juggy crack on short rib high up on left side of the crag.

2 **Shag** *   **10m VS 5a**

FA Ken Johnson 1960

The thin strenuous crack.

3 **Hangover Buttress Edge** *   **30m Moderate**

FA Jimmy Ness & B.Ellison 16 February 1946

The rib left of the roofs. Above the oak, a groove leads to an easy left-slanting ramp.

 Route 2 ★   25m HS 4b

FA Ian Clough, R.Mason, P.Hannon & O.Cook 1 September 1957

Popular and polished. Climb diagonally left to the left end of the roof, then traverse right between the roofs to finish up a chimney.

Cross 3 ★★   25m Difficult

FA unknown 1950s

Good well protected climbing, following the finely positioned left-slanting slab beneath the large roof.

TRICOUNI BUTTRESS

NN 1484 6864 **Alt:** 100m   15min

Two slabby walls about 100m right of **Hangover Buttress**. **Approach:** Head directly up the hillside on a vague path from one of the passing places.

 Corner ★★   15m Severe 4a

FA unknown 1950s

Tricouni Overhang ★   10m VS 5a

FA unknown 1960s

Tricouni Slab ★   25m Very Difficult

FA unknown 1960s

Good climbing up the prominent right-angled diedre.

Black Slab Edge ★   25m VS 4c

FA Ken Johnson 1960/61

Climb the steep right edge of the slab by a groove. Step right to finish up short wall.

TRICOUNI RIGHT

NN 1489 6866 **Alt:** 110m    15min

The slender clean slab 50m up right of the main crag.

Fly Direct ★★   25m HVS 5a

FA Ken Johnson (aid) 1960/61; FFA Ray Treadwell & Richard Shaw 2 June 1983

Start at small rib beneath the main edge. Move up to pass a small overlap on the left, then more easily up the edge of the slab.

Parisian Walkway ★   25m E2 5b

FA Ramsay Donaldson (solo) 8 June 1983

Bold climbing up the centre of the slab. Climb past a thin section to mid-height break, step left and finish at the same point as 1.

DUNDEE BUTTRESS

NN 1505 6863 **Alt:** 110m **W** **SE** 9min

The small slender buttress immediately left of **Calvary Crack Buttress**, just above the descent path.

1 Dundee Weaver ★   20m HVS 5a

FA R.Gray & G.Low 1965; FFA Klaus Schwartz & Alec Fulton 1967

Trend leftwards to a small platform, then tricky moves up the crack just right of the edge to easier ground.

2 Promises ★★   20m HS 4b

FA Ken Johnson 1960

Direct up into the niche, finishing by the crack on the right.

3 Dark Horse   20m E1 5b

FA Steve Hill (on-sight solo) 1988

Line 2m right of the arête.

4 Wren's Delight ★   25m Difficult

FA Ken Johnson 1960/61

Climb lower slab to slanting heather ledge, then slabby wall above on 'wart-like' holds.

AFTER CRAG

NN 1504 6867 **Alt:** 150m     12min

A fine open crag just above and left of the **Calvary Crack Buttress** descent path.

Approach: Follow the path up the left side of **Dundee Buttress**, then up slightly leftwards.

1 Kraut ★★   20m E1 5b

FA Klaus Schwartz, Sammy Crymble & Blyth Wright (aid) 6 April 1969; FFA Kenny Spence 1970

Strenuous well protected climbing up the left-slanting fault on the steeper sidewall.

2 Cross Examination ★   20m E1 5b

FA Steve Kennedy, Eileen Blair & Colin Moody 7 June 2022

Climb the steep crack with an interesting move onto the buttress edge, finishing up 3.

3 Afterthought ★   20m HS 4a

FA unknown 1960s

Climb over a bulge, then follow a fault leading into shallow heathery scoop.

4 After a While/Rubberface ★   20m HVS 5a

FA Stevie Abbott, Ed & Rona Grindley 16 June 2003

Climb the corner and shallow groove, stepping right onto the slab. Climb this to left end of sloping ledge, move right along this, finishing leftwards. 4a *Direct* ★ E2 5b.

5 Rubberface Direct ★★   20m E1 5b

FA Klaus Schwartz & Loch Eil Outward Bound 15 June 1968; Direct: Ed Grindley 2 March 1976

Climb over a bulge (crux), step left onto the slab and continue direct to the left end of a sloping ledge, finishing directly.

6 Sauer ★   20m VS 4c

FA Ed Grindley, Fiona & Davy Gunn & George Reid 28 March 1982

Direct up the slab just right of a mossy streak.

Dave 'Cubby' Cuthbertson on his own route Soap Suds (page 220).

CALVARY CRACK BUTTRESS W S

NN 1510 6861 **Alt:** 100m   7min

The largest of the lower buttresses. The left face is split by prominent diagonal cracks, easily recognized by a large Scots pine sprouting high up with two further prominent Scots pines halfway up the left side of the front face.

Approach: From the long layby, follow the good path up the right side of Pinnacle Burn for 60m, then a good path traversing leftwards, crossing the burn, then direct up to the base. This avoids the boggy more direct paths starting further left (west).

Descent: A steep path down the left side, skirting round the left side of **Dundee Buttress** near the base.

1 **The Old Wall** ★★★   40m VS 4b

FA Ike Jones & party 1963

Steady open wall climbing on good holds. Protection is noticeably lacking. Start up a wide left-slanting heathery crack high up the left side of the base. Climb the initially vegetated crack for 6m to just past a tree. Move right and up before heading across the wall through *Storm* to finish on the right edge of the wall level with the pine on *Storm*.

2 **Storm** ★★★★   85m HVS 5a

*FA Ian Clough & Terry Sullivan (3 PA) 3 May 1959;
FFA Allan & Jenny Austin 1962; True Finish: John Taylor
& Klaus Schwartz 1976*

A well protected classic with fine situations, giving the best route of its grade on the crags. Omitting the final crux pitch by escaping out right from the tree belay to finish up *Heatwave* gives a great three star VS. Start a few metres up left from the very toe of the buttress, beneath a left-slanting ramp, just left of a holly and a birch.

  1 **25m 4b** Follow the ramp to belay on a ledge at the base of a long diagonal crack.

  2 **30m 4c** The wide crack to a wonderfully positioned belay on the large pine.

  3 **30m 5a** Ascend the shallow groove on the right to cross the bulge (crux) then much easier and trending

slightly right up the wall above. 2a *The True Finish*, **E1 5b moves out left from above the crux and follows the shallow corner capped by a block overhang.

3 Heatwave/Vampire Combo ★★   **90m HS 4b**
FA Ian Clough, J.Pickering, R.Henson, P.Brocklehurst & R.Porteous 22 May 1959

A good link. Start just left of the toe of the buttress, as for *Storm*.

1 25m 4b Follow the ramp to belay on a ledge at the base of the long diagonal crack of *Storm*.

2 5m Traverse the horizontal break right to belay at the base of a left-facing groove.

3 30m 4b Climb the groove then continue up left-slanting cracks close to the right edge of the wall. The original route ascends the left wall of the gully at 4a.

4 30m 4a Finish up slabs above, trending left to a short groove and direct above.

Ewan Lyons seconding Karen Latter on the second pitch of Storm.

4 Vampire ★★   **85m HS 4b**

FA Ian Clough & Eddie Buckley 21 April 1959

Varied and enjoyable. Start 20m right of the buttress edge, at the same point as *Fang*. Scramble up left to belay behind small ash trees.

1 25m 4b Climb left slanting flakes left of the initial capping roof, past a small rowan at 10m, then traverse left beneath first pine. Continue left and down to belay on oak at good recessed bay.

2 30m 4b Climb the groove above, then continue up left-slanting cracks close to the right edge of the wall.

3 30m 4a Finish as for *Heatwave*.

5 Fang ★★★   **45m E2 5b**

FA Bill Skidmore, Peter MacKenzie & Jim Crawford (aid) July 1963; FFA Ed Grindley & Ian Nicolson 19 April 1976

Bold open wall climbing. Start beneath an open left-facing groove down and right of the open slab of *Vampire*.

1 25m 5b Go up a steep initial groove past an overlap then move steeply out right on good holds. Move up then pull leftwards along a diagonal crack to belay on a ledge.

2 20m 5a The groove above, turning the roof on the right.

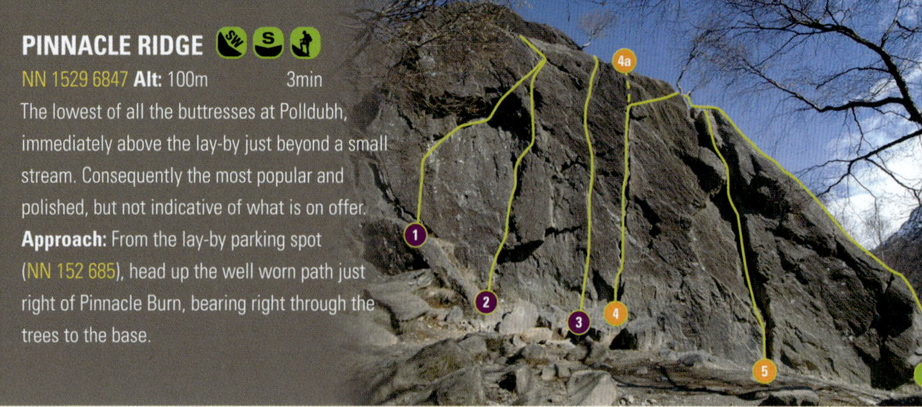

PINNACLE RIDGE

NN 1529 6847 Alt: 100m   3min

The lowest of all the buttresses at Polldubh, immediately above the lay-by just beyond a small stream. Consequently the most popular and polished, but not indicative of what is on offer.
Approach: From the lay-by parking spot (NN 152 685), head up the well worn path just right of Pinnacle Burn, bearing right through the trees to the base.

1 Soap Suds ★   12m E4 6a

FA Dave Cuthbertson & Ian Sutherland 1981

Start at the left side of the wall beneath a curving overlap. Undercut rightwards and pull over the overlap at a good hold on the lip. Step right to gain better holds. Bold.

2 The Sugar Puff Kid ★   12m E4 6a

FA Dave Cuthbertson / Gary Latter (both led) 15 May 1985

Ascends the wall midway between 1 and 3.

3 Chalky Wall ★   12m E4 6a

FA Dave Cuthbertson & Murray Hamilton April 1977

The hard Polldubh test piece. Start 2m left of the prominent vertical crack of 4. Up past a flake with a hard move to gain holds in the start of a diagonal crack. Pull directly and somewhat blindly onto the rounded slab above.

4 Clapham Junction ★★   10m VS 5a

FA unknown 1950s (aid); FFA Ken Johnson 1964

The obvious crack, hand traversing right at the top. Technical and well protected. The 4a *Direct Finish* pushes the grade up to HVS 5a.

5 Severe Crack ★   9m VS 4b

FA Jimmy Ness & Alan Burgon 25 May 1950

The crack with a perched block near the top proves harder than its name would suggest.

6 Hodad ★   10m HVS 5b

FA Ike Jones & John Grieve 1967

The left-slanting diagonal crack on the second tier. Make a long reach right to finish.

7 Pinnacle Ridge ★★   45m Severe 4a

FA Jimmy Ness & Dr Donald Duff 1947

One of the most popular routes at Polldubh, with a polished first pitch. Start at the toe of the buttress.

1 **20m 4a** Climb near the left edge of the slab, first left onto the edge (good diagonal finger crack for protection) then back right and up to a small birch. Continue up the easy angled scoop/jam crack to a terrace. Step right and belay at good crack immediately below another birch.

2 **25m** Climb up then along the top of the large flake on the right (past a third birch) then up rightwards to follow a good crack up rough slabs to the top. The fine open scoop above the large flake can also be climbed at the same standard.

Variations: *Staircase* 15m S 4a (*FA Jimmy Ness & Alan Burgon 28 Sept 1950*) Start 6m right of the normal start and follow, wait for it, *"staircase like holds"* to the first tree. *Tip Toe* 25m HS 4a (*FA Jimmy Ness, M.Hutchison & R.Corson 23 Sept 1950*) Start immediately right of *Staircase*. Go up the slab to a small foothold, and (you guessed it!) tip toe left to good hand holds at 10m and traverse left. Either climb a niche or the crack to its right leading to the terrace. Poorly protected on the lower slab.

PANDORA'S BUTTRESS

Lies immediately above and to the left of **Repton Buttress**. The buttress has a slightly overhanging front face with twin diagonal leftward slanting cracks. Two rock tongues extend down on either side of the cracks into the trees.

Approach: From the lay-by parking spot (NN 152 685), head up the well-worn path just right of Pinnacle Burn then cross the burn and head up through the trees to the base.

① Phantom Slab ★★★   25m VS 4c

FA Terry Sullivan & Ian Clough 3 May 1959

Excellent sustained slab climbing, well worth the trouble of searching it out. Start by climbing the first two pitches of 2. Descend a few metres to belay on a horizontal oak tree beneath the right side of the slab. This point can also be gained by fairly straightforward down climbing or a short abseil from just left of the main pitch on 4. Traverse diagonally leftwards and follow a line up the left edge of the slab, finishing by moving slightly rightwards at the top.

② Pandora ★★   65m HS 4b

FA Ian Clough & Eddy Buckley 20 April 1959

Good varied climbing.

1 **25m 4b** Climb the left of two rock tongues below the diagonal cracks to a large ledge.
2 **20m 4a** Continue up the rib to a tree on the right wall of a large corner.
3 **20m 4b** Up the corner to a large ledge then slabs on the left.

③ Tomag ★★   30m E3 5c

FA Murray Hamilton & John Fantini 1981

Strenuous and sustained climbing taking the finely situated parallel cracks across the overhanging wall. Gain these from a short groove, and follow them round the arête on fist jams (large cam useful) then cut back right to the top.

④ Flying Dutchman ★★★   60m Severe 4a

FA Terry Sullivan & Ian Clough 3 May 1959

Deservedly popular with an exposed and sustained second pitch. Start on the lowest rocks, to the right of *Pandora*.

1 **27m** Climb the crest, or heathery grooves just to the right to a terrace.
2 **24m 4a** Go up slabs left of a dièdre to traverse diagonally left on good footholds to a short corner in the left side of the roof. Pull over the roof, up a crack and round a rib to belay on a capacious ledge.
3 **9m** Follow the ridge directly or scramble up easy ledges leading left.

④a Direct Finish ★   6m VS 4c

FA unknown 1960s

From the belay ledge, follow the thin rightward slanting crack.

REPTON BUTTRESS S 6min

NN 1524 6857 **Alt:** 140m

Slightly below and immediately to the right
of **Pandora's Buttress**.

Approach: From the lay-by parking spot (NN 152 685)
head up the well-worn path just right of Pinnacle Burn
then cross the burn to the base.

 **Three Pines** ★　　　　　**30m Severe 4a**

FA Terry Sullivan, Eddie Buckley & A. Flegg 17 March 1959

A popular route. Climb the rib immediately right of the
central gully to the pines. Follow the groove behind the
central tree moving right under the roof with interest to a
platform. Continue by a crack.

 **Three Pines Variations** ★　　**30m Very Difficult**

FA unknown 1960s

Start 6m right of the original route and traverse up left
to the rib which is followed to the pines. Escape out left
then back right.

2 Right Wall ★　　　　　　**30m Very Difficult**

FA unknown 1950s

Start on the right of the crag. Climb up to the right end of
the overhang, step left onto it and up to a tree. Traverse
3m to another tree and through yet another tree into a
cleft. Up behind this.

LITTLE BUTTRESS SW S 15min

NN 1526 6873 **Alt:** 170m

A slabby crag with a large Scots pine at the left end of
a mid-height ledge. It lies just up and right of **Repton
Buttress**, about 100m left of **Pine Wall Crag**.

Approach: As for **Pine Wall Crag**, traversing left 100m.

Descent: Down the right side of the crag.

 **Spike Wall** ★★　　　　**55m Very Difficult**

*FA Ian Clough, R. Henson, P. Brocklehurst & R. Porteous
28 June 1959*

A popular and delightful climb. Good value for the grade.
Start near some boulders.

1 30m Climb the slab leftwards past a rock
scar (spike now gone!), then traverse left
almost to the edge. Climb a recess and the
ridge above to the big ledge with the pine.

2 25m Climb the slab above on small quartz holds
to finish on the rounded crest of the ridge.

 **Spike Direct** ★　　　　　**30m Severe 4a**

FA Ken Johnson 1960

Climb a groove just right of the rib for 5m then move right
towards the rock scar. Continue straight above the rock
scar, gaining the ridge by a crack on the right to reach a
block.

THE ALP

Hidden from the road above and well right of **Pinnacle Ridge** and just beyond the origin of the Pinnacle Burn lies a large sheltered flat area of grass with two particularly fine contrasting crags, together with a strangely popular instantly-drying micro-crag.

Approach: From the lay-by parking spot (NN 152 685) head up the well-worn path keeping just right of the Pinnacle Burn, passing a boggy section just before the end of the burn. Continue past the 12m high smooth steep slab of **SW Buttress** to gain the Alp. **Pine Wall Crag** is just up through the trees on the left, **Styx Buttress** immediately in front.

SW BUTTRESS 15min

NN 1537 6866 **Alt:** 180m

A micro-crag, but very quick drying. It lies just above the origins of the burn.

Descent: By path down the right (east) side.

1 Tear ★★   **12m HS 4b**

FA Ken Johnson 1963/4

The vertical crack just left of centre with a technical but very well protected crux past the bulge.

2 Scratch ★   **12m VS 4c**

FA Ken Johnson 1963/4

Climb slightly rightwards up the slab, from just right of 1.

3 SW2 ★   **12m E2 5c**

FA unknown early 60s

Start up the short corner just right of 2 to the diagonal break. Step right to a foothold and ascend thin cracks.

4 SW Diagonal ★   **15m HVS 5b**

FA Ken Johnson 1963/4

Diagonal left-slanting crack with one hard move low down.

5 Tee   **12m VS 5b**

FA Mike Hall (solo) 1971

Follow a direct line just right of the diagonal to finish up crack. Again, one hard move.

Tear, Penny Dunbabin climbing.

PINE WALL CRAG 15min

NN 1541 6869 **Alt:** 190m

A prominent 60m high ridge facing the glen, clearly
distinguished by a large Scots pine at two-thirds height.
Descent: Traverse leftwards on a path crossing a small
burn then descend a short shallow V-gully leading left
of the watercourse facing out, i.e. east, to large boulder
field leading back to the base.

1 Eigerwand ★★ 45m HS 4a

FA Ken Johnson 1960s

Start up left from the base beneath a large tree on a
terrace at 8m. Climb up to the tree then a small gully
on the left for 3m. Move right and follow a direct line,
finishing up the centre of the superb headwall.

2 Pine Wall ★★★ 65m HS 4a

FA Jimmy Ness & Alan Burgon 1 June 1950

Excellent exposed climbing on perfect rock, making it
one of the best climbs of its standard on the crags. Much
variation is possible, although the described line gives
the best climbing of the grade. Start at the toe of the
crag, just left of *The Gutter*.

1 **35m 4a** Move up to a dièdre and climb this to a platform at 12m. Move left and cross a bulge then climb

the immaculate slabs to the right of the rounded
ridge to belay on large ledge where the angle eases.

2 **15m** Follow either the ridge or grooves left of
it to belay beneath the steeper headwall.

3 **15m 4a** Pass an overlap on either side to reach
a small recess just left of the crest, and
continue more easily up this to finish.

3 The Gutter ★★★ 65m Difficult

FA unknown 1940s?; after cleaning Klaus Schwartz 1976

A classic with fine situations – far and away the best
route of its grade on the crags. Start at the lowest point
of the crag.

1 **30m** Climb crack (awkward bouldery start) to a
ledge at 12m. Continue in the same line by a
deeper crack to belay just above a birch sapling.

2 **15m** Continue up a shallow groove just left of the
crest of the ridge to belay on the big pine.

3 **20m** Move up left past the left end of a
ledge to a good flake handhold on the wall
then go out right to finish easily up the fine
deep crack in the crest of the ridge.

STYX BUTTRESS

NN 1545 6869 **Alt:** 190m 15min

Immediately right of **Pine Wall Crag**, with a typically steep side wall and slabby front face.

Descent: Make a short step down into the open gully bounding the left side of the crag then easily down this.

1 Tobe Hooper ★★   20m E3 5c

FA Stevie Abbott & Ed Grindley 21 August 2004

Well-protected on superb rock. Start at the upper left end. Climb easily up to small tree, then left trending line to above bulge. Step right and climb tiny groove and wall above.

2 Ascension ★★   30m E2 5c

FA Klaus Schwartz & Jim Mount 19 May 1973; FFA Nick Colton & Rab Carrington 5 June 1976

Good well-protected climbing, following the obvious left-slanting fault. Follow direct line, then over a bulge to gain and follow the fault. Starting up the diagonal chimney/offwidth gives a logical *"character building"* approach.

3 Black Friday ★★   25m E5 6a

FA Klaus Schwartz & Blyth Wright (aid) 13 May 1969; FFA Murray Hamilton & John Fantini 1981

A well protected struggle up the prominent overhanging cleft splitting the centre of the left face. Climb up rightwards past a sloping ledge to the base of the cleft fault. Ascend this strenuously to easier climbing up the final wider cleft where the angle eases. Slow to dry.

4 Resurrection ★★★   25m VS 4c

FA Ian Clough & A.Lakin 5 April 1959

A Polldubh classic – sustained and well protected, taking the tapering ramp in the centre of the crag. Climb the slabby ramp with the crux at the narrow middle section, finishing by a slightly easier wide fault leading to a fine pine tree belay. Scramble to finish.

5 Curse ★   30m VS 5a

FA John Taylor & Alan Kimber 1976

Start 3m right of the ramp of 4. Follow gangway round left side of initial roof onto slab, then obvious natural line round left side of roof. Finish more easily above.

6 Breakheart Pass ★   30m E2 5c

FA Tom Ballard 18 May 2005

Start beneath low roof just right of 5. Climb up to and through roof on good holds and up to the main roof. Cross this with difficulty into a small hanging flake leading to heathery ledge above. Finish more easily.

7 Damnation ★★★   30m VS 4c

FA Ian Clough & John Alexander 28 June 1958

Excellent varied climbing, improbable looking for the grade. Protection is good where it matters. Go leftwards up a ramp, until it is possible to step up right to a rounded vertical rib leading to an overhang. Cross this slightly leftwards on good holds and continue directly up the fine cracked slab to finish.

8 Jericho Rose ★   30m E2 5b

FA Ed Grindley & C.Phillipson 8 June 2004

Up steep wall to a gangway leading left. Cross roof at its widest point to small ledge. Step up thinly to a pocket then finish directly.

11 That Hollow Feeling ★   9m E1 5b

FA Bob Hamilton & Steve Kennedy 19 April 2009

Climb the slab then pull steeply up right into a V-shaped recess. Pull through overlap at small block, then up slab to lower-off from small Scots pine.

12 Right Wall ★★   50m Very Difficult

FA unknown 1960s

A long pitch on good clean rock. Near the right edge of the crag is a clean slab. Start near the left edge of this. Climb diagonally left to the edge of the slab and follow this, passing some pine saplings and keeping right of a recess to finish by a fine crack past another sapling.

9 Iche ★★   30m HVS 5a

FA Terry Sullivan & Ian Clough 11 April 1959

Start just left of the diagonal heather groove. Climb leftwards up the slab and cross the overhang 2m from its right end. Move left to a thin crack and follow this to finish up the slab above.

10 Fidelity ★★   30m VS 4c

FA Ian Clough & Terry Sullivan 11 April 1959

Start right of the broken central fault. Climb direct to a holly, then leftwards to finish up slab and fine crack. Starting up the left-slanting slab further right is HVS 5a.

Karen Latter on the immaculate upper slab of Damnation.

ROAD BUTTRESS ⓢ 🧗 10min

NN 1548 6862 **Alt:** 110m

200m right of **Pinnacle Ridge**.

All routes cleaned 2021.

Approach from 150m east of the lay-bys below **Pinnacle Ridge**.

1 Sidewalk ★   30m Severe 4a

FA unknown 1960s

The blunt left rib, bypassing the roof on its left.

2 No Entry ★   25m VS 5a

FA Klaus Schwartz & Rebecca Morrow 3 June 1976;
Direct Start Ed Grindley 22 April 1978

Start just left of the groove of 3. Climb up to a curving crack and follow it to a heathery scoop.

3 The Web ★★   20m E2 5c

FA Klaus Schwartz & Blyth Wright (A2) 29 May 1969;
FFA Ed Grindley & Willie Todd 5 June 1976

The open groove with an obvious overhang halfway up. Layback round the first overhang on good holds and cross the top crux bulge directly.

3a Variation The layback crack of 4 can be gained by starting up 3 at ★★ E1 5b.

4 Wee One ★   25m E3 6a

FA Jim Mount & Brian Chambers (A2) 1973;
FFA Murray Hamilton & Kenny Spence 1980

Start just right of 3. Hard bouldery moves lead into a shallow left-facing groove. Up this and finish by a layback crack.

5 Withering Crack ★★   20m E3 5c

FA Klaus Schwartz & Jim Mount 11 October (aid) 1972;
FFA Ed & Cynthia Grindley 1978

The jam crack splitting two overhangs on the right side of the crag. After a bouldery start difficulties soon ease just above the first overhang.

6 Atree ★   20m VS 5a

FA Ed Grindley 31 August 1978

The groove to the right of 5 contains a tree. Reach this by a tricky wall and finish by a slab.

SECRETARIES' BUTTRESS

NN 151 688 **Alt:** 200m 🌥 S 🧗 25min

High up the hill, above and left of **Calvary Crack Buttress** and about 200m left of **High Crag, Lower Tiers.** Easily recognised from the road by a three tier steep left wall split by two oblique faults, with a slabby frontal face.

Approach: As for **Calvary Crack Buttress** then follow the path up the left side of the small subsidiary crag to the flat top of the main buttress. Continue slightly left and up past the left side of a further smaller buttress.

Descent: Down either side of the buttress.

❶ Ring of Fire Right-Hand ★★★ 25m E3 6a

FA Murray Hamilton 1984

Fine open wall climbing with a gymnastic start. Start at the top left end of the buttress. Swing athletically right to a large flat hold above the lip of the initial roof and make a hard move to good holds and up to a ledge. Break out right and follow an obvious line up the wall leading to the top crack on *Vincent* and finish up this.

❷ Pablo ★★ 25m E6 6b

FA Gary Latter 23 October 2016

Line up the right side of the top tier. Start up the fault of 3 for 5m to a projecting boss on the left. Pull round this leftwards into the hanging left-slanting groove and peg runner. Move left and up to good hold in quartz recess, then leftwards to join 1. Pull out rightwards and over bulge to rejoin that route and climb directly above to join *Vincent.* Pull straight up the wall above and continue directly up rightmost of parallel cracks to finish at a good flake.

3 Secretaries' Crack ★★ 25m Difficult

FA Jimmy Ness & Alan Burgon 11 May 1950

The fault splitting the second and third tiers of the crag gives one of the few deep chimney lines hereabouts. Follow this to a ledge on the front wall. Stays dry in the rain. Escape off right, or finish up the second pitch of *Secretaries' Direct*.

4 Verbatim ★ 18m Very Difficult

FA Gary Latter (solo) 11 October 2016

The right edge of the wide dyke of 3. Move up rightwards to good holds and continue up the rib.

5 Autumn Leaves ★ 20m VS 4c

FA Gary Latter (solo) 11 October 2016

Start at the same point as 4, but step down right to follow line of good holds rightwards to gain diagonal crack. Move up leftwards and trend left to finish up the right side of the rib.

6 Tilted ★ 25m HVS 5a

FA Gary & Karen Latter 23 October 2016

Start as for 4, but step down and follow the diagonal crack rightwards to finish at a tiny rowan.

7 Plagiarists ★★ 20m E4 6a

FA Gary Latter & Ewan Lyons 11 October 2016

The diagonal fault that splits the first and second tiers originates as a jam crack splitting the initial roof. Climb this through the roof and step right and up to ledge. Move up onto the wall and climb directly then move leftwards to a good jug just before prominent diagonal crack. Finish quite boldly with difficulty to the apex of the wall.

8 Last Word ★ 20m HVS 5a

FA Ian Clough 22 May 1959

Start at the lowest rocks. Follow left-trending ramps, then head direct to the right end of the right-slanting fault. Poorly protected.

9 Just Passing ★ 25m E1 5b

FA Ian Taylor & Neil Brodie 22 June 1989

Another sparsely protected pitch. Climb direct to the break of *Vincent*, then continue up the wall above via a pair of thin cracks to gain the arête. Finish easily up this.

10 Vincent ★★★ 55m E3 5c

FA Dave Cuthbertson & Ian Sykes 5 August 1981

Superb climbing on beautiful rock, taking a diagonal line across all three tiers. Start at the toe of the buttress.

1 **30m 5c** Climb ramp to join the start of the prominent diagonal crack. Follow the crack to its end at wide right-slanting fault. Make a long reach to get established on the wall above, then continue steadily to protection in diagonal crack near top. Move leftwards just below slabby edge to belay in recess.

2 **25m 5c** Pull round with difficulty to good diagonal crack on the wall. Follow the crack with interest until a traverse can be made to a ledge, then finish up the crack on the right.

11 Secretaries' Super Direct ★★★ 50m E1 5a

FA Alec Fulton & Klaus Schwartz 28 July 1969; p1 Klaus Schwartz & Jim Mount 19 May 1973

A fine exposed line heading for the left edge of the slabby face. Start below the corner of *Secretaries' Direct*.

1 **20m 4c** Move left across the steep slab to cross the overlap near the left edge and up this to the first ledge system.

2 **30m 5a** Follow the left-slanting line boldly to the second ledge. Cross the gap and finish up easier slabs.

12 Secretaries' Direct ★★★ 75m Severe 4a

FA Ian Clough & Eddie Buckley 21 April 1959

Excellent climbing, giving one of the best lines of its grade in the glen. Start below the shallow left-facing corner in the front face.

1 **15m 4a** Climb corner to the first ledge system. Move right 4m to belay below crack.

2 **30m 4a** Follow the central crack on superb quartz holds to the second fault, then slabs above.

3 **30m 4a** Climb the left-slanting ramp right of the edge, then continue up slabs near the crest of the ridge above.

13 Twitch ★ — 15m E2 5b

FA John Cunningham & party 29 October 1969

The thin slab midway between the second pitches of *Super Direct* and *Direct*. Unprotected, the difficulties easing as height is gained.

14 Right Wall ★ — 30m Very Difficult

FA unknown 1960s

Start from the first terrace. Climb the slab near its right end by the line of least resistance. Poorly protected. Further lines further right can be climbed at the same grade.

NAMELESS CRAG 35min
NN 1518 6866 **Alt** 310m

Lies at the same level, and immediately to the left of the skull of **High Crag, Upper Tier**.

Approach: As for **Secretaries' Buttress**, cutting up the right side of that crag.

Descent: Down the right side.

1 Electronic Omission ★★ — 20m E1 5b

FA Nathan Adam & Garry Campbell 11 June 2020

The left fault. Climb the left side of the quartz studded wall into a right-facing corner. Go up this to the roof and climb the excellent crack above to a short corner and the top.

2 Anode ★ — 20m E1 5b

FA Jamie Skelton & Morag Eagleson 1 June 2020

The right fault. Move up the lower wall to a small bay below the crux groove. Jam or layback this to the top.

3 Triode ★★ — 20m E5 6a

FA Dave Cuthbertson & Andy de Klerk 19 May 1987

Thin bold climbing up the blankest section of the face. Start beneath a shallow left-facing groove. Up this for about 6m then move right to a crack leading to horizontal break. Nut placement in slot above. Move left along the horizontal then up to a quartz hold. Continue on small holds to a series of undercuts heading out left to finish. Easy for the grade.

3a Triode Direct Finish ★★ — 20m E5 6a

FA Jamie Skelton & Morag Eagleson 1 June 2020

Technical, but only half a grade harder than the original. Stuff small gear in the overlap, step right and commit to the short headwall above.

 **Diode** ★★★   20m E2 5c

FA Brian Sprunt & C.Hill 29 July 1977

"Great well protected technical climbing." A thin crack splits the right end of the face. Climb this direct passing a narrow roof at mid-height.

 **Lightbulb** ★★   14m VS 4c

FA Morag Eagleson & Jamie Skelton 1 June 2020

The obvious crack-line. Climb up and onto a small ledge before stepping left into the crack. Follow this over a small bulge to the top.

HIGH CRAG   25–30min

NN 153 688 **Alt:** 300m

LOWER TIERS

The biggest crag at Polldubh, composed of three tiers of slabs. The lower tier is scrappy, mostly obscured by trees; the second tier 70m high and much better. The routes are a little bit dirtier than the more popular shorter routes lower down, but not unduly so.

Descent: Down the left side, skirting left (west) round a short undercut slab near the base.

 **Cervix** ★   30m VS 4c

FA Klaus Schwartz & Brian Chambers 11 October 1969

Start high up on the left wall above the gully and overlooking the subsidiary hanging slab down left. Go up the initial wall by a left-slanting crack. Finish up the fine steep slanting chimney.

2 Crag Lough Grooves ★★   135m E1 5b

FA Terry Sullivan & Ian Clough (3 PA) 23 March 1959;
FFA Allan & Jenny Austin 1962

Varied climbing, with a hard crux section through the steeper final tier. Start at the bottom left of the middle tier. Scramble to overhangs at 10m.

1 30m 5b Move up and traverse right across three small ribs to gain left-slanting dyke. Follow this to belay on small tree on grassy terrace. Pulling right round the roof into the dyke is well-protected E2 5c.

2 40m 4c Follow the crack up the edge of the slabs finishing by easier slabs on the right to a thread belay on the second terrace.

3 15m 5b Walk to the right a few metres. Cross the initial bulge and follow a prominent rightward trending gangway to a small stance around the rib.

4 20m 4c Follow steep groove above to slabs and a ledge.

5 30m Finish up easy slabs.

3 Kinloss Grooves ★ — 65m VS 4c

FA Ian Clough & Terry Sullivan 11 April 1959

Start at a tree below the first break in the overhang right of *Crag Lough Grooves*, left of a red wall.

1 **33m 4c** Climb to a small niche below the overhang at 9m, exit right and up to another niche. Continue up the slabs to a ledge.

2 **32m 4a** The slabs above, climbed anywhere, to the second terrace.

4 Autobahnausfahrt/Enigma ★ — 165m VS 4c

FA Klaus Schwartz & Brian Chambers (1 PA) 2 September 1969; pitches 4 & 5 Terry Sullivan & Ian Clough 11 April 1959

The longest route in the glen with pleasant and varied situations.

1 **35m 4a** Climb slabs near the right side of the base of the lower crag to the first terrace.

2 **15m** Climb the block and bulge above to a tree ledge.

3 **45m 4b** Surmount the overhang 5m above and head for a small groove and up this to the second terrace.

4 **35m 4c** Walk right and climb the steep slabs at the right end of the overhanging base, passing a heather groove at 8m on its left, to belay at twin birches.

5 **35m** Continue straight up over a bulge to easy slabs, climbable anywhere.

UPPER TIER

NH 1522 6883 **Alt:** 305m

Easily recognisable from the road, the steep left side of the crag bears a remarkable resemblance to a skull. The severely overhanging walls in the centre of the front face, either side of *Sky Pilot* give good, though limited hard **bouldering** above a wide grassy terrace. Worth knowing about for a rainy day – if you can bear the walk in.

Approach: As for **Secretaries' Buttress**, cutting up the right side of that crag to gain the left end of the terrace.
Descent: Down the left side.

5 Sky Pilot ★★★ — 30m E5 6b

FA Dave Cuthbertson 1981

The centre of the overhanging 45 degree wall is breached by a hanging crack, forming a block near the lip. Climb up and out on improving holds to the lip and step left to gain easy slabs. Bouldery.

6 Slatehead Slab ★★ — 30m E2 5b

FA Dave Cuthbertson & Gary Latter 27 October 1985

Right of the right eye (as you're looking at it) of the skull is a fine slab above a tree with a short wall at its foot. Gain the base by scrambling in from top right. From the tree traverse left then up the wall to pull left onto the slab at good side pulls. Follow the obvious line right across the slab to a good pocket at a break. Continue past another break to the top.

⟨📷⟩ Gary Latter on the superb first pitch of Vincent (page 229), Secretaries' Buttress. Photo Karen Latter

BLACK'S BUTTRESS  30min

NH 1546 6894; **Alt:** 310m

Situated at the same level and well to the right of **High Crag, Upper Tier** about 100m above **Pine Wall**. The buttress is composed of a vegetatious lower tier marked by a clean left edge. Immediately above, separated by a wide flat grassy terrace (obscuring much of the crag from below) is a smooth 30m slab of immaculate rock.

Approach: Follow the **Pine Wall Crag** descent path (up the left side). Continue up rightwards on a vague path.

Descents: Traverse right and down steep grass to right, or abseil from slings and maillons.

1 Zelos ★   60m Very Difficult

FA P1 Alec Fulton & Brian Chambers May 1969;
P2 Klaus Schwartz 15 February 1970

Climb a line close to the left edge.

2 Seven Fours ★★   30m HVS 5a

FA Ed & Rona Grindley 31 May 2009

Climb up to the nose just left of 3 and move up to small overlap. Step right and climb direct up the centre of the slab.

3 Shergar ★★   30m VS 4c

FA Ed Grindley, Cynthia Grindley, Pete Long & Gordon Higginson 13 September 1981

Move easily up to step right onto ledge, then direct up thin crack leading to easier climbing up hanging left-facing corner.

4 Land Ahoy ★★★   25m E3 5b

FA Ed Grindley & Davy Gunn October 1981

Brilliant climbing, sustained and unprotectable as far as the crack. Start beneath the centre of the smooth slab. Climb up to large flat hold at 4m, step left then climb direct to the crack in the upper half of the wall, then more easily up this.

5 Centrepiece ★★   25m E6 6b

FA Kev Howett / Gary Latter (both led) & Dave Griffiths 11 October 1987; FA sans PR Mark Garthwaite 1994

Superb fingery climbing directly up the centre of the slab. Start as for 4 to the large flat hold. Step right to tiny L-shaped hold (1m below pegs). Make hard moves up to pegs in thin horizontal break, then further hard moves past these to incipient crack and poor RPs. Gain a good flat hold just above, then continue more easily.

6 Kaos ★★ — 30m E3 5c

FA Klaus Schwartz & P.Logan (3 PA) 7 July 1968; FFA Kenny Spence 1980

Delicate and fingery with a poorly protected crux. Follow the obvious thin vertical crack then up and left with difficulty to gain a further crack and small ledge at 10m. Continue more easily in the same line to the top.

7 Desmo ★ — 25m E2 5b

FA Pete Long, Cynthia & Ed Grindley & C.Higginson 13 Sept. 1981

Climb the thin crack just right of 6, step right into hanging crack and up this to a recess near the top.

8 Crybaby ★★ — 25m VS 4c

FA Klaus Schwartz & Moira Horsburgh 29 June 1970; FFA Klaus Schwartz & Loch Eil party 1972

The prominent right-slanting cracks. Follow cracks past a small triangular niche to a ledge. Finish up the crack on the left.

9 Soho ★ — 25m E2 5c

FA Ed Grindley & Johnny MacLeod 17 September 2009

Good well-protected climbing. Start a few metres right of 8. Climb the wall, then direct up thin crack, then easy slabs on right to finish. 9a *Heulsuse* ★ HVS 5a moves left to finish as for 8.

10 Poeme a Loup ★ — 20m E2 5b

FA Ed & Rona Grindley, Johnny MacLeod & Peter Duggan 13 September 2009

Bold but steady climbing in the upper section. Start 3m right of 9. Climb up to protection in small overlap, then continue direct.

11 Oraisons ★ — 20m E3 5c

FA Ed & Rona Grindley 3 June 2009

Climb delicate slab to gain thin vertical crack leading to easier slabs.

12 Kyanite ★ — 20m E2 5a

FA John Taylor & Klaus Schwartz 1976

Unprotected climbing up blunt right rib of the crag.

Ewan Lyons almost at the first protection on the rather run-out Land Ahoy

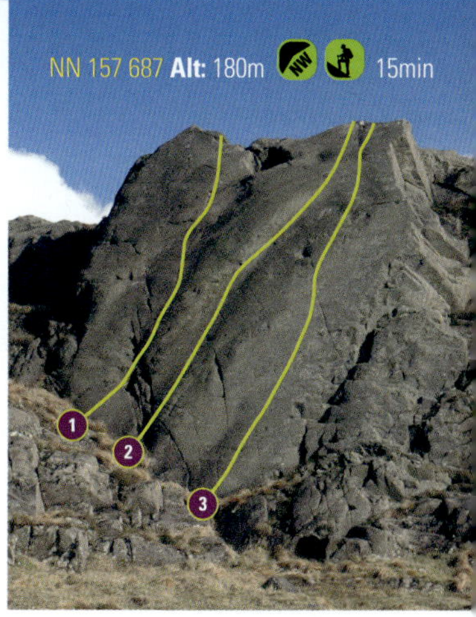

SCIMITAR BUTTRESS

The furthest right (eastmost) of the Polldubh crags, approximately 300m right of **Road Buttress** and slightly higher, on the rocky spur of Mam Beag (small breast).
Approach: From the small bridge 200m up the road from the lay-bys head slightly right up the hillside to the base.
Descent: Either scramble easily up or down the ridge to easy ground then down either side.

1 Wanderlust ★   35m Very Difficult

FA Ian Clough 22 April 1959

Climb the wide right-slanting fault at the right end of the crag then follow an obvious left-trending line across the top of the wall. Finish up a short rough slab.

2 Razor ★★   20m HVS 5a

FA Brian Sprunt & Andy Slater 1978

Start at the right end of the crag. From the base of 1 climb straight up to join 3. Follow this for a few moves then traverse left along a shelf to finish up the flake crack in a fine position.

3 Diagonal Crack ★   20m VS 4c

FA R.Wilkinson & Doc Pipes (7 PA) 8 April 1958;
FFA Alec Fulton & Brian Chambers 1968

Steep climbing up the prominent right-slanting crack. Well protected on good holds.

4 Nutcracker Chimney ★★   25m HS 4b

FA R.Wilkinson & Doc Pipes 8 April 1958

Good well-protected climbing up the shallow flared chimney at the upper left side of the crag.

UPPER SCIMITAR BUTTRESS

A short steep isolated gritstone-like slab 150 metres up the ridge from the lower crag. All three routes are unprotected.
Approach: Continue up past the lower crag for a further few minutes.

1 Sweet Little Mystery ★   10m E4 6a

FA Dave Cuthbertson (solo) June 1984

From the bottom left end of the crag, ascend diagonally rightwards to a steepening. Make an awkward step up to good break and finish directly.

2 Jahu ★★★   10m E6 6a

FA Dave Cuthbertson (solo) July 1984

Climb the right trending scoop in the centre of the slab in its entirety. Very thin and committing, with a definite crux at mid-height.

The right side of the slab is taken by 3 *Where the Mood Takes Me ★* E5 6a.

CAR PARK AREA

Between Polldubh and the road end car park are a number of crags, all on the lower slopes of Sgurr a' Mhaim, on the south (right) side of the road. **Whale Rock** is the first to come into view, across the river from where the road passes under a pair of distinctive Scots pines. Further on opposite the car park lies, not surprisingly, **Car Park Crag** itself, at 60m high, the largest steep and continuous crag in the glen.

WHALE ROCK    15min

A slabby front face, steepening and increasing in height towards the right end, where it forms a series of discontinuous scoops.

Approach: Cross the river at the bridge then the stream (Allt an t-Snaig) low down, near where it joins the river and follow a diagonal line direct to the crag from here.
Descent: Down either side of the crag, by steep ground.

NH 1621 6851 **Alt:** 170m

1 The Fascination Trap ★   25m E1 5c

FA Dave Armstrong & Dougie Borthwick 4 July 1984

A diagonal left to right line across the slab, with a bouldery start. Just right of the heather gully is a thin crack in the slab. Up this to ledges then follow the obvious slightly rising traverse line with a further hard move across the slab to reach good holds leading into the top of 2. Continue in the same line to the top right of the crag.

2 Earthstrip ★★   25m E2 5c

FA Dave Armstrong & Andrea Wright 7 September 1983

The central, widening crack line in the slab, easing towards the top.

3 Run for Home ★★   20m E5 6b

FA Kev Howett 28 February 1985

Thin sustained climbing on small edges up the diagonal

hairline cracks starting 3m right of 2 joining that route at half-height. Many small RPs required.

4 Hold Fast, Hold True ★★★★　　　　25m E10 7a
FA Dave Macleod (headpointed) 16 December 2002;
Jules Lines (solo) June 2013

A death defying line up the pristine wall. Start at the right edge of the small ledge (poor skyhook possible). Climb directly up the faint rib through a desperate technical and sustained crux to gain a tiny finger flake (no protection). Continue with further difficulty to a hard finishing move to gain a line of good edges leading leftwards into 2 – 4a *Hold Fast* ** E9 7a. Continue up the wall to reach a flake jug, then follow flakes up and right to finish.

5 Femme Fatale ★★★　　　　25m E8 6c
FA Dave Cuthbertson (headpointed) 27 July 1986

Very serious and technical climbing up the bulging scoops on the steepest section of the crag. Directly over the first bulge with difficulty (runners in opposition used to protect this – cam on ledge on left, and small RPs low down in crack on right) to a no hands rest and skyhook placement in first scoop. Move right and blindly place HB #2 & HB #1 in the thin crack to the right. Up rightwards into the second scoop, to gain a knee bar in the base of the flake (RP #1). Finish up the flake. The route awaits an on-sight ascent.

6 Just a Little Tease ★★★　　　　25m E5 6b
FA Dave Cuthbertson June 1984

Excellent, well protected climbing up the scoop and twin ragged cracks at the point where the crag bends around the hillside. Place a high runner on the right and a second opposing runner on the boulder at the foot of the previous route. Make difficult moves across the scoop to gain a good hold and protection. Attain a standing position on this (rest possible) before following the cracks which lead strenuously to the top.

7 *Midgiematosis* ★, E2 5c follows a groove and cracks just right; 8 *Strategic Midge Limitation Talks* ★, E3 5c a thin crack climbing through a Caledonian pine sapling with interest.

Just a little Tease, Andy Nelson climbing.

NH 1712 6883 **Alt:** 260m

CAR PARK CRAG  40min

The large buttress overlooking the car park, on the slopes of Sgurr a' Mhaim. It presents the highest vertical face of any of the crags in the glen, though others give longer, slabby routes, separated by grassy terraces.

Approach: From the car park, cross the river at a weir or boulder hop if the water level is suitably low. Otherwise, especially after heavy snowmelt in the spring (or for those with hydrophobia) cross the river by the footbridge about a kilometre further downstream, as for **Whale Rock,** then head steeply up the hillside directly to the crag. A pleasanter approach is to cross the River Nevis at the head of the gorge (either by boulder hopping or wading just upstream). Head up a good drovers track which zig-zags up the left of the crags then contour right (west) round the hillside for 200m to where the track levels out at a large flat terrace.

1 Gobstopper Groove ★★　　　　　60m E2 5c

FA Dave Armstrong, Al Murray, Murray Hamilton & Pete Whillance April 1981

Steep, well situated climbing. Start at the right side of a large tilted roof.

1　**35m 5b** Climb the crack and groove to a large depression. Exit on the right and up a quartz band to a terrace.

2　**25m 5c** Move onto the rib on the left and climb it by a shallow groove and thin crack.

2 **Quality Street** ★★★★ 70m E3 6a

FA Klaus Schwartz & Brian Chambers (HVS/A2) 20 June 1970;
FFA Pete Whillance & Murray Hamilton April 1981

Superb, sustained well protected climbing up the
tram-line cracks on the left side of the smooth wall in the
centre of the crag. Hard for the grade. Start on a raised
ledge behind a large rowan, directly beneath the cracks.
Cleaned 2023.

1 **40m 6a** Climb the easier initial crack which
leads to a ledge on the left, beneath the twin
cracks. Follow these (crux) to pull out left on
good holds. Traverse right to better holds in the
shallow, left-facing groove. Continue up this to
easier ground and a small ledge. Nut belay.

2 **30m 5a** Continue up the shallow right-
facing corner then slightly left and directly up
the easier slab. Tree belays further back.

Iain Small starting up Quality Street.

STEALL AREA

"... where crenellated crags tower"
– W.H.Murray, The West Highlands of Scotland, 1968
This section contains all the crags in and around the
Steall meadows, including the gorge. All are approached
from the car park at the end of the road (NN 168 692).

Both **Gorge Crag**, nestling in the trees and **Wave
Buttress**, directly above with a prominent quartz patch
in the centre, can be clearly seen from the car park as
can a number of crags on the hillside above.

EAS AN TUILL
THE NEVIS GORGE

This gorge has been described as one of the finest examples of its kind in Britain. *"The Nevis gorge, taken alone, has no counterpart in this country and is internationally famous. Its Himalayan character arises from a peculiar combination of crag and woodland and water, which is not repeated elsewhere in Great Britain."* – W.H.Murray, *Highland Landscape*, 1962.

From the car park at the road end, a path follows the true right bank of the river. The gorge itself extends for about 1.5km, the river level falling some 130m in that distance with the rocky side walls covered in native pine, birch, oak and rowan. Above the gorge the scene changes to one of Arcadian grandeur – the transition to flat meadowland is truly stunning.

GORGE CRAG

 20min

NH 1742 6905 **Alt:** 230m

The first crag encountered in the gorge, easily seen from the car park. A slabby left wall sweeps round into a steep and imposing front face, with a couple of corner systems bounding the right side.

Approach: Along the path, the crag squats in the trees about 30m above the path, above a wooden boardwalk.

Descent: Either by abseil from trees at the top of *Plague of Blazes* or *Travelin' Man* (slings & maillons usually in situ) or contour left (north) from top of crag and down easy angled ground just before the stream.

1 Brief Candle ★   30m E2 5c

FA Ed & Rona Grindley 30 September 2007

The rib. Climb steep wall, stepping left and up left-facing corner to ledge on left. Continue easily until rib steepens, then by a thin crack on the right to a ledge. Finish up slab above, stepping right then direct to clump of trees.

2 Plague of Blazes ★★★   30m E2 5b

FA Ed Grindley, Fiona Gunn, Noel Williams & Willie Lawrie 27 May 1982

An excellent pitch meandering up the centre of the slab. Start 5m up and left from the toe of the crag, at the leftmost of two thin crack systems running up the slab. Follow the zig-zag crack for 12m to a flake. Move left up into a recess; step right and up a slab to the final wall. Up this on good holds to finish at a Scots pine.

3 In the Groove ★★   39m E3 5c

FA Dave Cuthbertson & Gary Latter May 1982; p1 Gary Latter & Ian Campbell 24 July 1983

The diagonal steepening groove, trending right onto the front face, finishing in an exposed position.

1 **12m 5b** The shorter, right crack in the slab to the ledge.

2 **27m 5c** The groove above, and the subsidiary right groove past an obvious block (crux). Easier rightwards up ramp.

4 Travelin' Man ★★★   36m E2 5c

FA Dave Cuthbertson & Gary Latter 22 May 1982

Finely positioned climbing up the groove system splitting the rib between the left and front faces.

1 **12m 5b** Climb the cracked groove just right of the toe of the buttress to a ledge and belay.

2 **24m 5c** The groove above, stepping left (crux) to a crack in the slab. Up this and the easy ramp above.

> Scramble up right over some boulders (or avoid by a detour round right) to gain the base of the following:

5 Cosmopolitan ★★★★   30m E5 6b

FA Dave Cuthbertson & Gary Latter 20 – 21 May 1982

A fierce, well-protected technical test piece taking the wonderfully positioned hanging finger-crack in the headwall of the front face. Start below and left of a hanging left-facing groove in the centre of the overhanging wall.

1 **15m 6a** A bouldery start (crux) leads to a handrail (F #2 at right end) leading out right to the groove. Follow this, exiting out rightwards to a spacious ledge and belay.

2 **15m 6b** Gain the thin diagonal crack from the right with a hard initial move and up this to good holds just below the top. Brilliant.

6 Conscription ★★   20m E1 5b

FA Dave Cuthbertson & Gary Latter 19 May 1982

The obvious wide crack up the right side of the wall. Climb the crack then move right under a bulge to step right onto a ledge.

6a Confrontation ★★   30m E2 5c

FA Jamie Skelton & Morag Eagleson 8 June 2020

Superb exposed climbing up the steep wall above *Conscription*. Start as for that route to the point at which it goes horizontally right. Climb up and slightly left across the wall to gain a layback crack at the right end of the big ledge system. Continue up wildly on big holds to a ledge and finish either directly or via the left-leaning crack.

7 All Our Yesterdays ★★   20m E1 5b

FA Gary Latter & Dave Cuthbertson 19 May 1982

Good well protected climbing, often dry even in the rain. The corner crack, stepping out right onto a ledge.

8 Mini Cooper ★★   30m HVS 5b

FA Willie Jeffrey, Noel Williams & Pete Hunter 11 June 1983

Start beneath hanging groove at the left side of lower tier. Cleaned 2024.

1 **15m 5a** Gain the groove steeply and follow it to terrace.

2 **15m 5b** The fine square-cut corner

Ian Sheridan pulling through the crux section of Travelin Man.

9 Gorgeous ★★★   20m E7 6c

FA Jules Lines (headpoint solo) 20 May 2021

The left side of the arête gives a masterclass in the genre of pure arête climbing. Gear and better holds appear at half height.

THE RIVER WALLS     20min

NN 1747 6893 **Alt:** 200m

Extending left for about 100m from about 100m beyond **Gorge Crag**. Steep slabs and walls rise abruptly from the river. The start of the routes may be inaccessible when the river is high. The most obvious feature is a prominent steep rectangular hanging slab split by thin vertical cracks with a shorter, wider left-trending crack on the left. Both routes start from the boulder choke at the bottom left of the hanging slab rising from the water.
Descent: From the top of 1, scramble leftwards down grassy gangway; from the top abseil diagonally from block.

Harri Durban on a somewhat tilted Plague of Blazes – don't worry, it's a tad slabbier than it appears here! (page 241). Photo: Will Jones

1 Liquidator ★★   12m E1 5b

FA Paul Laughlan & Martin Macrae 4 May 1985

Traverse rightwards across a scoop to the base of the left trending crack. Follow this to grassy gangway.

2 Aquarian Rebels ★★★   25m E4 6a

FA Dave Cuthbertson & Gary Latter 23 April 1985

Excellent climbing up the thin cracks in the centre of the slab, directly above a beckoning pool. Cross the scoop to gain the leftward crack of *Liquidator*. Arrange protection a short way up this then traverse right to follow thin cracks to a finish up the final steep crack. A double set of small cams is useful.   Cleaned 2021.

WAVE BUTTRESS   40min

NN 1763 6899 **Alt:** 320m

Clearly visible from the car park on the slopes of Meall Cumhann above and slightly right of **Gorge Crag**. A relatively exposed position ensures a fast drying time and often a midge free haven when the gorge is unbearable.

The crag is generally slabbier (75–85 degrees for the most) than most of the crags hereabouts, and split into two buttresses by a heather-filled gully. The left buttress has two obvious crack-lines: the left one *On the Beach* fading out at a horizontal crack at mid-height, and the continuous left to right diagonal crack of *Crackattack*.

Approach: Follow the main path through the gorge into the Steall meadows then a steep zig-zagging path up the hillside left of the wide, open gully, left of the **Meadow Walls**. Alternatively, the old drovers track, avoiding the precipitous section of the gorge, affords a slightly more direct approach. This branches off left from the tourist path just over 100m beyond where the path has been blasted across a stream and rises initially before contouring around the hillside above the gorge to link up with the zig-zag path from the meadows below Wave Buttress.

Descent: Either by abseil, or by an indistinct path along the top and down the right side.

1 First Wave ★★   30m E2 5c

FA Ed Grindley & Noel Williams 20 April 1982

The left edge of the crag has a narrow right to left ramp with a series of ledges and short walls above. Start at some quartz blotches. Boulder diagonally rightwards to the foot of the ramp. Up this and follow a direct line to finish up a steep wall on good flakes.

2 On the Beach ★★★★   30m E5 6a

FA Murray Hamilton & Rab Anderson April 1984

Bold open wall climbing with spaced protection in the upper half. From the pedestal follow the crack-line moving slightly rightwards to the horizontal break. Step left and follow the shallow runnel above to better holds at some quartz. Continue more easily above.

3 The Edwardo Shuffle ★★★   30m E6 6b

FA Dave Cuthbertson 20 June 1985

Bold and sustained climbing up the rippled wall; a logical natural line. Start midway between the two cracks. Trend slightly leftwards to a tiny scoop then slightly right to the centre of the small overlap (HB 4 on side, pulling to left, in shallow horizontal above). Step left then make thin and committing moves to the horizontal break and protection (RPs #2 & #3). Move left along break and finish as for 2.

4 Ground Zero ★★   30m E3 5b

FA Ed & Cynthia Grindley, Noel Williams & Fiona Gunn 9 May 1982

Varied open wall climbing, with a bold lower section. Low in the grade. Start just left of the diagonal crack. Climb directly up the wall past some thin flared cracks (RP #4 behind a good hold at the steepest section), moving slightly right on good holds to the base of a diagonal crack/ramp. Ascend the crack in the groove past an awkward bulge and finishing up the quartz staircase above.

5 Crackattack ★★   30m E3 6a

FA Ed Grindley & Barry Owen 14 August 1983

The diagonal crack. Well protected. Crux at the top.

6 Bewsey Crack ★   30m HVS 5a

FA Ed Grindley & Noel Williams 23 May 1983

The shallow fading groove runs into a diagonal crack. Follow this into a sentry box near the top. Exit this on good holds.

7 Think Vertical ★   30m E3 6a

FA Ed Grindley & Barry Owen 9 August 1983

A fine natural line though it also happens to be a natural watercourse and therefore slow to dry. Start at the lowest point of the buttress, 6m right of the gully. Climb cracks leftwards into a right facing groove. Up this to a ledge and the steepening groove above (crux) to the top.

8 Walter Wall ★★   30m E4 6a

FA Kenny Spence & Spider McKenzie 28 April 1984

Serious though steady wall climbing up the shallow depression in the centre of the wall. Start at the toe of the buttress. Up this a short way then easily right across a scoop to good holds above. Boldly climb the wall above with no protection to a good horizontal break and protection. Hard moves past this lead to a shelf. Exit rightwards.

Iain MacDonald on the superb steady mid-section of Crackattack.

9 Edgehog ★★★ 30m E3 6a

FA Ed Grindley & Noel Williams 26 April 1982

No prizes for guessing which route this refers to. Gain
the arête from the right, and follow it to a large ledge at
its top. Finish easily up the slab. Better protected than
first impressions would suggest.

10 Teenoso ★ 30m VS 4c

FA Ed Grindley 1 June 1983

The corner crack. A bit grassy, but worth doing anyway.

James Sutton starting up the spectacularly-positioned Edgehog. Photo Mike Hutton

SPREADEAGLE BUTTRESS

Situated a few hundred metres above and right of **Wave Buttress**. The crag is characterised by a prominent arching stepped roof/corner system on the left, sweeping over a steep square lower wall, which tilts back to form a steep slab in its upper section. There is a short corner on the right, running out at a wide ledge halfway up.

40min

NN 178 689 **Alt:** 450m

Approach: As for **Wave Buttress** then by a vague path diagonally right, crossing the open gully just right of that crag.
Descent: By an easy gully and down the rocky shelf on the right side of the crag.

wall above, keeping just left of the arête. This leads with increasing difficulty past a poor horizontal RP #2 slot (Tri Cam #1? also) to hard moves pulling over onto a sloping ledge. More easily leftwards up a slab.

① Slip Away ★★   30m E3 6a

FA Dave Cuthbertson & Gary Latter June 1982

1 **18m 6a** The clean cut corner crack with the difficulties past an obvious smooth section on the right wall. Easier up the wide crack to a spacious belay ledge.

2 **12m 5b** Step left off the ledge and trend leftwards to finish up the final section as for 3.

③ Spreadeagle ★★   30m E4 6a

FA Dave Cuthbertson & Gary Latter June 1982

The shallow hanging groove in the centre of the wall. Start left of the groove beneath an obvious scoop/depression. Up into this, and make a hard move rightwards at a horizontal hairline crack to better holds at the base of the groove and finish up this. The upper groove is often wet, though usually avoidable.

② Chiaroscuro ★★   30m E7 6b

FA Kev Howett & Andy Nelson 10 May 1988

A varied pitch with much contrast; a hard and strenuous lower section leading to a bold upper half. It follows a fairly direct line up the blunt rib bisecting the groove of 3. Start beneath the scoop just left of the prominent corner-crack of 1. Up the scoop then climb leftwards past a series of vertical slots to an obvious projecting block on the arête. Climb into the hanging groove of 3 to break out left onto an obvious hanging block. Continue up the

④ The Singing Ringing Tree ★★   30m E5 6a

FA Dave Cuthbertson 9 July 1983

In the centre of the wall left of the hanging groove of 3 is a short crack, running out at a bulge. Start beneath this. Up the crack to a good nut slot near the top and pull right (sling on large spike hold, RPs in thin crack down and right to hold it in place) on improving holds to an obvious flake crack. Much easier up the wall above, trending leftwards. The difficulties are short lived and concentrated on the initial steep start.

NN 178 690 **Alt:** 500m

 40min

BLADE BUTTRESS

Above the right side of **Wave Buttress**, opposite and slightly higher than **Spreadeagle Buttress**. The buttress gets its name from the prominent hanging blade-like arête forming clean cut grooves on either side. Round to the right is a narrow, clean slab, sometimes referred to as a pillar.

Approach: As for **Wave Buttress**, continuing up the hillside, initially by an indistinct path, which veers off rightwards to **Spreadeagle Buttress**.

Descent: Make a 30m abseil from sling & mailllon on large block up and left from 1.

The first two routes start in the open bay beneath the blade. Start by scrambling up into this.

1 Cruisability ★★★　　　　　　　**20m E5 6b**

FA Gary Latter 12 July 1986

The stupendous overhanging groove up the left side of the blade. Gain and climb the flake crack leading to the roof. Pull through this with difficulty, using a good incut hold over the lip on the left. Continue up the groove, pulling out right to finish. Strenuous and well protected.

2 Ugly Duckling ★　　　　　**30m E2 5b**

FA Kev Howett & Alan Moist 20 May 1985

Good climbing with a bold start, following the arête and crack up the left side of the narrow slab. Start at a flake belay at the base of the wall directly below the thin crack. Up to roof, traverse left to the arête, and up this (crux) to gain a good ledge. From the right side of the ledge climb directly up the crack to the top. Friend belay next to a large flake well back.  Cleaned 2023.

3 Flight of the Snowgoose ★★　　　**30m E6 6a**

FA Kev Howett, Gary Latter & Dave Cuthbertson 21 May 1985

Thin bold slab climbing in its upper reaches. Flight would not be advisable. It climbs the centre of the narrow slab after sharing a common start with the fledgling 2. From near the top of the crack traverse right across the wall then up to small ledge (tied down skyhook runners). Ascend the wall with increasing difficulty to better holds just below the top. Belay well back. Cleaned 2023.

THE SPACE FACE

NN 178 690 **Alt:** 550m

An imposing crag on the right side of the open gully above **Blade/Spreadeagle** characterised by a distinctive frontal face seamed with a series of criss-crossing cracks.

Approach: As for **Wave Buttress** continuing up the hillside initially by an indistinct path then direct up the right side of the open gully.

Descent: Scramble back a short way then traverse leftwards and down the easy gully.

1 The True Edge ★★★   35m E5 6b

FA Kev Howett & Alan Moist 10 July 1986; Direct (as described) Gary Latter & Andy Nelson 17 June 1988

Spectacular climbing up the hanging arête in a stunning position. Very well protected after an initial bold start. Follow the obvious crack-lines up the centre of the wall avoiding the big easy flakes on the lower arête. Using good underclings gain a good flake and finger-locks in the base of the bottomless crack in the upper arête. Swing round and layback up the overhanging left wall of the arête to gain a good foothold in the crack. Swing back round passing a further tricky section which enables a good horizontal break to be reached. Saunter to top. The main difficulties can be avoided by climbing the finger-crack on the right at E4 5c.

Niall McNair tussling with Cruisability.

GALAXY BUTTRESS   25min

Beyond the **Meadow Walls**, and higher up the slope facing towards Steall Hut and the waterfall is a distinctive clean cut hanging slab above a long roof. Broken areas of rock lie on either side of this buttress.

Approach: Directly up the slope, leaving the path midway between the right end of the Meadow Walls and the wire bridge across the river.

Descent: Down the shallow gully/rake midway between the buttress and a narrow slab at the left side of the crag.

1 Short Man's Walkabout ★★★ 45m E5 6b

FA Dave Cuthbertson & Gary Latter 7 July 1983

Stunning, well protected climbing up a thin crack in the centre of the hanging slab. Start by scrambling to beneath a short right-facing corner and thin crack splitting the centre of the roof.

1 **33m 6b** Up to a PR at the back of the roof (F #1.5 halfway out) and cross this with difficulty to a good fingerlock over the lip. Up the crack (no hands rest possible out right) to a ledge and belay.

2 **12m 5a** Gain the scoop above, from the right.

STEALL HUT CRAG  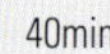 40min

NH 1764 6830 **Alt:** 320m

This impressive crag lies not surprisingly on the hillside behind Steall Hut to the right of the waterfall. Though mostly permanently dry, it is also a strong contender for the midgiest crag in Scotland – be warned! *Stolen*, *Leopold* and *Steall Appeal* are *"virtually permanently dry"* in Summer and can be climbed in heavy rain, but the finishes of the others take a while to dry out.

Approach: Cross the wire bridge at Steall and head diagonally up the hillside behind the hut.

1 Lame Beaver ★★★ 25m E7 6b

FA Kev Howett (2 PA) 31 May 1985; FFA Dave Cuthbertson 25 May 1987

A sustained pitch with sparse protection, breaching the left side. Start at the left end of the wall, about 2m from the left edge. Climb up past a shield of rock, heading for an obvious hold in the apex of the niche above (good small cam & R #3 behind this). Undercling the roof system rightwards with difficulty to gain good underclings in the niche on the right. Pull up and left using a good hidden pocket to jugs (crux). Superb climbing gains and climbs the easier flake crack in the headwall.

2 The Fat Groove ★★★ 30m F8a

FA Dave MacLeod 18 May 2011; Steallworker Dave MacLeod 12 August 2011

The compelling diagonal corner/overlap system. Continuing to the LO on 6 gives the stupendous *Steallworker* ★★★★ 40m F8b.

3 Stolen ★★★★ 25m F8b

FA Dave MacLeod 30 July 2007; Direct Start: Dave Macleod 2012

Brilliant and permanently dry climbing up the overhanging walls and bulges left of 4. Difficult climbing leads leftwards to the big tooth (no-hands rest). Follow the undercut groove above with more cruxes to a rest and yet another tricky section to gain the finishing headwall. Best F8b in Scotland?

4 Trick of the Tail ★★★ 25m F7b+

FA Mark McGowan 1 June 1989; retro-bolted May 2011

A stunning pitch up the diagonal crack/groove.

5 Arcadia ★★★★ 25m F8a

FA Gary Latter (red pointed) 20 September 1993; retrobolted October 2023

Two prominent diagonal cracks offer superb sustained climbing. Both share a common start. The leftward slanting crack, finishing up the final twin cracks.

6 Leopold ★★★★ 25m E7 6c (F8a+)

FA 1970s; FFA Murray Hamilton (protection placed on abseil & pink pointed) 21 July 1992

The wider right-slanting crack with a deviation out right on good undercuts at the obvious shield of rock. Follow good holds to pull out right into the crack. Follow this on sloping holds past two cruxes until possible to pull rightwards along a series of undercuts. Pass a final crux above a resting jug to easier ground. Continue more easily above. Bolts, then in situ wires, pegs and a cam.

7 Ring of Steall ★★★ 25m F8c+

FA Dave MacLeod 1 August 2007

Brilliant technical climbing, the completion of a longstanding project, bolted in the early nineties. Follow 6 to just beyond its crux, then move left to a big undercut. A heinous boulder problem leads to the base of the leftward slanting crack which leads with slightly easier climbing to the top.

7a Fight the Feeling ★ 22m F9a

FA Dave Macleod 2012

A more direct finish to 7. From the base of the crack beyond the crux, move up and right with desperate climbing on tiny edges.

8 The Gurrie ★ 15m F8a+

FA Dave MacLeod September 2010

The groove to the overlap, before making hard moves up and left to better holds.

9 Steall Appeal ★ 15m F8b

FA Malcolm Smith 19 September 1993

"I believe this is Scotland's hardest sport route" – Smith, 1993.

A hard boulder problem start followed by a powerful undercut move leads to some slightly easier climbing. Finish by lowering from the final bolt.

Tower Ridge. Photo Dave Cuthbertson, Cubby Images.

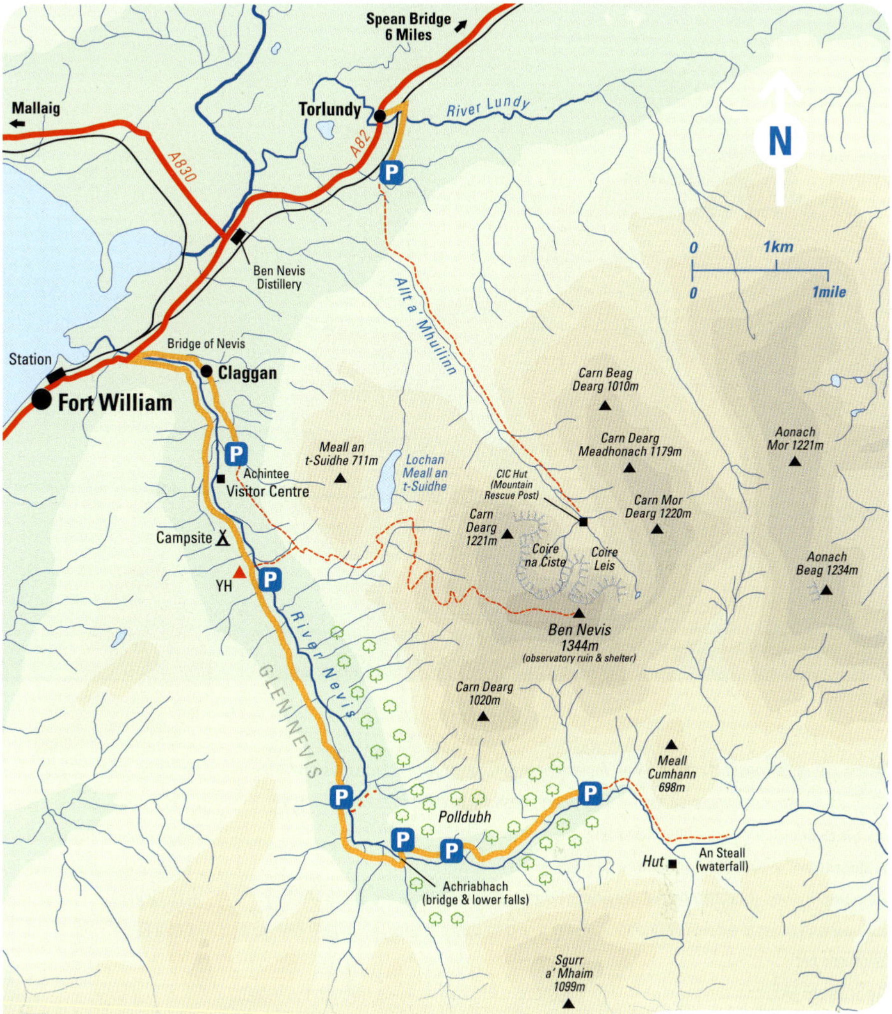

BEN NEVIS (THE BEN)

'Mountain with its head in the clouds' is perhaps the most likely possible translation of the name Beinn Nibheis. With the summit wreathed in clouds on average 200 days per year this interpretation certainly rings true. Avoid the summit if you can, as it is a bit of an eyesore with discarded litter left behind by the hordes of tourists that trudge up the pony track.

"Nevis, the most massive, malevolent, most elevated lump of rock on these islands, is itself an island, humping hideous flanks from endless bogs, hard to equal for hidden depths of character." – Jimmy Marshall

In the Groove, Scottish Mountaineering Club Journal, 1971

Access: Turn east off the A82 at Torlundy, 1.2 miles/2km north of the A830 Mallaig turn-off and cross the railway bridge. Turn right and follow track for 0.5 mile/0.8km to a large car park.

Approach: (A) From the car park, follow the steep path diagonally up through the woods to the upper (guides) car park by the dam. Continue less steeply up the left (north) bank of the Allt a'Mhuilinn (stream of the mill) leading to the CIC hut in around 1hr 40min.

(B) On foot from Fort William, either follow the pony (tourist) track starting from either Achintee or opposite the Youth Hostel in Glen Nevis, cutting off left (north) at the halfway lochan and following a path contouring round the hillside to the CIC Hut in 2hr 15min.

(C) Alternatively, cut through the distillery opposite the A830 Mallaig turn-off, cross the railway and follow a path up the right (south) bank of the Allt a'Mhuilinn for 500m, cross a small bridge and head steeply up a path to join up with the usual approach at the dam.

*"I've never seen a bigger hill,
I don't suppose I ever will,
But Everest is bigger still,
The very thought would
make you ill."*

– Doug Benn

The separate cliffs are described from **right to left**, as encountered on the approach.

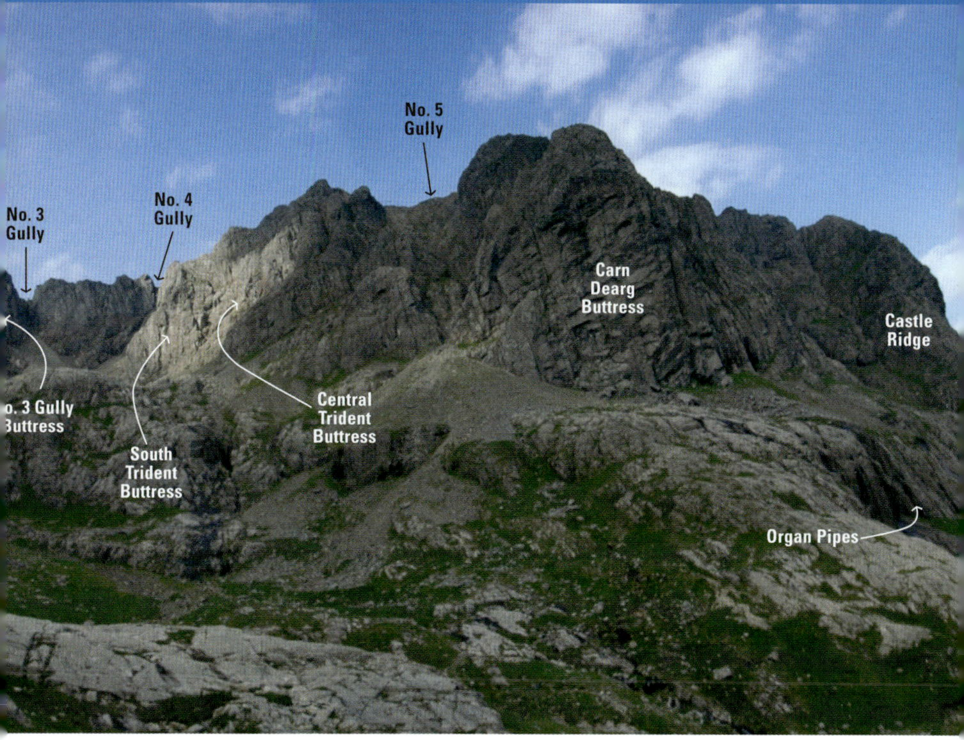

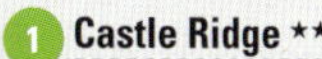

1 Castle Ridge ★★   270m Moderate

FA Norman Collie, William Naismith, Gilbert Thomson & M.W. Travers 12 April 1895

2hr

The northmost and easiest of the great Nevis ridges. Although not matching the quality of the 'big three' classic ridges (*Tower, Observatory and North-East Buttress*), being both lower down and further north and west, it comes into condition much more readily than the others, particularly in the spring when the Orion Face is still in winter condition. Start at a tongue of rock above 'the lunching stone', an obvious large boulder on the path from Meall an t-Suidhe. Wander up slabs to get established on the main ridge or gain the ridge by an obvious traverse above **The Organ Pipes** from the **CIC Hut**. Once established on the ridge itself, like all the other Nevis ridges the correct route is well marked by crampon scratches. Follow the easy angled ridge then a clean cut left-slanting corner leading to easy bouldery ground. A steep blunt tower is the crux starting by an awkward corner before moving up right to climb wide crack in the steep nose. A short chimney above marks the last of the difficulties. Continue along the crest then a large tilted slab to gain the final boulder slopes of Carn Dearg. **Descents:** Either **(A)** traverse a short way south west to avoid the North Wall and descend the tortuous boulder field down to the south end of Lochan Meall an t-Suidhe, or **(B)** from the summit of Carn Dearg contour south to gain the zig-zags on the touroid path, or **(C)** descend by *Ledge Route* on Carn Dearg Buttress.

THE ORGAN PIPES 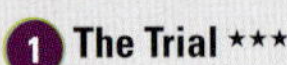 1hr 45min

NN 1647 7223 **Alt:** 715m

The long low crag situated below **Carn Dearg Buttress**, 10 minutes up right from the CIC hut.

① The Trial ★★★ 45m E5 6a

FA Wills Young, Andy Tibbs & Andrew Fraser 19 June 1988

An excellent pitch. Right of the main watercourse and at the right side of the crag is a dry area of buttress. At the right side of the steepest part is a corner which starts from a grass ledge. The route climbs the cracked arête on the smooth wall left of the corner. Start directly below the arête. Climb the clean wall to the arête, then the arête.

CARN DEARG BUTTRESS (RED CAIRN)

The showpiece cliff, the front face contains a series of slabs and overlaps, steepening on the right side of the buttress. Round the edge the north-west face presents a clean 100m high vertical to slightly overhanging wall with a fair concentration of high standard modern routes all finishing at a convenient central abseil point. The front face receives the sun until early afternoon; therefore an early approach is worthwhile.

Approach: Head directly up from the CIC Hut towards Coire na Ciste then follow a path cutting diagonally right across the screes to the base of the buttress. 15–20 minutes from hut.

Descents: Traverse leftwards down *Ledge Route* and down the lower reaches of **Number 5 Gully**, which often holds considerable quantities of snow well into the summer. This can be bypassed by an obvious fault on the left (west) side (about Moderate) or by continuing across a good path on the continuation of the upper shelf and down the lower reaches of Coire na Ciste.

Alternatively, in situ abseil anchors are in place on the belays of *Titan's Wall* (2 x 30m), which provides fast descents for routes from *The Bat* rightwards. 60m ropes reach ground on stretch!

① Ledge Route ★★ 450m Easy

FA J.Napier, R.Napier & E.Green 9 June 1895

An exhilarating route, the easiest 'climb' on the mountain, it is in reality no more than a simple scramble. Start up **Number Five Gully**, the large gully defining the left side of the buttress. This usually holds much snow until well into the summer. If so, climb a fault line up the buttress to the right. Traverse out right on the first prominent shelf (often damp – crux) and continue round this until beneath a shallow slightly left-slanting scree-filled gully. This leads to another wide shelf. Follow this rightwards until it eventually leads to the easy angled crest of the buttress. Finish up this airily to the plateau.

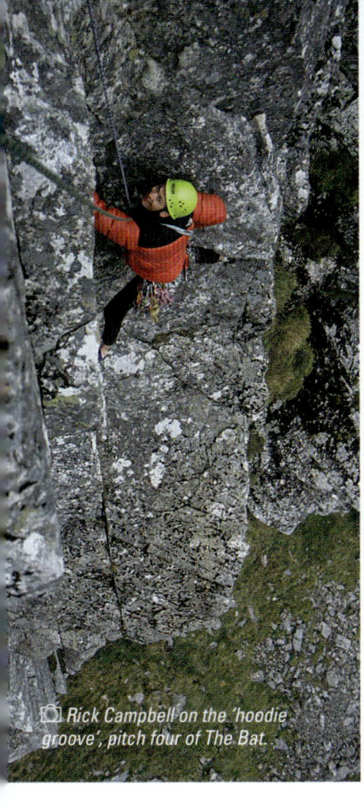

Rick Campbell on the 'hoodie groove', pitch four of The Bat.

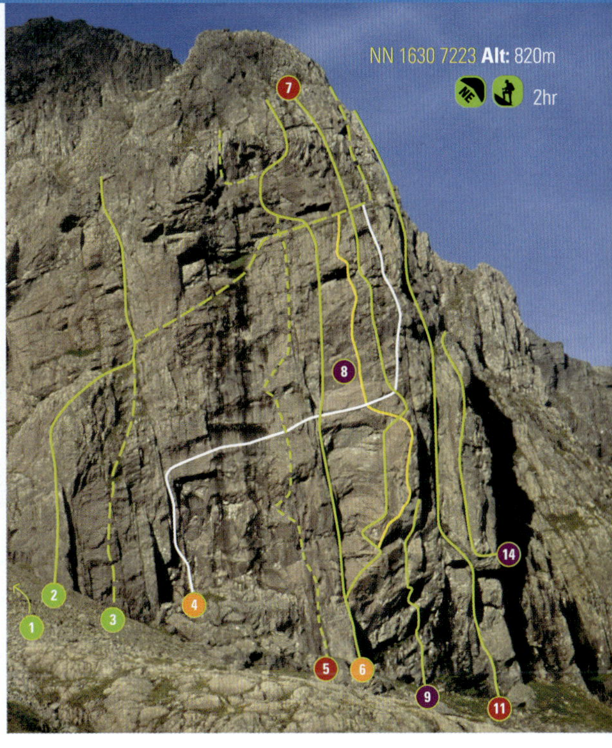

---

2 Route 1 Direct ★★　　　　　　　**180m Very Difficult**

FA Albert Hargreaves, Graham MacPhee & H.Hughes 17 June 1931; Direct Start Robin Plackett & W.Campbell 31 August 1941

The conspicuous chimney cleaving the left edge of the buttress. It provides a traditional struggle, saving its crux for the final pitch. Start to the left of the lowest rocks bounding the lower left flank of the buttress.

1　**30m** Climb just left of the right edge to belay by blocks.

2　**45m** Climb directly then a groove on the left to its top. Move right round the edge and up diagonal break then the wall above to belay on a second ledge.

3　**35m** Scramble to the top of the subsidiary buttress and walk right to the base of the chimney.

4　**25m** Grovel up the chimney (not as ferocious as it looks) then by a grassy groove to a recess. Go up the right wall to a belay.

5　**25m** Regain the chimney and climb it to belay on ledge above.

6　**20m** Continue up to gain the final chimney. This leads with interest to large block-strewn terrace. The path of *Ledge Route* lies just above.

3 Route II Direct ★★★　　　　　　**235m Severe 4a**

FA Brian Kellett & W Arnot Russell 9 June 1943; Direct Start Brian Robertson & George Chisholm 19 May 1962

A superb high level rising traverse across the buttress. Committing for the grade – there is much steep ground both above and below! Start at a cairn on a large wet mossy lump (spring) right of *Route 1 Direct*.

1　**30m** Climb up the centre of a smooth slab to a small ledge, traverse right 1m to a smooth wall and up a small slanting corner. Traverse left to a stance. Continue up a small black crack to a flake belay.

2　**15m** Climb straight up for 15m to a large block below a groove.

3　**10m** Climb the groove and move out right to a shattered ledge.

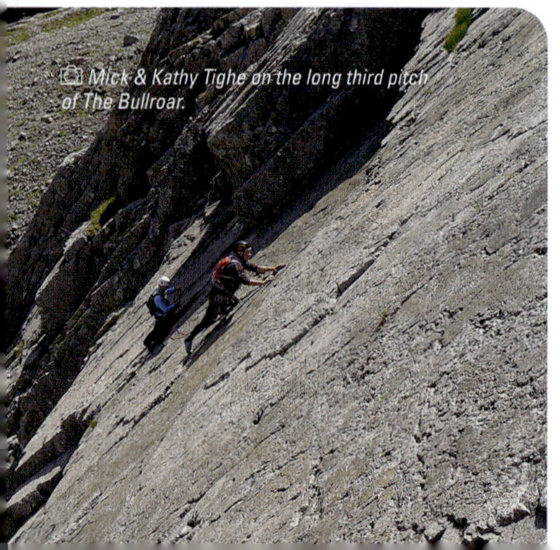

Karen Latter nearing the end of the long third pitch of The Bullroar.

4 25m Continue up broken ground to belay at the base of the chimney of *Route 1 Direct*.

5 15m Follow the chimney to a large chokestone at 10m then traverse out right on the slab to a small belay stance.

6 25m Continue out right to a large flake beneath the great overhangs.

7 10m Traverse the flake for 6m then up the rib to an inconspicuous thread belay low down.

8 40m Continue traversing the prominent fault to a platform on the edge of the buttress.

9 30m Scramble up the edge.

Mick & Kathy Tighe on the long third pitch of The Bullroar.

10 35m Follow the groove above mainly on the right wall to gain the crest of the buttress.

④ The Bullroar ★★★★ 300m HVS 5a

FA Jimmy Marshall & James Stenhouse 30 May 1961

A long route following a committing rising traverse line across the centre of the buttress. Excellent positions with considerable exposure – not for timid seconds! Start 30m right of *Route II Direct* at a bottomless groove 3m right of a right-facing corner with some boulders at its base. The wet streak on the main slab dries early in the day but returns by mid to late afternoon. The first pitch is slow to dry; thereafter the climbing until the end of the traversing is on immaculate rock. Many choose to continue easily out right and abseil down *Titan's Wall*, rather than continue up the dirtier and awkward route finding on the upper pitches.

1 45m 5a Climb thin crack into the bottomless groove and up this to a large flake. Move out left into the upper section of the *Direct Start*. Follow good holds up the left edge to belay at a thread 3m beneath the long overlap at the top of the central slabs.

2 20m 4c Traverse right along the slab to belay in crack forming the right edge of the roof.

3 40m 5a Traverse right across the slab, dropping down to a peg runner. Continue with a slightly descending traverse leading to a belay at the top of pitch 3 of *Centurion*.

4 **25m 4c** Climb the crack above then traverse right to a belay under the overlaps.

5 **30m 4b** Traverse right under the overlaps on excellent small flake holds to easier ground. Climb this to a large terrace above the chimneys of *The Bat* and *Sassenach*.

6 **10m** Traverse left from the left end of the terrace to an area of shattered rocks beneath an undercut groove.

7 **20m 4c** Climb the left-slanting groove to large ledge beneath undercut roof.

8 **45m** Negotiate the awkward slot to good holds and continue up the groove and continuation to grass ledge.

9 **50m 4c** Continue in the same line to large ledge with loose blocks.

10 **15m** Finish more easily up the slabby wall above the broken edge on the left.

Direct Start ★★   40m E1 5b

FA Will Hurford & M.Hancock 13 May 2001

An excellent sustained well protected pitch. Climb the obvious right-facing corner left of the original route to gain the belay at the end of pitch 1.

5 Torro ★★★★   225m E2 5c

FA John McLean, Bill Smith & Willie Gordon (1 PA) 25 July 1962; FFA Ian Nicolson & Ian Fulton June 1970

Excellent sustained climbing following a good natural line up the slabs left of *Centurion*. Start just left of the rib forming the left edge of the *Centurion* corner.

1 **30m 5b** Climb the steep initial groove to a flake then the groove above to a larger flake. Climb the right side of the flake and step back left and continue up groove to belay at a flake.

2 **30m 5b** Up a fault above then left and up a slim groove to PR on slab. Traverse left on a lip round the arête and up leftwards to belay on a ledge.

3 **25m 5a** Move up and right to a flake, step right and up rightwards up a slab to belay at an old relic.

4 **20m 5c** Follow a faint crack for 6m then step up left onto a higher slab. Climb up the edge of the slab to the roof, step left and cross this (crux) leftwards. Climb the groove above to a belay.

5 **35m 5a** Follow the fault for 5m then traverse right across the slab to a crack. Follow this to an overhang which is climbed trending leftwards, then by a groove to a grassy stance.

6 **20m 4b** Continue up the fault to a grass ledge beneath the long band of overhangs.

7 **30m 5a** Move up to the overhang then out left onto a steep slab, heading out left to another overhang. Step off a detached flake and delicately traverse left onto a large slab. Continue easily up rightwards for 12m to belay beneath the final overhangs.

8 **35m 5b** Traverse left into a steep little corner and follow this to a large grass terrace at the top.

Andy Stewart on the penultimate pitch of Centurion. Photo Dominic Oughton

6 Centurion ★★★★　　　　　　　　**200m HVS 5a**

FA Don Whillans & Bob Downes 30 August 1956

A magnificent route based around the great central corner – the classic line on the cliff. So named because the overlapping slabs on the right reputedly resemble the armour plates of a Roman centurion! The second pitch is slow to dry, as the corner faces west, though protection is plentiful – larger gear to 3" useful.

1　**15m 5a** Climb the left wall of the corner by an awkward crack to fine stance on top of the rib.

2　**35m 5a** Traverse into the corner and climb it to a belay on a slab in an overhung bay. A brilliant pitch – the holds and protection just keep coming!

3　**25m 4b** Traverse left onto the edge. Climb easy grooves until level with the lip of the big overhang then step back right onto the lip and move up to a stance.

4　**20m 4c** Move back into the corner. Traverse left up across the wall on flakes then climb cracks left of the arête to a block belay.

5　**50m 4a** Climb slabby grooves in the same line past a block then continue up easier ground to join the _Route II_ traverse. Trend diagonally left to belay on small muddy ledge.

6　**30m 5a** An intimidating exposed pitch. Move up to the overhang then out left onto a steep slab, heading out left to another overhang. Step off a detached flake and traverse left onto the slab on good hidden holds. Saunter rightwards up the fine rough slab above to belay beneath some blocks.

7　**25m 4c** Traverse right for 6m and climb a spiky arête to a bulge. Surmount this, step left into an easy groove and climb this to the terrace at the top.

7 King Kong ★★★★　　　　　　　**275m E2 5c**

FA Brian Robertson, Fred Harper & Jimmy Graham 1+2 Sept 1964: FFA gained from The Bat: Norrie Muir & Ian Nicolson June 1970

A fantastic long sustained route following an improbable and intricate line through the overlapping slabs right of _Centurion_. The original contrived start is seldom climbed (slow to dry) and the route is described by the more usual approach via _The Bat_.

1 **15m 5a** As for *Centurion*.
2 **15m 5b** Climb 6m up the corner then traverse right across the pink slab to belay at a perched block.
3 **35m 5c** Step left from the belay. Make a hard move up to gain a steep slab. Move up leftwards to gain an obvious line of weakness through the overlapping slabs and continue directly to reach an impasse at a large overlap. Undercling down and right to gain a thin crack springing from the lip of the overlap. Follow this diagonally rightwards across the slab above to belay at the right bounding rib.
4 **35m 5b** Climb directly up to gain a vertical crack in the red wall above. Climb this then trend up and right across the wall (bold) and up to belay at the top of *The Bat* corner.
5 **40m 5b** Move up from the belay then traverse left for 5m to the crack system left of *The Bat*. Climb this over a series of overlaps to move left to belay on a small grass ledge.
6 **25m** Climb easily by a grassy bay to a vertical wall then traverse left and up to a block belay beneath an overhanging wall capped by a small roof.
7 **40m 5a** Follow the crack through the roof and the continuation corner on the left to belay beneath a corner. *Route II* crosses here.
8 **40m 4c** Climb the corner, swing across to a spike then move left and continue up to a spike belay.
9 **30m** Continue slightly right then up to a grassy groove, moving left then up to finish.

Rick Campbell nearing the top of pitch 4 of King Kong.

8 Trajan's Column ★★★★   305m E6 6b

FA Rick Campbell & Gary Latter 7 September 2004
(pitch 2: 20 September 2003)

Fantastic varied climbing up the front face of the pillar right of the main pitch of *King Kong*.

1 **35m 5a** As for *King Kong* to the large perched block belay.

2 **25m 6a** Move left, up then back right to a position above the belay. Continue right and up to a perch at the base of the main slab and ascend straight up until it is possible to pull round an arête on the right onto a sloping ledge in a recess. (Spike belay on *King Kong* 5m higher.)

3 **30m 6b** A fantastic wildly-positioned pitch with devious wandering climbing and spaced protection. Step down right then make hard fingery moves to gain good holds in the huge 'boot flake' and more easily up this to a small overlap. Make committing moves out right and up rightwards to a good incut jug then straight up past a small useful flake to a rest on the right arête below a smooth uninviting groove. Traverse hard left into a parallel groove, which is followed to a long thin overlap. Commit right (scary) into the right groove (gear at last!) and make a hard stretch up right to a good hold on the arête. Continue back left to below the left side of the final overlap and pull through this, keeping close to the left rib leading to easier ground and the belay common to *King Kong* & *The Bat*.

4 **40m 6a** Traverse out left along the obvious fault line (*The Bullroar*) 10m beyond the crack of *King Kong* to the crack splitting the slab. Pull over the overlap and climb the slab to a slim smooth hanging groove in the steep wall. Lean out rightwards (crux) to gain two good incut jugs then directly up the fairly sustained groove above, belaying just above the final capping roof.

5 **15m** Climb easily up slightly left to belay at the base of a small right-facing corner near the right end of the big roof system.

6 **30m 5c** Climb the corner then move out left into a slim groove and up this then the fine rib to belay on a good ledge above.

7 **50m 4c** Move out right and up over vegetated ledges to a loose wet corner. Climb this to pull out right onto clean solid rock at a wide blocky crack. Move up left and scramble up leftwards to a block belay on ledge.

8+9 **80m** Move out right and climb layback/jam cracks up a slabby corner. Progressively easier scrambling leads up leftwards to the top of the buttress.

9 The Wicked ★★★   107m E6 6b

FA Gary Latter & Rick Campbell 29 July 2000;
6 July 2001 & 22 September 2001

Follows a very well protected series of cracks up the left wall of *The Bat*. Pitches 2-4 all overhang by 30 degrees. Start immediately beneath prominent twin finger cracks about 40m right of *Centurion*.

1 **20m 5b** Climb the lower slab direct, stepping left up a tiny groove to pull onto the first diagonal shelf on good holds. Belay at the base of the rightmost of a pair of cracks.

2 **12m 6b** Make a long reach off undercuts to a good spike hold then pull up into the crack and layback this past good flat holds on the right to a sloping hold at the top. Finish with difficulty to belay on the shelf above.

3 **20m 6b** Step up left into the crack and follow it with difficulty to good holds at the top. Pull over on good flakes then climb a crack in the arête, gained delicately from the right. Belay on the next sloping shelf 5m higher.

4 **25m 6b** Climb the flake crack in the corner on the left (*The Bat* descends here) then move right along a wide crack (Camalot #4 useful). Climb the crack on good holds with some hard fist jamming near the top (F #4) to a good flake at the top of the steep section. Continue more easily up the right-slanting groove to pull round the arête then easily up the shelf to belay near the base of *The Bat* corner.

CARN DEARG BUTTRESS

2hr

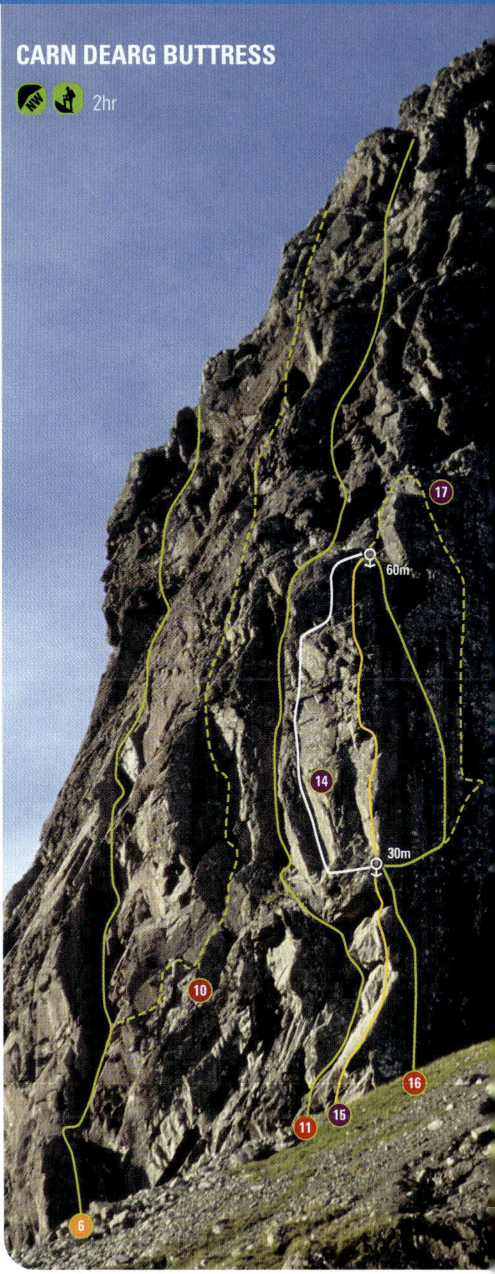

5 30m 6b Climb the corner for 5m then move left and arrange protection in the crack above. Step back down and hand traverse left along the ledge to step down into a short left-facing groove. Move up left to a good jug and arrange protection (F #0.5) in the crack above. Return to the centre of the ledge. Climb straight up on small edges then traverse left along a horizontal crack past a good undercut (crux) to gain better holds. Continue directly on improving holds and protection to pull out slightly leftwards at the top to belay as for *The Bat*. Either abseil down *The Bat* in two rope lengths or continue up the groove above as for that route, traversing right along the terrace to abseil down *Titan's Wall*.

10 The Bat ★★★   **275m E2 5b**

FA Robin Smith & Dougal Haston (some aid) September 1959; Independent finish Dougal Haston & Jimmy Marshall Sept 1959

A fantastic ultra-classic, restoring national pride. Immortalised in print and film, it still retains its once fearsome reputation.

1 15m 5a Climb *Centurion* pitch 1.

2 35m 5b Follow the *Centurion* corner for 6m then traverse right across a pink slab to a perched block. Continue moving right along a shelf to a block belay.

3 25m 5a Descend to the right for 3m to enter a bottomless groove and climb a short wall to a triangular slab. Follow the V-groove above then trend right along slabs to a belay beneath the left edge of the deep chimney of *Sassenach*.

4 15m 5b *The Hoodie Groove.* Climb a steep shallow groove on the left of the chimney then enter the main corner.

5 30m 5b Climb the corner to the overhang then launch into the wide corner crack above. Continue boldly to a ledge and belay, consciously trying not to emulate Haston: *"…as a black and bat-like shape came hurtling over the roof with legs splayed like webbed wings and hands hooked like a vampire…"* – Robin Smith

6 35m 4b Climb the groove to the left end of a large terrace. Either traverse right and abseil down *Titan's Wall*, or:

7-9 120m 4b Finish up the line of grooves above, belaying as required.

⑪ Sassenach ★★ 260m E2 5c with 2 PA (E3 6a free)
FA Joe Brown & Don Whillans (some aid) 18 April 1954;
FFA Steve Wilson & Colin Read August 1969;
Patey Traverse: Tom Patey, Bill Brooker & Mike Taylor 7 Oct.1953

An old-fashioned neglected classic following the great chimney-corner line running the length of the right side of the buttress. A wild roof. So named, as an understand-ably miffed Scottish team shouted up *"English Bastards!"* as Brown & Whillans were completing the route. Below and right of the corner there is a large slab of rock leaning against the face. Start just to the left of this.

1 25m 4c Climb the sloping mossy ledges for 4m until possible to step right onto a nose. Climb this on good holds then traverse easily left and up wide corner crack to a stance and belay at the top.

2 20m 5c Continue up the corner with interest to the overhang. Traverse left (often wet – two slings for aid or free at 6a) then up the grooves above,

moving leftwards to a stance and belay. *"Most sensational pitch I've done."* – Don Whillans

3 10m Up easily to the bottom of the corner.

4 15m 5b Climb the chimney past a tight constriction (crux) and belay where the angle eases. *"Good old-fashioned climbing"* – beware of several large loose spikes.

5 35m 5b Continue up the chimney to a large grassy terrace.

6 20m Move right across the terrace and scramble up easy rock to the foot of a V-groove capped by an overhang.

7 35m 4b Climb the groove for 10m then move out left onto a ledge. Continue up the crack above to enter another groove with a wide crack.

8 15m 4c Climb the groove and step right at its top.

9+10 85m 4b Climb the grooves above.

Note: The often wet crux pitch can be avoided by climbing pitch 1 of *Titan's Wall*, E2 5b then a 12m VS 4c pitch, the *Patey Traverse*, which traverses left round the arête and across slabs to a ledge at the base of *The Banana Groove*. Step down and left to a grass ledge at the base of the main chimney.

⑫ Antonine's Wall ★★★　　　　　　**50m E6 6a**
FA Iain Small & Tony Stone (both led) 30 July 2011

A good long sustained pitch on immaculate rock up the right wall of the *Sassenach* corner. Drift right to close to the right arête, then step left and up a bit leftwards to the centre of the wall, continuing fairly straight up to a depression in the wall. Finish up a crack/seam to the top.

⑬ Calgacus ★★★　　　　　　**60m E6 6b**
FA Iain Small & Rick Campbell 12 September 2009;
Pitch 1 Rick Campbell September 2006

The stunning arête right of the *Sassenach* chimney; well seen from the base of *The Bat* corner. The climbing is balancy, intricate and (apart from the initial arête) overhanging. Small nuts and as many tiny cams as you can carry. Bold.

1 15m 6b From the belay at the base of *The Banana*

Groove step left onto the slabby arête. Move precariously up this with runners placed (even more precariously!) in a seam on the right to where a bulging wall bars access to easier ground above. Hard moves are made up the wall to belay.

2 **30m 6b** Deceptive climbing up a slightly lichenous faint weakness right of the edge leads up to a scarred niche just left of *The Banana Groove*. The bulging rock above is avoided by bold and hard moves leftward around the arête onto the slabby left face of the arête. Continue delicately up this to a ledge on the edge itself.

3 **15m 6b** Move back onto the left-hand face again where more hard moves lead onto a step on the arête; side-step the final bulge in the same manner to a belay well back near the top of *Sassenach* chimney.

14 The Banana Groove ★★★　　　　107m E4 6a

FA Murray Hamilton & Rab Anderson 21 August 1983

Right of *Sassenach* are two prominent hanging grooves. This climbs the leftmost groove in its entirety.

1 **35m 5b** Climb pitch 1 of *Titan's Wall* to belay near the left arête.

2 **12m 4c** Traverse left round the arête and cross slabs to belay at the foot of the groove.

3 **45m 6a** An exhilarating pitch. Climb the corner above. Move up and right into the groove-crack and continue straight up the crack-line and ensuing shallow corner to pull out left onto a small sloping ledge.

4 **15m 5c** Continue up the groove to reach easier ground. Gain the spike abseil point on the right at the top of *Titan's Wall*.

15 Agrippa ★★★　　　　80m E5 6b

FA Pete Whillance & Rab Anderson 29 August 1983

The stunningly positioned arête forming the left edge of *Titan's Wall*.

1 **30m 5c** Climb the slabby rib for 15m to a small overhang. Pull over to a sloping ledge and up the groove above to a shelf below another overhang. Straight up to the ledge and

belay on *Titan's Wall*. A bit artificial.

2 **25m 6b** Move up left and climb a thin crack in the arête. Pull right to better holds and up to a large block. From a standing position on the block move round onto the left wall of the arête and up with difficulty to a good ledge and belay.

3 **25m 5c** Climb a slight groove on the right for a metre then up left across a wall to gain a flake in the arête. Climb this and continue on good holds, moving right to the spike belay on *Titan's Wall*.

16 Titan's Wall ★★★★　　　　80m E3 6a

*FA Ian Clough & Hamish MacInnes (A2) 19 April 1959;
FFA Mick Fowler & Phil Thomas June 1977*

Steep, varied climbing (a modern classic), precursor to a clutch of short high standard routes in the vicinity. Start 5m right of the left edge of the wall.

1 **35m 5b** Follow cracks to an overhang near the edge. Cross this and continue up the crack-line slightly rightwards to a long narrow ledge. Move left along this to in situ belay above the huge roof.

2 **45m 6a** Traverse right along the ledge to the crack splitting the centre of the wall. Launch up this with continuous interest, soon easing near the top. Continue to a spike belay at the top of the wall. An excellent well protected pitch.

17 Boadicea ★★★　　　　100m E4 6a

FA Willie Todd & Alistair Cain June 1989

The main pitch climbs the well protected thin crack up the impending wall right of *Titan's Wall*.

1 **35m 5b** Climb pitch 1 of *Titan's Wall* to belay near the left end of the long narrow ledge.

2 **20m 5a** Traverse to the right end of the ledge. From the foot of the crack on *Titan's Wall* follow a curving crack to a stance on *The Shield Direct*.

3 **40m 6a** Return left to the crack, which now becomes vertical. Climb it, thin at first to reach an easier section by some flakes. Continue with difficulty up the smooth section above to a belay.

4 **5m 5a** A short descent to the left leads to the abseil point at the top of *Titan's Wall*.

COIRE NA CISTE

This deep recessed coire containing twin tiny green lochans lies between **Carn Dearg Buttress** and the prominent projecting *Tower Ridge*. Between **Number 5 Gully** (bounding the left side of Carn Dearg Buttress) and **Number 4 Gully** lie the fine **Trident Buttresses**. **Approach:** From the CIC Hut head directly south-west up the slabby buttress into the corrie then bear right and up the lower scree slopes of **Number 4 Gully** to the base of the cliffs. 45 minutes from the hut.

CENTRAL TRIDENT BUTTRESS

A steep rounded wall of excellent rock in its lower reaches, though care should be taken on blocky ground higher up.

Descent: Traverse left and down scree and easy slabs bounding the left side of the buttress or make a 45m abseil from block (sling & maillon in situ) near the top of *Steam*.

Sheila Van Lieshout on Cranium.

1 Steam ★★ 90m HVS 5a

FA Stevie Docherty & Bobby Gorman Summer 1970

Start at an obvious right-facing corner groove near the left side of the face.

1 **25m 4c** Climb the corner to overhang which is passed on the left to a stance and PB.

2 **30m 5a** Climb the wall for 6m then move left to a corner. Climb this to a leftward-sloping ramp and belay in a greasy corner.

3 **35m 4c** Traverse diagonally up and right until a mantelshelf left onto a sloping ledge can be made (PR). Continue left and climb a steep wall to finish.

2 Heidbanger ★★ 80m E1 5b

FA Norrie Muir & Ian Nicolson 9 June 1970

Start 6m left of the base of the obvious crack system splitting the face.

1 **35m 5a** Climb a bulge moving right to the top of a steep groove at 6m and climb a short corner onto a band of slabs. Traverse rightwards to an arête and belay in a cave.

2 **30m 5b** Climb the hanging crack above then continue direct up the wall on fine incut holds to pull onto a ledge above. Continue up onto the next ledge and move out left to belay in a short corner.

3 **15m 4b** Climb the rib above on good holds then easily leftwards to a block belay/abseil on the left.

2a Cranium ★★ 20m E1 5b

FA Norrie Muir & Arthur Paul 19 June 1977

The prominent deep crack in the arête leading directly into the cave provides a good direct start to *Heidbanger*. Some large cams (up to 4") useful.

3 Metamorphosis ★★★ 100m E2 5b

FA Stevie Docherty & Davy Gardner August 1971

The best line on the cliff on superb looking rock. Start just left of a prominent overhanging crack at the right end of the face.

1 **35m 5b** Climb rightwards to join the crack below the bulge. Surmount this and continue to below a corner. Go right and up to another corner (PR) which is followed to a ledge and PB.

2 **25m 5b** Continue past a recess to a ledge. Go right along the ledge and up to PB beneath an obvious flake (poor stance).

3 **25m 4c** Climb the flake and wall above then trend right to easier ground.

4 **15m 5a** Finish by a corner and wall above.

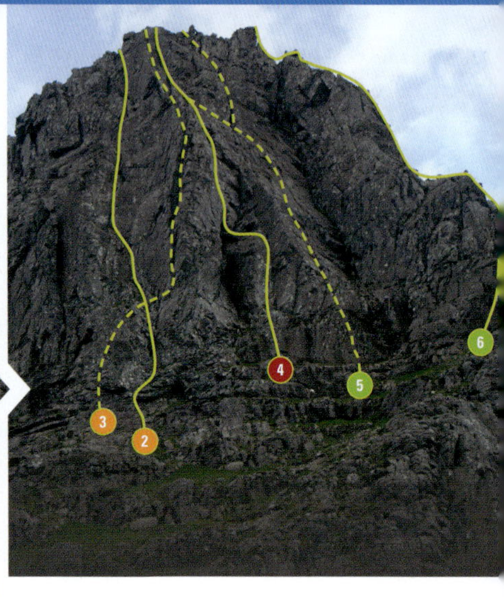

NN 162 720 **Alt:** 940m

2hr 20min

SOUTH TRIDENT BUTTRESS

The leftmost and most defined of the buttresses.
Descent: Traverse left and down the loose scree-filled
Number 4 Gully, which bounds the left edge. There is
a marker post at the top.

1 1944 Route ★★★   125m Severe 4a
FA Brian Kellett (solo) 30 July 1944

A superb route and a logical approach to the fine routes
on the upper tier.

1 **30m** Start by scrambling up to the base of the
steep wall about 30m right of the lowest rocks.

2 **25m** Go up left of the steep chimney
then trend left to a belay ledge.

3 **25m 4a** From a large leaning block climb to the base
of the second leaning groove from the left. Traverse
right to belay beneath the rightmost (fourth) groove.

4 **33m** Climb into the groove then by a crack in
the right wall to gain a system of ledges
which are followed to a short steep corner.

5 **12m 4a** Finish by the corner.

The following routes are all situated on the Upper
Tier of the buttress, gained either via *1944 Route*, or
scrambling in easily from the left.

2 Sidewinder ★★   75m HVS 5a
FA Jimmy & Ronnie Marshall & Andy Wightman June 1964

Look for a triple-tiered corner rising steeply leftwards
across the face 15m right of an obvious flake-chimney.
Scramble from the ledge to the foot of the corner.

1–3 **4b 5a 4c** Climb the corners by cracks in three
pitches (15m, 10m and 20m) to gain an easy slab.

4 **30m** Continue directly above by the continuation
of the crack and a large flake to gain the crest
of the buttress below the final tier.

3 Strident Edge ★★   75m VS 4c
FA Norrie Muir & Dave Regan July 1972

Originally dubbed 'Pink Dream Maker', but renamed by
the second ascentionists, thinking they were making the
first ascent. The prominent sharp arête on the middle tier
gives a pleasant and straightforward climb despite its
fearsome appearance. Start 6m left of *Sidewinder*.

1 **15m** Climb rightwards to belay in the corner.

2 **36m 4c** Move airily out right and climb the
steep cracked groove splitting the left
side of the arête to belay on the crest.

3 **24m 4b** Finish up the left edge of the arête.

4 Devastation ★★   80m E1 5b

FA Colin Moody & Andy Nelson 12 July 1995

Fine varied climbing up the steep cracked groove round right of *Strident Edge*. Named after the numerous trundles on the first ascent! Start beneath the second corner right of that route.

1 **40m 5b** Climb the corner until it ends at the flake; move left to a niche above the overhang. Climb the steep crack above.

2 **40m 4a** On up.

5 The Slab Climb ★   80m Severe 4a

FA Brian Kellet (solo) 30 July 1944

The crack up the left side of the slab. Start directly beneath the base.

1 **20m** – Ascend diagonally leftwards then traverse right and climb a short wall to the base of twin cracks.

2 **40m 4a** Climb the slab near the right crack towards an overlap, traverse left into the left crack and follow

it to a ledge. Continue up to negotiate a steep chimney-slot with difficulty then easier grooves above. The crux chimney can be avoided by a flake crack above an overhang on the left, finishing up the final straightforward groove, as for *Devastation*.

3 **20m** – Continue more easily leftwards to finish.

6 Pinnacle Arête ★★   150m Very Difficult

FA Harold Raeburn, Dr William & Jane Inglis Clark 29 June 1902

Fine climbing up the right edge of the buttress, overlooking the steep north wall and a good finish to *1944 Route*. Start from the right end of the ledge running beneath the middle tier. Move over sloping ledges for 3m then up an awkward corner on the right. 10m of difficult rock leads to easier ground. Trend left to the crest and climb to the base of a steep wall which is climbed direct on good holds. One further short steep section leads up to the narrow shattered crest and the final tier of the buttress which gives scrambling to the summit plateau.

Strident Edge.
Sheila Van Lieshout climbing.

NUMBER THREE GULLY BUTTRESS

NN 160 715 **Alt:** 1050m 2hr 40min
Just left of **Number 3 Gully** with the obvious groove of
The Knuckleduster clearly visible from the CIC Hut.

Approach: From the CIC Hut head up into Coire na Ciste
as for the Trident Buttresses then head up scree to the
base of the cliff. 1 hour from hut.

Descent: Down the right bounding scree-filled
No. 3 Gully, or the more straightforward **No. 4 Gully**
(NN 158 717) 300m further right (north-west) along the
rim, which contains a metal indicator post at its top. Both
retain much snow well into the summer. Alternatively the
top of *Ledge Route* (large cairn) lies 400m further north
than **No. 4 Gully** and will always be clear of snow when
the cliffs are in summer condition.

1 The Knuckleduster ★★   120m E1 5b

FA Jimmy & Ronnie Marshall 4 September 1966

Great climbing up the groove on the steep right side of
the buttress, though slow to dry. Start by scrambling up
to belay at the base of the groove.

1 **40m 4c** Follow the groove to belay
beneath an overhang.

2 **15m 5a** Turn the overhang by a slab on
the right to belay on the outer edge.

3 **35m 5b** Regain the groove by traversing
left along horizontal ledge. Continue by
a crack on the right wall to a ledge.

4 **30m 4c** Climb the wall on the right to gain the
large platform of *Number Three Gully Buttress*.

2 Last Stand ★★   90m HVS 5a

FA Doug Hawthorn & Arthur Paul 2 July 1984

Steep and exposed climbing up the arête right of 1 *The
Knuckleduster*. Start below the prominent groove of 1.

1 **30m** Follow a diagonal line of weakness
up and right to the foot of the arête.

2 **40m 5a** Climb by thin cracks on the right
side of the arête to a ledge and belay.

3 **20m 5a** Continue directly by cracks, cross a bulge
then finish more easily up further cracks.

3 Sioux Wall ★★   90m HVS 5a

FA Ian Nicolson & George Grassam 4 September 1972

An excellent route on good rock, following an obvious groove
line just left of the centre of the steep face up right of 1.

1 **25m** Follow a diagonal weakness up and right. Climb
past a large rock fin and up to a square-cut niche.

2 **30m 5a** Step left onto the steep wall. Climb this to
a crack then move left onto a small ramp which
moves back right to the base of the obvious
corner/groove. Bridge up this to below a roof.

3 **20m 5a** Climb the steep crack above (hard to
start) to a small overhang. Pull directly over
this on large holds to a ledge and belay.

4 **15m** Climb the continuation crack to a large
platform. Scramble to the top of the buttress.

DOUGLAS BOULDER   2hr

NN 167 719 **Alt:** 800m

A good direct approach to the base of *Tower Ridge*. Being one of the lowest lying buttresses on the mountain it is relatively quick drying and is often clear of the cloud when many of the other cliffs are shrouded in mist.

Approach: Follow a path heading south from the CIC Hut for 15 minutes to the base.

Descent: Abseil (slings in situ on blocks) or down climb down the back into the Douglas Gap and continue up *Tower Ridge*.

Direct Route ★★ 215m Very Difficult

FA William Brown, Lionel Hinxman, Harold Raeburn & William Douglas 3 April 1896

A good approach to *Tower Ridge*, following the groove splitting the centre of the front face. Start at the base of the lowest rocks. Climb easily up the large open groove until it steepens into an open chimney. Follow this for about 60m to a good ledge. Traverse right along this then ascend steep broken rocks to the top.

TOWER RIDGE 2hr 15min

Tower Ridge ★★★ 600m Difficult

First Descent Jack, Edward, Bertram & Charles Hopkinson 3 Sept 1892; FA (in winter conditions) Norman Collie, Godfrey Solly & Joseph Collier 30 March 1894

A well trodden classic, the most popular of the classic ridges. Once the base is gained the way is obvious – just follow the crampon scratches! Approach by skirting left round the base of the Douglas Boulder and cutting up easy broken ground on the right to an easy-angled loose fault leading up to the Douglas Gap. Ascend a polished 15m open chimney to gain the crest. Follow this easily to a short steep wall and outflank this by a ledge leading up rightwards. A further easy-angled section leads easily to another short step, the Little Tower. Climb this by a choice of lines on good rough rock then another long almost level section leads to the Great Tower. This is again outflanked, this time by the Eastern Traverse – a metre wide ledge running out left round the corner. This leads to a deep chimney capped by a huge chokestone. Climb the chimney (15m) then continue out left then steeply on good holds to the top of The Great Tower. Follow the fine narrow crest which leads to the highlight – Tower Gap. Drop down awkwardly for 3m to the base of the gap (or jump the gap if you're mad enough!) then more easily up the other side. Continue easily, turning the final steepening by a ledge and groove on the right leading abruptly to the summit plateau.

OBSERVATORY RIDGE   2hr 15min

NN 168 716 Alt: 920m

The long narrow buttress projecting north from the summit sandwiched between two of the most famous winter routes around – *Zero Gully* and *Point Five*.

Approach: From the hut follow a good path up beneath the base of the Douglas Boulder, continuing slightly leftwards up scree to the base. 30 minutes from hut.

3 Observatory Ridge ★★★ 450m Very Difficult

FA Harold Raeburn (solo) 22 June 1901

Completes the trilogy of classic ridges with excellent clean continuous rock, particularly in the lower section. The hardest of the ridges, especially the lower third. Mostly of Difficult standard, especially the upper section (after about 180m), where the standard and angle eases considerably. The best route is well-scratched and obvious. Start at the lowest rocks. Scramble easily rightwards then traverse right to the right end of a grassy terrace (65m). Continue just left of the crest by slabs and walls to steeper ground which is outflanked easily on the right. Follow cracks and grooves to gain the easier angled crest of the ridge which leads to the plateau.

NORTH-EAST BUTTRESS

This is the huge buttress projecting north-east from the summit plateau. Its north-west flank contains **Minus Face** and **Orion Face** – a great sweep of slabby buttresses rising up out of the usually snow-filled depths of the huge **Observatory Gully**. Separated by **Minus One Gully** and **Minus Two Gully**, the Minus Faces are numbered from right to left starting from the pronounced **Zero Gully** which separates the Orion Face from the broad *Observatory Ridge*. Routes below the First Platform are fairly quick drying with little run-off.

Approach: From the hut follow a good path up the right side of the burn up towards Coire Leis, which leads underneath the base of North-East Buttress. For the Minus and Orion Faces bear off right and up to the base of the chosen cliff. 20-30 minutes from hut.

Descent: For routes finishing on or near the First Platform, descend the easier lower section of **North-East Buttress**. For routes ending on the plateau, the quickest descent is to follow the path south-east down steep blocky ground (marker posts) to the abseil posts at the bealach at the top of Coire Leis and descend easy scree slopes (path) back down into Coire Leis.

NN 170 718 Alt: 850m    2hr 10min

4 North-East Buttress ★★★ 300m Very Difficult

FA Jack, Edward, Bertram & Charles Hopkinson
6 September 1892

Another great classic ridge. Start up left from the base of the ridge and traverse left beneath the lower rocks to gain a wide ledge which leads up right to the crest of the ridge. This leads to the First Platform, which can also be gained more interestingly by either *Raeburn's Arête* or *Bayonet Route*. Follow the crest of the ridge to a steepening and turn this by a shallow left-slanting gully. Trend rightwards up short chimneys and walls to gain a broad sloping shelf on the crest – the Second Platform. Continue up the narrow well defined ridge until a smooth overhanging wall bars progress. Outflank it either by a corner with large steps on the right, or more direct by following a right-slanting ledge to cross a bulge on good holds. The infamous Mantrap lurks a short way above. Attack this direct then continue by the '40 foot corner' which can be climbed direct or avoided on the left. Follow a small gully above which leads to easier rocks and the summit plateau.

5 Raeburn's Arête ★★★   230m Severe 4a

FA Harold Raeburn, Dr William & Jane Inglis Clark 30 June 1902

Fantastic climbing up the clean arête formed by the north and east faces of the buttress. Start directly beneath the arête, at the lowest rocks.

1 **20m 4a** Move up to a black overhang, bypass this on the right and continue to belay on a grass ledge.

2 **35m 4a** Climb the arête to belay on a ledge.

3 **40m 4a** Traverse 6m right then climb up to regain the arête as soon as possible.

4–6 **135m** Continue, keeping as close to the arête as possible to the First Platform.

6 Bayonet Route ★★   160m Very Difficult

FA Graham MacPhee & A.Murray 30 September 1935

Start from a large grass platform roughly midway between *Raeburn's Arête* and the right bounding chimney of the buttress.

1 **45m** Ascend a rib of rough rock veering slightly left to a grass niche at 20m. Traverse left across a rib and continue up a grassy groove to a belay.

2 **20m** Go out left onto the rib and follow it to beneath the left edge of the main overhang.

3 **25m** Gain the rib on the left and follow it until a right traverse leads into a grassy bay. Continue up the rib on the right of the bay.

4 **30m** The rib above.

5 **40m** Climb a corner to pull out left just above a prominent square-cut overhang. Continue by easier ground to gain the crest of the buttress. An ascent including the following two variations gives an excellent combination:

6a Direct Start ★★   35m Severe 4a

FA Jimmy Marshall & James Stenhouse August 1959

From the left end of the grass platform, follow the prominent corner, join the original route beneath the overhang.

6b Main Overhang Variation ★★   35m HS 4b

FA Brian Kellet (solo) May 1943

Bold and committing. From the grassy groove, climb a slab to a point beneath a V-notch in the main overhang. Gain this by cracks, exiting on the right. Trend left above the overhang to regain the original route.

THE MINUS FACE

NN 168 716 Al**t**: 950m  2hr 10min

The north-west flank of North-East Buttress contains The Minus and Orion Faces – a great sweep of slabby buttresses rising up out of the usually snow-filled depths of the huge almost corrie-like **Observatory Gully**. Separated by **Minus One Gully** and **Minus Two Gully**, the Minus Faces are numbered from right to left, starting from the deep pronounced **Zero Gully** which separates the Orion Face from the broad *Observatory Ridge*. All the routes finish at various levels on *North-East Buttress*, which can either be ascended or descended to finish.

MINUS TWO BUTTRESS

7 Left Edge Variation ★★★   80m VS 4c

FA N. & Bob Richardson & Alastair Walker 11 June 1988

An excellent start to *Left-Hand Route* providing better and more sustained climbing than the original lower pitches. Starts as for the Clough variation, but climbs the edge overlooking **Minus Three Gully** instead of the big groove.

1 **30m 4b** From the foot of the V-groove move out onto the left rib to climb up and left to a belay at flakes in a niche on the edge of the buttress.

2 **40m 4c** Descend right a metre or so then move up and right to climb a bulging groove then a further groove on the very edge of the buttress to a fine stance in a little bowl below an imposing wall.

3 **10m 4b** Pass this wall on the left then go up and right to belay above the crux of the original route. Continue as for the parent route.

8 Left-Hand Route ★★   270m VS 4c

FA Brian Kellett, Robin & Carol Plackett 20 June 1944

Good climbing up cracks splitting the front of the
prominent raised crest of the buttress.

1+2 60m 4a Ascend the cracks splitting the
    front of the prominent raised crest of the
    buttress to gain a ledge at the base of a
    slab to the left of some stepped roofs.

3 10m 4b Descend a slab from the left end of the
    ledge and traverse left to a second slab which
    is climbed on small holds. The slabs above
    the ledge can also be climbed direct at 4c.

4 40m 4b Climb the slab keeping close to the left
    edge, then a groove to step right to belay.

5–8 160m Continue easier up more broken ground, better
    on the left, to gain the crest of *North-East Buttress*.

MINUS ONE BUTTRESS

9 Minus One Direct ★★★★   315m E1 5b

FA Bob Downes, Mike O'Hara & Mike Prestige (1 PA) 11 June 1956;
Serendipity Variation: Ken Crocket & Ian Fulton 27 August 1972;
Arête Variation: Stevie Abbott & Noel Williams (1 PA) 28 Aug 1983;
FFA Willie Jeffrey & Noel Williams 6 July 1984

A fantastic route *"of almost alpine proportions"* on
superb rock throughout. When combined with the
Serendipity and *Arête Variations* it provides one of the
finest routes of its grade in the country. Start at the
lowest point in the centre of the buttress.

1 25m Scramble easily to belay
    under a left-facing corner.

2 25m 4b Climb the corner then exit
    right onto a glacis and belay.

*Jake Oughton on Serendipity Variation (pitch 5) of
Minus One Direct. Photo Dominic Oughton*

3 **45m 4c** Ascend a shallow groove in the wall above to a block at 6m. Pass this by a crack on its right (it is also possible to climb on its left) then climb by short walls, moving left to a niche. Step right and climb to the top of a vast plinth.

4 **30m 5b** The crux pitch of the original route; a sustained and committing lead on perfect rock. Climb straight up for 5m, then step right onto the nose above the overhang and climb it until it is possible to follow a ramp rightwards to finish at a block belay on a platform overlooking the final chimney of Minus One Gully.

5 **25m 5b** Traverse up and left until a hard move leads to a recess (crux). Climb this and continue to belay in a small grassy niche.

6 **20m 4c** Step left into a further recess and climb its slabby left wall. Climb the steeper rocks above until it is possible to break out onto right-trending slabs. Belay at a stack of detached blocks.

7 **20m 5b** On the left is a prominent slab capped by a long narrow overhang. Gain a foothold in the centre of the slab (crux), move up to the overhang (or climb straight above the belay to gain the traverse line – straightforward and more logical!) then traverse left on underclings to the arête. Cross the overlap with difficulty to a small exposed stance.

8 **40m 4c** Climb a crack in the crest above then continue more easily to a groove. Bridge strenuously up this to reach the great terrace.

9 **40m 4b** Climb to the top of a 12m pinnacle then continue easily to a leaning pedestal.

10 **45m** Finish up the finely-situated knife-edge arête to join *North-East Buttress* above the Second Platform and finish up that route.

THE ORION FACE

A vast mainly broken V-shaped slabby face extending from **Minus One Gully** rightwards to **Zero Gully**.

NN 169 717 **Alt:** 950m  2h 20min

10 **The Long Climb** ★★★   410m VS 4b

FA Jim Bell & John Wilson 14 June 1940

Aptly named, this offers the longest vertical face climb in mainland Britain, of almost alpine stature. (Though *The Chasm* on Buachaille Etive Mor is longer, much more sustained and an altogether bigger day). An extensive snow patch usually lies in the basin until late in the season. It is usually better to skirt round the rocks on the left side of this to gain the base. Start to the left of the base of Zero Gully, by scrambling up to a large platform.

1 **50m** Go up an easy-angled ochre rib left of Zero Gully to a small platform.

2 **45m 4a** Follow a rib leading up from the left end of the platform to the base of the distinctive Great Slab Rib. Step round the rib on the left and move up to its base, or climb either directly up the rib or the groove on its right (both 4c).

3 **30m 4b** Traverse right onto the crest of the Great Slab Rib and follow fine parallel cracks to belay in a recess. An excellent pitch.

4 **45m** Head out and up right then climb easily to The Basin.

5 **40m** Continue easily across The Basin heading up to the base of the distinctive Second Slab Rib, which bounds the top right side of The Basin.

6 **40m 4b** Climb the slabby left edge of the rib, bypassing a steepening on the left (4c direct), then a short awkward wall either on the left or right to a belay.

7–10 **160m** The original route trends up slightly left (harder ground lies further left), aiming for the base of another 60m high Great Slab. Avoid this on the right by a line leading to a niche then a short difficult pitch gains the top of the slab. Continue easily to gain the crest of *North-East Buttress* and finish up this. Much variation is possible in this upper section.

Daviot
Huntly's Cave
A82
Loch Ness
A939
Loch Duntelchaig
Duntelchaig
A9
Grantown-on Spey
B851
Brin Rock
The Camel
A938
A95
B862
A95
A82
Stac Gorm
B862
Aviemore
A9
B970
Fort Augustus
Comic Crags
Farrletter Crag
Kingussie
Cairngorm Mountains
Newtonmore
Creag Dhubh
Laggan
A86
Dirc Mhor
Loch Laggan
Spean Bridge
Binnein Shuas
Dalwhinnie
Loch Ericht
Loch Ossian
A9
B846
Pitlochry
Loch Rannoch
B846
Weem Hill Crags
Ballinluig
Aberfeldy
Craig a Barns
A826
Dunkeld
A827
A822
Loch Tay
A822
Killin
Glen Lednock
A822
Perth
A82
A85
Glen Ogle
A85
Comrie
Crianlarich
Lochearnhead
Loch Earn
A84
Crieff
A822
Bennybeg
N
0   8km
0   5mls

CENTRAL HIGHLANDS

This is by far the largest area covered in the guide. The dominant rock type within the southern and central Highlands is schist. Within this area lie two of the most significant Highland outcrops – Creag Dhubh and Craig a Barns. Both played a leading role in the rising standards throughout the decades of the 70s and 80s. Binnein Shuas and the more recently developed great slash of Dirc Mhor offer the only climbing of note on micro-granite. Also within this area lies Glen Ogle (a definite contender for the least aesthetic glen in the whole of the Highlands), where the trend towards ever shorter and harder pitches has led to a cornucopia of miniature sport climbs. In a similar vein, though in an altogether pleasanter environment, the short steep buttresses on the thickly wooded slopes of Weem Hill provide the best mid-grade sport climbs thus far in the country. Other isolated sport venues amidst fine locations include the unusual conglomerate crag The Camel at the south end of Loch Duntelchaig just south of Inverness.

Accommodation: See also Northern Cairngorms introduction. **TIC** in Pitlochry (01792 472215; www.visitscotland.com); Information Centre in Aberfeldy (01887 829187; www.locuscentre.org).

Youth Hostel: Pitlochry (☎ 01796 472308; www.hostellingscotland.org.uk)

Bunkhouses Pitlochry Backpackers (☎ 01796 470044; www.pitlochrybackpackershotel.com); The Bunkhouse, Aberfeldy (☎ 07849 689386; www.thebunkhouse.co.uk); Dalwhinnie Old School Hostel (07960 174462; www.dalwhinniehostel.weebly.com).

Campsites: Comrie Croft (01764 670140; www.comriecroft.com); West Lodge Caravan Park, Comrie (☎ 01764 670354; www.westlodgecaravanpark.com); High Creagan Caravan Site (☎ 01567 820449; www.ukcampsite.co.uk) by Killin; Invermill Farm Caravan Park, Dunkeld (☎ 01350 727477; www.invermillfarm.com); Aberfeldy Caravan Park (☎ 01887 822103; www.aberfeldycaravanpark.co.uk); Faskally Caravan Park (☎ 01796 472007; www.faskally.co.uk); Milton of Fonab Caravan Park (☎ 01796 472882; www.fonab.co.uk) both by Pitlochry; Blair Castle Caravan Park (☎ 01796 481263; www.atholl-estates.co.uk), Blair Atholl.

Club Huts: Raeburn Hut (SMC; NN 636 909) on A889 between Dalwhinnie and Laggan, The Cabin, Balgowan (JMCS; NN 638 944) between Laggan and Creag Dhubh.

Amenities: Two places particularly worth a mention for refreshment and easily accessible from the A9 are the Blair Atholl Watermill (☎ 01796 481321; www.blairathollwatermill.co.uk; open April-October) and Ralia Café with tourist information (☎ 01540 670066) just south of Newtonmore turn-off.

"The one way to know mountains, and by knowing enter into possession, is to walk through them and climb on them. That double act is needed. In normal weather, one of the first intimate things that Scottish mountain country reveals to each explorer is the 'stone-glint' of its skeleton – the white light reflected off wet rock. Everywhere the bare bones break surface, exposed in slabs, protruding in ribs, outcropping in crags, or lifting up as whole mountains."

– W H Murray,
Scotland's Mountains
(Scottish Mountaineering Trust, 1987)

GLEN LEDNOCK

A couple of easily accessible roadside crags set in a peaceful pastoral glen. The crags on the west side of the glen are schist whilst those opposite, though shorter, are composed of fine quality granite and dry much quicker.
Access: The town of Comrie lies on the main A85 Perth – Lochearnhead road, 7 miles/11km west of Crieff. From a prominent right-angled bend at the west end of the town next to Deil's Cauldron restaurant, turn up the minor road sign-posted towards Glen Lednock and follow this for 3.9 miles/ 6.3km to park (NN 7445 2695; 56.418021, -4.0367833) either side of the second cattle grid, for Creag na h-Iolaire. For **The Great Buttress**, continue for a further 0.3 miles/0.5km to car park at the end of the public road at Coishavachan (NN 7429 2728; 56.420941, -4.0395337).

CREAG NA H-IOLAIRE
(EAGLE CRAG)

A good collection of granite crags with a sunnier aspect than the schist crags across the glen.
Access: Park by the side of the road just before a cattle grid 0.4 miles/0.6km beyond the house 'Funtulich'.
Approach: Head directly across the easy angled slopes to the crag.

THE LOW WALL

The short wall low down on the left side of the crags.
Descent: Either walk left 50m from the top of the crag, abseil from trees or scramble down a grassy slanting rake/chimney fault (Diff) leading to the base of 5.

1 **Fringe Activity ★**   15m HVS 5a
FA Kev Howett & Janet Horrocks 14 July 1999

Start up the lower slabby rib, then the crack directly above. Step left into a shallow scoop. Pull round left onto a small sloping ledge, finishing up the rib above.

2 **Deuxieme ★**   15m VS 4c
FA Ian Duckworth (solo) 1979

The corner on the upper left side of the buttress. Start up a rib then step right and climb the corner past some saplings.

3 Premiere ★★   15m VS 5a

FA Neil Morrison, Rhoddy Stewart & Bryan Hogg 6 May 1979

The corner at the left end of the buttress. Climb this until it is possible to traverse right to a finely situated finger crack on the front of the buttress. Finish up this.

4 Junior's Jinx ★★   15m E1 5b

FA Ian Duckworth & Neil Morrison 25 May 1979

Gain the final finger crack of 3 by climbing direct up the left edge of the buttress.

5 Strategies for Survival ★   15m E1 5b

*FA Neil Morrison, Ian Duckworth & Bob Duncan
23 September 1979*

Start up the grassy descent rake to pull out left onto the arête. Finish up this or the finger crack further left.

Oddball, Very Difficult, takes a series of corners and ledges leading to the top end of the rake.

6 Get a Grip ★★   12m E2 6a

FA Neil Morrison 19 September 1979

The slim groove in the rib left of 6. Climb the lower wall and groove, with the crux reaching a good layback flange near the top.

7 The Strangler ★★   12m E2 5c

FA Ian Duckworth & Neil Morrison 25 May 1979

The recessed corner. Bridge up the corner, pull through a tricky bulge to finish up the easy upper groove.

8 Sultans of Swing ★★   12m HVS 5b

FA Rhoddy Stewart, Bryan Hogg & Neil Morrison 6 May 1979

The prominent right-slanting diagonal crack leads to a slight niche. Finish slightly rightward up the wall above.

8a Direct Start ★   10m E1 5b

FA Chris Calow & Bryan Hogg 1979

Steep start but on good holds.

9 Power Pinch Direct ★   10m E4 6a

*FA Gordon Lennox, D.Alexander & Craig Adam 30 May 1995;
Direct: Martin & John McKenna 23 July 2016*

Climb the arête to the good jug. Place small cam and RP in the right hairline crack. Climb wall above on small edges to the finishing jug

10 Cranium Wall Direct ★   10m E2 5c

FA Gordon Lennox & Craig Adam July 1994

The centre of the wall, climbing through the L-shaped niche past a PR and finishing slightly rightward.

11 Cranium Wall ★   10m E1 5b

FA Ian Duckworth & Chris Calow 1980

Start at the right end of the steep orange wall. Direct up to an overhang on the right, move right to a good jug and finish direct just left of the tree.

Deuxieme, Karen Latter climbing.

CENTRAL WALL

NN 7474 2713 **Alt:** 290m   10min

Some 50m up rightwards from the right edge of **The Low Wall**. All the routes belay well back from the top.
Descent: Abseil from trees down either end of the crag or traverse steep grass leftward for 100m then cut down to follow a vague path skirting back down to the base.

1 Black September ★★   20m E3 6a

FA Neil Morrison 29 September 1979

Excellent sustained well protected climbing. Start at the highest point of the wall. Climb to a good hold at 4m then span left to a good incut. Move left again then direct to a ledge. Climb the wall just left of the initial thin crack to a reasonable flake. Step right and finish up the fingertip crack.

2 Jessicated ★   20m E2 5c

FA Ian Duckworth & Neil Morrison 10 June 1980

Start beneath twin diagonal cracks just left of the shallow corner of 3. Climb the cracks to the ledge then pull through the bulge near its right end. Climb the wall to finish up a shallow corner.

3 Tigger ★   25m HVS 5a

FA Neil Morrison & B.Hogg 6 May 1979

A wandering natural line. Start below the corner in the centre of the wall. Up this to the ledge at half-height. Traverse down left on the ledge to gain a slab on the left side. Follow this delicately up and left to finish up a slim corner.

4 Disco Duck ★★   20m E3 6b

FA Neil Morrison 2 October 1980

A bouldery start leads to fine climbing up the left edge of the wall. Start 2m right of the shallow corner. Hard initial moves lead (sometimes) to better holds and a horizontal break. Move right along this then break leftward through the overlap and climb the left side of the upper wall on good diagonal slots. Move diagonally left through a break in the overhangs to finish up 2.

5 Disco Dancer ★   20m E3 6b

FA Ian Duckworth May 1980

A good slightly easier companion to 4. Start directly beneath a small vertical slot 4m up. Boulder past the slot then make a hard move left to a junction with 4. Pull through the overlap then shimmy up the diagonal crack and up on good holds to gain the grass ledge. Finish easily up a slim corner crack.

HIGH WALL

A further 60m diagonally up the hillside from the **Central Wall** is an acutely overhanging wall lying above an easy-angled ramp.
Descent: A 25m abseil from trees (belays also) at the top of the crag.

1 Diamond Cutter ★★★★   30m E3 6a

FA Kenny Spence & Spider Mckenzie June 1981

Brilliant well protected climbing up the flake crack and hanging corner at the left side of the wall. The best pitch in the glen. Start a short way up from the base of the

wall. Up diagonally left to the diagonal crack of 2 and up this to a ledge. Move up to the flake crack splitting the bulge on the left and up this to an easing below the corner. Finish spectacularly up this (crux) mainly by its right arête in a superb position.

2 No Cruise ★★   **30m E4 5c**

FA Spider Mckenzie & Kenny Spence June 1981

The wide diagonal crack gives a sustained and strenuous pitch. Take some large cams or hexes. Start as for 1 to gain the crack and follow this to a bulging V-groove just short of the top. Pull out rightward round the bulge then up and right.

3 Indoor Storm ★★   **18m E5 6b**

FA Matthew Thomson 6 October 2018

A good direct line with a wild finish above good gear. Climb onto the block ledge as for 4, then move into the short diagonal crack above and crank up the wall from a keyed-in block (micro-cams) to a spike. Move into the

crack of 2, then climb the exciting nose directly above on angular holds and blind slopers. Scramble up to tree.

4 Sidewinder ★★   **25m E4 6a**

FA Kev Howett & Dave Douglas 18 June 1992

A strenuous pitch up the centre of the crag. Start 7m further up the large ramp from the other routes. Move left onto the top of a large protruding block. Step left into the crack of 2 and up this to beneath a prominent fin. Climb direct using holds on either side of this then continue up the right slanting groove to a rest near the top. Swing out left on a sharp protruding hold to finish up a short jam crack.

5 Echo Box ★   **12m E4 6a**

FA Matthew Thomson 6 October 2018

Start near the top of the ramp below a low nose and high niche. Gain the nose (hollow) and move right and up to jugs. Climb past the break to the roof on hidden holds and gear, pull through the roof and final niche direct to the abseil tree cluster.

THE GREAT BUTTRESS

NN 7351 3252 **Alt:** 310m   E   35min

The biggest of all the crags in the glen with a good selection of extremes.

Approach: Walk along the private road through the locked gate towards the dam for just over 1km, then down the left fork to cross the river at a bridge. Walk back along the west bank of the river, then head diagonally right up the hillside, skirting a boulder field beneath the base of the crags. A more direct approach is possible, but this involves climbing over a deer fence and wading the river (wellies) – 20 min. **Descents:** Either abseil from trees (45m), or walk a long way right.

1 The Chancers *   45m E3 5c

FA Ed Grindley & E.Brookes June 1975

The overhanging corners and grooves just left of 2.

1 **30m 5c** Climb corner until level with a gangway on the left. Swing up right to the foot of a steep, thin corner crack and up this to a large ledge and belay.

2 **15m 5c** Climb the open corner above to the top of the prow and finish directly by a steep crack.

2 The Great Crack **   45m E3 5c

FA Ian Conway & party (1 PA) 15 May 1974;
FFA Ed Grindley & Ian Duckworth May 1975

Good climbing up the prominent wide crack.

1 **20m 5b** Ascend the corner and crack through the overhang, continuing until moves up and right lead to a belay.

2 **25m 5c** Climb the wide crack above.

3 Wendy's Day Out **   45m E4 6a

FA Guy Muhlemann & Simon Richardson (1 PA) 20 July 1985; FFA Tom Prentice & Guy Muhlemann late 80s

Sustained strenuous climbing up the hanging groove.

1 **20m 5a** Climb the crack then make a difficult move left to reach a sloping ledge. Traverse a metre further left then up the corner crack.

2 **25m 6a** Up the impending wall behind the stance and over a small overhang to make a difficult exit into the smooth slab above. Up the overhanging groove on the left to reach 3 below its final crux and up this more easily to finish.

3a Direct Start *   10m E5 6b

FA Michael Gardner & Ross Cowie 26 August 2021

Climb crack for about 3m, before a rising traverse left to beneath roof. Climb over the roof (crux) and continue up the wall to gain 3.

BENNYBEG

NN 8621 1888 **Alt:** 40m SE 1min

A sunny and very accessible crag with a good range of short low grade sport routes, many suitable for beginners and kids.

Access: The crag lies just east of the A822, just south of Crieff. From the south follow the A822 beyond Muthill for 1.1 miles/1.8km to park in car park (NN 8614 1887; 56.348433, -3.8438936) at the north end of the Plant Centre. From Crieff, the parking is just under 1 mile/1.6km south of the outskirts, at the end of a long straight.

Approach: Follow the path for less than 50m to the first sector.

LEFT SECTOR

1 Route Minus One ★		8m F3+
2 Route Zero ★		8m F4+
3 Bill and Benny the Flower Pot Men ★★		8m F3
4 Bennydorm ★		8m F5
5 Driven around the Benny'd ★		8m F5+
6 Benny and the Banshees ★		8m F4

---

4 **No Place for a Wendy** ★★★   40m E3 5c

FA Murray Hamilton 8 April 1978

"All in the feet - lovely bridging with plenty of no hands rests." Good climbing up the grooves in the right arête of the buttress. Scramble up right to belay on a slab beneath the prominent groove. Climb the initial dirty groove and step left into the main groove which is followed to the top.

5 **Pole-axed** ★★   15m E4 6a

FA Rick Campbell & Ian Taylor July 1993

Just up and left of **Great Buttress** is a prominent small buttress with a short isolated hanging arête. The base is easiest to approach by abseiling down the line of the route. Start up a short groove on the left then move right and up the arête strenuously.

7	Benny Hill ★	8m F5+
8	United Colours of Bennyton ★★	8m F3
9	Scorchio ★★	8m F5
10	Route Ten ★★	8m F4
11	Benny Goodman - King of Swing ★★	20m F5

RIGHT SECTOR

1	Beggar's Belief ★	10m F4
2	New Beginning ★	12m F5
3	Beguille ★★	12m F5
4	The Spanner ★	12m F5
5	Benny's Black Streak ★★	12m F5+
6	Lady Willoughby ★★	12m F4

7	Beg to Differ ★	12m F5+
8	Beg'tastic ★	12m F5
9	Benny Lane ★	12m F6a
10	Ally's in Wonderland ★★	12m F3+
11	The Beg Issue ★	12m F4
12	The Beggar ★★	12m F6a
13	Beggar's Banquet ★★	12m F6a
14	The Smiddy ★★	12m F6a
15	Begone ★	12m F5
16	Beg Pardon ★★	30m F5+

GLEN OGLE (HIGH GLEN)

Glen Ogle is the open 5 mile/8km long glen running south-east to north-west from the A84/A85 junction at Lochearnhead to the A85/A827 junction at Lix Toll, some 2 miles/3km south-west of Killin, close to the west end of Loch Tay. It is a strong contender for the least aesthetic glen in the entire Highlands, with a steady procession of trucks and caravans trundling up the road. A number of small schist crags situated on both sides of the glen provide a wide range of short bolted routes of all grades from F5+ upwards, together with a handful of traditionally-protected routes on Creag nan Cuilean. The rock on the east tends to be easier angled, and more importantly, receives the sun.

Nesting restrictions: From March – July peregrine falcons and buzzards nest on some of the crags. If present, move to another crag well away from the nesting site.

Access: Travelling from the south via Lochearnhead the majority of the crags are approached from a small rough parking spot on the west side of the A85 road opposite the viaduct 2.2 miles/3.5km beyond the village. The remaining crags, **Creag nan Cuileann** and **Mirror Wall,** are approached from a smaller lay-by on the east side of the road 0.4 miles/0.6km further.

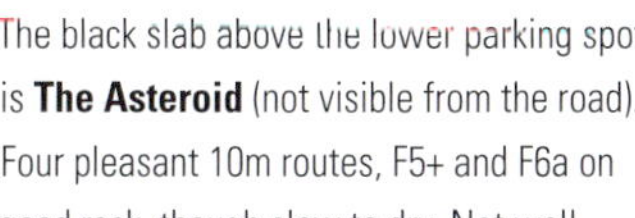

MEALL BUIDHE (YELLOW HILL)

The hill near the top of the east side of the glen.

NN 573 270 **Alt:** 400m   20min

THE ASTEROID

The black slab above the lower parking spot is **The Asteroid** (not visible from the road). Four pleasant 10m routes, F5+ and F6a on good rock, though slow to dry. Not well equipped, with single bolt anchors at the top.

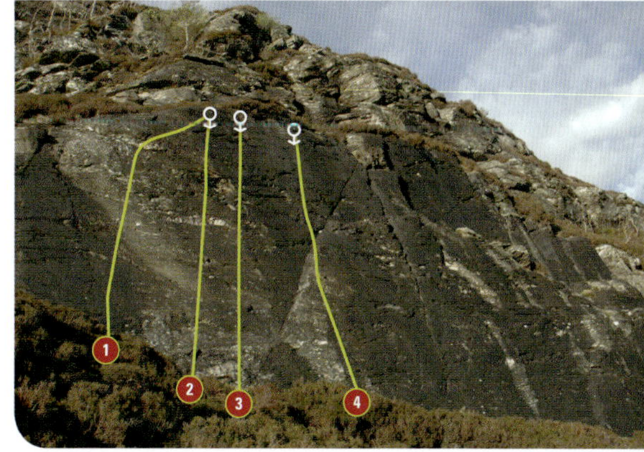

NN 568 273 **Alt:** 350m

CREAG NAN CUILEANN (CRAG OF THE PUP)

SW · 15min

The largest crag high up on the east side of the glen. Fortunately (or unfortunately, depending on your viewpoint) the traditionalists got here first.

Approach: From the lay-by at the top of the glen on the east side of the road head diagonally right up the hillside to the crag.

Descent: All the trad routes finish on a large grassy ledge just short of the top. Traverse right along this and down the right side of the crag.

1 Idiot Wind ★★　　　　　　　　　10m F7b

FA Rab Anderson 11 October 1998

The steep little wall at the left end of the crag just right of a short, roofed chimney. 5B to LO.

2 Merlyns Flight ★　　　　　　　　30m E3 6a

FA Kev Howett 24 August 1992

Climbs the prominent break through the roof at the left side. Start just right of a short hanging corner formed by a block under the roof. Exit out left on a horizontal crack and just before its end climb the crack above onto the slab. Finish directly up the wall.

3 The Bigger They Fall ★★　　　　　25m E5 6c

FA Rick Campbell & Rab Anderson (both led) 13 June 1992

A powerful well protected start leading to a bold headwall. Start 5m left of 5. Climb to the bulge and struggle up the thin crack (awkward to place R #2 at the start) to its top. Move up left to a small roof (F #0 or #0.5), pull over and continue to a quartz niche (F #1.5 in pocket below) then wobble up the pocketed headwall to the top.

4 The Harder They Come ★　　　　30m E4 5c

FA Tom Prentice & Kev Howett 3 August 1991

Climb a short crack immediately left of the central fault to join the fault at a holly. Pass the tree then traverse a ledge left into the centre of the wall and climb a very thin crack with difficulty to a deep horizontal slot on the left. Continue direct, exiting out left.

5 Pruner's Groove ★　　　　　　　30m E2 5c

FA Graham Little, Tom Prentice & Kev Howett 11 May 1991

The central fault. Climb up leftward to the right end of the roof. Cross this on good holds, foot traverse a good break left to the groove and finish up this.

6 **Mind Bogle** ★　　　　　　　　　　30m E5 6a

FA Kev Howett 3 August 1991

The roof and quartzy scoop right of the central line of 5. Quite serious. Start under the stepped roof right of 5. Up the initial wall to beneath the roofs. Pull out left through the arête of the lower roof to gain the base of a prominent white break that leads left. Up to join 7 at the crack. Step hard left into the break and follow it with difficulty to pull onto the wall just right of 5. Continue direct about 3m right of 5 through the bulges near the top. Friends essential.

7 **Poison Ivy** ★★　　　　　　　　30m E3 5c

FA Kev Howett & Tom Prentice 11 May 1991

The prominent crack-line splitting the wall on the right. Start below the roofs below the crack just right of 6. Up into the roofs and out left to the base of the crack. Climb it to a small shelf near the top, exiting out right onto grass ledge.

> The following nine sport routes are all located at the right side of the main wall. The first three routes all start from a boulder at the base, directly beneath the centre of the aspens at 10m. Roof and wall is 8 *Fight or Flight* F7a+; roof and bulge immediately right of the blocky groove 9 *Slaphead* F7a.

10 **Fat Chance** ★　　　　　　　　　10m F6c

FA George Ridge 24 September 1998

The short thin crack in the bulge just right of the blocky groove.

11 **Fight the Flab** ★　　　　　　　10m F6c+

FA Rab Anderson 24 September 1998

The roof left of 12 then the short blunt slabby nose. Approach from the right as for 12. **Take great care** with the block under the line – do not stand on the left/lower block.

12 **Let it All Hang Out** ★　　　　　12m F6c+

FA Rab Anderson 30 August 1998

Thug through the roof close to its widest point then the slab above to a LO in the trees.

13 **Happy Campus** ★　　　　　　　12m F6c+

FA Rab Anderson April 1999

Line through roof immediately right of 12.

14 **Hang On!** ★★　　　　　　　　　12m F6c

FA Janet Horrocks 19 September 1998

The small roof just right of 13, finishing up the featured wall and slab above.

15 **Step on It**　　　　　　　　　　　12m F6a

FA George Ridge September 1998

Start right of 14. Climb the wall, moving up left into 14 to finish.

16 **Life in the Fat Lane** ★　　　　　12m F6b+

FA Rab & Chris Anderson 29 August 1998

Short steep crack at right side of wall. LO on the heather ledge, beyond the rowan.

> Round the edge and up the slope a little is a short blunt arête.

17 **Chasing the Bandwagon** ★　　　10m F6a+

FA Colin Miln 3 July 1998

Line up wall left of the arête.

18 **Reaching the Limit** ★　　　　　10m F6b+

FA Rab Anderson 4 July 1998

Climb on and just left of the arête, sharing the first two bolts and LO of 17.

19 **Clutching at Straws** ★　　　　　10m F7a

FA Rab Anderson 20 June 1998

Aptly named. The steep leaning side-wall and arête climbed on its right side.

Set back to the right is a pale clean slab, with a diagonal line of aspens cutting up left from the centre.

1 Dazed and Confused   15m F6a
FA Janet Horrocks 29 August 1998

The leftmost route on the recessed section, using a tree to gain the mid-height ledge.

2 Having a Little Flutter ★★   15m F6c+
FA George Ridge 29 August 1998

The wall to the left of the thin crack in the headwall. Crux is clipping chains.

3 Ceuse Jimmy ★★   15m F6c
FA George Ridge 30 August 1998

The line up to, then following a prominent thin crack in the headwall.

4 Kinmont Times ★   16m F6a+
FA George Ridge 30 August 1998

Left-slanting crack on a recessed section, finishing on the left edge as for 1. Direct to the LO is 6b+.

5 Lichen Virgin ★   12m F6a+
FA Janet Horrocks 30 August 1998

The hollow flake, wall and groove bounding the right side of the recessed section.

6 Loose Living ★   12m F6a
FA George Ridge 1998

The prominent shallow groove and direct continuation.

7 Ghost Trail ★★   12m F6b+
FA Rab & Chris Anderson 14 June 1998

The pale streak on the right edge of the wall, sharing the same LO as 6.

About 80m right of **Creag nan Cuileann** at the same level as the upper (rightmost) wall lies **Bourneville**, a long 7m high vertical pocketed wall. Twelve routes, mostly F6b except (from left) 1 F6a; 2 F6a+; 6 F6a+ and 7 F6b+.

THE WEST (DARK) SIDE

A good introduction to small, dirty, usually wet and midgy crags …
sorry, I mean Scottish sport climbing. It can only get better!

SGORRACH NUADH
(NEW PRONGED/PEAKED/CLIFFY)

This is the hill on the west side of the glen with the old dismantled railway and viaduct running parallel to the road, beneath all the crags.

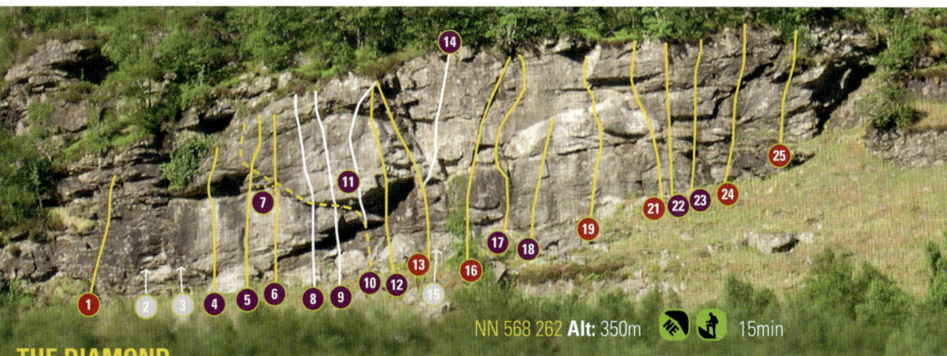

NN 568 262 **Alt:** 350m   15min

THE DIAMOND

The left side of the buttress has the best concentration of hard routes on any of the crags.

Approach: Follow a prominent narrow track diagonally down to a small wooden bridge across the stream then steeply up to the viaduct. Cross the fence and follow a path diagonally up rightwards to the crag.

The minuscule route left of the main wall is 1 *Midge Patrol* F6b. Routes 2 and 3 are projects.

4 Easy Over ★   12m F7a

FA Rab Anderson 8 May 1993

The black streak above the boulder to a tiny groove onto the shelf. Pull over the roof on large holds, move leftward and finish up slab to LO.

5 Digital Quartz ★★   15m F8b

FA Iain Pitcairn 1993

Start beneath a prominent quartz boss. The lower wall is the technical crux, the headwall F7b+.

6 Cease Fire ★★   15m F8a+

FA Dave MacLeod September 2001

From the cave sustained crimping leads to a rest at a quartz hold. Excellent moves on finger pockets (crux) lead to a jug rail and the top.

7 The Link ★   15m 7c+

The easiest line up the main face. Follow 6 to the horizontal traverse line then move left and finish up 5.

8 Spiral Tribe ★★   15m F8a

FA Duncan McCallum 1993

Follow the line of BRs direct from the first BR on 10.

9 Off the Beaten Track ★★★   15m F8a

FA Paul Thorburn 1993

The right line on the main wall.

10 Children of the Revolution ★★   20m F7b

FA Rab Anderson 12 July 1992

Start at the right side of the wall. Follow the fault leftward to the large ledge beneath the roof. Pull over the roof on good holds and follow a direct line up the headwall to LO.

11 Chain Lightning ★★   **15m F7b+**
FA Rab Anderson 1993

The short hanging groove in the blunt right arête of the main wall, past a shelf at ²/₃rds height.

12 One in the Eye for Stickmen ★   **15m F7a**
FA Neil Shepherd 13 June 1993

The cracked grooves 3m right of 11. Start just left of the prominent white streak near the base of the crag.

13 Old Wives' Tail ★   **15m F6b**
FA Neil Shepherd 12 June 1993

Climb *Metal Guru* to the second BR then follow the left-slanting ramp.

14 Metal Guru ★★   **15m F6c+**
FA Rab Anderson 20 July 1992

The thin crack up the vertical wall right of the main face.

15 is a project.

16 Sugar and Spice ★   **15m E3 6a**
FA Rick Campbell & Rab Anderson 4 June 1992

The original line on the crag, before the boltnasties arrived! Start directly beneath the wide slanting crack/ pod type thingy.

17 Gross Indecency   **12m F7c**
FA Rab Anderson 3 July 1993

The prominent short V-groove at 4m then the wall 2m right of the crack of 16.

18 The Trossachs Trundler   **10m F7c**
FA Malcolm Smith 1993

The thin V-slot 2.5m right of 17.

19 After the Flood ★   **12m F6c**
FA George Ridge 24 May 1993

The line of resin bolts right of 18.

20 is a project. PR and BRs midway between.

21 Arc of a Diver ★★   **12m F6c**
FA Rab Anderson 5 June 1993

Left-facing groove through the left end of the lower roof.

22 Climb and Punishment   **12m F7b+**
FA Rab Anderson 1994

Bouldery start then powerful above ledge.

23 Wristy Business ★   **10m F6c+**
FA Rab Anderson 1993

The groove to a ledge then the steep wall above.

24 Raspberry Beret ★   **10m F6b+**
FA Chris Anderson 4 July 1993

The featured wall starting from the ledge.

25 *Ship Ahoy* F6b is a scruffy wall at the right extremity.

CONCAVE WALL   NN 566 266 **Alt:** 360m

🧭NE 🧗 25min

A very steep crag high up on the right. From **The Diamond** traverse right past a number of smaller crags for a few hundred metres then directly up the hill from the right edge of the last crag. 10 minutes further.

1 Northern Exposure ★   **12m F6c+**
FA Dave Redpath 3 August 1998

Start up 2 then cop out onto the vertical wall.

2 Arms Limitation ★★★   **18m F7b+**
FA Mark Garthwaite July 1999

The prominent overhanging arête climbed on its left side.

3 Snipe Shadow ★★   **10m F8b**
FA Dave MacLeod April 2004

The similarly overhanging wall to the leftop end 8b.

4 Embrace My Weakness ★   **8m F7c+**
FA Dave Redpath 8 August 1998

The left micro-route – originally bolted for a warm-up!

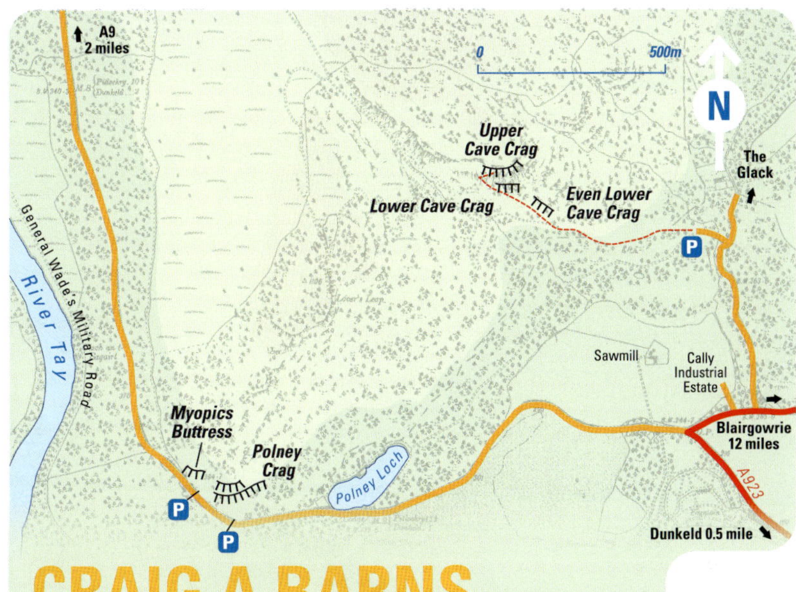

CRAIG A BARNS

This is the thickly wooded hillside overlooking Dunkeld (fort of the Caledonians), the 'gateway to the Highlands'. There are two main contrasting crags, the steep and imposing overhanging walls of Upper Cave Crag, catering mainly for the extreme climber and the slabbier walls of Polney Crag, offering a fine selection of routes in the lower and middle grades. The rock is schist and dries relatively quickly after rain.

Accommodation: Invermill Farm Caravan Park (☎ 01350 727477; www.invermillfarm.com) at Inver (NO 016 421), on the south side of the A9, 0.6miles/1km west of the main turn-off. Numerous hotels and B&Bs in and around Dunkeld. Wild camping, discreetly, is possible in woods up and left of Lower Cave Crag. There is an open cave just up the hill, about 200m up the track left of Myopics Buttress.

Amenities: Bank with ATM in square. Range of shops in main street, including a supermarket and numerous cafés. Late opening village shop in Birnam (1.3 miles/2km from Dunkeld). There is a good café in the Birnam Arts & Conference Centre. Bar meals available in many hotels – try the Taybank hotel (often live folk music) or the Royal Dunkeld Hotel in the main street opposite the chip shop and baker's. Public toilets at the car park at the north end of town.

"The pass into the Highlands is awfully magnificent; high, craggy and often naked mountains present themselves to view, approach very near each other and in many parts are fringed with wood, overhanging and darkening the Tay, that falls with a great rapidity. After some advance in this hollow, a most beautiful knoll, covered with pines, appears in full view and soon after the town of Dunkeld, seated under and environed with crags, partly naked, partly wooded, with summits of great height."

– Thomas Pennant, A Tour in Scotland, 1774

POLNEY CRAG

NO 012 432 **Alt:** 180m   1-5min

The large sprawling dome-shaped crag immediately above the old A9 road. The crag provides the best and most readily accessible low-mid grade routes in the area with a good selection, particularly in the Very Difficult to E1 range. This is the most popular crag in the Highlands, particularly at the weekend and on summer evenings.

Access: Follow the A923 through Dunkeld and continue straight on for 0.9 miles/1.4km past the turn-off towards Blairgowrie. Park on the wide verge on the right (east) side of the road (NO 0100 4307; 56.569200, -3.6122704), between the two northmost crags **Myopic's Buttress**

and **Ivy Buttress**. Best to turn in the gated track entrance, with vehicles facing south. From the north follow the A923 (signed Blairgowrie) for 2.6miles/4.2km to park on the verge beyond the forestry track.

Approach: Straight up the path behind the parking spot to the crag. For routes on **Ivy Buttress** continue round the road for a few hundred metres then cut up directly by steep path.

Descent: For routes on the left side of the crag (*Kestrel Crack* leftward) traverse left and round by a path leading down a steep muddy gully back to the base. For the routes on the **Upper Buttress** scramble down a shallow gully about 30m right of the top of *Ogg's Hindquarters* (pine and birch tree at top) to gain the start of the path above the finish of *Kestrel Crack*. There is an easy (though dirty and slightly loose) gully down the centre of the main section of the crag – **Hairy Gully** (Moderate). If intending to do a number of routes here it may be quicker to set up a fixed abseil down this. A broken path of sorts heads right along the top of the crag then down the right end of the crag – **take particular care in the wet**. In situ slings and maillons have been left on the pine trees at the top of 11 *The Groove*, 17 *The Creep* & **Hairy Gully** – please use these if intending to abseil. There are also slings & maillons on the trees at the top of the first pitch of 3 *Ivy Crack* & 4 *Poison Ivy*.

1 Psoriasis ★★　　　　　　　　　　**10m E3 6a**

FA Alan Taylor (solo) 1979

Hard initial moves lead to small cams in horizontal break.
Gain the angular niche above to finish direct.

2 Hot Tips ★　　　　　　　　　　**15m E5 6b**

*FA unknown 1969; FFA Dave Cuthbertson & Rab Anderson
11 June 1980*

The prominent hanging groove in the arête. A hard
bouldery start to get established in the groove, which
after a few hard pulls, soon relents. Finish much more
easily past a huge cracked block.

3 Ivy Crack ★　　　　　　　　　　**30m VS 4c**

FA Paul Brian (solo)1958

The prominent slabby corner.

  1　15m 4c Tricky initial moves lead to steady
    climbing up the corner crack to a tree belay.

  2　15m – Climb easily up a series of
    blocks directly above the belay.

4 Poison Ivy ★　　　　　　　　　　**15m VS 5a**

FA Robin Campbell (solo) 1961

A good well protected problem up the short right-facing
corner down and left of the main slabby corner of 5.

5 Consolation Corner ★　　　　　　**30m Very Difficult**

FA Paul Brian & Robin Campbell 1957

A good first pitch. Start round on the right side of a large
block pinnacle just right of a huge beech tree.

  1　15m Climb a short wide crack on the right side of
    the block to the top of the pinnacle (or 5a gain the
    same point from the start of 4). Step left and up
    the corner to a tree belay on a ledge on the left.

  2　15m Direct by two ribs and occasional rock
    above. Tree belay on the left at the top.

UPPER BUTTRESS　　NO 0108 4312 **Alt:** 160m

A series of disjointed crags directly above **Ivy Buttress**.
Approach: Via any of the routes on **Ivy Buttress** or
the left end of the **Left Section** or by scrambling up the
narrow muddy gully at the left end of **Ivy Buttress**.

1 Left-Hand Crack ★★　　　　　　**12m E2 5c**

FA Neil MacNiven 1961 (2 PA); FFA John Mackenzie 1974

The steep corner towards the left end of the buttress
immediately above the path. Abseil or lower from a small
beech at the top of the difficulties (sling & maillon in situ).

The following route is situated just above the path above a maze of jumbled boulders.

2 Hogg's Hindquarters ★★　　　　**30m Very Difficult**

FA Robin Campbell & Paul Brian 1959

Fine varied climbing. The route starts from 'Duncan Ogg's Hole', reputedly the hide-out of a local infamous cattle rustler. Start from the top of the boulder. Bridge up the overhanging groove (yes, the grade is right!) then step right onto the hanging slab and continue past a juniper bush. Traverse leftward into the centre of the slab. Follow the obvious line, pulling out slightly left then back right at the top. Tree belays further back.

MAIN CRAG
LEFT SECTION　　NO 0107 4308 **Alt:** 140m

3 Anon ★　　　　**35m Difficult**

FA unknown

The easiest worthwhile line on the crags. Start left of the pointed block at the base of 4.

1 15m Follow diagonal cracks rightwards, then direct up slab to a ledge, then traverse left to beneath the well defined square-cut corner.

2 20m Climb the corner (easier than it looks) to finish slightly rightward.

4 Kestrel Crack ★★　　　　**35m Severe 4a**

FA Robin Campbell & Paul Brian 1957

The slab and wide hanging flake in the centre of the recessed bay. Start beneath deep cracks above a pointy boulder at the left side of the slab. Climb these, then rightwards to ledge and thread runner. Pull onto the wall and climb the flake above (large hexes useful) which leads to a large ledge (possible belay). Continue easily up the slabby corner above.

5 Kestrel Variations ★　　　　**35m HVS 5a**

A link up of previous variations.

1 25m 5a Ascend left-facing corner bounding the right of the slab to gain the base of the huge flake. Move left and climb the outer edge of the flake to the bay.

2 10m 5a Move right and climb cracks and a slim corner up the front face of the buttress.

James Strongman on the airy and spectacular Hogg's Hindquarters.

Deziree Wilson on the short but action-packed Left-Hand Crack.

6 Katie Morag's First Steps * 40m VS 4b

FA Ben Ankers & Carol Fettes 1992

Good open climbing. Start in the back of the bay, just up from 7. Climb the wall just left of the corner and thin crack above, passing just left of small tree to below a long overlap. Continue for 3m, then traverse right and up the left side of a brown slab to move right to finish up the shallow groove, as for 8.

7 Twisted Rib * 45m Very Difficult

FA Robin Campbell & Peter Lightbody 1957

Wandering open climbing, starting up the right defining rib of the slabby recess.

1 **15m** Follow the initial rib (poorly protected, but on good holds) to a ledge. Move right and up short steep corner to a huge flake belay on large ledge.

2 **30m** Traverse a long way rightward on disjointed ledge then ascend a slab diagonally rightward to a grassy bay. Finish up the groove right of the grass.

8 Beech Wall ** 40m HS 4b

FA Robin Campbell & Paul Brian 1959

Start, not surprisingly, behind the beech tree. Climb the short steep initial groove then move right and up the second groove, with a hard move to good holds (crux). Continue more easily directly to steep headwall (possible belay on grass ledge). Finish up fine shallow left-facing flake groove.

9 Piker's Progress Direct * 35m HVS 5a

FA Ian Rowe 1969

Start below a break in the left side of a line of overhangs, 10m right of the beech tree. Climb the initial wall past a large flake then move leftward up the steep wall (crux) to a break in the overhang formed by a huge block. Pull strenuously right into a short slabby groove then traverse right to the prow to finish up the rib.

10 The Way Through ** 30m E2 5b

FA Kenny Spence & Russel Sharp July 1967

Good steep climbing up the centre of the double roofed wall. Start in the centre of the wall. Climb to the lip, pulling diagonally right to a slab beneath the second roof. Break through this on good holds at its widest point (good slot for gear on lip) then continue to cracked blocks beneath the prow. Pull over the overhang above at a thin crack then finish easily.

The following three pitches all share a common start, at the prominent break at the left side of the large overhang near the base.

11 The Groove ★★   30m VS 5a

FA Robin Smith (solo) 1960

Move up to beneath the roof then traverse a good break leftwards to a groove cutting through the left side of the roof. Make awkward moves up right past a PR (crux) onto a slab on the lip. Move up the cracked groove above then move up diagonally leftwards to cracked blocks. Continue direct up the fine right-slanting groove above.

12 The Rut ★★★   30m VS 5a

FA Neil MacNiven 1960

Excellent sustained well protected climbing up the short hanging groove above the common start – the best line on the crag. Finish by stepping left into the final section of 11.

13 Wriggle ★★   35m VS 5a

FA Robin Campbell & Paul Brian 1959

Follow the common start through the break in the roof then make an exposed traverse on good pockets above the lip of the roof rightward to a blunt rib. Finish up this.

14 Twilight ★   30m E1 5b

FA Dave Cuthbertson & Rob Kerr 2 November 1980

A fine direct line cutting through the traverse of 13. Start below the holly bush at the right side of the long low roof. Up a groove just left of the holly to a roof. Pull out left through the roof to the slab of 13 then quite boldly direct up the blunt rib climbing through a bulge at the top.

15 Holly Tree Groove ★★   30m Very Difficult

FA Robin Campbell & Paul Brian 1960

The diagonal right-slanting ramp line. Start just right of the holly bush. Climb a short awkward wall on the right to step up left into a short groove. Move up and follow the ramp to a ledge at its end, finishing up the open chimney above.

16 Recess Route ★   25m Very Difficult

FA John Proom 1960

The wall right of 15. Start on a small ledge with a block, below a faint rib. Climb the wall first on the left then right to reach a large grassy recess. Finish up the final open chimney of 15.

17 The Creep ★   30m HS 4a

FA Robin Campbell & Paul Brian 1960

Poorly protected climbing, taking a parallel line right of 16. Start just right of 16. Move up into the short corner then climb rightward up the slab to a small sloping ledge beneath the right edge of the first overlap. Move left through the overlap on big holds then up and right to the right side of a grassy recess. Finish up the easy-angled shallow groove above.

18 Dynamo ★   30m VS 5a

FA Rab Anderson & Dave Cuthbertson 10 June 1980

Fine varied climbing. Start at the left side of a long grass ledge at the base of the wall, gained by walking left from the base of 19. Follow a shallow right-facing groove in the prominent black seepage streak to a foot-ledge on the right below the overlaps. Pull through the overlap at a slight weakness (crux), step left then direct up the wall to finish up a steep crack just right of the final shallow grooove of 17.

Karen Latter nearing the top of The Groove.

Eilidh Milne nearing the top of the best pitch on the crag The Rut.

19 Cuticle Crack ★★   30m Severe 4a
FA Paul Brian & Robin Campbell 1960

Popular and polished, climbing the prominent deep crack 3m left of **Hairy Gully**. Follow the crack moving right along the top of a large protruding block. Finish directly.

RIGHT SECTION

20 Bollard Buttress   30m Very Difficult
FA Robin Campbell & Paul Brian 1959

Poorly protected climbing up the wall just right of **Hairy Gully**, starting 5m left of 21.

21 Bollard Buttress Direct   30m HS 4a
FA Paul Brian & Robin Campbell 1960

Protection is only just adequate. Start at a thin crack with 'Bollard Buttress' painted on the rock. Follow the crack then a shallow groove. Finish up a rib leading slightly rightward.

22 Springboard ★   35m VS 4c
FA Robin Campbell & Paul Brian 1960

A good though wandering and sparsely protected route. A short way right of **Hairy Gully** is a jumbled mass of boulders piled up against the crag. Start just left of these, at the lowest point of the wall. Climb first leftward then back right and up to a deep horizontal break under the overlap. Hand traverse this right and up into the short brown groove. Up the corner (unprotected) moving out left under the roof at the top to a slab. Climb left up an overhung recess to a slab and finish rightward up this.

22a Springboard Direct ★   30m E3 6a
FA Dave Bathgate (3 PA) 1960; FFA Derek Jamieson 1978

A technical and perplexing problem, previously severely undergraded (E1 5b!). Follow 22 to the overlap. Make hard moves to become established in the slim groove above. Fiddly protection.

23 The Chute ★   30m VS 5a
FA Robin Campbell (solo) 1961

Start at the base of the brown groove of 22. Step up then move right immediately to follow a slim groove to a quartz bulge. Cross the bulge with difficulty to a tiny ash tree then move up rightward, traversing right round the top of a large flake to finish up easy slabs.

24 The End ★★★   35m VS 5a
FA Robin Campbell (solo) 1960

Excellent varied climbing. Start at the base of the narrow clean slab on the lower tier, immediately right of the jumbled boulders.

1 **15m 4c** Up the left side of the slab under a shallow left-facing corner. Climb the corner to a bulging nose, up this and the delicate slab above to ledge.

2 **20m 5a** Break through the central weakness in the bulging wall above (crux) then direct up the slab to step right to below the break in the largest part of the overlap. Pull through this and finish easily up the fine slabs.

25 Barefoot/The Beginning ★★   35m E2 5c
FA Pitch 2 Dave Cuthbertson, Derek Jamieson & Murray Hamilton 1976 (previously top-roped); Pitch 1 Dave Cuthbertson & Pete Hunter 31 August 1980

A logical combination with both pitches serious. Start in the centre of the narrow slab, just right of 24.

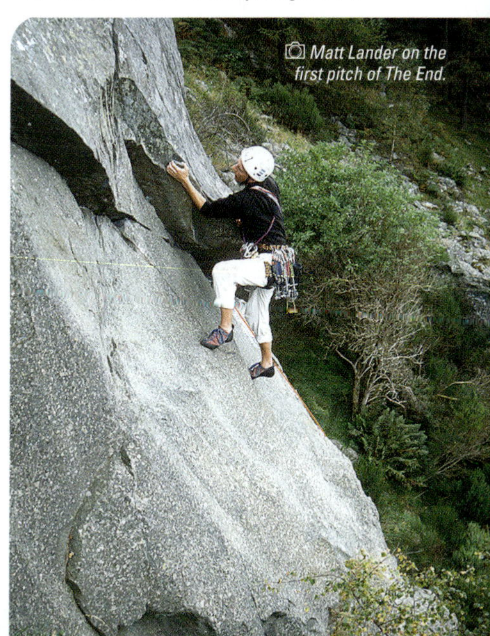

1 **15m 5c** Easily up the slab to a very thin crack. Move left then back right to better holds on the blunt arête near the top.

2 **20m 5b** Climb the short right facing tapering groove then left and across a steep wall onto the slab. Up this directly to the roof, crossing this at a prominent break. Continue more easily.

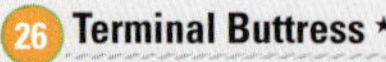

26 Terminal Buttress ★　　　　　**35m VS 4c**

FA Paul Brian & Robin Campbell 1959

Start at the corner near the right side of the lower slab.

1 **15m** Up the corner to ledge.

2 **20m 4c** Climb through the bulge just left of the weakness of 24 (crux) to gain the recess above. Step right and up a crack in the overlaps to finish up slabs.

27 Spirochaete ★　　　　　**30m VS 4b**

FA J.Cameron & C.Norris 8 December 1969

The last worthwhile route on the crag before it degenerates into the hillside. Start at the right side of the crag, behind a large birch tree. Climb a short right-facing layback groove then traverse out right to a small beech tree (runner). Move back left and finish quite boldly up the slab.

MYOPICS BUTTRESS

NO 0097 4313 **Alt:** 130m  1min

A short steep buttress very close to the road with a bunch of short action-packed sport routes.

Access: Drive round the bend heading away from the main crag (north) and park on the right, just before the start of a private track.

Approach: Zig up left then back right to the crag.

1 Phantom　　　　　**15m F7a+**

FA Unknown

The leftmost line, trending right at the top to a LO beneath small roof. Poorer rock.

2 The Chopping Block ★★　　　　　**15m F7b**

FA Neil Shepherd 15 September 1996

A direct line breaking through the roof at its widest point.

Follow good holds up to the roof. Cross this using a good flange to gain a good spike hold (crux) on the lip. A good quartz pocket above soon leads to more reasonable climbing up the headwall, to share the same LO as the following route.

3 The Vibes ★★　　　　　**15m F7c**

FA Ian Cropley 1990

The central line with a hard crux moving out left to underneath the roof. Climb the lower wall with a long reach right to good holds, to hard moves up left to the main roof. Pull through the roof on good holds. Easier up short final wall to LO.

3a Vibe Heads ★★★　　　　　**15m F7b+**

An excellent combination, giving the best route on the crag.

4 Granola Head ★　　　　　**15m F7c**

FA Duncan McCallum 1990

The right line. Cross the initial roof with difficulty then up to a good ledge beneath the final overhanging arête. Climb this with difficulty to an abrupt finish.

CAVE CRAGS

Two steep crags in a fine secluded setting high on the hill, much quieter than the nearby Polney Crag and with a finer outlook.

Access: Leave the A9 and follow the A923 through the historic town of Dunkeld, complete with its own cathedral. About 0.3 mile/0.5km beyond the town, turn right towards Blairgowrie then take the second un-metalled track on the left (by a post-box, signposted The Glack). Follow very rough track for 0.4 miles/0.65km to turn left into the large Cally Car Park (NO 0236 4395; 56.577291, -3.5909633).

Approach: From the car park follow the obvious well worn forest track traversing the hillside westwards, until a narrower path forks off on the right (just before the forest track starts to descend). Follow this to just beyond large fallen yew tree then cut up the hillside just before the stream. **Lower Cave Crag** lies 30m right of the 'cave', a man made structure supposedly erected for one of the Duke of Atholl's 'paramours'! For **Upper Cave Crag** cross the stream (next to the cave) and follow the steep path directly up the hillside, re-crossing the stream

Dave Cuthbertson on an early on-sight of The Vibes.

just beyond a large mossy boulder, to follow the path steeply up to the left end of the crag.

EVEN LOWER CAVE CRAG

NO 0198 4395 **Alt:** 170m   SW   7min

A small crag just above the approach path.

Approach: The crag is just over 100m beyond the fork from the wider track, 150m before **Lower Cave Crag**.

1   Arpeggios ★   15m F6a+
FA Karen Latter 12 February 2023

2   Just for Laughs ★   15m F6b+
FA Alan Leary, Martin Macrae & David Cassidy July 6 July 1987; retrobolted February 2023

LOWER CAVE CRAG   10min

NO 0149 4381 **Alt:** 220m

A dank and often overgrown crag in the trees overlooking the path, though with a few good exposed routes well worth seeking out.

Descent: Scramble up through the boulders and trees to the base of the upper crag then left and down the approach path for that crag, or abseil from trees from slings and maillons at top of *The Hood* or *Cherry Tree Ridge*.

The obvious hanging finger crack and roof at the left end is *The Civer* ★ E3 6a; an early testpiece.

Karen Latter pulling round the rib on the final section of The Hood.

1 The Hood ★★   **35m VS 4c**

FA Robin Campbell & partner 1959

Good exposed climbing, staying dry in the rain. Start near the right end of the crag just left of where the face changes direction.

1 **20m 4b** Climb the wall to the centre of a long roof then traverse right along a juggy break beneath the roof and pull round the arête onto a ramp, which leads back left. From the top of the ramp step right to gain ledges and easy ground leading up to a large roof. Bypass this on the right to gain a large block belay on the right edge beneath an overhanging corner.

2 **15m 4c** Step up onto the slab and traverse left onto the arête. Step up to the base of the prominent overhanging crack and swing out left to finish up a slabby rib.

2 The Hood Direct ★   **35m HVS 5a**

FA Brian Robertson & partner 1963

Start just round the arête from the normal route at a short leaning corner cutting through the overhanging walls at the base.

1 **20m 4c** Go up the short corner past the lower roofs to below a bigger roof. Traverse awkwardly left to gain the base of the ramp of the normal route. Follow this to the left then step right onto easy ground at a large roof. Bypass this on the right to a ledge.

2 **15m 5a** Follow 1 onto the slab and climb leftward to beneath the overhanging crack. Finish strenuously up this crack on painful jams.

3 Fuck Face ★   **35m HVS 5a**

FA Dougal Haston 1959

A good direct line up the smooth corner right of 2.

1 **20m 4c** Gain the corner from the left and climb it by thin bridging to pull onto a large slabby area.

2 **15m 5a** Pull onto the steep wall on the right (PR) and climb through the bulge with difficulty. Finish up the easier upper wall, trending right to finish.

4 Cherry Tree Ridge Direct ★   **30m Severe 4a**

FA J.Hockey & John Proom 1969

Traverse right on narrow path past obvious gully to base of broken rib. Climb the lower stepped rib and fine short corner to a large ledge near the top. Finish up the tricky short steep arête on good holds.

UPPER CAVE CRAG

NO 018 438 **Alt:** 260m    15min

Regarded by many as one of the best 'roadside' crags in Scotland. Once dry, much of the crag is often climbable even in heavy rain, with the sport routes on the central wall usually climbable well into the autumn.

Descent: There is a well worn path down the left side of the crag, scrambling past a birch tree at the upper left end of the crag. Another possible descent is down the centre of the slabby vegetated face left of *Flook* on the right side of the crag at about Diff. standard. At the bottom head left (facing out) across a mossy slab to the base of the corner of *Flook*. Avoid in wet conditions. A 25m abseil from the pines at the top of *Gnome* or *Tombstone*, or 30m abseil from tree at the top of *Mousetrap* (slings & maillons in situ) are other options.

① The Ramp ★   40m HS 4b

FA Paul Brian 1957

Bounding the left side of the steep section of the crag is a large ramp, split by a large ledge at mid-height.

1 **15m** – Climb the ramp by a crack on its right side to a spacious belay on the ledge. A fine easy pitch on its own – about Difficult. (Abseil descent from tree).

2 **25m 4b** Continue up the corner crack in the back of the ramp, which widens and steepens before a ledge/pinnacle thingy. Make steep moves up the wall above (often dirty – crux) to a heathery finish.

The following four routes all start from the large platform half way up *The Ramp*.

② Lily Langtry ★★   20m E2 5b

FA Mal Duff & Rab Anderson 21 November 1978

Good fairly well protected climbing up the shallow groove up the left side of the wall. Climb the initial groove past a small sapling, then either continue up the groove above, or easier, step left and up the wall to reach a recess under a bulge. Cross the bulge rightwards to finish.

③ Voie de l'Amie ★★★   30m E3 5c

FA Dave Cuthbertson, Rab Anderson & Mal Duff September 1979

A fine contrasting pitch, quite bold in the lower half. Start just left of the scoop of 4. Direct up a shallow groove to a bulge. Hand traverse diagonally right on improving holds to the ledge on 4. Finish with interest up the superb thin pocketed crack above.

④ Hang Out ★   25m E3 5c

FA M.Forbes & G.Miller (5 pts aid) 1972; FFA Murray Hamilton, Dave Cuthbertson & Derek Jamieson 1976

Climb boldly up and right (wires or tiny cams in flake) to gain the scoop and follow this more steadily to long ledge on right. Move right along this to finish up the final easy corner of 7.

⑤ The Pied Piper ★★★   50m E3 5c

FA Dave Cuthbertson & Mal Duff 24 April 1980

A brilliant left to right girdle of the crag, crossing some impressive ground.

1 **30m 5c** Follow the horizontal crack (2 PRs) into 7, step down then continue rightward to the larch (possible belay). Move up rightward to good holds then out right. Reverse down a short way to a good flake then follow the obvious line across the big open groove of 10 to belay on the lip of the lower roof of 12.

2 **20m 5c** Finish up the top pitch of 11.

⑥ Lady Charlotte ★★★★   30m E5 6a

FA Dave Cuthbertson & Mal Duff April 1980

The classic hard route, taking the intricate pocketed wall the full length of the left side of the crag. Start on good ledge a few metres up *The Ramp*, beneath a tiny right-facing groove. Climb the wall right of this and step left to a good foot ledge at its top. Continue up the pocketed wall above, slightly leftwards, to good holds and protection in the obvious horizontal break. Gain the hanging flake above and follow it leftwards to pull into the final section of the scoop on *Hang Out*. From the ledge above, finish up the superb thin crack in the headwall.

6a Lady Charlotte Direct ★★   30m E6 6b

FA Dave Cuthbertson 12 May 1987

Considerably harder than the original. From the top of the hanging flake, span right to a good pocket, then make difficult moves to gain a hidden peg runner in tiny groove. Continue directly to the large ledge above.

7 Rat-catcher ★★   30m E3 5b

FA Alan Pettit & Ken Martin (A3) 1969;
FFA Dave Cuthbertson & Murray Hamilton 1976

Bold steady climbing with little protection and some suspect rock. It follows the shallow open groove up the wall left of the larch. Direct up the wall and leaning corner, turning the bulge on the right to PR at 12m. Move slightly right then back left to pegs in the horizontal break. Continue in the same line to the top.

8 In Loving Memory ★★   30m E6 6b

FA Ken Martin (A3) 4 October 1969;
FFA Dave Cuthbertson & Martin Lawrence 1981

The short but action-packed overhanging wall above the larch, starting up that. Monkey up the tree to its uppermost branches. Step onto the wall and up past a PR (rock in small pocket just above) to poor holds in a small

recess (RP #3 in short pale crack up on left). Continue to a good jug. Head up leftward to finish up a short tiny right-facing corner.

9 Morbidezza ★★★   30m E5 6a

FA Dave Cuthbertson 24 June 1979

The blunt hanging arête in the centre of the crag, just left of the open groove of 10. An optional (bit contrived & crux – 6b boulder problem) start up the overhanging right rib of 10 leads to a scoop. Move left into the groove of 10, up this for a few metres then left to the arête (possible rest in tree out left). Up this by a crack to a large flat hold at its top then finish up the left side of the large cracked block.

The central groove just right of the larch tree is the often wet 10 *Mousetrap* ★ E5 6b.

11 Warfarin ★★★   40m E2 5c

FA Derek Jamieson, Grahame Nicoll & Mal Duff 5 September 1978

A surprisingly balancy route meandering up the centre of the crag. Start below a hanging right-facing groove, 3 metres left of the central crack of 12.

  1 20m 5c Boulder up the steep wall to the groove, up this (PR near top) then move up leftward into the groove of 10. Up this for a few metres to traverse right on the lip of the roof to a hanging stance on 12. N & PB.

2 **20m 5c** Up to the roof and some old PRs then follow good line of holds diagonally right under the roof to a tricky finish.

12 Rat Race ★★★★   30m E4 6a

FA Neil MacNiven (to top roofs)/Brian Robertson & John McLean (on bolts) 1963; FFA Pitch 1 Murray Hamilton with 1 PA on new direct top pitch 1976; FFA complete Mike Graham 1978

The showpiece of the crag with a sustained first pitch. Often climbed as a superb single pitch. Start beneath the pea-pod in the centre of the crag.

1 **15m 6a** Climb the wall just left of the pod to a good 2" cam slot. Continue up the crack with a hard section to gain the niche (good no-hands rest). Jug hauling above leads to a hanging stance on the lip of the roof.

2 **15m 5b** Direct to the roof then out rightward for 6m to a good flange in the roof (old PRs). Cross the roof leftward to finish up a short wall. Belay far back on pines.

THE SPORT WALL

13 Ultima Necat ★★   17m F7b+

FA upper section: Mark McGowan 8 July 1987; independent start: Neil Shepherd 14 August 2013

The leftmost line of bolts, squeezed in between *Rat Race* and *Marlene*. A sustained and bouldery start, sharing some holds on *Marlene* to join that route just before its big undercut, then direct up the wall above.

14 Marlene ★★★★   20m F7c

FA Dave Cuthbertson 3 August 1986; left start - as described Steve Lewis 13 September 1986

Very sustained with a bouldery start; no desperately hard moves. Follow the middle line of bolts, moving right past the 4th bolt with a long reach to the base of the diagonal crack. Up this and pull blindly up left at its top to LO.

14a Hamish Teddy's Excellent Adventure ★★★   25m F7b+

FA Duncan McCallum, Johnny May & Rab Anderson (all redpointed!) 1992

The most popular sport route in Scotland and justifiably so, tackling the finely situated right arête of the wall on good holds. Follow *Marlene* to the start of the diagonal crack, move up right and finish up the left side of the arête to LO beneath the capping roof.

15 The Silk Purse ★★★★   25m F8a

FA Graeme Livingston 12 May 1987

The classic hard sport route – a long stamina pitch. Follow the right-most line of bolts with a hard move left past the 4th bolt, then up to join 14 just before the start of the crack. Up the crack to its end then make hard moves up and right to some barn-door-layback moves up the scoop to reach good quartz pockets and LO just above.

15a Ching ★★   20m F8a

FA Gordon Lennox 3 April 2006

The hardest most direct line, squeezed into the final remaining gap. Climb *The Silk Purse* to the second bolt, then move out left and take a direct route cutting through the horizontal

Martin McKenna on the crag classic Rat Race.

break of *Marlene* to rejoin *The Silk Purse* half-way up the left-slanting diagonal crack. Finish as for that route.

15b Silk Teddy ★★★   F7c
FA Dave Cuthbertson 1990s

An obvious link-up, combining the first half of *The Silk Purse* with the top arête of *Hamish Teddy's*. Move slightly right from the fourth bolt and up to good rest in niche, then pull direct to finish up the upper section of *Hamish Teddy's*.

16 Baled Out ★★   15m F7a+
FA Ian Taylor 2015

The rightmost line; can also readily be linked into the finish of 14a at F7b.

17 Squirm Direct ★★   20m E3 5c
FA Dave Bathgate (5PA) 1960;
FFA Murray Hamilton & Alan Taylor 1976

The right edge of the central bolted wall has a prominent groove leading to a bulge. Gain the groove by swinging left on good holds from the lower right-trending groove. Up

this to a crucial PR at the bulge, pull through this and pull out left to a recess (old PRs). Move out rightward up easier ground to gain the belay ledge on the ordinary route.

18 Squirm ★   30m E1 5c
FA Neil MacNiven (1 PA) 1960;
FFA Murray Hamilton & Alan Taylor 1976

Good climbing on the main pitch with a short well protected crux and fine airy finish. Start 10m right of 16.

1 **15m 4b** Up easy broken ground to a large belay ledge.
2 **15m 5c** Climb the green wrinkled slab with difficulty past a PR (crux) to easier ground. Head out left and finish spectacularly up the right side of the arête on huge flakes.

19 Corpse ★★   30m E2 5c
FA Neil MacNiven (3 PA) 1960;
FFA Murray Hamilton & Alan Taylor 1976

Good climbing with a short well protected crux. Start just left of the big open corner. Easily up to the base of the shallow right-facing groove. Up this (difficult to protect) to a glacis on the right (PRs). Move out to the right end of this and finish up the short strenuous crack sprouting from its right end.

The following five routes all finish on broken ground about 10m below the top. Either abseil from thread anchor; scramble up easy vegetated groove, or in dry conditions scramble diagonally rightwards down the slabby face to gain the base of *Flook*.
High Performance & *Death's-head* start from the back of the huge boulder/pinnacle-like feature, by scrambling up from inside the 'cave'.

20 Coffin Corner ★★   25m HVS 5a
FA Pete Smith & Ferranti Club members 1960

The large left-facing corner near the right side of the crag. Easy ground leads to the base of the corner. Follow this past some threads to pull out right at its top. Block belay just above. Easier ground remains.

The classic hard Upper Cave Crag sport route, The Silk Purse. Grant Farquhar displaying his stylish tartan trews on an ascent in the late eighties. Photo Rick Campbell.

Coffin Corner, Sam Forrest nearing the top.

21 High Performance ★★   15m E3 6a

FA George & R Farquhar 1960;
FFA Dave Cuthbertson (on-sight) 10 October 1978

Gymnastic climbing through the roof just right of the big corner. Scramble up leftwards to the roof. Climb the crack in the roof (overhead protection) to a good jug on the lip and PR just above. Step right and up shallow groove past a possible no-hands rest (a cheval!) to finish.

22 Death's-head ★★   15m E1 5b

FA Dave Cuthbertson, Alan Taylor & Murray Hamilton Oct 1976

Follow the cracked groove, turning the bulge on the right. Step back left and up the short left-slanting groove to finish on good holds. Protection is reasonable, though fiddly to place.

23 Crutch ★★   30m Severe 4a

FA Andy Wightman 1959

Fine well protected climbing and the best route of the grade on the crags. Start 5m right of the 'cave'. Up past a small beech tree, then left along shelf into the corner. Up this steeply on good holds, then pull directly past flake to finish up crack at left side of orange wall.

24 Marjory Razorblade ★★★   25m E3 5c

FA Dave Cuthbertson & Murray Hamilton March 1977

The characteristic offset S-shaped crack in the steep wall right of *The Crutch*. A thug's delight.

1 **15m 5c** Climb steeply on good holds directly up to the base of the crack (possible belay). A combination of jamming and laybacking may lead to a belay on slabby ledges above.

2 **10m 5a** Step left across slab to a short steep left facing groove. Up this to easier ground.

Round to the right is a wide vegetated slab. Overlooking this is a shorter steep back wall, with a holly bush low down in the centre.

25 Flook ★   30m Very Difficult

FA Pete Smith 1960

The dièdre bounding the right edge of the slab. Near the top, traverse left onto large blocks and finish up the short corner above the large roof.

Tim Miller approaching the
crux of Morbidezza.

26 Tumbleweed ★   25m E2 5b

FA Derek Jamieson & Dave Cuthbertson June 1976

Scramble up the gangway for 10m to below the holly bush. Up to a ledge beneath the holly then follow a gangway/ramp up leftward to the base of the corner. Step right and climb a shallow groove and wall rightward to the top. Sparsely protected. So named as the first ascentionist plummeted from the finishing moves clutching a handful of heather!

27 Summer Days ★★   30m E3 5c

FA Dave Cuthbertson & Rab Anderson November 1978

Start up the cracked arête, as for 29, then the upper diagonal break leftwards leading to the holly bush. Pass this with interest to finish directly up the strenuous shallow groove in the wall above.

28 All Passion Spent ★★   25m E4 6a

FA Ewan Cameron (2 bolts); sans bolts Mark McGowan 1987

Powerful climbing up the steep pocketed wall right of the holly bush. Climb the lower wall to gain the large block just right of the holly. Pull out right and climb with difficulty past 2 pegs then on good holds into the shallow groove above, finishing slightly rightwards.

29 Gnome ★★   25m E1 5c

FA Brian Robertson 1960; FFA Mick Couston 1974

The right arête of the wall. Start at the foot of the arête. Climb by cracks on good holds up the left side to move right to a ledge beneath short open groove. Up this with interest past a PR to top.

29a Laughing Gnome ★★   25m E4 5c

FA Dave Cuthbertson & Duncan McCallum September 1980

Exhilerating sparsely protected climbing up the sharp knife-edge upper arête. Climb the lower cracked arête of *Gnome* to the ledge. Arrange good nut in pocket, then pull up then left round the edge and up this to finish on good holds.

30 Tombstone ★★   18m E2 5c

FA Dave Cuthbertson & Mal Duff 10 October 1978

Steep and sustained – low in the grade. The prominent groove right of 29 gives steep jamming and bridging. Scramble up to large ledge at the base of the crack. Follow the crack to overhang, cross this and continue on good holds (peg on right wall) in the same line to finish.

High Pitched Scweem, Weem Rock. Sadie Renwick reaching the sanctuary of the final juggy top out.
Photo Dave Cuthbertson, Cubby Images

WEEM HILL CRAGS

A fine varied selection of quick drying, mainly sports crags set in the thickly wooded hillside above the village of Weem, overlooking the River Tay and Aberfeldy in the heart of rural Perthshire. The crags have a sunny aspect, particularly in the spring and late autumn when there are no leaves on the trees.

A peaceful place to climb. The rock is generally an excellent schist, often encrusted with many garnets in its upper reaches, giving good friction. The majority of the routes are all well bolted.

Access: Leave the A9 at Ballinluig and follow the A927 west for 10 miles/16km to Aberfeldy. Turn north at cross-roads at the west end of the town (signposted Kinloch Rannoch) and follow the B846 across Wades Bridge over the River Tay to reach the village of Weem after 1 mile/1.6km. For **Weem Rock** and **Aerial Crag**, park in car park (NN 8439 4981; 56.625823, -3.8859295) on the right (north) just before (east) of the church. For **Manyana Wall**, **Secret Garden Crag** and **Easter Island Buttress**, continue through the village for a further 0.3 mile/0.5km turning right (signposted Castle Menzies), taking the right fork to a car park (NN 8396 4970; 56.624729, -3.8928843).

Amenities: Aberfeldy has all the usual requirements – a supermarket, some cafés, and several pubs. The Watermill (down on right before (east of) crossroads/traffic lights - signposted) is fantastic bookshop/café/gallery, with great coffee & tasty cakes. For supplies of chalk, climbing gear, etc. there is a small outdoor shop, Munros in Bridge End, just east of the cross-roads (☏ 01887 820008).

Accommodation: The Bunkhouse, Aberfeldy (☏ 07849 689386; www.thebunkhouse.co.uk). Masses of hotels, guest houses and B&B establishments in the area. Information Centre in the Square, Aberfeldy (01887 829187; www.locuscentre.org). Aberfeldy Caravan Park (☏ 01887 822103; www.aberfeldycaravanpark.co.uk) is at the east end of the town. Wild camping is more of a problem as the surrounding area is heavily farmed.

It is advisable to locate and stay on the approach paths described or take a machete as the undergrowth is very dense. Deer ticks are a bit of problem but only seem to be encountered if you stray from the paths.

Paul 'Stork' Thorburn on the second ascent of *Saving Up for a Rainy Day*, Aerial Crag (page 315).

WEEM ROCK 15min

NN 8443 5012 Alt: 200m

An excellent crag with two contrasting faces. The left side of the crags sports a couple of wildly steep lines, overhanging at about 20 degrees. Further right the front face has an excellent slab with some good long pitches, together with a number of shorter routes further right.
Approach: Walk up the steep road at the side of the church for just over 100m to a wide grassy track on the right, opposite the entrance to the last two houses. Walk along this for 50m then cut up left (prominent Scots pine ahead) to gain the main Weem Forest Walk. Follow this right for 200m, cutting back left at a hairpin. After about 100m, just before some steps on the path, cut steeply up right on a small path through the undergrowth. Then scramble steeply over fallen trees and rocks to reach the left end of the crag.

SIDE WALL

The acutely overhanging left wall.

1 High Pitched Scweem ★★★   **18m F7a+**

FA Neil Shepherd 13 April 1997

Excellent steep climbing. Climb the easy lower slab to ledge. A strenuous fingery start soon leads to a line of good holds and flakes, finishing direct.

2 The Screaming Weem ★★★   **18m F7b**

FA George Ridge 13 April 1997

Excellent sustained climbing up the central line of resin bolts left of the blunt arête. Difficult moves around the 3rd bolt and a long reach for a good hold on the lip of the roof.

2a Manga Manga ★★   **18m F8a**

FA Ross Kirkland 20 July 2017

Thin technical moves up the arête. Climb 2 to split at 3rd bolt. Use triangle shaped side pull at base of crack to move right to arête. Climb direct finishing up 2b.

2b The Last Gasp ★★   **15m F7a+**

FA Gary Latter 1998

A right finish to 2. Climb that route to the 5th bolt, then move right around the arête to finish up the centre of the wall.

3 The Last Temptation ★★   **25m E2 5c**

FA George Ridge & Janet Horrocks 10 April 1997

The hanging right-facing groove up the right side of the overhanging wall. Climb a short steep crack (crux) into

the groove. At the top, pull over the small roof on good holds. Step right and move up right to the LO of 5.

4 The End of Silence ★★   **15m F7b**

FA Colin Miln 29 March 1997

Technical fingery climbing up the narrow lower wall, soon relenting in its upper half.

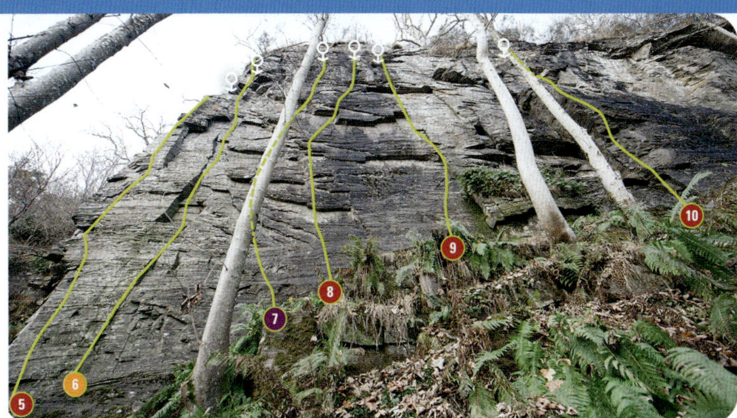

FRONT FACE

5 The Real McKay ★   24m F6a+

FA Dougie McKay June 1997

Finely situated climbing up the right side of the arête with difficult moves past the 2nd bolt. Continue much easier up the groove, finishing up a pleasant cracked groove.

6 Back to Basics ★★   23m VS 4c

FA Grahame & Mel Nicoll & Bill Wright 26 August 1997

The line of cracked grooves a few metres right of the left arête. At the overhang move right and climb another groove to finish.

7 The Long Good Friday ★★★   20m F6c+

FA Isla Watson 25 May 1997

The line up the left side of the slab with the crux passing a prominent triangular slot just above half way.

8 Confession of Faith ★★★   20m F6c

FA Janet Horrocks 12 April 1997

Excellent sustained climbing, taking the central line with the difficulties just above mid-height.

9 Mannpower ★★   20m F6b

FA Dave Pert 12 April 1997

The prominent thin intermittent crack at the right side of the slab. Start up the wall just left of the corner (crux) then follow the intermittent crack, passing the small roof with difficulty.

10 Staring at the Sun ★   18m F6a

FA George Ridge 10 October 1998

The arête just to the right of the vegetated corner.

Karen Latter on the fine crack of Back to Basics.

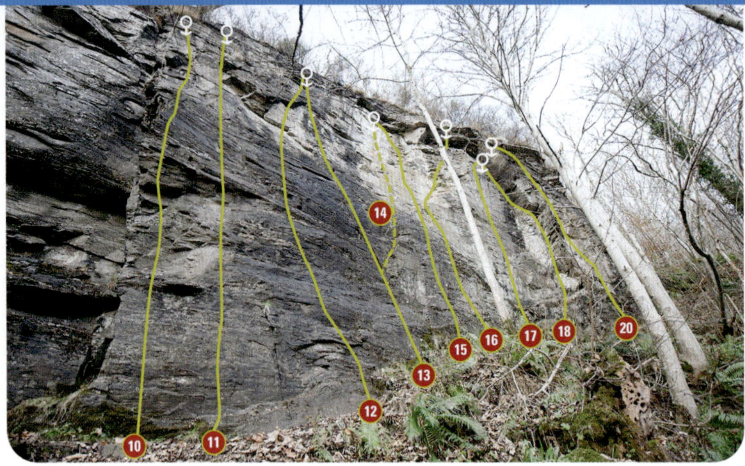

11 The Soup Dragon ★   18m F6a

FA Janet Horrocks 20 September 1997

The first line right of the arête of 10.

12 Scooby Snacks ★   18m F6a+

FA George Ridge 21 August 1997

Start up the centre of the broken slab, trending right at the overhangs to LO shared with 13.

13 One Step Beyond ★   15m F6a+

FA Isla Watson 24 August 1997

Start beneath a ledge at 3m. Climb up past the ledge then direct up the crack in the slab; steep moves to gain the LO.

14 Down to the Last Heartbreak ★   15m F6a+

FA George Ridge 30 June 1998

Climb 13 to the 2nd bolt, then step right and direct up the streaked wall.

15 The Trial of Brother No. 1 ★   15m F6a+

FA Colin Miln 7 September 1997

The centre of the scooped wall to a LO just beneath a prominent curving overlap at the top.

16 Lap Dancing   15m F6b

FA Rab Anderson 13 September 1998

A mossy slab, then blunt arête, surmounting the roof at the top.

17 The Llama Parlour ★   10m F6c

FA Janet Horrocks 28 September 1997

Sustained and fingery climbing up the steep technical wall 4m left of 18.

18 The Protection Racket ★   10m F6a+

FA George Ridge 21 August 1997

Strenuous climbing up the prominent left-facing groove at the right end of the face with the crux where the angle eases.

19 Mutant Stars ★   12m HVS 5a

FA Gary & Karen Latter 6 July 2022

The slim groove in the arête.

20 Lighten-up ★   12m F6a

FA Rab & Chris Anderson May 1999

A useful warm-up. The left-slanting flake leading to a LO up left. Awkward to strip.

21 Crowing at the Enemy ★   10m F6b

FA Rab Anderson 10 October 1998

The rib and wall right of the flake.

22 Bark Bacherache ★   10m F6b+

FA Rab & Chris Anderson May 1999

The rightmost route.

AERIAL CRAG    30min

NN 8460 5021 **Alt:** 300m

A short buttress situated almost directly above **Hanging Rock**. A good open crag with a fine outlook from the top. The two pitches up the left wall are very steep and stay dry in the rain.

Approach: As for **Weem Rock** to the hairpin, continue straight on for 100m and go up the right side of the 5m high bouldering wall and out right until a large oak tree. Head directly across broken ground, past the left side of the steep 10m high **Hanging Rock**, continuing steeply up an open glade, cutting out right about 50m above some dead elms just above a band of small outcrops running up the hillside.

1 Communication Breakdown ★　　12m E3 6a

FA Kev Howett & Lawrence Hughes 4 August 1997

The thin discontinuous crack at the left end of the steep left wall, finishing up an awkward V-groove at the top.

2 Saving Up for a Rainy Day ★★★　　18m E5 6a

FA Gary Latter & Rick Campbell 31 August 1997

Excellent sustained well protected climbing taking the obvious challenge up the centre of the wall. Climb the steep finger crack to a break at its top. Move out left along the break and up to good holds then climb the flange to a good nut placement in quartz near the top of the flange. Climb the wall above on thin slots to gain good flat holds at a small ledge. Stand on this and finish more easily.

2a Thunderstruck ★　　18m E4 6a

FA Jamie Skelton & Morag Eagleson 21 July 2020

The right arête. Start as for 2 to a jug at the top of the finger crack. Stretch up towards the next break and protruding ledge at the bottom of the arête. After mounting the ledge exposed turbo jugs on the very edge lead to the top.

3 Pawn Channel ★　　10m E1 5c

FA Lawrence Hughes & Kev Howett 15 August 1997

The fine crack up the right side of the steep wall, finishing up the easier crack in the arête.

4 Kissing the Witch ★　　10m F6b+

FA Janet Horrocks July 1997

The left-most line of bolts up the wall.

5 Static in the Air ★　　10m E3 5c/F6c

FA George Ridge May 1997 (with bolts); FA sans bolts Kev Howett & Lawrence Hughes 16 August 1997

The thin hanging crack-line on the left side of the front face. Gain the crack from the left by stepping off the boulder and follow it quite boldly near the top to finish past a block and sapling. Tree belay above the ledge.

6 Cracking Good Reception ★　　15m HVS 5a

FA Kev Howett & Alastair Todd 18 August 1997

Follow the deep crack immediately right of 5.

7 Strong Signal ★　　15m HVS 5a

FA Grahame Nicoll & Kev Howett 2 August 1997

The shallow groove in the centre of the front face just right of 6. LO from huge bolt near top.

8 Moving the Aerial ★　　15m E1 5b

FA Kev Howett, Grahame Nicoll & Alastair Todd 4 August 1997

Start just right of 7. Climb up and move right round a blunt rib into a shallow groove. Up this then move out left and up past the left side of a small overlap to finish.

The following crags are all best approached from the castle car park.

MANYANA WALL   10min

NN 8380 4988 **Alt:** 180m

A small buttress, with some reasonable easier routes.

Approach: Follow the less obvious path directly behind the car park steeply past two zig-zags, then cut left to follow the path along the back edge of the walled garden to its end, then cut under a fallen tree. Head steeply up the hillside for 200m, then traverse right to the crag.

1 Tomorrow Never Comes ★　　　　　**12m F6b**

FA Neil Shepherd 2007

The right side of the arête.

2 Sometime Soon ★　　　　　**12m F6a**

FA Derek Armstrong 2007

The central line.

3 Don't Do Today What You Can Do Tomorrow ★

　　　　　10m F6a

FA Neil Shepherd 2007

Right line, pulling through niche in roof on good holds.

Adam Russell on the fingery testpiece The End of Silence, Weem Rock (page 312)

SECRET GARDEN CRAG

NN 8375 4993 **Alt:** 200m

A good steep face with a prominent roof running across the right side of the crag at mid-height.

Approach: As for **Manyana Wall**, but instead of traversing right, continue directly up (avoiding fallen trees and a boulder directly beneath the crag) for just over 100m to gain the right end.

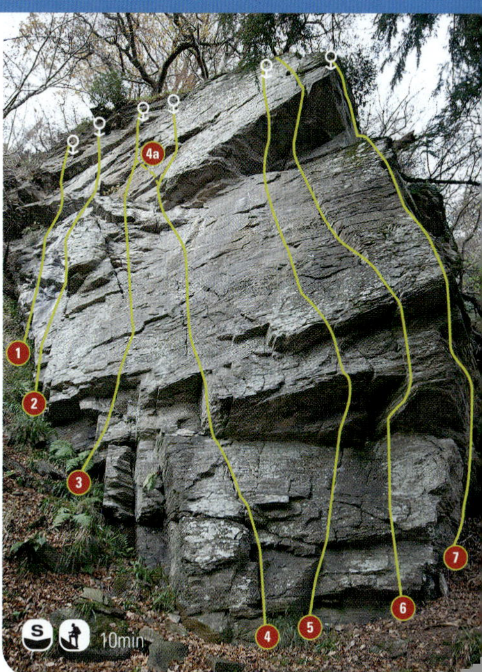

1 100 Ways to be a Good Girl — 12m F6b

FA Janet Horrocks 5 April 1997

The grey streak at the left end, cutting through three small roofs. Dirty rock and slow to dry.

2 Batweeman ★ — 15m F6b

FA Isla Watson 22 March 1997

The cracked left-facing corner and technical wall above. The best of the easier routes to warm up on.

3 Forbidden Fruit ★ — 15m F6b+

FA Isla Watson 12 April 1997

Start just right of the arête, 3m right of the groove. Climb on sloping holds to a ledge then with interest above.

4 Faithless ★ — 15m F6c

FA George Ridge 30 March 1997

The left side of the highest section, breaching the left side of the mid-height roof. 4a *The Missing Link* ★ F6c moves diagonally left past a gold bolt to finish up the crux headwall of 3.

5 The Watchtower ★★ — 15m F6c

FA Colin Miln 2 March 1997

The central line through the roof.

6 Caledonia Dreaming ★ — 15m F6c

FA Neil Shepherd 9 March 1997

Start just left of the arête. Follow the crack just left of the right arête with spectacular moves on good holds through the body-length roof.

7 Don't Knock the Block ★ — 15m F6a+

FA Janet Horrocks 21 February 1998

Right arête of front face. Up to the first bolt on 8, step left to the arête. Follow the arête to a ledge, step left above the roof and finish directly.

8 Brass Monkeys — 8m F6b

FA George Ridge 8 March 1997

The short wall at the right side of the crag. Can be a bit green.

EASTER ISLAND BUTTRESS 10min

NN 8382 4988 **Alt:** 180m

A good little crag about 50m right of **Manyana Wall**. There are rather a lot of bolts on the tiny triangular-shaped lower wall (about twenty-five!)

Approach: As for **Manyana Wall**, continuing right about 50m on a vague path/jungle bash to the base.

1 Right in the Face ★   **12m F6b**

FA Rab Anderson 13 July 1997

The right arête, swinging left to the LO of 2.

2 The Republic of Scotland ★★   **8m F7a+**

FA Colin Miln 20 April 1997

The line of bolts up the centre of the wall. Excellent climbing on good edges.

3 President Shhean Connery ★   **8m F7a**

FA Colin Miln 20 April 1997

Rather closely spaced bolts just a tad close to 4. The 4th bolt is about 3m up!

4 Left on the Shelf ★★   **20m F6c+**

FA Rab Anderson 22 June 1997

Follow the left line of bolts to the top of the initial wall, step left past a couple of good pockets (bolt on slab above) and pull over the roof on good holds to finish up the easier arête in a fine position.

The leftmost line is 5 *Motion Sickness* F7a, breaking left out of 4, through the roof and up narrow wall.

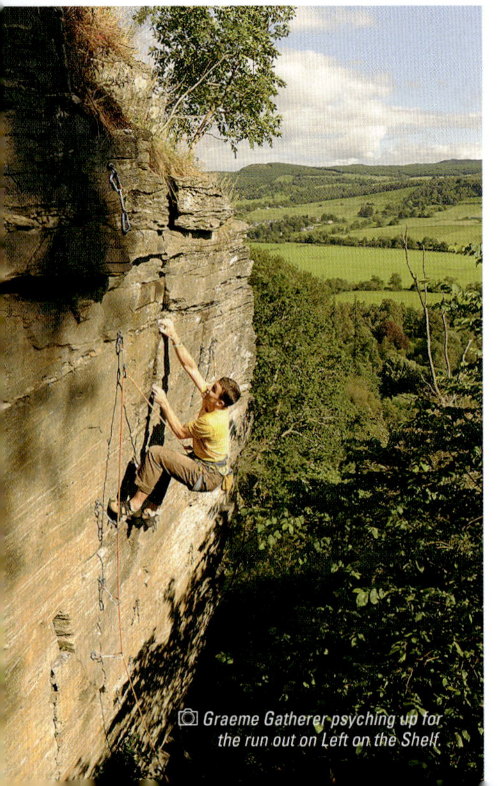

Graeme Gatherer psyching up for the run out on Left on the Shelf.

DIRC MHOR
(THE GREAT SLASH) (w) (hiker) 1hr 30min
NN 591 861 **Alt:** 560m

This impressive defile lies about 5 kilometres (3 miles) west of Dalwhinnie. It is a narrow kilometre-long slot lined with cliffs, particularly on the east side, where, at their northern end, they culminate in **Sentinel Rock**, an impressive 60 metre barrel-shaped buttress. The rock is micro-granite, similar to but of even better quality than the nearby Binnein Shuas. Both **Sentinel Rock** and **Ship Rock** dry quickly, though many of the other crags seep for several days.

Access: Follow the A889 for 1 mile/1.6km north of the distillery at Dalwhinnie to park on the east side of the road, in small car park (NN 6362 8624; 56.947220, -4.243431) just right of quarry entrance, opposite a track signposted 'Old Drovers Road to Feagour'.

Approach: Follow the track which leads to a new house after 1km. Skirt left on path below house, then cut up to gain an ATV track which follows the Allt an t-Sluic westwards. Follow this *"yellow brick road"* criss-crossing the burn then the burn itself heading up into the Dirc Mhor. Avoid the rough going boulder-choke in the base of the Dirc by cutting up left to contour into the base of **Sentinel Rock** on a good path.

SENTINEL ROCK

The showpiece, a stunning barrel-shaped 60m tall buttress standing guard over the lower reaches of the Dirc.
Descent: From the boulder-strewn terrace above the crag continue up more broken ground over the top of the small hill then head left (north) and return round to the base of the crag. Alternatively, a 50m abseil from an in situ sling and crab on block on the top right side of the crag leads diagonally rightward to the third grass ledge, or 60m abseil from thread at top of 3rd pitch of 3 *Fanfare*.

1 Working Class Hero ★★★   85m E2 5c

FA Allen Fyffe, Keith Geddes & Ado Liddell 1 August 1980

The ramp and corner up the left wall. Start at the base of the large slanting slab.

1 **15m 4b** Climb the big slab to belay at its top.
2 **30m 5c** Gain and climb the small right-facing corner in the overhanging wall above which leads to the clean cut ramp/corner line. Follow this spectacularly to a niche and exit out right onto a large ledge cutting across the front of the buttress. A stunning pitch.
3 **40m 4c** Step up to the large ledge which leads diagonally right across the face to a corner. Up this to the top of a pinnacle. Traverse diagonally right using a thin crack and from its end traverse back hard left. Easier rock leads to the top.

② The Man with the Child in his Eyes ★★★★

85m E6 6b

FA Kev Howett & Graham Little 24 June 1995

Stunning climbing up the left arête of the front face. Start as for *Working Class Hero*.

1   **15m 4b** Climb the big slab to a belay.

2   **30m 6b** Climb a small groove in the wall just right of the corner of *Working Class Hero* for 4m until it is possible to span out right around the arête to gain a big flat-topped flake. Move up to join *Fanfare for the Common Man* at the left end of its belay ledge. Up the arête past a quartz fin where hard moves gain a prominent hold in the arête. Continue up the thin crack on the left side of the arête with sustained difficulty until an exit left near the top gains a belay at the left end of the large ledge.

3   **15m 6a** Follow slabby rock up and left to gain the thin crack near the arête. Up this to belay.

4   **25m 4c** Trend leftward to the base of the left trending ramp. Follow the crack directly above to broken ground. Scramble rightward to reach the boulder strewn terrace.

◎ Ewan Lyons & Karen Latter on the main pitch of Working Class Hero.

③ Fanfare for the Common Man ★★★   **90m E5 6b**

FA Kev Howett & Graham Little 16 July 1994

Brilliant climbing up the left side of the front face. Start as for *Working Class Hero* at the lowest point of the crag.

1   **25m 5a** Climb the easy angled slab for 10m (as for *Working Class Hero*) to a point where a prominent flake crack breaks the right wall. Ascend this then traverse right along a fault to step up onto the right of two sloping ledges. Move left to take an awkward belay at the junction of the two ledges. Climbing direct from the fault to the ledges via a short steep diagonal crack is 6a.

2   **25m 6b** Step up onto the higher ledge then move left to the base of a slight groove. Up this to a thin horizontal crack. Pull up bulging rock with increasing difficulty to a hairline horizontal crack then make committing moves left and

up to gain better holds. A sequence of good
holds lead to the large ledge. A superb pitch.

3 20m 5c Climb easy rock just right of the belay to gain a
right-trending ramp shared with *Working Class Hero*.
Follow this to the start of a less distinct left-trending
ramp. Ascend this to a deep incut hold below bulging
rock. Go directly over the bulge to a thin flake then
move left to gain a flange. Move left and up to ledge.

4 20m 4a Climb a left-trending stepped
groove then scramble back right and up
to gain the block strewn terrace.

④ Waiting for the Great Leap Forward ★★

60m E7 6b

FA Iain Small & Tony Stone 10 August 2013

A direct line up the centre of the buttress, with a very
technical section through the lower roof and bold wall
climbing in the upper section.

1 40m 6b Climb the initial few moves up the flake of
Positive Earth before moving leftwards to below the
widest part of the roof. Powerful moves across the
roof gain a good hold at the lip. Pull up and slightly
right to gain a standing position. Pull ropes through
to reduce rope drag above. Continue fairly directly to
a small ledge below a steep compact wall and small
knifeblade peg. Climb up and make a reach to clip
the peg. Small edges and smears lead out leftwards
then up to a blind reach to a good left sidepull.
Further technical climbing leads back rightwards
to easier angled ground beneath a slim right-facing
corner with a small roof at its top. Pull out leftwards
from the top of the corner and follow breaks steeply
up and slightly left to gain the large belay ledge.

2 20m 6b Climb easy rock just right of the belay to gain
a right-trending ramp shared with *Working Class
Hero*. Follow this to the start of a less distinct left-
trending ramp and ascend this to deep incut hold
below bulging rock (as for *Fanfare*). Pull up and right
to a vague groove (good cam down and right). Using
crimps on the right pull up and reach out leftwards
to sloping holds and use these to pull over onto
easy ground. Continue up to blocks above to belay.

⑤ The Scent of a Woman ★★★   95m E6 6b

FA Kev Howett & Graham Little 30 July 1994

The central line up the front face completing a great trilogy.

1 25m 6b Climb the flake as for *Positive Earth* then
continue up the overhanging corner cutting through
the roof to a rest above the lip. Move right then up
the pod on the right side of a big block to a pedestal.

2 30m 6b Step right and up into a short, shallow
right-facing groove. Pull out left to gain a thin crack
which leads up to join the left-trending diagonal
crack of *Positive Earth*. Follow the continuation
of the diagonal line up left on good holds then
traverse diagonally left to under a prominent
isolated roof in the centre of the face. Climb
directly up the two-stepped overhanging wall
above, with hard moves over the final lip then
move up right to gain a rock ramp which leads
to a belay at the right end of the large ledge.

3 20m 6b Move left up a short slabby wall to a right-trending ramp. From this follow a less distinct left-trending ramp (common with *Fanfare for the Common Man*). Move right and up to a diagonal crack in the bulging wall then move up and right to gain a sharp edge. Make difficult contorted moves up to reach an undercling (F #0). Undercling rightward to gain a poor crack in a bulge. Strenuous moves on scarcely adequate holds lead up to a belay ledge.

4 20m 4a Climb a short slabby corner then broken ground leading to the terrace.

6 Positive Earth ★★   85m E1 5c

FA Martin Burrows-Smith & Allen Fyffe 23 August 1980

A line up the right side of the face heading for the big corner. Start at the toe of the buttress.

1 20m 5a Climb the corner until it steepens then escape up the right wall onto a large slab. Follow a line diagonally up the wall above to belay at large prominent blocks.

2 20m 5c Move slightly left into a niche. Climb the thin diagonal crack which leads leftward with a hard bouldery move to reach a good small triangular ledge. Up the wall above on good holds leading into the base of the corner.

3 15m 4a Climb the corner and exit left under the roof to belay on the large ledge on the front of the buttress.

4 30m 4c Finish up the fine vertical crack above.

7 Holy Smoke ★★   80m VS 4c

FA Alasdair 'Bugs' McKeith & party 23 August 1966

An excellent superbly positioned top pitch. Start at the lowest of three grassy slabs, above and right of the lowest rocks. The first (5a) pitch of *Positive Earth* would provide a fine start.

1 20m 4b Climb a short wall then rightward to the large blocks of the first belay of *Positive Earth*. Continue up the wall above by a crack on the right to the third grassy slab.

2 25m 4c Climb the corner then move left onto the wall of the buttress. Continue up to belay on small ledges on a slab below the big corner.

3 35m 4c Climb the left arête of the corner to the roof then go left to the large ledge. From the top of the pinnacle above, hand traverse up and rightward in a wild position until a narrow ledge leads right. Directly up then left to a slab. Follow the crest to the top. **7a** A more direct single pitch version is *Holy Moly* ★ VS 4c.

SEA OF SLABS

NN 590 860 **Alt:** 570m

The inaptly-named more broken area of rock extending rightwards from **Sentinel Rock**. Slow to dry.
Descent: Make a 50m abseil from nut & maillon (take a spare 4 or 5 nut to replace).

1 Scorched Earth ★★   45m E1 5b

FA Janet Horrocks & Kev Howett 25 June 1995

Ascends the clean-cut slab sandwiched midway between two corners. Start below the corner between two small rowans.

1 25m 4c From the start of the big corner scramble horizontally rightward along the grass ledges then the slabby wall leading up and right to an easy slab.

SHIP ROCK

NN 589 859 **Alt:** 600m · NW SW · 1hr 40min

Two hundred metres further south up the Dirc is a short steep projecting buttress above an area of easy angled slabs and grass.

PORT WALL 

Belay on Fs and Ns on large block 10m back from the top.

Approach: All routes are approached from the left. Scramble up easy ground then climb a short corner and arête (Severe 4a) which leads to the left end of the wall.

Descent: Scramble up heather then walk right (south) to the top of the descent gully, or abseil (35m) from in situ anchor down big corner on far right of buttress.

1 Spurlash ★★   15m E4 6b

FA Kev Howett 28 August 1994

The deep central crack. Start off the left end of the ledge. Move left under a small roof to gain a flake. Pull over the roof rightward to gain the crack, which leads with

Traverse right and up to a chokestone belay beneath the open chimney on the right side of the slab.

2 **20m 5b** Climb a crack-line up the centre of the slabby wall then large knobbles to the easier central slab. Follow this slightly left to near its left arête just below the top of the crag. Step right into a short ramp and make a hard exit. Belay at small rock outcrop 15m further back.

difficulty to a shelf and large flake. Pull up the wall above and finish up the fine deep crack in the pale coloured rib.

2 A Deep Green Peace ★★   15m E4 6a

FA Kev Howett & Tom Prentice 11 August 1994

The thin crack immediately right of the deep crack of 1. From the left end of the ledge follow a very thin crack diagonally right into a short corner. Up this then make strenuous moves out right to gain and climb the thin crack to the shelf. Stand on the shelf and step left to the big flake of 1 and finish up this.

3 Murderous Pursuit ★★   15m E3 6a

FA Kev Howett & Janet Horrocks 13 August 1994

The shallow scoop in the right side of the wall just left of the arête. Climb up to a vertical quartz-studded 'bore-hole'. Gain the thin flake crack above and follow it to a jug at its top. Pull up and right to a small flake then step right to pull onto the shelf. Finish up the cracks above.

4 Breaking the Wind ★★   15m E4 6b

FA Kev Howett & Lawrence Hughes 13 August 1995

The arête. Start up 3 to just past the 'bore-hole'. Reach right round the arête to gain a crack. Follow this into the short hanging groove, exiting onto the easier wall above.

STARBOARD WALL 

⑤ Close to the Wind ★★   **30m E5 6b**

FA Gary Latter (on-sight) 9 July 1996

Excellent well protected climbing. On the overhanging right (starboard) wall of the crag is a very prominent crack directly above the right side of the blocky ledge. Climb the crack on good holds to a good undercut flake at half-height. Continue with difficulty up the thin finger crack above, which leads to an easy-angled slab. Finish more easily up the rib (just right of 4) to belay a long way back.

SHINING WALL

NN 589 858 **Alt:** 610m (w) 1hr 40min

One hundred metres beyond **Ship Rock** is a further projecting buttress of clean white rock which forms the left side of the only descent gully on the entire east side.

① Carry on up the Khyber ★★   **45m E1 5b**

FA Graham Little & Kev Howett 30 July 1994

The fine left arête. Climb a stepped corner leading right to a heather ledge. Move up the wall on the left to flakes near the arête. Climb just right of the arête to gain easier angled rock, then cracks up the left side to reach

easier rock. Finish by a steep little corner containing a worryingly jammed block.

② Nature's Raw ★★   **45m E2 5b**

FA Kev Howett & Grahame Nicoll 28 September 1997

Excellent climbing up the grey wall right of 1. Scramble up to underneath the short corner immediately right of that route. Climb the corner passing a perched flake with care to gain a sharp flake forming the top of the corner. Step immediately right into the centre of the face and climb up, then right to gain a slight recess near the top right side. Finish onto the easy slabby rock at a junction with 1, then up the final blank looking corner, which is well protected and covered in good holds.

③ Here's Johnny ★★★   **40m E3 6a**

FA Tim Miller & Jamie Skelton 26 June 2020

The right-slanting crack. When the crack peters out, traverse left to a junction with 2 and finish up it.

DESCENT GULLY WALL

Alt: 660-680m   (w) 1hr 40min

Two walls on the right of the gully. Clean rock, but slow to dry.

① Bogart Corner ★★   **55m E2 5b**

FA Hannah Burrows-Smith & John Lyall 15 June 2006

A striking grey roofed corner above a bay, gained by scrambling.

1 **40m 5b** Climb black rock at a break right of a quartz dyke and go boldly up into the corner. Follow the corner and go round a roof on the left to a ledge.

2 **15m** Finish up a short wet wall and easy heather.

② Bournville ★   **30m E1 5b**

FA Grahame Nicoll & James Hotchkiss 13 August 1995

Further up the gully is a beautifully clean brown wall just left of a low mossy cave. Start from a good ledge and climb the centre of the wall until possible to move right and up to a ledge. Move left and up to finish.

NN 467 824 **Alt:** 500m

1hr 45min 1hr 15 min

BINNEIN SHUAS
(SMALL, PEAKED MOUNTAIN)

A fine cliff offering a good range of routes with an idyllic outlook. The cliff has a beautiful remote feeling despite its accessibility. Being further east it can often be dry and sunny here when nearby Glen Coe & Ben Nevis are enshrouded in cloud (but has a more westerly climate than Creag Dhubh and the Cairngorms). The rock on the left side of the crag is composed of micro-granite and feldspar, while the right (Far East Sector) is composed of horizontally bedded mica schist, similar in character to nearby Creag Dhubh.

Access: Along the A86 Spean Bridge – Newtonmore road. Park at a large lay-by on the south side of the road at (NN 4328 8304; 56.912093, -4.5758393) just east of a concrete bridge over the River Spean, about a mile/1.6km west of the west end of Loch Laggan. This is 14.2 miles/22.7km west of the A889 Laggan turn-off (4.2 miles/6.7km west of the Creag Meaghaidh car park) and 14 miles/22.4km east of Spean Bridge. There is additional parking 0.2miles/0.3km further west.

Approach: Cross the bridge and follow a good Land Rover track through a large gate and continue to a fork at the top of a rough steep section. Continue right, curving round the south flank of the hill to the shore of Loch na h-Earba. From just before the bridge a rougher boggy path heads diagonally across the hillside on the north side of the loch to reach the cliff (45 minutes).

Descent: Follow a vague path heading left above the top of the cliffs then a well-defined path skirting beneath the **West Sector**.

WEST SECTOR

From the left the first crag of any height is a fine 25m rectangular wall seamed with lots of vertical cracks.

 **Comraich ★**   30m E1 5b

FA Andy Tibbs & Alastair Matthewson 14 August 1993

Start up a pale streak 5m left of 3, midway between twin cracks. Ascend the streak, moving right at 6m. Continue up the right edge of the hollow, stepping left to finish up deep twin cracks. The direct start is 1a *Cloudberry*, E3 5c.

2 Bearberry Groove ★   30m E2 5c

The crack-line immediately left of 3.

3 Blaeberry Grooves ★★   75m VS 4c

FA Graham Hunter & Dougie Lang 17 September 1967

An excellent well protected main pitch up the deep crack system in the centre of the rectangular wall.

  1 **30m 4c** Climb the crack which lies back in its upper section. Belay on the heather terrace beneath a

crack system in the centre of the hanging slab.

2 45m 4a Climb the steep wall on good but well spaced holds to a good spike at the top. Pull onto the slab and continue easily by the central crack. Belay just beneath the top.

4 Medusa ★   30m HVS 5a

FA Andy Nisbet & Jonathan Preston 20 May 2008

Start immediately right of 3. Move right, up and back left to gain and climb the crack-line which is just right of 3.

5 Gorgon ★   75m E1 5b

FA John Mackenzie & N.Fraser 1969

5m right of *Blaeberry Grooves* is a deep slot which peters out at half height.

1 30m 5a Climb the offwidth slot with interest, continue direct in the same line to the heather terrace.

2 45m 4b Start just left of right-slanting overhung ramp. Climb wall on good holds, then crack up slab easily to top.

THE FORTRESS

The centre of the buttress above a large grassy terrace has a huge capping roof. Approach by scrambling up a grassy ramp and short corner containing a rowan to gain the left end of the terrace.

1 Kubla Khan ★★★   107m HS 4b

FA Graham Hunter & Dougie Lang 25 June 1967

Fine climbing up the pale wall just right of the prominent vertical dyke at the left side of the cliff. Start just right of the dyke.

1 15m 4a Climb the cracked lower wall directly underneath the prominent white streak to a good ledge.

2 45m 4b Follow the cracks up the slab just left of the pale streak, transferring right to a further set of cracks up the pale streak to a thread belay below a huge boulder (possible abseil descent, 50m ropes gain easy ground on left). A brilliant pitch.

3 12m Walk up heather on the right to the wide grassy terrace.

4 35m Up a boulder on the left and finish

up pleasant easier slabs. Tied off clump
of grass or sprig of heather for belay!

2 The Rubaiyat ★★   70m E1 5b

FA Gary Latter & Judy Hartley 30 August 1996

A direct line up the right edge. Start down and right of
Kubla Khan beneath twin parallel cracks.

1 **20m 4c** Climb direct, passing a tiny rowan sapling
near the top to pull onto a heather terrace.

2 **50m 5b** Climb an easy niche which leads to a flared-
crack in the slab. Follow this past a thin section
low down (poorly protected crux) and continue in
the same line to a prominent right-slanting break.
Shuffle right along this and continue in the same line
to finish up a wider crack. Move out left to belay as
for *Kubla Khan*. Either continue up this or abseil off.

3 The Keep ★★   95m E1 5b

FA Dougie Lang, Graham Hunter & M.Main (3 PA) 16 July 1967

A good route with a finely situated main pitch up the left
edge of the steep central section of **The Fortress**. Start

from the right end of a grassy ramp leading up to The
Garden (rowans at left end).

1 **10m 4b** Up the short steep hand/fist
crack to a block belay on a ledge.

2 **35m 5b** Climb the edge on good holds to the
prominent twin cracks splitting the steeper wall
above. Up these with interest and continue
more easily in the same fault line. An easy slab
leads to a belay on a small grass ledge.

3 **50m 4a** Move out leftward and up the
centre of the easy unprotected slab.

4 Siege Engine ★★★   60m E7 6c

FA Iain Small 12 October 2016

A mega endurance trip taking the huge diagonal fault in
the left side of huge roof. Well protected but sustained
with the crux near the top.

1 **30m 6c** Climb easily up the short slab to gain the
diagonal undercut feature leading out left to *The
Keep* and a good rest on the arête. Arrange gear
and launch rightwards along the diagonal ramp/

flake. Continue to a stopping place below a roof and an in-situ thread runner. Step right and climb the steep groove (crux) to eventually gain a welcome jug just before the lip. A further sting in the tail leads to a belay on the slab above.

2 30m 4b Climb easily up the slab above.

4a Dun Briste ★★★　　　　　55m E8 6c

FA Dave MacLeod July 2017

The obvious direct finish to 4, taking the large roof and headwall above its traverse. Follow 4 to the thread. Arrange good cams in the roof. Make a very powerful stretch across the roof to gain the short flake-crack on the lip (protection). From the top of the crack, make a hard lunge (crux) to reach the slabs. Climb easily up the slabs above to finish. Well protected and excellent F8a+ climbing.

5 Storming the Bastille ★★　　　80m E3 5c

FA Mark Charlton & Alan Moist (1 PA) 17 May 1984;
FFA Mark Charlton 1980s

Paul 'Stork' Thorburn on the first ascent of Greatness and Perfection, the inaugural hard route on The Fortress.

The spectacular offwidth chimney. *"Holds and gear are friable up to the roof, but it's easy. Gear then is excellent."*

1 40m 5c Make a hard entry into the bomb bay crack and follow it to the roof. Back and foot the chimney through the roof to a steep slab, which is followed to good finishing holds.

2 15m – Climb the easy corner to ledge and tree belay.

3 25m 5a Follow diagonal crack and slab above.

6 Greatness and Perfection ★★★★　　40m E7 6c

FA Paul Thorburn & Rick Campbell June 1998

A stunning route breaching the centre of the roof. Start in the centre of the buttress 3m right of *Storming the Bastille*. Pull over the bulge into a scoop on the left then a gain very shallow left-hand runnel and follow it to good nut placements at 12m (very serious). Continue with difficulty to the roof (large Fs) and undercut this rightward. Climb the thin diagonal finger crack with a fiendishly hard move to reach better jams above the lip. The easier upper crack leads to a nut belay in a bay.

7 Isinglas ★★★★　　　　　　40m E7 6b

FA Iain Small 29 August 2016

An excellent varied climb taking the large flange feature in the centre of the wall. Climb a delicate slab to reach the roof at the base of the flange. Undercut rightwards to reach a welcome peg runner (crucial wire above). Make a hard move over the bulge and follow the corner/ramp with slightly less difficulty to its top. Move right, then up under the large roof, then step left with a technical move to reach a jug in the break at the left edge of the roof. Make hard technical moves to get established on the slab above (crux) and continue easily to a belay at a flake on the heathery ledge above.

8 Ardanfreaky ★★★　　　　　105m E3 5c

FA Mark Charlton, John Griffiths & Alan Moist 27 May 1984

"Fantastic positions and climbing, with an exciting traverse at the top with tons of air below your feet. Very exposed; but well protected and not pumpy – low in the grade." The first two pitches can readily be run together,

omitting a cramped belay. Start below grassy cracks at the right side of the huge roof.

1 **20m 5c** Climb the line of grass (yes, there is a crack in there somewhere) to move right into the corner and up this to a ledge. Your mate's pitch!

2 **20m 5c** Pull over the roof above on good holds into a corner. Traverse left under the large roof to a curious drooping flake. Step up to the roof and continue traversing in a fine exposed position into a short hanging corner. Struggle up this to move left on to a ledge.

3 **40m 5b** Follow the diagonal crack rightward then slabs above in the same line to a tree belay.

4 **25m 4b** The corner crack to the top.

9 Wild Mountain Thyme ★★   45m E5 6b

FA Neil Craig/Gary Latter (both led) & Rick Campbell 31 July 1995

Right of the central roofs is a steep wall with three prominent cracks. This follows the right-slanting diagonal one. Amble up an easy slab to a crack and along this to an undercut. Move up to better holds then make a hard move back right which soon leads to easy ground. Finish up the easier vertical crack over the bulge as for the next route.

10 Bog Myrtle Trip ★★   40m E4 6b

FA Rick Campbell 6 July 1995

The deep crack on the right side of the wall. Follow the crack with difficulty to some good finger slots then up the excellent hand crack to a bulge. Cross this on good holds then trend slightly right to belay at the back of the terrace

EAST FACE.

1 The Fortress Direct ★★   155m HVS 5a

FA Tom Patey, Robin Ford & Mary Stewart (1 tension traverse) 16 May 1964; Lower part Rab Carrington & Jimmy Marshall July 1970

Described starting up the *Variation Start*, which gives the best line. Start on the steep right wall, about 20m up right form the toe of the buttress beneath a short left-facing groove with a roof low down.

Mike Reed starting up the superb crack of Delayed Attack.

1 **55m 5a** Climb corner and undercut left into an easier groove until possible to traverse left onto a large grass ledge. Follow twin cracks and a diagonal crack leading to the right end of **The Garden**. Belay at the right end of the wall - medium-large nuts/cams useful. This pitch can readily be split by belaying at the base of the twin cracks.

2 **45m 5a** Move up rightwards into triangular niche with a crack rising from its apex. Climb the fine crack, step right at the top, then direct to large ledge.

3 **55m** Finish up the edge above.

2 Delayed Attack ★★★   45m E3 6a

FA Geoff Goddard 30 July 1983

The superb right-curving crack up the right side of the face. Follow the crack with a hard section over the overhang and up a small groove. This leads to a short wide hand and fist crack in a fine position to the top. Very well protected.

EAST SECTOR

This is the area of rock right of **Hidden Gully**, an aptly-named left-slanting feature splitting the much steeper **The Fortress** from the generally slabbier **East Sector**. The gully only becomes apparent from below.

① Ardverikie Wall ★★★★   **187m HS 4b**

FA Graham Hunter & Dougie Lang 24 June 1967

"This route can be considered a classic. Its appearance belies its grading and, with the exception of the first [originally 12m] pitch and some lichen on the rib, is on perfect rock throughout and with magnificent situations. The pioneers can think of no other route of the same grading of comparable quality." – Doug Lang
Scottish Mountaineering Club Journal, 1968

A classic route on immaculate rock, possibly the best route of its grade in the country. Expect other parties at weekends. It follows a direct line up the face right of Hidden Gully. Start 8m left of a boulder which forms an arch. Harder and more direct variations may be made.

1 **45m 4a** Climb the rib on good holds leading to a point overlooking a short corner on the right. Move back left and continue steeply up the rib then by an immaculate slab slightly rightward and back left to ledge and flake belays. A brilliant pitch.

2 **27m 4a** Move right and up the beautiful flake groove to its top. Step up right to a ledge then climb cracked wall to large flake belay at base of obvious thin right-slanting crack.

3 **40m 4b** Gain and climb the crack (crux) for a few metres, then more easily up an ill-defined rib leading to easy-angled left-slanting grooves which lead easily to a large flake belay in a shallow scooped area.

4 **25m 4a** Trend leftwards to gain right-slanting flake/overlap and follow this, stepping left at its top over small bulge. Continue easily to back of terrace.

5 **50m** Easy slab to a block belay on the left.

③ Behind Enemy Lines ★★★   **40m E2 5c**

FA John Aisthorpe & Masa Sakano 27 September 2020

The well protected crack above the ramp. Originally climbed in three pitches. Start up the easy ramp and belay below the vertical finger crack (Camalot 3 or 4). Climb the finger crack and follow a widening crack leftwards and then traverse further left until jugs. Climb straight up the unlikely overhang above on good holds, with gear in hidden breaks. Belay on the ledge under the roof (Camalot 3 and many small to hand-sized cams useful). Climb a very short easy traverse pitch left to **The Garden**.

Ardverikie Wall. Martin Grant nearing the top of the crucial third pitch.

1a Flypaper ★★ 50m HVS 4c

FA Graham Hunter & Dougie Lang 24 June 1967

A very worthwhile pitch, providing a fine alternative to the easy final pitch of *Ardverikie Wall*. Traverse easily along the grass terrace for 50m to the base. Ascend the prominent pink streak on good holds to a flake on the right. Climb the wall above the flake on small holds to reach and finish up a layback crack.

FAR EAST SECTOR

About 70m beyond *Ardverikie Wall*, the rock changes from the micro-granite to horizontally bedded schist with large pegmatite intrusions. Only the lower 50m is good clean rock; abseil anchors in situ at top.

Gary Latter nearing the top of the first pitch on the first ascent of Thru With the Two Step. Photo Karen Latter

① Dancing in the Dark ★★★   50m E5 6a

FA Gary Latter & George Beeton 1 June 2023

A superb direct line on impeccable rock up the right side of the smooth wall.

1 30m 6a Climb up to the left side of the large sloping ledge. Pull left on slopers above lip of roof to good holds on left (1" cam). Direct to good fingerlock at base of left of twin hanging cracks, and up these to the left end of small roof. Move right then up crack/arête to good holds in break. Continue direct, then leftwards on huge holds to belay on small ledge. Yellow/blue Camalot. A double set of cams useful.

2 20m 6a Easily above, then direct up the bulging wall to a good flat hold. Make a long span right to good holds, then rockover onto ledge above. Trend right, then left to top.

② Thru With the Two Step ★★★   55m E2 5c

FA Gary & Karen Latter 21 April 2023

The hanging right-facing curving corner. Start 6m right of the large embedded block.

1 20m 5c Climb easy flakes to ledge, then the flake and corner leading rightwards to ledge.

2 35m 5c Traverse right, then trend up leftwards on pegmatite, continuing to slabby right-facing corner. Climb this to large ledge on left at its top. Make committing moves up rightwards (crux), then move right and up on large pegmatite holds. Move out left to finish on good holds to the abseil point.

③ Soft Shoe Shuffle ★★   55m HVS 4c

FA Graham Hunter & Dougie Lang 24 May 1968

Good climbing on fairly steep rock. Start at the right end of an overhang.

1 25m 4c Either traverse left above the overhang, or harder direct on big holds through the right side, then go up to gain a flake crack leading to the right end of a roof. Continue up the wall above on huge holds, then make a short thin traverse right before moving up to a flake. Make a thin move up right to another flake at the left side of a slab.

2 30m 4c Traverse left to climb a steep pegmatite band trending slightly left on good holds to belay below a grass cornice.

NN 671 958 **Alt:** 370m   10–20min

CREAG DHUBH

With over 200 routes this is far and away the largest and best roadside crag in Scotland. The rock is horizontally bedded mica-schist. In the past the crag unjustifiably earned the moniker 'Creag Death'. This was nothing to do with the climbing, but a reference to the proliferation of rotting sheep carcasses that once littered the base. Thankfully, sheep haven't roamed (or leapt from!) the hillside for at least the last couple of decades. In addition to a range of bold run-out routes, the majority of routes on offer are adequately or indeed well-protected, particularly with a good range of cams.

Many of the routes dry quickly, with any seepage lines obvious from the approach. It is often a good retreat when it's wet further west around Nevis and Glen Coe and it's often possible to climb here most of the year. Although there are numerous easier routes, the best climbing and by far the highest concentration of routes is from HVS upwards, though there is still enough of interest in the lower grades, with a fine selection of VSs. Double ropes will also be found useful, as most of the routes involve abseil descents. A good overview of the crags can be gained from the field opposite the parking spot.

Access: Creag Dhubh overlooks the infant River Spey and the A86 Spean Bridge – Kingussie road, some 3.3 miles/5.3km south-west of Newtonmore (4.2 miles/6.7km east of Laggan). Park in a lay-by on the north side of the road (NN 6735 9573; 57.033529, -4.1873626), opposite the westmost of two small ox-bow lochans.

Accommodation: There are a couple of good camping spots beneath the crags, one on flat ground beyond the stream 300m west of the parking spot, the other 80m up right of the parking spot, in line with **Sprawl Wall**.

Approach: Go through the gate at the car park and follow the path which soon rises steeply, scrambling over a boulder field near the top to gain a horizontal path just beneath the crags. **The Great Wall** is just up to the left, with the **Lower Central Wall** closer on the right and **Lower Right Wall** just beyond. **Sprawl Wall** lies well up and right of here, above a large boulder field. The continuation of the crags just left of **The Great Wall** soon becomes **Waterfall Wall**. Higher up and left lies the initially-hidden **Little Rock**, with **Bedtime Buttress** the furthest left, with a prominent large projecting roof high on its right side, well seen in profile.

Crags are described as logically encountered on the approach, first from **The Great Wall** rightward, then leftward.

THE GREAT WALL 

NN 6713 9581 **Alt:** 380m 10min

The showpiece of the crags, in its centre a superb 40m gently overhanging wall containing many of the best routes on wonderful folded contorted schist.

Descent: By abseil from in situ anchors - see topo.

1 King Bee ★★★    85m VS 5a

FA Dougal Haston, James Moriarty & Arthur Ewing April 1965

Excellent climbing with a short well protected crux. The classic of its grade on the crag. Start at a prominent rib at the left end of **The Great Wall**.

  1 **25m 4c** Up the slabby right side of the rib to a small tree then left and up a shallow corner to a tree. Traverse right under a roof and climb to belay at a small tree just above.

  2 **35m 5a** Trend diagonally left to a tiny groove in the rib. Make a difficult move over the bulge (good protection) and continue to a belay on a quartz slab beneath a roof. Either abseil off, or:

  3 **25m 4c** Move out right to break through the roof just left of a large perched block. Up vegetation right-wards to belay on ledge. Traverse to abseil trees.

1a King Bee Direct ★★   20m HVS 5a

FA Kenny Spence & R.Gough September 1967

An excellent exposed direct on the first pitch. From the first small tree, climb up the left side of the arête then step right (F #3 in break) and pull spectacularly through the roof on good holds to a belay just above.

2 Erse ★★   60m E3 5b

FA Pitches 1 & 2 Dougal Haston & James Moriarty May 1965 (aid on P.1); FA Pitches 3-5 Dougal Haston & Joy Heron May 1965; FFA unknown

A serious first pitch and aptly named, as that's what you would land on from a great height! Start just left of *Brute*.

1 **25m 5b** Climb a small shattered groove crossing some overlaps then boldly up the wall above to a V-notch in the roof. Cross this on good holds to belay on large flake just above.

2 **35m 5a** Ascend the wall and crack to a ledge beneath a roof. Turn this by a bulge on the left and continue up a groove.

3 Jump for Joy ★★   60m E2 5b

FA Dave Cuthbertson, Bob Duncan & Alan Taylor August 1981

Good climbing up the wall between 2 and 4.

1 **30m 5b** Climb *Brute* corner, then direct up the black streak in wall above (bold) to an old peg under roof. Pull rightwards over the bulge, then move up left into a shallow groove leading to the long ledge and bolt belay of 4 on right.

2 **30m 5b** Climb the wall and steepening groove to prominent V-notch at left end of the roof. Pull through this on good holds then climb directly up the wall to a tree.

4 The Brute ★★   40m VS 5a

FA Terry Sullivan & Nev Collingham October 1959

Good climbing on the lower pitches. Start at an open groove just left of a large birch near the left end.

1 **15m 4c** Ascend the groove then step right and up to a ledge.

2 **25m 5a** Move diagonally right under the roof to cross it by a crack at its smallest point (PR). Continue, trending leftward to belay at a pair of bolts.

5 Little Mother F****r ★★   60m E1 5b

FA Fred Harper & Bill March 10 October 1970

A good direct line cutting through 4. Start just right of that route.

1 **30m 5b** Climb flake crack in the right arête of the

Gary Latter on first ascent of Fou Rire. Photo Craig Lamb

corner to the belay ledge on 4. Continue up the wall, skirting the left side of the roof, then trend rightward then back left to the bolt belay on 4.

2 **30m 5b** Climb directly above into the open quartz groove, step out right to a shallow groove and peg runner and directly up this to the right end of the roof. Move out left above the lip of the roof, then follow direct line to the large abseil tree on the right.

6 Fou Rire ★★   60m E3 6a

FA Gary Latter, Craig Lamb & Scott Grosdanoff 10 March 2016; pitch 2 Gary Latter & Dave McKinney 25 June 2016

Good varied climbing with a well protected crux. Start midway between the fallen birch tree and 7, just right of a shelf at the base.

1 **35m 6a** Make bouldery moves past good incut hold to the second break. Move left along this, then up past a left-curving flake to large quartz recess. Pull out rightwards and direct up wall trending left to

beneath the roof. Cross this at a prominent V-slot (jugs on lip), and continue easily up right edge of wall to gain the right end of the belay ledge on 4.

2 **25m 5c** Climb directly up the wall to the roof and good cams in slot. Reach over to good small edge on lip then span left to jugs (2" cam in slot). Pull over to good small break, then more easily up wall above rightwards then trending left to tree.

7 Outspan ★★★   **45m E2 5b**

FA Robin Barley & Brian Griffiths 13 April 1971

Excellent bold climbing, steady at 5a after the initial boulder problem start, following a wandering line up the left side of the impressive main section of the wall. Start at the thin crack near the centre of the wall (8m right of the birch). Climb direct to an obvious handrail and follow this left for 4m. Ascend directly up the steep black wall on good holds (bold) to a horizontal break. Move right along this to a bulge (PR), pull over this and go up right to a

large flake, continuing diagonally rightward to join 14 on long ledge. Follow this easily leftward to tree belay.

8 The Fuhrer ★★★   **45m E4 5c**

FA Dave Cuthbertson & Ian Duckworth 17 June 1979

An excellent serious pitch up the left side of the wall. Start at a thin crack near the left side of the wall directly beneath a prominent plaque of rock at 6m. Up the wall to a good rounded spike then the quartz scoop diagonally left to gain the plaque. Stand on the top of this then traverse right and up to the PR on 7. Pull over the bulge and go up right to a large flake. Climb the quartz wall diagonally leftward to a broken ledge, finishing up the steep shallow groove which leads to tree belay.

9 Colder than a Hookers Heart ★★   **45m E5 5c**

FA Dave Cuthbertson, Gary Latter & Calum Henderson 23 October 1985

Very good but serious climbing, taking an eliminate line based on 8. Climb to the spike on 8. Move slightly right to a small overhang (poor F & RP #1 on right), pull over this and directly up the wall above to the PR on 7/8. Trend left and up the overhanging wall to a jug at the base of the quartz wall. Up this going slightly left to a crack in the arête above the girdling break. Climb the crack past a dubious block and trend right to the tree belay.

10 Harder than your Husband ★★   **40m E5 6a**

FA Grant Farquhar & Ian Marriott 21 May 1988

The most serious pitch on the wall, taking a fine central line. Start beneath quartz streak 2m left of 11. Climb direct until forced right onto the ledge on 11, then continue up the bulging wall to a good finger jug about 3m below the traverse into the niche. Break out left and climb the quartz bulges up into the slanting niche (protection). Continue to the girdle ledge then move left to finish up 8.

11 The Final Solution ★★   **40m E5 6a**

FA Steve Monks & Willie Todd May 1987

A serious pitch just right of the centre of the wall. Start at the foot of the boulder to the left of 12. Climb the initial wall via a series of bouldery moves to stand on a ledge. Step left and climb directly up the bulging wall to gain a small niche at 15m – F #2. Climb directly out of the niche and up the wall passing through the second niche of 12 traverse to the tree belay at the top.

12 The Hill ★★★★　　　　　　　　　45m E2 5b
FA Kenny Spence & John Porteous September 1967

A tremendous route, serious in the lower section, then very exposed. Start by a large flat block beneath the rust coloured streak 7m left of 14. Climb past some spikes and up bulging wall to a poor peg, then up to good protection in the flake above. Continue over bulge into tiny groove/niche above, then traverse left rising slightly to another slight niche, then direct to gain the right end of a ledge. Continue straight up to finish up the last few moves of 14 to the tree belay.

12a Over the Hill ★★★　　　　　　　40m E3 5c
FA Dave Cuthbertson & Rob Kerr 24 October 1980

An excellent sustained direct pitch. Climb 12 until the tiny groove/niche above the small overlap. Continue up and right to good small spike, then traverse left 3m and climb difficult thin crack and short wall to ledge. Finish as for *Inbred*.

13 Under the Skin ★★　　　　　　　40m E3 5c
FA Gary & Karen Latter 11 September 2016

Bold climbing leading to a well protected crux, taking the obvious line between 12 and 14. Start midway between the two. Go directly past incut flakes to 1.5" cam slot at 5m. Move left, up, then back right, then direct up the wall, crossing the small roof in its centre to good holds above. Make a hard move from good horizontal break (1" cam slot), then slightly rightwards. Pull through the right end of the roof above the ledge on good holds and follow the fault diagonally leftwards, then direct up the wall to a thread belay.

14 Inbred ★★★　　　　　　　　　　55m HVS 5a
FA Dougal Haston & T.Gooding October 1964

The original route to venture up **The Great Wall** and

one of Haston's best rock routes. A fine natural line and a good introduction to the crag. Start near the right end of the ledge, directly beneath a large spike.

1　**25m 5a** Straight up the crack system to a PR, then trend leftward to pull into the triangular niche from the left. Step down and move right, then direct pulling through a bulge (just left of a large quartz patch) to the left end of the belay ledge. BB and large block on right. A more direct line perhaps nudges the grade up to E1 5a.

2　**30m 4b** From the left end of the ledge follow a leftward trending line to a tree belay.

14a Inbred Super Direct ★★★　　　　40m E1 5b
FA Gary & Karen Latter August 2016

A superb direct pitch, well protected after the niche. Follow 14 as far as the triangular niche. Climb the crack above steeply on good holds then trend leftwards to gain right end of large ledge above. Pull through crack at right end of the roof on good holds, then follow diagonal fault leftwards, then direct up the wall past good horizontal breaks to the thread belay.

14b Inbred Direct ★★★　　　　　　　40m E1 5b
FA Dave Cuthbertson & Rab Anderson 11 November 1976

Follow 14a to the left end of the belay ledge, then direct for 2m and move right under a bulge. Cross this and continue slightly leftwards over a further bulge to finish up a crack leading to the rightmost abseil anchor.

15 Strapadicktaemi ★★★　　　　　　50m E1 5b
FA Dave Cuthbertson & Rab Anderson 10 September 1976

Sustained climbing on good holds – a little bolder than *Inbred*.

1　**25m 5a** Start up 14 for 5m then follow a thin crack cutting diagonally right until above the tree then direct, trending slightly left through bulge to the belay on 14.

2　**25m 5b** Traverse right 3m then climb the left-slanting crack to a bulge. Step left and continue up another crack, finishing direct onto cleaned ledge.

LOWER CENTRAL WALL

S · 10min

NN 6719 9583 **Alt:** 360m

The shorter wall starting 50m right of **The Great Wall**.
Descent: Abseil from trees at the top of all the routes.

1 Rib Direct ★ — 90m Severe 4a

FA B.Halpin, T.Abbey & S.Tondeur, date unknown

The left-defining rib of the crag with a fine upper pitch.

1 **10m 4a** Climb left-slanting flake crack
  to belay on the large tree.

2 **25m** Continue up the groove to belay
  at a tree under an overhang.

3 **20m 4a** Move up under the overhang and
  move left onto shelf, then climb the main
  groove leading to another tree.

4 **35m 4a** Move out right onto the rib and follow
  this in a fine position, climbing over a small
  pinnacle to finish up the slabby crest.

2 Ticket to Ride ★★★ — 25m E3 5b

FA Dave Cuthbertson & Alan Taylor 9 September 1976

Serious sustained climbing up the prominent black streak
just right of the obvious quartz patch. Low in the grade.
Start below the quartz streak. Climb up and right across
the quartz to a spike. Ascend direct to a small niche,
moving left through a horizontal quartz band to finish up
the steep wall.

3 Cunnulinctus ★ — 30m VS 4c

FA Ray Burnett & Alasdair 'Bugs' McKeith November 1965

The vegetated chimney with a couple of holly bushes at
10m. Often damp in the lower half, but not unpleasantly
so! Start 5m down right of the chimney. Climb leftwards
up shallow grooves to the holly, then the open chimney,
finishing up the wall on the left to tree.

4 Phellatio ★★ — 30m HVS 5a

FA Arthur Ewing & Ian MacEacheran May 1965;
Direct Finish (as described) Ken Crocket & Colin Stead 4 June 1972

A popular companion to *Cunnulinctus*! Good steady
climbing starting up the shallow left-slanting groove
9m right of the chimney of 3. Ascend the groove to
a ledge at 15m. Continue directly up the wall above,
passing through a niche to finish slightly leftward up
the headwall.

5 Fiorella ★★   30m HVS 4c

FA Fred Harper, Bugs McKeith & Jock Knight November 1965

Another fine sparsely protected pitch. Start beneath
shallow groove 6m down left of 6. Climb the groove,
moving right at small bulge, then back leftwards. Trend
rightwards to finish.

6 Mirador ★   30m Severe 4a

FA Ian MacEacheran, Jock Knight & Ray Burnett October 1965

Start at the base of a short groove on the left bounding
rib of the deep gully. Climb the groove then move right
and follow the rib to the top.

LOWER RIGHT WALL

NN 6723 9584 **Alt:** 360m

A good compact clean wall, with convenient abseil
anchors at the top.

1 Youthanasia ★   30m E1 5b

FA Sandy Allan & Andy Nisbet 19 July 2014

Trend right past a tree and up a shallow corner then
slightly right. Pull leftwards through the roof, then trend
rightwards to the abseil point.

2 Scraping the Barrel ★   30m HVS 5a

FA Fred Harper & Bill March April 1972

Despite the name, it's actually not too bad, now that
it's been cleaned. Well protected. Climb slabby ground
leftwards into the corner; follow this over a couple
of bulges.

3 Oh de Goo ★   25m E2 5b

FA Gary & Karen Latter 5 July 2015

Good climbing crossing 4. Climb the wall, moving up right
to join that route at a small but well defined spike. Move
up right into a small groove and reach the lip, then make
a long reach up leftwards to a good hold over the lip.
Trend leftwards up the slab to the abseil point.

4 Gouttes d'Eau ★★   25m E2 5b

FA Dave Cuthbertson & Rab Anderson June 1981

Varied climbing on great rock. Start 3m left of the spruce.
Climb the wall over a bulge to a small but well-defined
spike on the right. Traverse left across the slab to a
small groove on the arête, which is climbed steeply on
improving holds to easier ground.

5 Featherlight ★   25m E2 5c

FA Dave Cuthbertson & Rab Anderson June 1981

Start behind the tree. Ascend wall over an overhang then up to the roof. Surmount this, then easier above.

6 Thyme ★   25m VS 4c

FA John Lyle & Andy Nisbet August 1991

Pleasantly juggy. Starting from just right of the tree, climb a small arête then fairly direct on good holds.

7 Just in Thyme ★★   20m E1 5b

FA James Thacker & Kris Hill 2 July 2020

Start 4m left of 8 and climb directly to an area of quartz. Step up and right to move up to a small overlap (thread) which is passed using an obvious V-slot, before climbing the green headwall close to a thin vertical crack.

8 Man on Fire ★★   20m E1 5b

FA Mal Duff & Tony Brindle 10 August 1983

Good well-protected climbing up the hanging vertical crack. Trend slightly left up quartz fault, stepping up right into thin crack which soon becomes more defined.

9 The Sting ★   25m E1 5b

FA Tony Brindle & Mal Duff 10 August 1983

Climb lower wall directly into the hanging corner. From a short way up this, break out right across the blocky roof, then direct to finish.

SPRAWL WALL    15min

NN 6740 9594 **Alt:** 410m

The prominent overhanging black-streaked wall at the top right end of the crags above the large boulder-field bounding the right edge of the crags. The main (right-most) streaked section of the crag is the slowest of the crags to dry, taking a couple of days after heavy rain (check from parking spot).

Approach: From the parking follow a small path diagonally right to a flat knoll, from where the wall can be viewed. Continue diagonally right, negotiating the boulder field.

Gary Latter on the steep crux of Jump So High (page 342). Photo Karen Latter

Descent: Either walk off right (east) and down the right edge of the cliff or abseil from trees

1 Tree Hee ★★★   60m HS 4a

FA Hamish Small & Jimmy Graham April 1965

Good climbing up the left side of the slabby right wall of the buttress. Start round right from a short overhanging wall at the toe of the buttress.

1 **30m 4a** Traverse easily diagonally up left to a shallow left-facing groove near the edge. Up this to a good ledge on the arête, directly beneath a holly tree.

2 **30m 4a** Cross the overlap just right of the holly then trend slightly left up the slab Bypass the final overlap on the left.

1a Left Variation ★★   35m VS 4c

FA Ian MacEacheran, Jonnie Knight & Ray Burnett May 1965

The left edge of the upper slab. From the belay beneath the holly tree, pull through the bulge just left of the holly with difficulty (4b using holly), then follow the obvious left-trending line in a fine exposed position, trending left to gain the large birch at the top left of the slab.

1b Tree Hee Direct ★★   60m VS 4c

Good climbing up the centre of the slab.

1 **30m 4c** Follow the easy initial traverse of 1 to the left end of wide horizontal crack, then direct up the slab on thin cracks to belay on the birch.

2 **30m 4b** Climb directly up the slab to beneath capping roof, and cross this 5m left of its right end at a small rib. Finish leftwards to the large birch.

2 Giggling Slab ★★   55m VS 4c

FA John Lyall & Dag Bulmer 1985

A good reasonably protected direct line up the right side of the slab.

1 **10m** Start as for 1 and scramble up to belay from highest birch tree.

2 **45m 4c** Move out left and pull through the right end of the wide slot and direct up the slab to wide vertical crack. At the bulge, step right along good break, then direct up slab. Continue direct past easing in the angle, aiming for left-facing corner at top. Climb this on good holds, moving out left at its top. Finish direct to sling and maillon on birch tree. 50m abseil.

③ Jump So High ★★   70m E1 5b

FA Fred Harper, Alasdair 'Bugs' McKeith & Arthur Ewing May 1965; Direct Finish George Shields & Chris Norris (1 PA) early 1970s; FFA Murray Hamilton & Kenny Spence 1978

Good varied well protected climbing. Start beneath the large vegetated corner left of the short overhanging wall at the right end.

1 **25m 4a** Climb the corner to belay on holly.
2 **30m 5b** Follow the hanging slab leftwards to its end. Climb the thin left-slanting steep crack to a good ledge.
3 **15m 5b** Finish up the fine crack above, finishing round the left side of the corrugated roof.

④ Jump So High Direct ★★★★   35m E2 5c

FA Fred Harper, Alasdair 'Bugs' McKeith & Arthur Ewing (5 PA) May 1965; Direct Finish George Shields & Chris Norris (1 PA) early 1970s; FFA Murray Hamilton & Kenny Spence 1978

One of the best pitches on the crags, with superb sustained climbing on impeccable rock. Start beneath a shallow left-facing groove down right of the ledge at the base of the overhanging left wall. Climb left-slanting groove with a hard pull to better holds, then boldly up black wall to base of steep crack. Climb the crack, pulling out right at the top on good incut holds to good ledge (possible belay). Continue up further short steep crack, then bypass the left side of the corrugated roof to a sporting finish pulling onto heather. Take plenty of medium cams if climbing in a single pitch.

The following five routes all start from a large ledge at the base of the left side of the clean overhanging wall, gained by scrambling up rightwards then over some blocks to gain its left end.

⑤ Instant Lemon ★★   45m E3 5c

FA Derek Jamieson & Dave Cuthbertson 29 September 1980

Excellent serious climbing meandering up the left side of the wall. Start at the left end of the ledge.

1 **20m 5b** Climb the easy broken quartz flake then step onto the wall and follow the prominent handrail out right, then direct up the wall at its end to large ledge.
2 **25m 5c** Traverse left on superb holds for 5m, move up then follow a line of holds in a pale streak leftward to enter a groove which is followed rightward to finish up the right side of a corrugated overhang.

⑥ Instant Lemon Direct ★★   40m E5/6 6a

FA Dave Cuthbertson, Pete Hunter, Derek Jamieson & Cameron Lees 30 September 1980

A very serious start since the disappearance of the original peg and some holds – the only protection in first 10m are 3 skyhooks.

1 **20m 6a** Climb directly up the first white streak 3m to the right of the flake, to join the end of the hand traverse. Move out right to belay.
2 **20m 5b** Traverse left and climb a stepped ramp then finish diagonally rightward up a wall and slab.

⑦ Migrants ★★★   35m E5/6 6a

FA Gary Latter 27 September 2015

A direct line up the yellow wall just right of second black streak from the left. Start as for 6 to the point where it

Karen Latter on Tree Hee Direct.

moves out right (or much safer, start up 5 to the same point – E4 6a overall). Continue straight up on good holds (good nut runner in small slot up on right) to good long incut ledge. Continue up crack in black rock with plentiful protection to its top, then use twin side pulls to pull up left onto large hold at base of easier slab. Up slab to gain large fault midway between two small aspens. Move up to bulge, then out right to belay on small birch.

8 Yes, Yes *   30m E7 6b
FA Dave Cuthbertson & Joanna George September 1997

An eliminate, giving serious technical climbing direct up the wall right of 6. Start from a thread belay at the right end of the ledge.

1 **14m 6b** Climb the pale wall to a small overlap and step right into the niche on the right. Continue on good holds steeply for about 3m (protection in a succession of thin cracks to the left, the third being in the black wall above the prominent handrail which runs diagonally right across the wall). Stand on the handrail and reach left for a small hold and so to a poor PR in a little left-facing corner (peg replaced 2012). From the top of the corner, rock over and make a long reach left for a good finishing jug. Continue up and right to belay at the thin crack of 4.

2 **16m 6a** Climb a thin crack 2m left of 4 until it peters out. Go left and boldly up to a large recess and protection. Continue to finish up the right side of the corrugated overhang, as for 5.

9 The Meejies **   14m E5 6a
FA Gary Latter & Lawrence Hughes 10 August 2000

A direct line up the centre of the wall. Climb easily up the crack into the niche then up to the obvious handrail. Gain a standing position on this then continue with good protection to finish on good holds. Finish up any of the routes above.

WATERFALL WALL   15min
NN 6700 9576 **Alt:** 360m

The steep impressively-situated walls flanking the waterfall (Cadha an Feidh).

Descent: By abseil from in situ anchors – see topo.

1 Tip Off **   35m VS 4c
FA Dave Bathgate & Ian MacEacheran March 1965

Good well protected climbing, taking the obvious right-facing groove/ramp beneath the left side of overhung recess high up. Climb directly up into the corner/ramp and up this moving left at the large roof. Continue up steeper ground, finishing through prominent V-notch.

2 Hayripi **   40m HS 4b
FA Dougal Haston & Mike Galbraith June 1965

The shallow groove and ramp. Start just down and left of a birch on the grass ledge.

1 **30m 4b** Climb through grass ledge, then easily up veg-etated wall and shallow corner to prominent triangu-

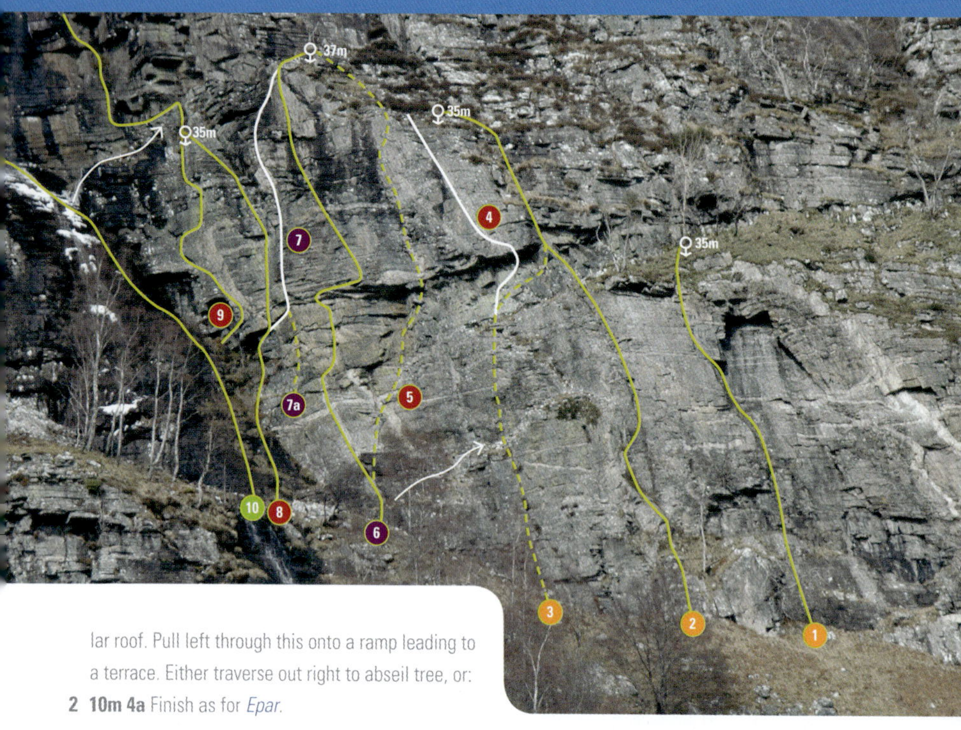

lar roof. Pull left through this onto a ramp leading to
a terrace. Either traverse out right to abseil tree, or:

2 10m 4a Finish as for *Epar*.

3 Epar ★   40m VS 4c

FA Dougal Haston May 1965

Start beneath the right end of the upper roof, beneath a
prominent wide flake crack.

1 10m 4a Climb groove to good ledge beneath
the flake crack. This point can also be gained
by scrambling along ledges from the left.

2 20m 4c Gain the flake crack with a steep
start, and follow it on good holds. Walk
along ledge to belay beyond roof.

3 10m 4a Finish up the left-slanting hanging
groove above. This pitch is slow to dry,
due to lots of grassy ledges above.

4 Face Value ★★   25m E3 5c

FA Dave Cuthbertson & Rab Anderson June 1981

A good pitch with spectacular moves through the initial
roof. Belay on the ledge at the top of the second pitch
of 3. Climb easily up to the roof about 2m left of its right
end. Pull left through the roof and move out leftward

above the lip to a good foot ledge and protection.
Continue on excellent rock up the wall above, left of a
prominent quartz vein. Tree belay up on left.

5 Independence ★★   40m E3 5c

FA Dave Cuthbertson, Rob Kerr & Calum Fraser 5 April 1981

An excellent pitch. Start beneath a shallow groove at the
right side of the steep wall. Climb easily up right trending
grooves to pull round a rib onto a small ledge beneath a
roof. Climb up to an old PR over the roof, cross this and
continue quite boldly on reasonable holds to a small
ledge and protection. Continue up the wall, stepping out
right to finish.

6 Acapulco ★★★   42m E4 5c

FA Dave Cuthbertson & Rab Anderson June 1981

Excellent climbing forging a line up the centre of the wall
with a serious first pitch. Start at the base of the wall.

1 12m 5b Follow the left-trending quartz
seam to a good ledge. Unprotected. This

pitch can be avoided on either side.

2 **30m 5c** Climb the short corner on the right then direct up the wall. Trend slightly leftward to a prominent flange over the roof then move right to a prominent block on the roof. Pull over to a good ledge. Trend leftward up the wall to a large ledge then direct from the right side to finish. The block can also be gained directly at 6a – not quite as spectacular.

7 Bratach Uaine ★★★ 35m E4 6a

FA Graeme Livingston & Andy Cunningham August 1987

Well positioned climbing up the blunt left arête of the wall. Follow 8 to the roof, traverse right along the prominent break to a good jug on the arête and pull through the roof just to the right (crux). Continue steadily up the right side of the arête.

7a Snotrag ★ 35m E6 6b

FA Grant Farquhar & Clare Carolan May 1989

A short, though hard and committing start takes the arête all the way, to join the ordinary route at the crux roof. Much more sustained. Boldly climb up the arête to gain a large 'fin' and protection. Continue with some *a cheval* moves to join 7 at the jug.

8 Wet Dreams ★★★ 30m E2 5c

FA Dave Cuthbertson & Rab Anderson June 1981

A brilliant well protected jug haul. Steep and sustained. It takes the big left-facing corner right of the waterfall, bounding the left side of the very steep **Waterfall Wall**. Start by scrambling up slabby ground (often wet) to a ledge at the base of the corner. Climb up to the first roof and pull through this (crux). Continue up the corner and through a second easier roof then trend left to belay at a large aspen.

9 Niagara ★★ 65m E2 5c

FA George Shields & Chris Norris April 1971

A very exposed line through the slabs and roofs right of the waterfall. If the first pitch is avoided by starting up 10 the route is HVS. Start at two parallel flakes 10m up the waterfall slab.

1 **25m 5c** Traverse across the flakes and pull up to a ledge beside some quartz. Move left to another flake, up it then return right to the top of the quartz. Climb a short green groove then trend left to ledges with trees.

2 **10m 4c** Pull out left onto a recessed slab at the right end of a smooth overhanging wall. Climb the slab to trees, traverse right through these and up to a ledge.

3 **30m 4c** Spectacularly hand traverse left to avoid a roof and climb the arête above.

10 Oui Oui ★★ 80m Difficult

FA Ray Burnett & Alasdair 'Bugs' McKeith October 1965

A fine route for water-nymphs, climbing under the waterfall with an excellent main pitch. Start to the right of the waterfall.

1 **20m** Climb grot and vegetation to belay in a flake crack beneath a steep corner on the right wall.

2 **35m** Continue up the slabby corner, (cleanest further left) passing behind the fall, to nut and PB on good ledge.

3 **25m** Continue diagonally up left past some perched blocks to a tree belay on a grass terrace. Abseil descent.

11 Brass ★ 30m Severe 4a

FA Allen Fyffe & Bill March April 1972

Good climbing following a wandering line up impressive ground, though slow to dry. A gully slants leftward from the slabby lower section of the waterfall to a level grassy section. Start left of here, level with a grass ledge in the centre of the wall. Traverse right to the ledge then trend leftward to the left end of the roof. Continue leftward passing its left end, finishing up a large groove.

LITTLE ROCK  20min

NN 6688 9572 **Alt:** 420m

Well to the left of **Waterfall Wall**, higher up the slope is an overhanging yellow wall with a stepped ramp-line running below it. The steepest crag here.

Descent: Traverse left and down the easy open gully, or 25m abseil from sling & maillon on thread.

1 Mistaken Identity ★ 45m VS 4b

FA Tony Brindle & Mal Duff 19 August 1983

Finely positioned climbing up the slabby right arête, though could do with more traffic. Start 4m right of the arête. Move up leftwards and follow a line just right of the arête. Belay beneath a short wall just short of the top. Scramble to finish.

2 Cross Leaved Heath ★★ 15m E5 6a

FA Chris Forrest 1992

Good sustained climbing. Start up the initial crack as for 3. From the good rail above the first finger-lock, move left then direct past good slots. Well Protected.

3 Heather Wall ★★ 15m E4 6a

FA Gary Latter 16 June 1985

Climb direct into finger crack and up this (crux) to good break and rest. Continue straight up, trending slightly right to finish.

4 This One ★ 15m E4 6b

FA Paul Thorburn & Gary Latter 1991

Another good sustained workout. Start beneath a couple of prominent quartz slots. Climb past these and place a small cam in the short vertical slot above. Make hard moves to gain a large protruding plaque directly above then climb straight up until a step right at a good thin horizontal break leads to some sloping holds to finish past a cracked block.

5 Ling ★★ 15m E5 6a

FA Gary Latter 20 September 2016

Start directly beneath a prominent wide vertical slot. Climb directly to the slot, then right to another V-slot and good holds above. Move back left and up into tiny left-facing corner. Move left to good quartz undercling beneath overlap, then direct on good holds to finish leftwards at the same point as 4. Very well protected.

6 Un Petit Mort ★ 12m E2 5b

FA Rusty Baillie & Martin Burrows-Smith 1976

Steep and strenuous climbing on huge holds up the widening crack at the right side of the wall.

7 Erica ★ 12m E4 6a

FA Gary Latter & George Beeton 17 May 2023

Line up narrow wall. Start 2m right of 6. Up on good slots, then direct (crux) to better holds. Move out left to finish direct.

Karen Latter on the finely positioned Mistaken Identity.

BEDTIME BUTTRESS

LOWER LEFT WALL  15min

NN 6678 9559 **Alt:** 370m

A more broken lower wall with a roof forming an alcove at the base.

Descent: Either abseil off, or scramble and up easy ground on the left.

1 Peigneur ★   **25m E1 5b**

FA Gary & Karen Latter 6 September 2020

A good direct line cutting through 2. Start 3m left of that route, at square cut groove. Up this to pull through roof at good crack to join 2 on the ledge. Step right and climb directly up the wall (left of 3) to finish on ledge directly beneath the abseil block.

2 Negligee ★★   **30m VS 5a**

FA Ian MacEacheran & Ray Burnett May 1965

Good climbing with an awkward start. Start at the left end of the initial roof. Climb to the roof and make an awkward move (crux) to gain ledge (large cams useful); continue up the left-slanting groove/flake, finish out rightwards to ledge.

3 Quickie ★   **30m HVS 5a**

FA Dave Cuthbertson & Rab Anderson 19 September 1977

Follow 2 to the ledge then move right about 3m and follow a line diagonally leftward to finish by a short crack.

4 Downtown Lunch ★   **80m HVS 5b**

FA Fred Harper, Arthur Ewing & Alasdair 'Bugs' McKeith May 1965

Good first and last pitches. Start at the right end of the initial roof.

1 **25m 5b** Step up right to a ledge with a small roof on its left then move back left to gain the main slab above (crux). Follow the obvious line, trending slightly leftward to belay at block.

2 **25m 4a** Scruffy ground leads to the right end of the terrace beneath **The Barrier Wall**.

3 **30m 5a** Go up the gully round the right end of **The Barrier Wall** to a ledge on the left then a groove. Step left onto the front face to an easier finish.

ledge. Continue up to roof, then traverse left, continuing beneath roof to exit out left.

 **2 Most Girls Do** ★★   35m E1 5a

FA Simon Richardson & Roger Webb 7 June 1985

The fine central crack. Climb out rightwards to base of crack and follow this, then direct to ledge below roof. Move right over block to finish up cracked wall.

3 Porn ★★   35m E1 5a

FA John Porteous & Mike Watson 11 April 1970

A fine left trending line, after a steep start. Start behind horizontal birch. Climb steep initial wall to short corner, then follow left-trending line of flakes to junction with 2. Continue to left end of roof, traverse back right on ledge and over block and finish as for 2.

LOWER RIGHT WALL   15min

NN 6680 9555 **Alt:** 370m

A prominent quartz-streaked slab down and right of the prominent huge beak-roof.

Approach: (A) From the parking spot, walk west along the road a few hundred metres to a gate 100m beyond the bridge over the burn. Follow a good track up the left side of the burn until opposite the crags. Go through another gate and follow a small path leftwards to cross burn at a small dam, then cross rough ground and up boulder field to the base. **(B)** Continue traversing left from the base of **Waterfall Wall**.

Descent: Abseil from tree on the terrace.

1 Cuckold ★★   35m E2 5b

FA Dougal Haston & James Moriarty May 1965;
FFA Pete Boardman & partner 1974

Good varied climbing with a serious lower section. Start beneath the more pronounced central diagonal crack, as for 2. Trend leftwards to a vague lower crack and follow this leftwards to good protection. Step right and move up then direct to corner on left. Move up left onto

THE BARRIER WALL   20min

NN 6678 9556 **Alt:** 410m

This is the upper tier of **Bedtime Buttress**, the furthest crag from the road, high up on the extreme left of the crags. The angle varies from just off the vertical to slightly overhanging, providing a fine range of short extreme pitches. Very quick drying - retains the evening sun, with clear views out to the west.

Approach: From the left end of the **Lower Left Wall**, cut up a short steep gully (knotted rope in situ), then diagonally right to gain the left end.

Descent: By abseil from in situ slings & maillons, or lower-offs on the trees at top of the shorter routes above the 'crevasse'.

The first 3 routes all belay on short wall 10m back.

1 First Offence ★★   20m E4 5c

FA Dave Cuthbertson & Rob Kerr 5 April 1981

Fine steep climbing. The original start up the gully is often wet. Starting up 2 is much better. Climb dirty broken rock to a sapling, then swing out right and move rightward to a good pocket. Go leftward to gain the prominent left-facing flake. Climb this to pull over bulge leftward to finish.

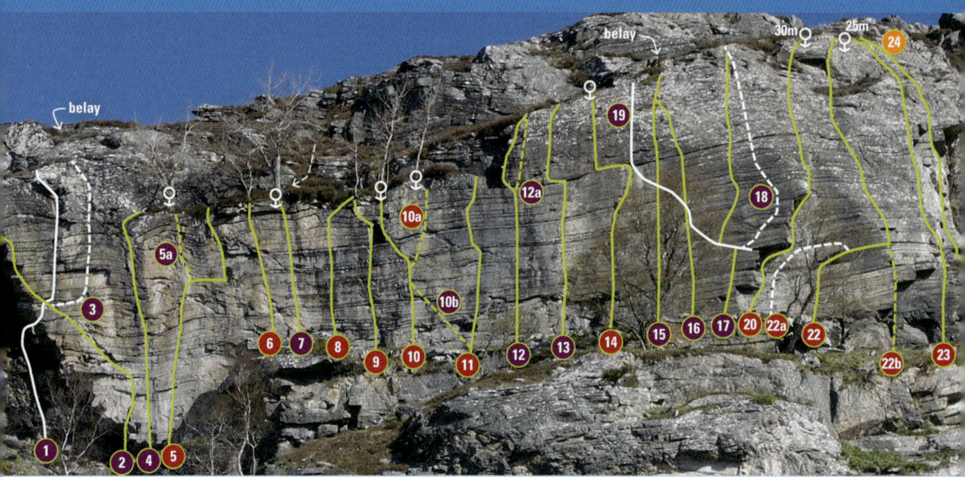

2 Zen ★★★　　　　　　　　　　　　　　20m E5 6a

FA Gary Latter & Masa Sakano 15 October 2014

Superb, well-protected climbing up the left edge of the wall. Start at the left end of the wall. Climb small left-facing corner and up to good wide break. Traverse left, then trend left up wall on good breaks to a triangular jutting block on the arête. Stand on this and climb directly up the wall above until a rounded boss about 2m from the top. Traverse thin break and swing left around the arête to good holds, then up to good finishing hold on arête. Mantel over onto broken ground.

3 The Atom of Delight ★★　　　　　　　18m E4 6a

FA Gary Latter & Robert Durran 21 March 2015

Sustained well-protected climbing up the wall between 1 & 4. Start up 2 and follow it up leftwards to the small hole. Move right and go up to small overlap, then make hard moves up wall above, moving rightwards to a good flange. Pull onto a sloping ledge, scramble leftwards, then direct.

4 The Art of Relaxation ★★★　　　　　15m E4 6a

FA Dave Cuthbertson & Roy Williamson May 1981

Sustained and well protected on good holds. Start just left of the prominent overhanging crack of 5. Up the wall to a good break on the left, move slightly right and continue up a crack to a ledge.

5 Case Dismissed ★★★　　　　　　　　15m E3 6a

FA Dave Cuthbertson & Rab Anderson 12 October 1978;
Direct Finish: Dave Cuthbertson & Rab Anderson June 1981

Superb strenuous and well-protected climbing, taking the steep hanging finger crack up the longest section of wall just left of the 'crevasse'. Direct up the wall to gain the crack and up this. At its top step up and right onto a foot ledge. Up the quartz crack on the left to a quartz overhang, finishing out left to a ledge. The 5a *Direct Finish* gives a more sustained route, perhaps nudging E4.

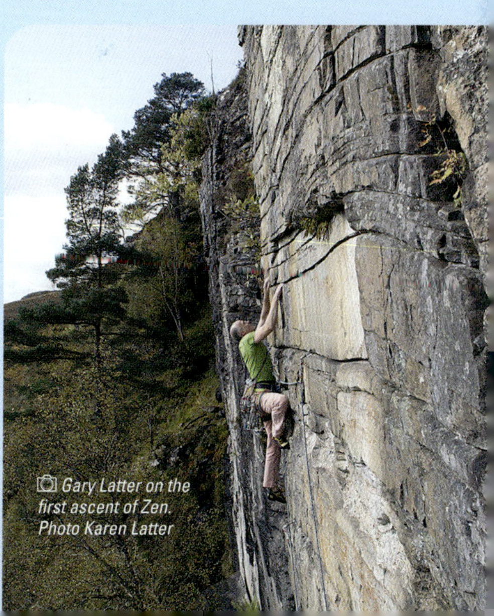

Gary Latter on the first ascent of Zen. Photo Karen Latter

6 Coda ★ — 8m E2 5b

FA Gary Latter & Alastair Robertson 18 October 2015

The left side of the wall beneath the large aspen. Start atop the left end of the 'crevasse'. Climb direct on good holds immediately right of the large quartz blotch, finishing up slot at top. Lower off from the large aspen.

7 Cabriolet ★ — 8m E4 5c

FA Gary Latter 1 October 2014

The wall beneath the large aspen. Start atop the 'crevasse'. Arrange protection in two breaks, then move up left to good flat hold. Continue directly (serious, but straightforward) to right end of small ledge and good protection in break. Pull up, then trend left on good holds to good break beneath overlap. Pull directly over this.

8 Ça Va? ★ — 9m E3 5c

FA Gary Latter 2 October 2014

The wall left of 9. Start atop the right end of the 'crevasse'. Climb direct up the wall to good finishing holds. Move right to pull over at the leftmost of three small aspens.

9 C'est la Vie ★ — 9m E3 6a

FA Martin Lawrence (solo) 26 April 1981

Climb thin crack into niche, exiting leftwards.

10 Hands Off ★ — 10m E2 5c

FA Martin & Dave Lawrence 8 May 1981

Steep and well-protected. Start 3m right of 9. Gain the left-slanting crack leading into the right end of the niche. Finish direct.

10a Carte Blanche ★★ — 9m E3 5c

FA Gary Latter & Masa Sakano 15 October 2014

The direct finish to 10. Climb that route to the large rectangular hole, then move up rightwards, then direct to good finishing holds in the small triangular niche.

10b Hands Off Right Start ★ — 10m E4 6a

FA Ian Taylor 2021

Start from below 11. Follow the thin diagonal cracks all the way leftwards to join the original.

11 Legover ★ — 10m E2 5c

FA Nick Sharpe & Alan Moist 17 July 1986

The wall right of the diagonal crack of 10. Climb the wall to a good break, then up and left to a good layaway, finishing up the left side of the open recess.

Rob Patchett on the classic Case Dismissed.

12 Subjugation ★★   15m E4 6a

FA Gary Latter 31 August 2014

Start at a good break and climb direct up the wall to good break level with base of niche at top. Climb the right edge of this to pull out rightwards over block at top.

12a Contagion ★★   15m E4 6b

FA Gary Latter 2 October 2020

The direct finish. Follow that route to the good break where it escapes left. Move up past large sidepull, then using sidepull on right make hard moves to good break, finishing at two short cracks onto ledge.

13 Galaxy ★★★   15m E4 6a

FA Dave Cuthbertson & Martin Lawrence 26 April 1981

Good well-protected climbing after a boldish start. Start just left of the hanging crack of 14. Climb the wall past a small plaque to good holds in break. Continue up thin crack above to good large break, then move slightly left and climb wall to finish up fine finger crack.

14 Ruff Licks ★★★   15m E3 5c

Dave Cuthbertson & Rab Anderson 19 September 1977

The fine hanging crack gives a good introduction to the harder routes on the wall. Start 2m right of the crack and climb up past short finger slot to gain the main crack. Follow this to horizontal break, move left along this to finish up a thin crack.

15 These Days ★★   18m E5 6b

FA Gary Latter & Robert Durran 21 March 2015

The wall right of 14. Start 3m right of the overhang at the base of that route, atop a small bouldor. Climb direct up the wall (good small cams on left) to the small ledge at the end of the traverse of *Ayatollah*. Continue directly to the circular quartz niche, then up to a large incut quartz hold. Step right and continue directly up slab to finish.

16 Twenty Years from Now ★★   18m E5 6b

FA Gary Latter 2 April 2015

Another fine well-protected pitch. Start 2m right of 15, just right of flat boulder at base. Climb the wall to the first small sloping ledge on *Ayatollah*. Continue direct, using sidepull in diagonal crack to gain good break, then direct to good deep break at top. Pull straight up onto slab above, finishing up easier final section as for 15.

17 Apathizer ★★   18m E6 6b

FA Graeme Livingston & Willie Todd September 1986

Well-protected fingery climbing directly up the centre of the wall. Start just left of the corner. Climb direct to the quartz slot on 19, then follow that route up and left to the break. Move up rightwards to reach some flakes. Move up and slightly leftwards to make hard moves into the horizontal break. Leave the break with difficulty to finish directly up the slab.

18 Aye ★★   18m E7 6b

FA Gary Latter (headpointed) 30 July 2014

The rightmost line up the wall. Well protected with many small cams. Start up 17 to the quartz slot. Step right to the arête, then up the wall to a thin break. Move up leftwards and make hard moves up and left to another break. Continue more easily up slab to finish.

19 Ayatollah ★★★   25m E4 6a

FA Dave Cuthbertson, Rob Kerr & Calum Fraser 6 April 1981

Excellent sustained climbing with improving protection. Start as for 20. Up the corner to roof, pull left round onto the wall then up to a good quartz slot (protection). Move left then up past two good breaks to prominent circular quartz recess. Continue in the same direction to gain good break at junction with 14. Move right and up wall and slab to finish.

20 Muph Dive ★★   25m E2 5c

FA Dave Bathgate & Dick Holt (3 PA) May 1965;
FFA Murray Hamilton & Adge Last 19 September 1977

The stepped right-facing corner near the right end. Climb the corner, finishing up fine slab.

21 Tyranny ★   25m E3 5c

FA Gary & Karen Latter 31 August 2014

A route squeezed in between 20 and 22. Start 2m right
of the corner. Climb the wall to gain the left end of the
niche, move right and climb directly up the wall from the
right side of the niche to large flat hold above. Continue
directly up wall and slab above.

22 Muffin the Mule ★★   25m E2 5c

FA Dave Cuthbertson & Dougie Mullin 14 October 1977

Good well-protected climbing up the wall right of the corner. Climb thin crack with difficulty to right end of recess
and good horizontal break. Traverse this rightwards to
ledge on right, then move up slightly leftwards to good
jug. Climb directly up the wall past some good breaks,
finishing up fine slab, crossing small overlap near the top.
The 22b *Direct Start*, E3 6a tackles the poorly protected
shallow overhanging groove.

22a Muffin the Mule – Left Start ★★   25m E1 5b

Lovely well-protected climbing and a good warm-up.
Start 1m right of the corner. Climb direct on good holds to
obvious horizontal break. Follow this rightwards, stepping
down to good wide break and a thread runner at its right
end. Continue as for the parent route.

23 Disaffection ★   25m E2 5c

FA Gary & Karen Latter 31 August 2014

The right arête of the wall. Well protected, after a bold
start. Start at the right end of the terrace. Climb onto
small incut ledge, then up to good ledge above. Continue
up the left side of the arête, then easier slab, passing
tiny Scots pine sapling to thread belay/abseil point.

24 Downtown after Lunch ★★   25m HVS 5a

FA Fred Harper, Arthur Ewing & Alasdair 'Bugs' McKeith May 1965

This is the top pitch of *Downtown Lunch*, a worthwhile
pitch in its own right. Start round the corner. Climb a
groove to make awkward moves out left onto slab near
top, trending leftwards to belay.

COMIC CRAGS

Alt: 300m | SE | 10-12min

Three small crags on the small hill of Creag Beag behind Kingussie, with many short lower grade trad and sport routes and bolted lower offs on all routes. Routes were cleaned, bolted and climbed by a number of local climbers from 2020 onwards: *Rafael Salazar, Kate Wilson, Sandy Allan & Alastair Wilson.*

Access: Turn off the main Newtonmore Road up the right side of the Duke of Gordon Hotel, taking the second left into a car park with toilets, in front of the Medical Practice (NH 7560 0074; 57.080806, -4.0540094).

Approach: Take a rising path at the back left of the grass, then follow the road uphill (signposted to Creag Beag Summit). Before the first house, turn left onto a path into the woods. Follow this for around 200m; 80m after a little zig-zag, turn right onto a smaller path leading directly up to the foot of the rightmost crag.

MARVEL BUTTRESS

NH 7513 0090 **Alt:** 300m

1 Thor – Son of Odin ★		15m F5+
2 Black Panther ★		15m F6a+
3 The Wasp ★		15m F6a
4 Ant-man ★		15m F6a
5 Captain Marvel ★		15m F6a+
6 Captain America		12m Severe 4a
7 Dr Strange		10m Severe 4a
8 Black Widow		10m Severe 4a
9 Hulk		10m Severe 4a
10 Iron Man		10m Severe 4a

DOUGLAS BOULDER

Alt: 300m

1 Micro Slab ★		7m F2a

3 bolts up a very short slab at the bottom left.

DC BUTTRESS

NH 7512 0087 **Alt:** 300m

Next buttress left.

1	Left Arête ★	18m F4+
1a	Batman ★	18m F6b
1b	Robin ★	18m F6b
2	Wonderwoman	18m Severe 4a
3	Superman	18m Severe 4a
4	Catwoman	18m Severe 4a
5	Aquaman	18m Severe 4a
6	The Flash	18m Severe 4a

ALL ABILITIES CRAG

Alt: 310m

Small (8m) very easy angled slab, with two sets of anchors, each beneath the small pine trees.

MANGA BUTTRESS

NH 7508 0084 **Alt:** 310m

Leftmost buttress, slightly higher up.

1	Pikatchu	15m F4+/5
2	Charmander	15m F4a
3	Dragon Ball	15m F4a
4	Pokemon ★	15m F3a
5	Doraemon	15m F3a
6	Mazinger Z	25m F3a

Route can be started just off the path at the very bottom of the crag. Weave through the heather on the cleanest rock to reach the base of the better upper section where a belay can be taken if desired. The upper section is very pleasant and tops out nicely (there is no lower off) to finish on the large upper ledge from where it is possible to walk off left. Tree belay with care. The upper section can also be climbed independently although the belay, which is well off the ground by this point, consists of only one hanger.

FARRLETTER CRAG  2min

NH 8266 0320 **Alt:** 260m

A very convenient compact schist crag in the woods just east of Loch Insh. Almost all the routes were originally climbed with often sparse peg protection in the eighties; all have now been retro-bolted by *Rafael Salazar*, with the exception of 8 *Too Farr for the Bear*.

Access: Turn off the B9152 Kingussie-Aviemore road at Kincraig and follow the road down past the north end of Loch Insh for 1.3 miles/2.1km. At the T-junction turn right down the B970 for 1.1 miles/ 1.8km to park considerately at the side of a wide forest access road on the left (NH 8280 0327; 57.105410, -3.9364860).

Approach: Follow well-worn path rightwards for 150m to the crag.

1 Farrplay ★   15m F6a+

FA Keith Geddes & Andy Cunningham 1987

2 Ceasefarr ★   15m F6b

FA Andy Nisbet & Keith Geddes August 1985

3 Farr ★   15m F6a

FA Rafael Salazar 2018

4 Private Farr ★   15m F6a

FA Andy Cunningham & Andy Nisbet July 1985

5 Farr Gone ★   15m F6c

FA Rafael Salazar 2018

6 Ausfahrt ★   15m F6b

FA Andy Cunningham & Martin Burrows-Smith 1990

7 Farrout ★★   15m F6b+

FA Andy Nisbet & Andy Cunningham July 1985

8 Too Farr for the Bear ★★   15m E4 5c

FA Andy Cunningham & Andy Nisbet July 1985

"*Steady and well protected.*" The vertical crack-line with a porthole near the base. Climb the crack past 2 peg runners, with a move right and back left near the top.

9 The Farrter ★★   15m F7a

FA Martin Burrows-Smith & Libby Healey 1990

9a **The Master Farrter** ★★ 15m F7a+

FA Alastair 'Plod' Ross & M.Sutherland 1988

10 **Farr Harder** ★ 15m F7b

FA Steve Johnstone 3 March 2009

11 **Yet so Farr** ★★ 15m F6c

FA Andy Cunningham & Andy Nisbet July 1985;
Right Finish: Rafael Salazar 2018

12 **Yet so Farrther** ★★ 15m F7a+

FA Martin Burrows-Smith & Stevie Blagborough 1990

13 **The Art of Coarse Climbing** ★★ 15m F7a+

FA Andy Nisbet & Andy Cunningham July 1985

14 **Midgie Farrt** ★★ 15m F7b

FA Alan Cassidy 1 September 2023

15 **Pick** ★ 15m 6a+

FA Rafael Salazar 2021

16 **And Mix** ★ 15m F6a

FA Rafael Salazar 2021

HUNTLY'S CAVE

NH 025 326 **Alt:** 300m N 5min

An excellent steep little crag situated in a pleasant open wooded gorge with a stream running through it. The rock is schist, generally vertical or overhanging, with the bedding plane dipping down at an angle forming lots of excellent incut holds and blocky roofs. All the routes are well protected.

Although it faces north it is often sheltered from the wind by a stand of mature larches just to the north. Unfortunately this sheltered location makes it very midgy for most of the summer months — much better as a spring or autumn venue. The cave that the crag takes its name from lies beneath a boulder about 50m downstream from the far left end of the crag. First ascent details for the earlier routes are not available as the crag was only written up in the early eighties, hence the somewhat uninspiring route names.

Access: From the south follow the A939 for 2.9 miles/4.7km beyond Grantown-on-Spey to park in a layby (NG 022 326) on the right (north-east) side of the road 0.3 miles/0.5km beyond a trio of farmhouses at Glaschoil. If approaching from the north, the layby is just before the end of the disused railway cutting running alongside the north-east (left) side of the road for about 500m (welcome to Cairngorms National Park sign).

Approach: Follow the mossy tarmac road for just over 100m, then cross a stile on the left and follow a path leading down for about 150m to arrive at some level slabs of rock at the top of the crag.

Descent: By a good muddy path zig-zagging down the left (facing out) side of the crag.

Routes described from right to left.

1 Hanging Groove   10m E1 5c

Start 2m left of the prominent cave-like block chimney at the right end of the crag. Ascend the wall to cross the roof at a small jammed block. Finish up a short crack and the hanging groove.

2 Huntly's Jam ★   10m E2 5b

FA Nick Sharpe 1987

Follows the arête right of 3. Cross the initial roof (crux) then follow the easier arête, finishing up its left edge.

3 Pete's Wall ★★   15m E2 5c

FA Pete Boardman 1970s

A good well protected pitch. Start just right of 4. Climb the initial wall to break through the long low roof at its smallest point to gain a good incut break (crux – reachy). Continue directly up the wall past further good breaks near the top.

4 Right Hand Groove ★   10m Very Difficult

Well-defined groove, finishing past a horizontal birch tree.

5 Slot Direct ★   10m Severe 4a

The leftmost of the three grooves.

6 Left Rib ★   10m Severe 4a

Start from a smaller subsidiary groove at the base of 5. Move left to a recess then past a pedestal. Step left round the edge and finish by cracks leading to the slot.

7 Dead Tree Wall ★   15m VS 4c

A bit of a misnomer as the tree is long gone. Start below the prominent roof a few metres down and left of 6. Climb the crack through the right end of the roof, hand traverse left a move then cross the roof and climb the wall just right of a groove. Continue more easily to a spectacular finish through the split roof left of the more prominent 5.

8 Central Crack ★★   20m HS 4b

Start as for 7. Traverse left beneath the roof to the arête, move up then back right. Trend leftward to gain and follow the deep crack to finish out leftward beneath the final roof. A *Direct Finish* over the roof to the right is possible at VS 4c.

9 Lime Street ★★   20m E4 5c

FA Pete Livesey late 70s

Very sustained and strenuous climbing up the innocuous looking right wall of 10, unfortunately slightly escapable near the top. Start off the blocks at the base of 10. Move up rightward to the small roof and pull through this to a huge hold. Continue up the flake, step left and up the cracked groove and the good hand crack which curves towards the top roof of 10. At the top of the crack move up rightward to layback up a line of small flakes just left of the arête. Finish more easily up the final wall.

10 Double Overhang ★★★   20m HVS 5a

Excellent climbing up the central left-facing corner system.

11 Bo-po Crack ★   20m E3 6b

FFA Martin Lawrence 3 October 1980

A well protected technical problem through the thin roof crack left of 10. Start 2m left of 10. Up the scooped wall to the crack then through the roof and up the wall to good ledges. Finish up the final crack of 13 to the rowan tree.

12 Diagonal ★★   30m VS 4c

Start up 10 to the roof.

1 **20m 4c** Traverse left underneath the roof and step out left (crux) and up into the left-facing groove. Follow this then continue out left and up to belay in a recess beneath the final roof.

2 **10m 4b** Move up above the belay then out right round the roof and easily to finish.

13 Diagonal Direct ★   25m E1 5c

Start a few metres right of 15. Ascend the green scooped wall and up the first corner of the normal route. Pull out right (possible belay under roof). Finish direct over the roof (crux) and the easier short wall above.

14 Slabby Groove ★★ — HVS 5a

A bit of a misnomer, though the groove is easier angled than many of the other routes. Start up 13 but move left onto the arête. Step left and up the left side of the arête then make a hard move into the deep cracked groove. Continue directly up past small ledges to finish up the wide crack of 15.

15 Cave Route ★★★ — 30m HS 4b

An excellent route amidst impressive surroundings, which can be split into 3 pitches. Start beneath the large bay towards the left end of the crag.

1 **18m 4b** Climb the easy initial corner to a large ledge. Traverse left 3m above the ledge then up blocks on the arête and move back right to thread belay on a large block in the 'cave'.

2 **12m 4a** Step up right and finish up the exposed wide bottomless crack – not as fierce as it looks!

Alternative Finish: Stick insects can squeeze through the narrow chimney running back from the 'cave' or the similarly thin slot at the base of the bottomless crack.

15a Cave Route Direct ★★★ — 10m VS 4c

A better way of climbing the middle section is to continue up the right side of the flake/corner above the first ledge. Very well protected.

16 Huntly's Wall ★★ — 25m E3 6a

FA unknown 1970s

The luminous yellow wall bounding the left-edge of the crag, with a spectacular finish through the capping roof. At the far left end of the crag is a short easy left-trending ramp. Start up the ramp for a few metres then move onto the wall on the right and up to the roof. Pull through this and up by the line of a thin finger-crack (crux) to gain a good ledge. Move easily up to a good horizontal break beneath the final roof. Pull out right onto the lip then use a good incut hold to reach up rightward to good holds. Finish past some keyed-in blocks.

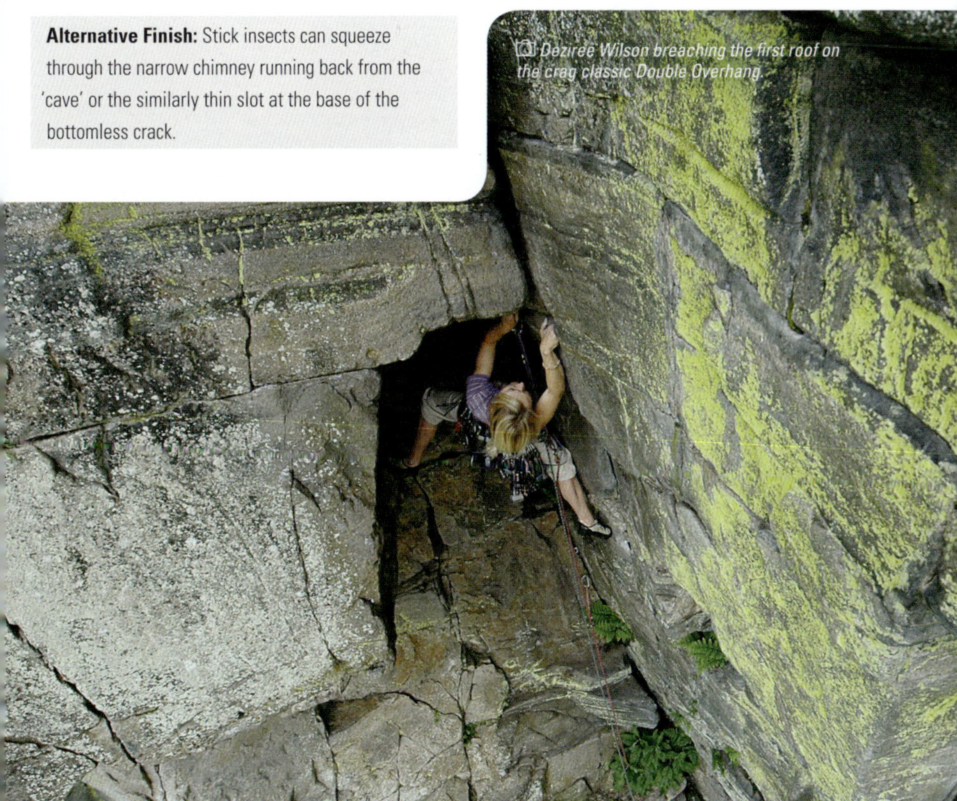

Deziree Wilson breaching the first roof on the crag classic *Double Overhang*.

DUNTELCHAIG (ROCKY FORT) NH 643 318 **Alt:** 240m 

A collection of steep gneiss crags overlooking the north-east end of Loch Duntelchaig, which lies to the east of the north end of Loch Ness. Good evening crags in a pleasant setting.

Access: Turn off the A9 just south of Daviot (7 miles/ 11km south of Inverness) along the B851 Fort Augustus road. Follow the road for 4.5 miles/7km to turn off right 0.2 miles/0.4km south of Inverarnie. Follow the single track C-class road turning left at the T-junction just beyond the River Nairn. Continue, passing through Dunlichity and along the shore of a small loch to a small unmetalled road on the left leading to a parking area on the left at Loch Duntelchaig (NH 6474 3207; 57.358961, -4.2500923).

DRACULA BUTTRESS (w) 10min

NH 6435 3179 **Alt:** 240m

The first large buttress on the approach, with prominent roofs on the left.

Approach: Pass a locked gate and continue down the forestry track on the east side of the loch for 450 metres then scramble up the left side of a mass of jumbled boulders to the left end of the crag.

Descent: Down the grassy ramp on the right.

1 Cyclops ★★ 30m E4 6a

FA unknown; FFA Duncan McCallum 6 August 1983

Just right of the lowest point of the buttress is an overhanging right-facing corner. Climb the corner to the lip of the roof and continue up the corner to a ledge. Climb more easily up the next corner to a belay. Either escape left or finish up easy slabs above.

2 Wolfman ★★ 30m E6 6b

FA Chris Forrest & Roy Henderson August 1991

The hanging groove in the arête. Start at the large block at the foot of 3. Head out left to the arête and pull over a bulge to the base of the groove. Up this with difficulty using a hidden hold on the left to pull out right to a good rest. Traverse right to join 3 in the middle of its roof pitch and finish up this.

3 Dracula ★★★ 30m E3 5c

FA unknown (A1); FFA Bob Brown & Fred Williams Autumn 1969

Superb athletic climbing through the spectacular roof crack. One of the hardest pitches in Scotland for its day.

1 15m 4c Climb the prominent corner past an awkward bulge to a fine belay perch below the roof.

2 15m 5c Follow the roof crack out left past some thin jams near its end to an easy finish up the front face. Very well protected and worth knowing about for a rainy day.

1 20m 5a Follow a left-slanting line of flakes (poorly protected) to the base of the corner then up this to a sloping stance under the roof.

2 10m 6a Make some precarious moves to gain excellent holds on the lip. Finish up the overhanging groove above.

4 Vampire ★★   30m E3 6a

FA Ado Liddell & Martin Burrows-Smith 1976

The next roofed corner above and right of *Dracula*. Start 15m right of *Dracula*.

5 Garlic ★★   30m E3 6a

FA Ado Liddell & Malcolm Campbell 1982

The corner system. Climb the groove, over blocks then move left using the arête to finish up the groove.

THE MAIN CRAG       15min

NH 6431 3169 **Alt: 240m**

A fairly rambling and generally uninspiring crag, sloping up from left to right above the jumble of boulders immediately right of **Dracula Buttress**.

Approach: Directly through the boulders from the track.

Descent: Follow a path rightward then cut down just beyond the Chasm Block, a large prominent block delineating the right extremity of the crag.

1 Misty Crack ★★ — 30m E1 5b

FA Allen Fyffe & Ado Liddell 1982

The roofed corners and crack bounding the right side of the main slab. Start at the lowest point of the crag. Climb the first corner to exit right, then the second corner, exiting left on to mossy slabs. Traverse right just above the lip of the roof then up (possible belay) to finish up the fine crack in the steep headwall. Tree belay. Abseil descent.

2 Monolith Crack ★ — 35m HVS 4c

FA Ian MG & Richard Frere (in descent) 24 June 1937

The huge corner in the upper crag directly above a large aspen. Start at a crack 8m left of the monolith.

1 15m Climb the short wide crack above a rowan tree then by vegetated slabs trending up left to belay on the aspen.

2 20m 4c Climb the offwidth crack in the corner with sustained interest (poorly protected) then an easy lichenous slab to a belay.

3 Monolith Recess ★★ — 30m E2 5b

FA Ado Liddell & party 1978

A huge flake sits against the face at the lowest point of the crag.

1 20m 5a Climb the left side of the block then the fine layback crack above. Continue up more broken ground to belay at the base of the prominent crack in the headwall.

2 10m 5b Attack the fine crack, finishing strenuously on good jams.

4 Drum ★ — 60m Very Difficult

FA Ian MG & Richard Frere 24 June 1937

Good varied climbing up the right side of the crag. Start a short way up the vegetated ramp that runs up under the right side of the crag, directly beneath a prominent tree on the face at 12m.

1 12m Climb a short right-facing corner over a chokestone then up to the tree.

2 13m Continue up the offwidth corner-crack above then out left along a large flake crack and up past a tiny pine to belay at the left end of large ledge.

3 15m Climb the crack above the right end of the ledge, soon easing.

4 20m Continue up the fault in the same line, crossing a chokestone steeply on huge holds.

The following three pitches start from a cluster of trees on the heather terrace at the top right of the crag.

5 Drumhead ★★ — 18m VS 4b

FA Ian Sykes, Pothecary, John Hinde, Grant & Bell 23 May 1962

Walk left then up a heather groove to belay at large block. Climb the fine flake crack through two overlaps then easily up slabs to a block belay.

6 Mica Arête ★★ — 20m Severe 4b

FA Richard Frere 1937

From near the left end of the terrace climb an easy groove then up the right-facing groove above. Ascend the V-groove on the left, then a short crack leading onto the arête. A further shallow groove leads to a final slab.

7 Mica Chimney — 20m Very Difficult

FA Richard Frere 1937

Follow 6 to the V-groove and continue up that to finish by a deep slot.

8 Top Corner ★★ — 20m HS 4b

FA R.Bell & R.Todd 19 April 1964

Gain the terrace then scramble up and right to a tree belay at the base of the prominent corner. Climb this direct, exiting left.

9 Mica Slab — 30m Difficult

FA Richard Frere 1937

The slab, though mossy, is also a useful descent.

THE SEVENTY FOOT WALL 15min

NH 6426 3159 **Alt:** 250m

The steep clean wall below the right end of **The Main Crag**. The central section is unfortunately frequented by abseilers from a local outdoor centre, witnessed by the two muddy streaks down this part of the crag.

Approach: Continue along the forestry track beyond **Dracula Buttress** for a further 200m to where the wall becomes visible and follow a small well-worn boggy path crossing a stream to the base.

Descent: By a well-worn path down the right side.

1 Slings ★★   **15m E1 5b**

FA Duncan McCallum 11 April 1980

Sustained climbing with a perplexing finish Right of a blocky crack right of the pillar is a thin crack. Climb the crack into a right-facing groove, up this to the roof and pull out right to finish.

2 Razor Flake ★★   **15m HVS 5a**

FA Duncan McCallum & John Mackenzie July 1978

The central wall has a huge sharp-edged flake above its base. Gain the right edge of the flake and hand traverse it leftward. Follow a crack to a ledge at half-height, finishing up the thin cracks above, moving left near the top to a muddy finish. The 2a *Direct Start* is 5b.

3 Seventy Foot Wall ★   **15m HVS 5b**

FA unknown, circa 1970

The next crack-line to the right. A serious start up the wide flake crack leads to a prominent break in the overhanging headwall.

The fine flake crack of Drumhead.

ZED BUTTRESS SE 20min

NH 6630 2937 **Alt:** 310m

Access: Turn off the A9 just south of Daviot and follow the B851 Fort Augustus road for 6.4 miles/10.4km, then turn right towards Dunlichity, parking on the left just beyond the bridge (NH 6686 2957; 57.337328, -4.2133748).

Approach: Head through the field to the bottom right corner, over a stile and down the right side of the deer fence. Cross the ditch, then follow a vague path steeply up through the boulders in the direction of the obvious crag visible on the skyline. Continue leftwards up a faint path over rough ground to arrive at the right end of the crag.

1a Rockness Power Link ★ 15m F7a+

FA Jamie Skelton & Andy Wilby 30 September 2023

Climb through the crux of 1 to OK rest, then move left via new bolt and finish direct.

1 The Rockness Monster Returns ★★ 16m F7a

FA Andy Wilby 2013

If you are near 6ft then its only 6c+.

2a Jobsworth Left 15m F6c+

Left variant, with a boulder section off the ledge.

2 Jobsworth 15m F6b+

FA Sue Wood 2013

3 Making Movies ★ 18m E1 5a

FA Allen Fyffe & Ado Liddell 1984

The right-slanting ramp to its top (unprotected), then move left along a sharp flake. Step right and up, trending leftwards to finish left of the obvious roof.

4 The Wild Man ★★ 16m E2 5b

FA Andy Nisbet & Andy Cunningham 14 June 1986

Climb the fine flake crack to a spike, then direct up corner past a small rowan. Cleaned in 2023.

5 Zorro ★ 16m F7b+

FA Andy Wilby 2013

6 Go Go Gadget Arms ★★ 16m F7c

FA Andy Wilby 2013

Direct via a very bouldery sequence to a jug and awkward but good rest. Continue slightly left and up via a confusing sequence to a rest then a final juggy pull through the roof.

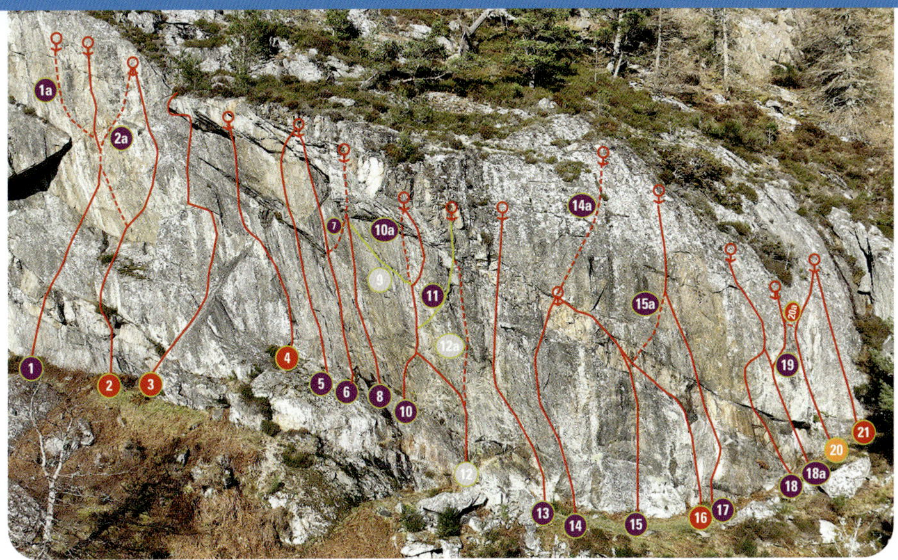

7 Gone in 60 Seconds ★★   15m F7c+

FA Andy Wilby 2013

Start up 6. At the jug pull right via a series of cracks to finish over roof on right.

8 Trick of the Tail 2 ★★★   15m F8a+

FA Andy Wilby 9 August 2015

Good sustained climbing after a hard start to finish up 7.

9 Open Project

10a The Force Direct ★★★   15m F8b

FA Calum Cunningham 10 October 2015

A harder finish. Climb the first 5 bolts of 10 then move left to a good hold, make strenuous moves to under and through the bulge.

10 The Force ★★★   15m F8b

FA Andy Wilby 2013

Steep wall to the left of the obvious arête.

11 The Fury ★   15m F8a+

FA Andy Wilby 30 August 2015

Climb the first 3 bolts of 10 then make hard moves right and up on to the arête to finish.

12 Dave Mac's Project   F9a/9a+

12a Project

13 Dave's Arête of Doom ★★   12m F7c

FA Andy Wilby 2013; rebolted 2023

Obvious arête to the left of the initial wall.

14 Little Minx ★★   12m F7b

FA Sue Wood 2013

14a Big Minx ★★   20m F7c

FA Jamie & Morag Skelton 9 January 2024

15 The Living Dead ★   12m F7b+

FA Jamie Skelton 12 October 2023

15a The Living End ★   20m F7c

FA Jamie Skelton 12 October 2023

16 Zed's Dead Baby ★   15m F6b

FA Dave Douglas June 2013

Start up the obvious corner in the quartz band. Follow this to the left to finish above the small right facing roof/corner. Best stripped by seconding the route!

17 Brick ★★★ — 12m F7c+
FA Andy Wilby 2013

Start at the same point as 16 and keep going straight up to a tricky finish!

18 It's High End 8! ★★ — 12m F7b+
FA Andy Wilby 2013

Hard crimping above the quartz band. Not for cold fingers!

18a Heather Bashing — 12m F7a+
FA Andy Wilby 2013

Start up a short corner to tricky moves out right. Easier climbing up the arête.

19 High End Bashing ★ — 12m F7c
FA Andy Wilby 2023

20a No Fatties Bashing — 12m F6a+
FA Andy Wilby 2015

A link up of 20 into 19 with 1 new bolt.

20 No Fatties Allowed ★ — 12m F5c
FA Sue Wood 30 October 2015

21 Rails of Disappointment ★ — 12m F6b
FA Sue Wood 30 October 2015

THE BLOCK 🧭 25min
NH 6628 2939 **Alt:** 360m

The overhanging prow leaning out over the hillside above **Zed Buttress**.

Approach: Skirt round the right side of **Zed Buttress**, then continue up until you can cut left following a vague path until below the steep ground under the crag. It's best to keep walking past the crag towards **The Needle** before turning back on yourself and up the heather to the base.

1 AirHog ★★★ — 12m F7b
FA Jamie Skelton & Tim Miller 4 May 2023

Sustained line tackling the soaring prow.

2 As Bare As You Dare ★★ — 12m F8a
FA Jamie Lowther & Jamie Skelton 6 June 2023

Climb 1 to its fourth bolt and make a big span right onto the nose feature. A hard boulder problem up the face above provides the crux.

3 Closed Project — 15m F8a+/8b

4 Closed Project

5 Less Pants, More Party ★★ — 12m F7b
FA Jamie Skelton & Tim Miller 4 May 2023

Punchy line of incut holds and edges on the right side of the buttress.

THE NEEDLE SE 30min

NH 6620 2937 **Alt:** 360m

Approach: Continue diagonally left up vegetated ground beyond **Zed Buttress** for a further 10 minutes. A pleasanter approach in summer may be made by continuing along the top of **Crag One**, then down the descent (35 min).

Descent: Down vegetated but not too tricky ground to the right.

1 Gold Digger ★★   25m E2 5c

FA Dougie Dinwoodie & Andy Nisbet September 1985

"Good, clean rock, well worth the uphill slog." The overhanging groove just left of the crest. Scramble up left from a ledge at the base of the prominent jutting nose to a large ledge. Climb the groove until possible to move right to join 2. Continue up by an *"apparently loose"* flake to climb the finishing flake-crack of 2 from its base.

2 The Prow ★★   30m E1 5b

FA Andy Nisbet & Simon Stewart August 1985

"Steep start and an exciting finish - the action is in the final 6m!" From the ledge at the base of the jutting nose, climb up between two hanging flake-cracks, then direct to a ramp leading left to the crest. Climb the crest to the last ledge, then make a long and scary move to reach up left for a flake-crack and very sensational finish.

CRAG ONE SE 25min

NH 6602 2922 **Alt:** 285m

The biggest and best crag, at the top left of the hillside.

Access: From the A9 just south of Daviot, follow the B851 Fort Augustus road for 7.5 miles/12km to a layby on the left (NH 6587 2833; 57.325726, -4.2292884).

Approach: Walk south down road for 150m, then follow tarmaced road towards the white house of Achneim. Turn right immediately after bridge, and follow ATV track snaking steeply up the hill. Follow small well-worn path rightwards along the top edge of triangular deer fence, continuing in the same line to a small cairn, then head down rightwards to the crag.

LEFT SECTOR

1 Keep Calm and Carry on Climbing ★

20m F6a+

FA Andy Wilby March 2010

2 Run the Gauntlet ★★★   25m F7a

FA Andy Wilby March 2016

3 Confessions of a Serial Climber ★★   20m F6c

FA Andy Wilby March 2016

Start about 15m right of 2, at the base of a black streak and follow this straight up. Tricky technical climbing.

MAIN CRAG 

NH 6602 2921 **Alt:** 350m

1 The Gangplank ★★   40m HVS 5a

FA Allen Fyffe & Andy Nisbet July 1985

The right-slanting ramp. Start at some blocks 5m left of
the corner of the recess. Climb cracks to below a shallow
corner, then the left-facing corner to gain the ramp.
Follow this over the bulge at its top, then traverse right to
another corner system which is followed to the top.

2 Snow on the Ben   15m F7a

FA Andy Wilby April 2012

3 Despicable Me ★★   15m F7a+

FA Andy Wilby April 2012

A strenuous start leads to a good hold and a rest, then a
reachy move to gain better holds and a puzzling finish.

4 'Owl at the Moon ★★★   15m F7b

FA Andy Wilby 22 June 2019

Very steep and pumpy jug pulling. Make sure you've
something left to clip the chain.

5 The Pink Wall ★★★   15m F7b

FA Andy Wilby 2011

Start up an overhanging corner to a good hold at the
base of the pink wall. Crimps lead slightly left then
right to a jug underneath the last few strenuous moves.
Awesomely steep but surprisingly good holds.

6 Dodged a Bullet ★★★   15m F7c+

FA Nick Duboust 2012

Start as for 5 for the first 2 bolts to the jug. Make a long
clip out right, then hard moves to gain the arête. Go up
this to turn it at about halfway onto the front face and a
welcome jug before the crux. More difficulties lie in wait
clipping the chain.

7 Brin It On ★★★★   15m F7c

FA Andy Wilby 2011

"Stunning, if very hard." Strenuous from the start and

doesn't let up. Start directly below the roof. Climb the
corner to a shallow slot, make a long move left to an
undercut, then run your feet round on nothing. Don't even
breathe as you pull the slack out to clip.

8 Whinging Consultants ★★★   15m F7b+

FA Murdoch Jamieson 2011

"Another stunning route." Very bouldery once the wall is
gained. Hard moves on sidepulls lead to a rest, then good
crimps and jugs to the top.

8a Snake in the Grass ★★   15m F7a+

FA Andy Wilby 2011

A slightly harder start to 9. Start as for 8 to the first bolt,
then follow a fingery handrail right past 2 bolts to join 9
at its 4th bolt.

9 The One and Only ★★★★   15m F7a

FA Dave Douglas 2011

The crag classic. The white overhanging wall on surprisingly incut crimps or jugs, amazingly sustained.

10 Third Time Lucky ★★ — 15m F6c

FA Andy Wilby 22 June 2019

Steady climbing to a tricky finish to clip the chain!

11 Treasure Island ★★ — 40m E2 5b

FA Richard McHardy & Peter MacDonald 1975

"Perhaps the best trad route at Brin?" A sensational route. Gain a right-slanting shallow groove system from the right and follow it to a large slabby area. Climb the slab just left of the corner above, or the corner itself, then continue up a ramp angling out right to finish near the fence.

12 Captains of Crush ★★ — 12m F6c

FA Pete Clarkson 1 June 2012

A good sustained technical climb with a stiff pull at the finish.

13 The Power of Three ★★ — 25m F7b+

FA Andy Wilby 2011

Three scoops with hard exits, and clipping the chain is perplexing. Hard to on-sight, with thin climbing followed by blind moves out of the scoops, but with hands-off rests in each. A different challenge to the rest.

14 Vagisil Overdose ★★ — 25m F7a+

FA Dave Douglas 1 May 2012

15 Christmas 1937 ★★ — 25m F7b+

FA Andy Wilby 2011

16 The Secret Garden ★ — 10m F7a

FA Andy Wilby 2012

Sustained moves until a good hold is reached at halfway, then it gets a bit technical with an interesting finish.

Access: Leave the A9 just south of Daviot and follow the A851 Fort Augustus road, for 8.1 miles to turn right almost immediately on entering the village of Croachy. Follow the narrow road for just over 1 mile/1.6km to park in small RSPB car park on the left (NH 6383 2805; 57.322608, -4.2629844).

Approach: The crags are not visible from the parking, facing down the strath. Cross the fence and head directly up the hill on a good initially boggy path, past the prominent Ruthven Boulder. Continue up the path to cross a fence 5 minutes beyond the boulder. Follow the fence leftwards over rough ground slightly downhill to another fence. **Frank Sinatra Walls** are visible in profile up right from here. Cross the fence and continue in the same line aiming for a cairn-shaped boulder on the skyline. Continue traversing at this line to arrive at the terrace beneath the **Upper Tier**.

TYNRICH SLABS

SE 20min

NH 6353 2727 **Alt:** 310m

A quick-drying two-tiered slab low down on the flank of the hill.

LOWER TIER

SE 20min

NH 6353 2725 **Alt:** 290m

1 Puff Ball ★★   **15m VS 5a**

FA Andy Nisbet, Brian Davison & Helen Geddes 3 September 1988

The prominent steepening crack splitting the centre of the slab.

UPPER TIER

2 Wrinkle and Crack ★   **20m Severe 4a**

FA N.Lawford, F.Adams & P.Savill 1 June 1982

The wide shallow crack, best started by climbing up to to

a hollow flake just right of a recess, then traversing right into the crack.

3 Fly Agaric ★★   15m E2 5b

FA S.Clark, Ben Sparham & Dave Porter 3 June 2004

Climb just right of a recess, then left to the left end of large ledge. Follow some *"weird erosion features"*, then a faint incut crack diagonally right to finish up the final moves of 5.

4 Angel's Wings ★   15m HVS 5a

FA Andy Tibbs & John & Linda Biggar 18 May 2007

Climb short wall, then follow a right-trending line leading to a crack immediately left of 5; finishing as for that route.

5 Scorpion ★★   15m VS 4c

FA N.Lawford, F.Adams & P.Savill 23 May 1982

Start at blocks on the terrace. The short groove and left-slanting crack.

6 Trumpet of the Dead ★★   15m E2 5c

FA Brian Davison & Andy Nisbet 28 August 1988; Direct Start: Jules Lines (solo) 1988

Start up the groove of 5, then take the flake crack out right. Where it ends, climb directly up the slab. The *Direct Start* ★ E4 6a ascends the tenuous slab beneath the flake crack.

7 Blewit ★★★   12m E2 5c

FA Andy Nisbet & Brian Davison 28 August 1988

Start 5m right of the blocks. Ascend the shallow crack, with the crux stepping up left to the finishing crack.

8 Slippery Jack ★★   12m E4 5c

FA Brian Davison & Andy Nisbet 28 August 1988

The faint crack just left of 9. Ascend lower wall, with thin moves over bulge to slabbier rock (unprotected). Continue up short diagonal crack, then follow obvious line of holds diagonally left with further hard moves to finish.

9 **Strewth** ★★ 12m HVS 5a

FA N.Lawford & F.Adams 16 June 1982

The wall and vertical crack, with the crux at the top.

10 **Horn of Plenty** ★ 10m VS 4c

FA John Lyall 1990

Gain a small flake by reachy moves from the left, then up to the horizontal break. Traverse the break with difficulty, finishing by easier cracks.

11 **New World Blues** ★ 12m E2 5c

FA Michael Barnard & Alan Hill 3 September 2022

Start as for 10 to the horizontal break, then directly up the intermittent crack above.

12 **Boletus** ★ 10m HVS 5b

FA John Lyall 1990

A boulder problem start to gain the break of 10. Cross this to finish up fine left-slanting layback crack.

FRANK SINATRA WALLS  25min

NH 6316 2734 **Alt:** 370m

Routes 1, 3 & 4 cleaned June 2020

A steep compact wall at the back of a small valley which cuts westwards from the top of **Tynrich Slabs** towards the summit of the hill.

Approach: Either up leftwards from the top of **Tynrich Slabs**, or continue up the path towards the summit of Stac Gorm for another 5 minutes beyond the fence, until a short overhanging crag can be seen on the left. Contour

Bobby cruising up the fine crack of Puff Ball (page 370).

left below this crag on rough heather to the crag, which lies just beyond yet another fence.

Descent: down to the right.

① My Way ★★   **20m E3 5c**

FA Paul Thorburn, Dave McGimpsey, Robin McAllister & Andrew Fraser 20 September 1998

Start beneath the recess. Climb up to its top left side. Pull out left onto a ledge, then direct up the slabby wall.

② Anything Goes ★★   **20m E6 6b**

FA Paul Thorburn & Alastair Robertson 15 May 2004

Excellent climbing, with wild moves through the central roofed recess. Start as for 1, then direct to the recess, pulling through the centre of the roof to finish up the very thin headwall.

③ Vagabond Shoes Direct ★★   **20m E1 5b**

FA Andy Tibbs & Davy Moy 23 September 2014; Original Start: Andrew Fraser, Robin McAllister & Dave McGimpsey 20 September 1998

Climb undercut layback flake, then a ramp up rightwards, stepping right into a small recess. Pull on a large *"worrying"* flake to finish up the excellent flake crack.

④ Let's be Frank ★★   **20m E4 6a**

FA Robin McAllister & Dave McGimpsey October 1998

"A brilliant route comprising of three distinct thirds." Start at the lowest point of the wall. Climb steep crack right of the curving roof, then a short groove on the left to the top of a flake. Place crucial good small RPs, then climb intricately up left to good holds. Continue up to and over a bulge and up a blank flake. Finish up slab. The 6a bit is short and next to really good RPs.

⑤ Frank's in a Trance ★★   **25m E4 6a**

FA Dave McGimpsey & Robin McAllister October 1998

Start up the steep crack as for 4, continuing to a bulging wall. Continue in the same line up diagonal crack through bulges to a small ledge. Step left onto the slab and traverse across it, then up to a horizontal break (unprotected, but only 5a/b). Finish up cleaned right streak.

CREAG NAN CLAG

THE CAMEL

NH 6003 2893 **Alt:** 280m      10min

A fine, steep sport crag of unusual conglomerate. It lies near the top of an open gully, which doesn't get much sun and can be quite breezy (keeping the midges at bay!). Take up to **15 quickdraws** for some of the routes.

Environmental restrictions: Due to the presence of rare lichen (the site is a SSSI), it has been agreed with SNH that no climbing takes place left of the first route described.

Access: From the south: Turn off the A9 just south of Daviot (7 miles/11km south of Inverness) along the B851 Fort Augustus road. Follow the road for 7.8 miles/12.4km to turn off right (north, signposted Loch Ruthven) just less than a mile beyond the prominent crags of **Brin Rock** up on the hillside. Follow the single track C-class road for 4 miles/6.4km to park on the roadside directly beneath the crag. **From Inverness:** Follow the A862 through Dores heading towards Errogie for 11 miles/17km to the south-western tip of Loch Duntelchaig then turn left (south) up the single track C-class road for 0.7 mile/1.1km to park directly beneath the crag.

Approach: Head straight up the hill to the crag.

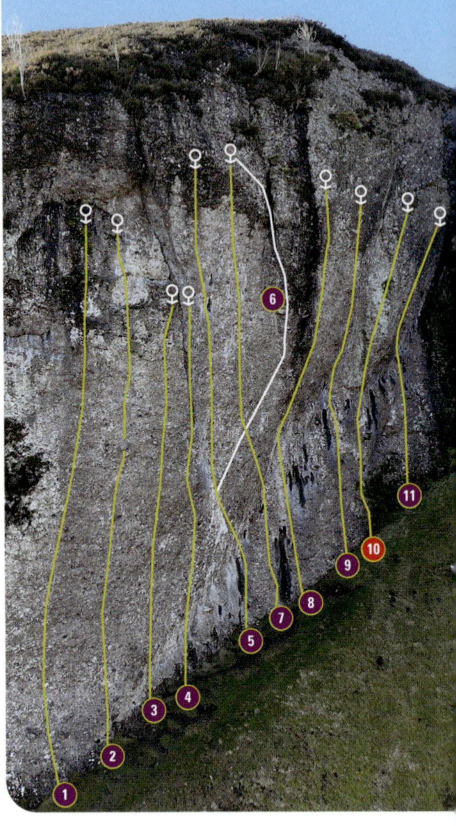

1 Inverarnie Schwarzenegger ★★   25m F7a

FA Neil Shepherd 11 August 2000

The leftmost line of resin ring bolts, easing after the crux at 20m.

2 Stone of Destiny ★★★★   27m F6c+

FA Neil Shepherd 12 October 1999

One of the best routes of its grade in the country and almost always dry. Follow a sustained line up the centre of the wall with the main difficulties centring around the biggest 'pebble' on the cliff.

3 Paralysis by Analysis ★★★★   25m F7a+

FA Neil Shepherd 7 August 2004

Superb sustained climbing up the continuously overhanging wall just right of 2.

4 There's Sand In My Pants ★★   25m F7a

FA Ali Robb 7 August 2004

Head for the huge hanging fissure at 10m which leads with interest, finishing more easily past a ledge.

5 Eyeballs Out ★★★   30m F7b+

FA Andrew Wilby July 2011

Climb 6 to the base of the ramp, step left then climb straight up the wall with superb sustained climbing.

6 The Final Straw ★★★   30m F7a

FA Neil Shepherd 5 August 2000

Stunning climbing up the huge hanging ramp. Make hard moves up the wall to gain the easier ramp. Continue up final groove and capping overhang to gain the LO.

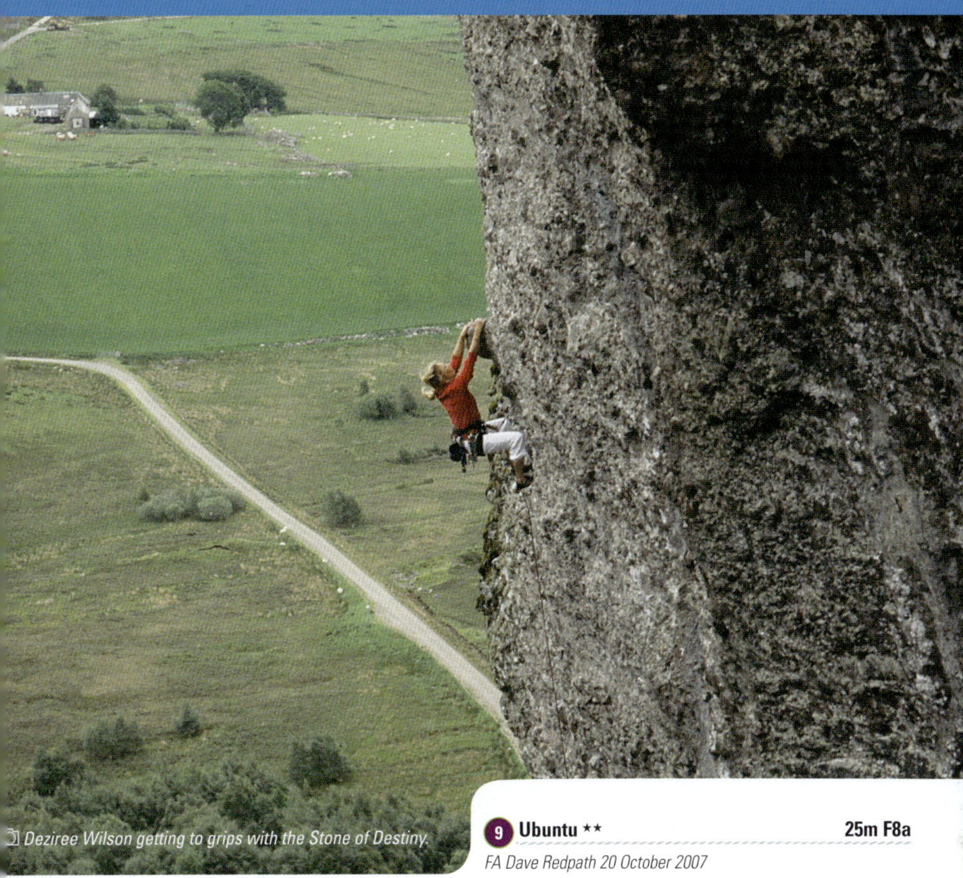

Deziree Wilson getting to grips with the Stone of Destiny.

7 **Giza Break** ★★ — 28m F7b

FA Neil Shepherd 24 September 2005

Line through ramp of 6 to share same LO.

8 **Death is a Gift** ★★★ — 25m F8a

FA Andrew Wilby July 2009

Superb sustained climbing. Climb the slab left of 9 to a hole in the steep wall, up to an undercut and reach through the bulge. Make a long move right to share jug on 9, then straight up.

8a **Gift Link** ★★★ — 25m F8a+

FA Dave Macleod July 2009

Up 8 to join 9 at the crux.

9 **Ubuntu** ★★ — 25m F8a

FA Dave Redpath 20 October 2007

Breaches the centre of the sweeping bulge. Start easily up the slab, then tenuous moves on undercuts lead right to a powerful crux off small edges to a handrail. Continue following the cool line of cobbles above.

9a **The Weakest Link** ★★ — 25m F7c+

The other obvious link up. Start up 9, finishing up 8.

10 **Two Humps Are Better Than One** ★ — 23m F6b

FA Neil Shepherd 30 August 1998

The prominent ramp near the top of the gully, with crux just below LO.

11 **Over the Hills and Farr Away** ★ — 20m F7a+

FA Neil Shepherd 31 August 2013

Very steep right-trending line.

Matthias Hausleber on the popular
Northern Corries classic Savage Slit (page 445)

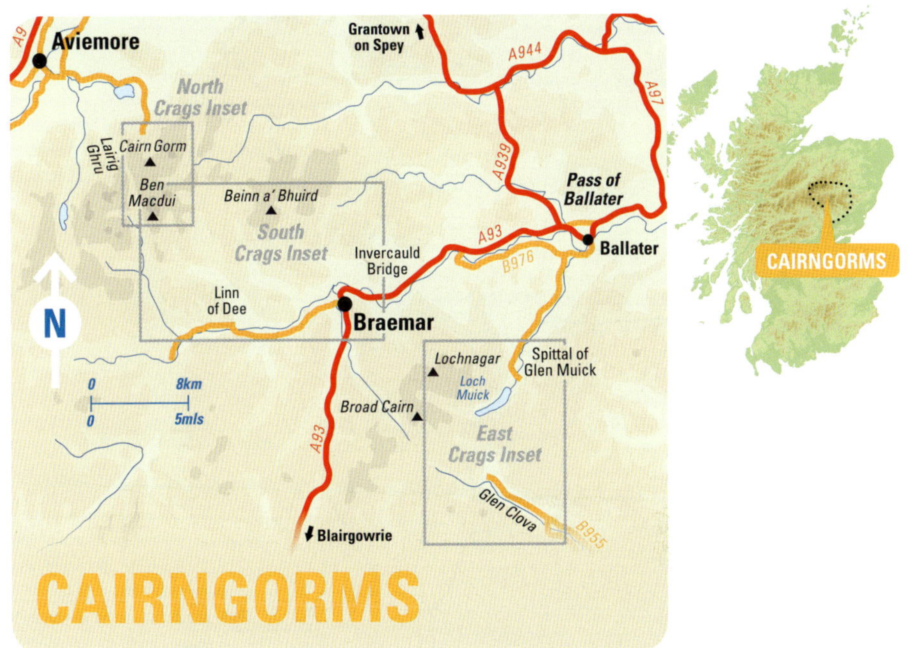

The mighty rampart of the high sub-arctic granite plateau of the Cairngorms forms
part of the Cairngorms National Park, covering 4,528 square kilometres - the largest
national park in Britain. Bounded on the west by the A9 Perth – Inverness road, the
high Cairngorms are divided roughly from north to south into three plateaux by the long
Lairig Ghru and Lairigh an Laoigh passes. These high, windswept and barren plateaux
include four summits over 1220m, forming the largest tract of high land in Britain. Within
this area are to be found some of the finest, and most remote, alpine-like corries in the
country. In contrast, ski developments high on the northern slopes of Cairn Gorm afford
convenient short approaches to a range of cliffs and corries. East of the A93 Glenshee
road (the highest trunk road in Britain, rising to 670m) and the River Dee lies a broad
rolling tableland dominated by the famed Lochnagar, in the heart of the Balmoral estate.
Further east lies the mighty Creag an Dubh-loch, for the rock climber the finest mountain
cliff in Britain with a frontage of over a kilometre of rock rising to almost 300m.

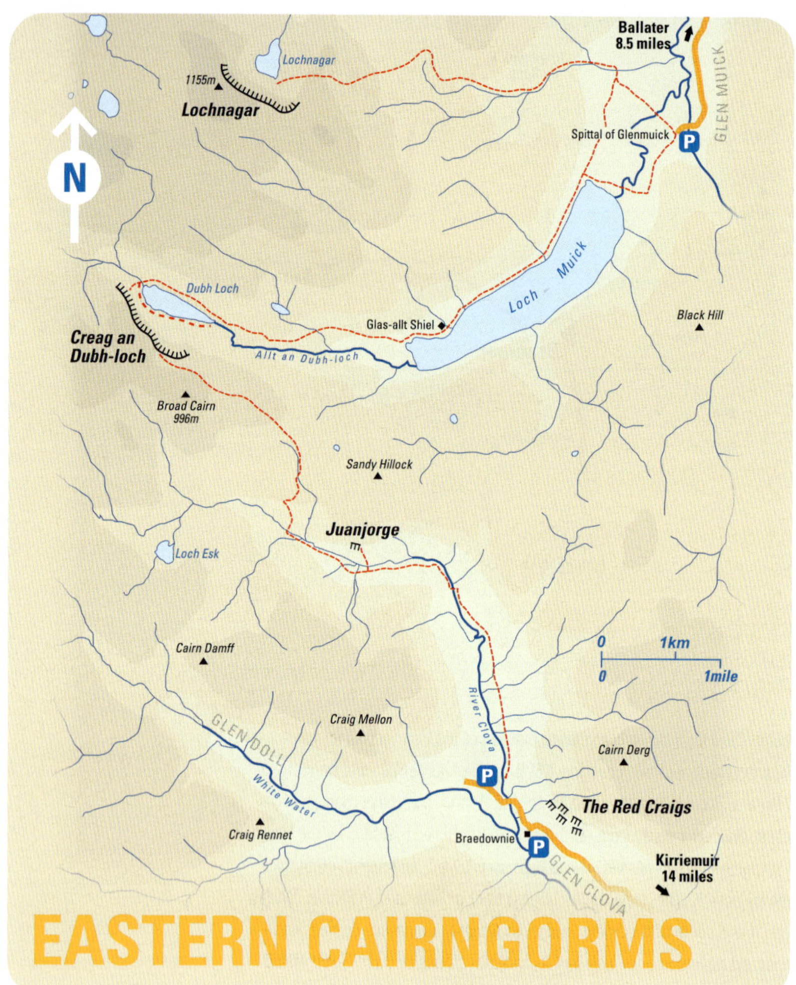

Accommodation: Ballater is the nearest town with numerous hotel and B&B establishments though there is a dearth of bunkhouses and Youth Hostels, except Ballater Hostel (01339 753752; www.ballater-hostel.com). **TIC** in town (☎ 01339 755306; www visitscotland.com).

Amenities: Fish & chip shop in Viewfield Road, near the square. Bar meals in many hotels. Lots of tourist cafés in Bridge (main) Street. Bank with ATM here also. Public toilets in the main square behind the church. Outdoor shop: The Outdoor Store (☎ 01339 753878; www.braemarmountainsports.com) in Bridge Street, with a café. Mountain bike hire: Bike Station Ballater (07552 169 272; www.bikestationballater.co.uk).

Campsite: Ballater Caravan Park (☎ 01339 755727; www ballatercaravanpark. com) at south end of town (signposted) at (NO 371 954). There are spots for wild camping in upper Glen Muick.

GLEN CLOVA

THE RED CRAIGS

Glen Clova lies in the southern edge of the Cairngorms and The Red Craigs provide many easily accessible, fast drying routes all within 10–15 minutes walk from the road. A south-west aspect makes it possible to climb on the crags for most of the year, though it should be remembered that they lie at an altitude of over 300m (1000ft). The climbing and setting is somewhat reminiscent of the Lake District (without the lakes and the crowds). The rock is diorite, a coarse grained form of granite varying from rough and rounded to smooth and angular.

Access: Follow the A90 north from Dundee turning off to Kirriemuir. Bypass the town and continue north on the B955 following the signs for 'the glens'. From Aberdeen the shortest route is to leave the A90 at Finavon and follow the B957 to Tannadice then by minor roads to Memus and Dykehead. On reaching The Glen Clova Hotel pass over the narrow bridge and continue up the glen for 2.9 miles/4.7km. Park on the verge on the left just beyond a long thin stand of larches and pines beneath the crags or a few hundred metres further on in the quarry on the right.

Accommodation: The Glen Clova Hotel (☎ 01575 550350; www.clova.com) serves good pub food and even better ale. For the impecunious, there is a capacious (sleeps 12 comfortably) but draughty howff under the boulders below the **South-East Crag**. There is also the Carn Dearg MC Hut just beyond Breadownie Farm at NO 286 757.

The Red Craigs, on the right (north) side of the road near the head of the glen, consist of six crags in all — the slabby **South-East Crag**, with the steep **Central Crag** just up to its left, the **Upper** and **Lower North-West Crag** and the two-tiered **The Doonie**.

Approach: Head directly up the hillside to the crags.

SOUTH-EAST CRAG

The eastmost (furthest right) crag on the hillside, with the best easier routes

Descent: Scramble off to the left and down the open gully running underneath the base of **Central Crag**, or abseil from slings & maillons on blocks.

1 Three J's Chimney ★ **12m Very Difficult**

FA Jack Scroggie, Jack Scott & John Ferguson July 1938

Just right of the left edge of the crag is a short chimney. A tricky start leads to better holds and protection. (The nearest city, Dundee, used to be famous for the 'three J's' – jute, jam & journalism.)

2 The Wildebeest ★★ **20m E4 6a**

FA Neil Shepherd & Kenny Clarke 10 August 1986

Good sustained climbing tackling the roofed niche in the upper right side of the wall. Start beneath a prominent orange coloured wall. Step off a boulder and climb directly up the wall to a ledge below the niche. Climb the niche, pulling through the roof on good holds to finish more easily.

3 Parapet Route Direct Start ★ **25m HS 4b**

FA unknown

Start above and right of the lowest point of the wall right of the vegetated gully. Follow a corner-crack fault line leading to the terrace.

4 Pilgrims ★★ **40m E1 5b**

FA Michael Barnard & Matthew Thompson 20 June 2010

Start as for 5a. Climb the initial corner and fine vertical cracks above to the foot of a hanging corner. Climb this past a wide section (crux), moving up left at the top to join 5 near its huge flake. Finish as for that route.

The following routes all start from a long grassy terrace, gained either from the right or by starting up the previous route.

5a Parapet Route Direct ★★ **40m VS 4c**

FA Ged & Ian Reilly 1970s

Start 4m further right than the normal start at another short shallow left-facing corner near the right end of the terrace. Climb cracks above, stepping left to a ledge beneath a prominent steep crack (possible belay). Climb up into the hanging corner finishing up the final section of the ordinary route.

5 Parapet Route ★★ **35m Severe 4a**

FA John Ferguson & Graham Ritchie 1 September 1940

Start just right of 6 beneath a short left-facing corner. Climb the corner and the short steep chimney just right then move diagonally left to a groove. Ascend the groove,

NO 297 756 **Alt:** 330m

15min

stepping right delicately to a crack leading to a large flake. Finish steadily up the right rib of the wall above.

 **Central Crack** ★★★　　　　　**35m HS 4b**

FA George Malloch & party 1950s

An excellent well protected pitch on superb rock up the shallow left-facing corner.

 **Flake Route** ★★　　　　　**35m VS 4b**

FA Bill Ward & John Ferguson 27 August 1939

Start beneath a short chimney at the left end of the terrace. Climb the chimney and step left and up over some blocks to a projecting block on the arête. Pull over this and up to a large flake. Move up diagonally right to a rounded boss then direct to finish on better holds.

CENTRAL CRAG 15min

NO 295 757 **Alt:** 360m

The overhanging wall with a distinctive orange patch above the descent rake for **South East Crag**.
Descent: Abseil from slings & maillons at top of crag.

> The first two routes start by scrambling up broken ground to belay below overhung niche.

 **West Side Story** ★　　　　　**20m E2 5b**

FA Graeme Ettle & Simon Stewart 23 August 1986

Climb strenuously up into the niche from the right, move up into the apex and pull out right onto the grey wall. Follow the line of holds to a flake crack on the right. Follow this, then continue up the wall above.

 **Bark at the Moon** ★★★　　　　　**15m E8 6c**

FA Tim Rankin & Guy Robertson (head pointed) May 2008

The stunning overhanging prow and crack up the left edge of the wall. Climb juggy hollow flakes just right of the arête to a good flat hold at a break from where protection can be arranged. Gain a sloper on the lip above and make a sequence of desperate slaps up side pulls to a good hidden hold at the base of the hanging groove. A further hard move up the groove leads to good holds; swing right then up to jugs and a rest. Move out left, then up to pull over at the crack. Excellent powerful climbing, well protected after a bold crux.

3 Mearns Wall ★★   30m E5 6a

FA Murray Hamilton & Spider Mckenzie 1983

An impressive pitch up the left side of the wall. Start by scrambling up the left side of large block to belay at its top. Climb direct up to the steep wall. From the ledge make committing moves up to the roof. Pull over this at a thin crack and follow for a few moves before stepping right to better holds below a wider crack. Climb this to finish.

4 Sun King ★★★★   25m E7 6b

FA Iain Small 2013

Superb well protected and sporty climbing with an exhilarating feel on the crux. Start up grey slab left of 5. Climb through overlap between two saplings to gain ledge below thin undercut flake. Climb this (camalot #2 in slot) and gain larger undercut flake leftwards, follow this (small and medium cams) to a niche below roof and a good rest. Climb directly above to gain a wide slot (camalot #1 & #2). Move hard right via undercuts to a stuck-on jug and mono pocket then hard right again to a hidden pocket and long reach to a crack in headwall. Climb ledges and short wall above trending rightwards. Belay 16m back in short steep wall. Double set of cams up to 2" useful.

5 Empire of the Sun ★★★   30m E4 6a

FA Dougie Dinwoodie & Jeff Hall 22 September 1986

Good well protected climbing starting up the orange patch. Climb slabs then directly up a crack in the loose orange rock to a slabby ledge. Gain the flake on the left then move up and left to another flake. Move up right along a crack until possible to pull onto small flat ledge (crux). Climb direct past large ledge and short wall above.

6 Sunset Song ★★   30m E5 6b

FA Iain Small & Jonny Clark 18 September 2005

A forceful route climbing the wall right of 5. Start below and right of the orange patch of rock. Climb the slab to an overlap, pull over and move left over shattered rock to gain a slabby ledge. Step right onto the impending wall and gain a flake. From its top make committing moves up and left to a small overlap. Pull out right to a short slot/crack. Final steep moves lead to a good hold from which balancy moves right allow a hanging corner to be gained. Finish up this to ledge and large thread.

6a Sunset Song Direct ★★   30m E6 6b

FA Iain Small 24 May 2009

Follow 6 to where the hanging corner can be gained from the left. Instead make a long undercut move to gain spaced holds leading up left to better holds.

LOWER NORTH-WEST CRAG

The prominent crag directly above the approach path.

Descent: Down a path at the right side of the crag.

1 Twenty Minute Route ★   45m Moderate

FA Jim Nisbet & John Ferguson March 1939

The broken rib bounding the left side of the main face. Start below a corner crack at the lowest rocks.

1 **10m** The corner to belay on the large terrace.

2 **10m** Left slanting crack to tree belay.

3 **15m** Climb direct behind the tree to gain a bulging slab or move out left onto the rib to gain the same position.

4 **10m** Cross the bulging slab and move out left to easy ground.

2 Wander ★   30m HVS 5b

FA W.Divers & Ken Sturrock 1957

Start beneath a prominent vertical crack 6m left of the corner. Go up the crack to a ledge then move left and climb the shallow corner to a tree. Make an exposed step out right and finish up a crack past ledges.

3 Wandered ★★★   30m HVS 5a

FA Neil Shepherd & Gordon Clarke 18 July 1982

Very good well-positioned climbing up the open wall left of the corner. Start up the crack as for 2 to the ledge. Follow the steep flake crack on the right to gain a large ledge. Move up into the recess on the left, step round the arête and finish up the exposed final wall on good holds.

4 The Beanstalk ★ 30m VS 4b

FA Tom Patey 1954

The great recessed diedre, now considerably harder since the disappearance of the 'beanstalk'! Easily up broken rock to a short chimney leading to the remnants of the beanstalk. Continue up wide slot above (crux) to large ledge. Move up and right past another tree to finish up the flake crack above. The 4a *Direct Finish* up the corner is ★VS 4c.

5 For a Handful of Beans ★★ 30m HVS 5a

FA Simon Richardson & Tim Chappell 24 Aug 2014

Good climbing up the centre of the wall. Climb direct up cracked wall to a niche and arrange protection in the wide left-curving crack above. Step back right and climb the wall above on good positive holds, trending left to a large rowan. Finish up crack above past a further rowan.

6 Proud Corner ★★★ 30m VS 4c

FA George Malloch 1950s

One of the best pitches on the crags, following an exposed reasonably protected line up the blunt right arête of *The Beanstalk* dièdre. Start beneath a pair of cracks on the arête. Follow either of the cracks or the arête to a good ledge at 10m. Climb large flakes up the

arête to a PR on a small triangular slab. Continue directly on good pockets to reach a small ledge, move right then directly by a tiny corner to finish.

7 Cinderella ★★ 20m E6 6a

*FA Graeme Ettle (headpointed with 1PA) 4 October 1987;
FFA Gary Latter & Kev Howett 6 July 1988*

A hard and very serious pitch up the leaning wall round to the right of 6. Start at the large pointed flake on the ledge left of 9. Climb the thin crack (PR) and the ramp leading left to a large flat hold (small shallow cams – very marginal). Stand on the hold (offset nut in shallow flared slot) and go up steeply to a good jug out right. Pull up into corner above, and continue more easily up this and directly above.

8 Taken by Force ★ 20m E2 5c

FA Neil Shepherd & Graham Woodtine 9 June 1984

The steep jam crack 3m left of the corner, finishing up easier headwall.

9 Monster's Crack ★★ 20m HS 4b

FA Andy Mitchell 1954

The prominent left-facing corner crack. Climb the lower corner past assorted shrubbery, finishing up the short steep corner crack on the right at the top.

10 Witch's Tooth ★★   20m E1 5b

FFA Murray Hamilton & Dave Brown 1976

Well protected climbing up the left side of the pillar. Climb the crack leading into a niche. Step left into a corner crack and follow this and the short wall above.

11 Cauldron Crack ★★   20m HVS 5a

FA Neil Shepherd & Colin Adam 14 July 1983

The overhanging recess up the right side of the pillar gives a good steep pitch on huge holds. Climb the lower corner to huge flake holds at the roof. Move out left then back right past a holly to gain a ledge. Finish up the short wall.

UPPER NORTH-WEST CRAG

A fine varied crag high on the slope just up and left of the top of **Lower North-West Crag**.

Descent: Follow a path up and left to a small cairn at the top of the descent gully bounding the left side of the crag, or abseil from sling & maillon on block at the top of *The Red Wall*.

Routes from right to left.

1 The Red Wall ★★★   35m E1 5b

FA B.Forbes (Aid) 1950s; FFA Murray Hamilton & Dave Brown 1976

Fine climbing up the central crack in the smooth red wall. Start beneath a large hanging flake.

1 **20m 5a** Climb the right side of the flake and hand traverse left along the large block. Pull over onto a huge sloping ledge. (The hanging flake crack just right gives a harder start - ★★ E1 5c with overhead protection). The easy open quartz chimney above leads to a belay.

2 **15m 5b** Step out right and climb the fine crack on immaculate rock to the top.

2 Kremlin Control Direct ★   35m E2 5c

FA Alistair Ross & Ged Reilly 1985; Direct – Simon Stewart & Catherine Smith June 1986

The leftmost crack-line in the smooth red wall at the lowest point of the crag. Start below some blocky overhangs at the left side of the steep grey wall.

Karen Latter on the finely-positioned Proud Corner.

1 **15m 5c** Climb up to a good hold, step right onto the grey slab and pull over the bulge to a spike. Continue past large sloping ledge above then easily to a flake belay.

2 **20m 5a** Move up left then back right beneath a small roof to the crack and finish up this.

3 Alder ★★   30m VS 4b

FA D.Brown 1951

Excellent airy climbing crossing some unlikely looking ground for the grade. Start in the right side of the overhung recess behind the large aspen (not an alder!).

1 **10m 4b** Follow wide cracks to a belay on a large ledge.

2 **20m 4b** Move out right then back left and finish directly.

4 Puddin' Fingers ★   25m E2 5c

FA Alec 'Tam' Thomson & Ian Shepherd (3PA) 1982; FFA Grant Farquhar, Graeme Ettle & Simon Stewart 1985

The steep flake crack on the right side of the overhanging wall. Start behind the rightmost aspen. Climb the steep initial flake crack continuing in the same line.

5 Sorcerer's Apprentice ★ — 25m E2 5c

FA Colin MacLean & Ged Reilly 1985

Sustained climbing up the steep crack in the right side of the overhung recess. Start up and left of the rightmost aspen below a steep flake. Climb the flake then move out right and up the crack passing a tiny rowan near the top.

6 The Supernatural Anaesthetist ★★ — 20m E4 6a

FA Grant Farquhar & Simon Stewart 1 October 1987

Sustained technical climbing up the hanging orange groove on the left side of the overhung recess. Start up 7 to the huge hanging flake. Make hard moves right into the groove and ascend this with interest. Near the top step out right then back left to finish at a projecting flake.

7 The Sorcerer ★★ — 20m E3 5c

FA Colin MacLean & Ged Reilly (in two pitches) 1985;
FA in one pitch Murray Hamilton 1985

Good well protected climbing. Start at the second aspen a short way up 9. Step right and up a short groove and slab to a large block. Undercut this leftward to good flakes in the corner and up to beneath the huge hanging flake. Undercut this spectacularly leftward then finish up an easier flake groove. Belay just beneath the top or further back.

8 A Vanishing Breed ★★ — 20m E6 6b

FA Grant Farquhar & Arthur Collins 17 July 1990

Steep powerful climbing up the wall high in the centre of the crag. Start at the base of the cleft of 9. Gain the wall via a crack, which is followed until it peters out. Move up and step right to a large sloping shelf and protection (2 RURPs in situ; R #1 down and right; Tri-cam #1 up and left in crozzly pocket). Climb the wall above bearing right to join 7 at the end of its flake.

9 High Level Traverse & Direct Finish ★★ — 40m HS 4b

Direct Finish D.Thomas, J.Fleming & 'Goggs' Leslie April 1950

A fine unusual route with two contrasting pitches. Start at the back of the overhung bay down and right of the huge overhung cleft-fault.

1 **20m 4b** Move up left to the base of the cleft and enter it with difficulty. Squirm more easily up it to spike and nut belay on ledge at its top.

2 **20m 4b** Traverse left across the exposed slab and round the corner to the base of a short narrow chimney. Move up this with interest then continue up the wide flake-crack to finish more easily.

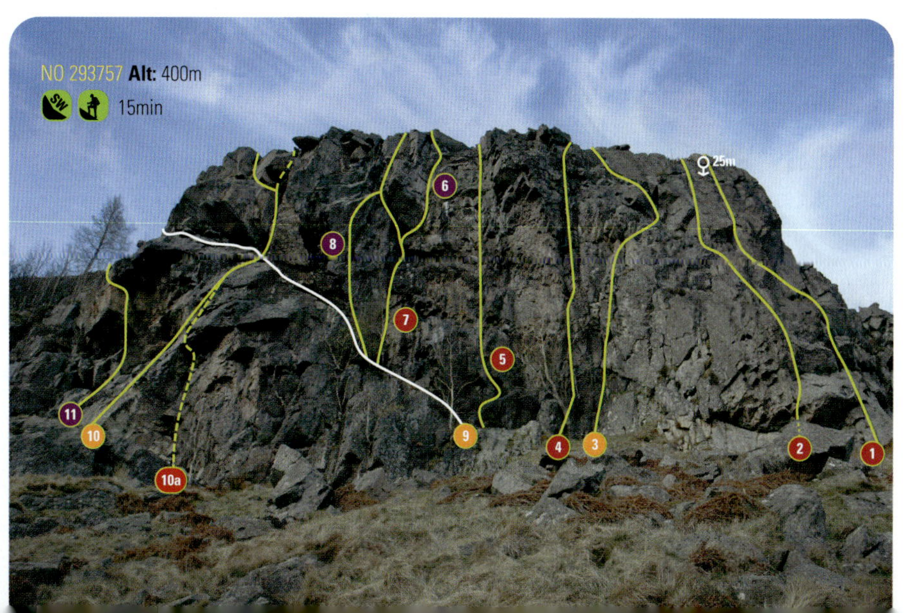

10a Zig-zag Double Direct★★   30m E2 5c

*FA Mick Tighe (pitch 1) 1972; FFA Simon Stewart
& Grant Farquhar September 1985*

An excellent well protected pitch. Start beneath the large
open corner of 10. Follow the steep crack-line through
the roof then out rightward as for 10. Continue easily
up the corner to negotiate the capping roof (crux) which
soon leads to good holds and the top.

10 Zig-zag Direct ★★   35m HVS 5b

FA Doug Lang 1960s; FFA Steve Scott & Ian Reilly 1974

A good sustained meander. Start at cracks in the slab,
down and right of the prominent cracked roof of 11.

1 15m Zig up and right across the slab to a large block
(PR). Pull over the block and climb the crack on the
right to belay on a pedestal above a deep cleft.

2 20m Climb the corner above then zag out left at a
prominent hand rail beneath the capping roofs.

11 Roman Candle ★★   18m E4 6a

*FA Davie Crabbe, J.Howe & Doug Lang (A3) 1964; FFA Murray
Hamilton, Rab Anderson & Duncan McCallum 24 September 1983*

A short steep athletic problem tackling the roof crack at
the left end of the crag. Up easy slabs to the right-facing
corner in the roof. Climb the corner and hand traverse out
to the lip (PR). Pull over strenuously to gain the
slab above.

12 Just Another Sparkler ★★   15m E3 6a

FA Murray Hamilton & Neil Shepherd 4 June 1985

The flakeline in the steep wall just left of 11. Strenuously
gain and climb the flake with difficult finishing moves
pulling onto the easy slab above.

*Malcolm Davies pulling through th
crux moves of Just Another Sparkler.*

THE DOONIE

The dome-shaped mass of rock on the left side of the hillside, above the quarry. The upper and lower tiers are split by a large left-slanting diagonal terrace.

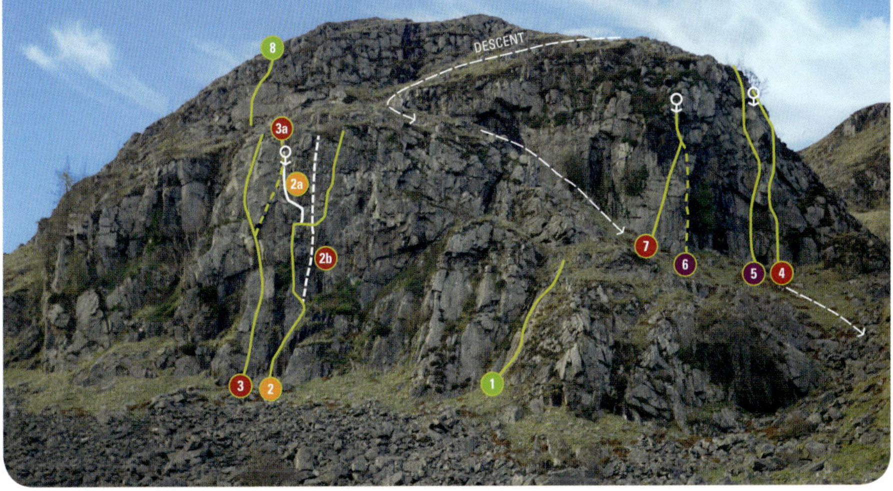

LOWER DOONIE   10min

NO 290758 **Alt:** 300m

The generally broken lower crag though with a fine clean wall near the left end.

Descent: Down the diagonal terrace then steep grass at the right side of the crag, or abseil from sling & maillon on large rowan 3m down from top of *The Furstenberg Finish.*

 ① Ant Slab ★   25m Moderate

FA Unknown 1950s

A good pitch up the easy-angled pink slab on the right side of the crag. Follow a diagonal crack up right then another back left into the centre of the slab. Direct up this to a block belay.

② Special Brew ★★   50m HVS 5a

FA George Malloch 1950s (sling aid); FFA Ged & Ian Reilly 1974/75

A good exposed route up the centre of the main face. Start below a pink right-sloping ramp, just right of 3.

1 15m 4a Follow the vegetated ramp then move up and right to belay at a peg.

2 20m 5a Move up and left to a PR at a small ledge. Climb the steep juggy cracks above to a short corner (PR). Traverse right to belay on the large ledge.

3 15m 4c From the right end of the ledge climb over the large detached block and continue in a fine position to the top.

② The Furstenberg Finish ★★   25m HVS 5a

FA Grant Farquhar & Simon Stewart 27 February 1988

A good left finish. From the short corner on the second pitch step right then up the wall to a shelf beneath a prominent V-niche. Traverse left with a tricky move to gain better holds in the groove on the left. Finish more easily up this.

② Special Brew Direct ★★   25m E3 5c

FA Nick Sharpe & Graeme Ettle 1985

A good well protected line cutting through the original route. Climb directly up the grooves above the belay to gain the large belay ledge on the right. Climb the wall directly above the left side of the ledge to finish.

3 Guinness ★★ — 40m E1 5b

FA Fred Old, George Malloch, Frank Anderson & Alex Ferguson (2PA) 1958; FFA Ged & Ian Reilly 1976

The crag classic on superb rock. Start 5m up and right of the toe of the crag directly beneath an overhanging beak roof.

1 **20m 5b** Gain and climb the grey concave slab leading directly to a good belay ledge.

2 **20m 5b** The black leaning corner above past a large ledge. Finish up the wide crack on the right .

3a Variation Finish ★★ — 20m E1 5b

FA Martin Hendry 1960s

Move up and right below the 'beak' overhang to finish up right slanting cracks.

UPPER DOONIE   15min

NO 291759 **Alt:** 300m

The sprawling walls directly above the steep grassy descent rake for the **Lower Doonie**.

Descent: Abseil from trees at the top of the crag or walk off left and down the diagonal grassy terrace.

Routes from right to left.

4 Dancin' in the Ruins ★★ — 35m E3 5c

FA Simon Stewart & Chris Cracknell 1986

A good varied first pitch up the right side of the wall right of 5. Start below a tongue of rock projecting from the roof.

1 **20m 5c** Climb the slab, move left and continue to a small niche (often wet) at the left side of the diagonal roof. Swing rightward through the roof on huge holds and continue steeply to belay at a small aspen.

2 **15m 5b** Move up and left then step right and finish up a right-facing groove just right of a steep crack containing loose blocks. Belay on a metal fence post.

5 The Whoremistress ★★ — 35m E4 6a

FA Grant Farquhar & Graeme Ettle 13 March 1988

An excellent long sustained pitch up the centre of the alcove. Start in the alcove below a square-cut recess with a small tree down and right of prominent evergreen bush. Up the recess and the V-groove above to a large diagonal shelf. Move up into the groove above with difficulty (poor PR). Climb direct up to the left side of the overhang and cross this using a prominent small flake on the wall above (crux). Step left into wide vertical crack and finish more easily to a ledge and tree belay on the right.

6 DRI ★★ — 30m E6 6a

FA Graeme Ettle & Grant Farquhar (headpointed) 19 May 1987

A technical and bold pitch with hard fought protection; up the blunt arête in the centre of the crag. Start beneath twin cracks in the right side of the arête. Climb the crack to a good ledge on the front face. Climb the arête on its left side passing a prominent side-pull to reach good holds near the top. Pull onto the slab above. Continue more easily up the wall above to finish up the final layback crack of 7.

7 Vindaloo Direct ★★★ — 30m E1 5b

*FA J.Cadger & J.Thomson (A2; 3 bolts) 1972;
FFA Simon Stewart & Grant Farquhar 1986*

A fine varied pitch with an exciting finale. Start beneath a crack 5m up and left of the blunt central arête of 6. Climb the initial steep crack, move left round the overlap and up the slab to a recess above. Traverse right round a block on huge holds then finish by a short fierce layback crack.

8 Larch Tree Wall ★ — 35m Very Difficult

FA unknown 1950s

Start at the top left end of the diagonal terrace. Climb the pocketed wall just right of the prominent lower crack then directly up the compact cracked wall above. Finish more easily up the upper rocks. The larch fell over in 1999.

JUANJORGE

NO 265 795 **Alt:** 500m   1¼hr 45min

A fine isolated wall of immaculate granite in the upper reaches of Glen Esk, which forms the upper right branch at the head of Glen Clova.

Access: Continue up Glen Clova past **The Red Craigs** for a further 0.6 mile/1km to car park (charges) and information hut.

Approach: From the car park at the head of the glen go back across the bridge and follow the track (on the east side of the River Esk) north through the forest. Just over 100m beyond the farm at Moulzie follow the track down left and cross the River Esk. Continue round the track for a further 1.5km then re-cross the river and head diagonally leftwards up the hillside to the crag.

Descent: Abseil from trees at the top.

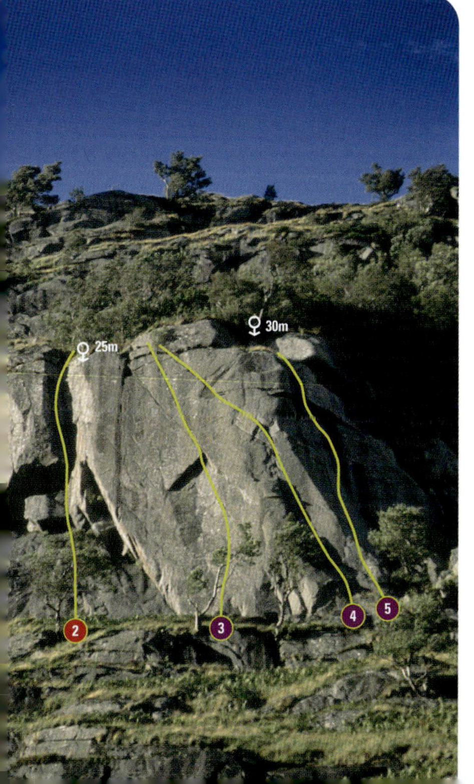

1 Granite Heids ★　　　　　　　　　25m E1 5b

FA Dougie Dinwoodie & A.Gunn April 1987

Start below the leftmost of two grooves, left of 2. Climb the groove until it is possible to move out left onto the edge (crux). Finish out right to a tree belay.

2 Rhiannon ★★　　　　　　　　　　25m E3 6a

FA Dougie Dinwoodie & Greg Strange 11 October 1986

The obvious corner line to the left of the smooth face. Climb straight up the slabby face below the corner. Pull over a bulge and go up a short groove to flakes right of the base of the corner. Climb the sustained corner to the top. Tree belay.

3 Time's Arrow ★★★★　　　　　　　30m E6 6c

FA Grant Farquhar & Clare Carolan 4 July 1995

The compelling central line through the triangular niche. Move up then step right to the base of a thin diagonal crack. Up this passing a Lost Arrow peg. Make difficult moves (crux) through the niche to gain the crack sprouting from its apex. Continue up the sustained crack to join 4 at its final moves.

4 Roslin Riviera ★★★　　　　　　　35m E4 6a

FA Murray Hamilton & Greg Strange 21 June 1983

Excellent well protected climbing. Climb the groove in the toe of the crag to its top then follow the left-slanting diagonal crack with a hard move to a resting place. Continue up the crack until a step left is possible to another crack, which leads after a couple of moves to a horizontal break. Finish above a small tree.

5 Ladies of the Canyon ★★　　　　　30m E5 6b

FA Kenny Spence & Murray Hamilton July 1983

Climb a short crack and step left into the scoop in the wall 5 metres right of 4. From the top of the niche make a hard move right to holds which lead left to a crack. Follow this through the roof to easier ground.

LOCHNAGAR (LOCHAN OF THE NOISY SOUND)

"Yet still are you dearer than Albion's plain.
England! Thy beauties are tame and domestic,
To one who has roved o'er the mountains afar;
Oh! for the crags that are wild and majestic,
The steep frowning glories of dark Loch na Garr."
– George Gordon, Lord Byron, 1806

The north-east corrie of Lochnagar presents an almost 1.5km long crescent of cliffs encircling the loch after which the mountain is named. Although there are substantial areas of grass and vegetation on the cliff including large areas of loose rock, especially in the gullies, there are also many excellent routes throughout a wide spectrum of grades, all on immaculate Cairngorm granite.

The correct name for the mountain should be Beinn Chiochan, or 'hill of paps', whilst the individual tops are Cac Carn Beag and Mor (little shit cairn & big shit cairn) – little wonder that the name was altered shortly after Queen Victoria purchased the estate!

Access: From Ballater head east out of the town over the River Muick. Turn right and follow the B976 south for 0.6 miles/1km then turn left at the bridge of Muick up the minor C-class single track road up Glen Muick for 9 miles/14km to the car park (pay & display) at the road end at the Spittal of Glenmuick.

Approach: From the car park follow the track for a few hundred metres to the visitor centre. Turn right (sign-posted) and head north-west across flat ground crossing the River Muick by a bridge to a path cutting through the woods. Continue by a good Land Rover track up first the left then the right side of the Allt-na-giubhsaich (stream of the pinewood) for just over 2.5 km to just short of the summit of the bealach. Follow a narrower but well constructed path west up the hillside for 1.5km passing the Fox Cairn Well to gain the col between the large flat mass of Cuidhe Crom on the left and Meikle Pap. Follow a small path cutting down into the corrie, which drops down until about 50m short of the lochan before rising up to a large monument from where all the routes can normally be viewed.

Descent: The fastest descent is by either branch of *The Black Spout*. The left branch lies 300m north-west from the top of *Eagle Ridge*, with a prominent large cairn about 50m from its top. This is relatively straightforward with an 'entertaining' through route beneath a chokestone near the base, just before reaching the main branch. If these are filled with snow continue north-east along the rim of the corrie, descending by easy slopes beyond **West Buttress**.

EAGLE RIDGE NO 247 856 **Alt:** 970m   2hr

① Eagle Ridge ★★★★   **250m HS 4b**

FA Jim Bell & Nancy Forsyth (1 PA) 24 July 1941;
FFA S & Mrs Thomson June 1944

"…for difficulty, narrowness and steepness altogether superior to any of the well-known Nevis ridges." – Bell
A majestic route, more popular than all the other rock routes in the corrie put together. It dries quickly and the rock is clean, making it possible (at VS) in the rain. Start beneath a prominent slabby V-groove just inside the open scree-fan of Douglas-Gibson Gully. Climb the groove then continue more easily up shallow gully, trending right to negotiate a chimney containing some chokestones (30m).Trend back leftwards, ascending a 10m inset corner on the right then steeper rock to regain the crest. Move up right to a recess then continue steeply, moving left to a *"splendid sentry box"*. Continue up the smooth arête, finishing by a small corner to a ledge on the crest. Ascend the 'whaleback' crest for 20m then a short slab corner leading to a knife-edge on the crest. The crux looms above – a vertical 4m wall split by a jam crack where a few moves (well-protected) lead to a good ledge. Continue along the almost level crest to a square-cut projecting overhang. Swing up onto the 'coping slab' above and mantelshelf from some cracked blocks into a V-recess finishing rightwards up easy-angled slabs to blocks just below the top.

Grant Kasprowicz on pitch 2 of Eagle Ridge. Photo Zelin Liu

BLACK SPOUT PINNACLE  2hr

NO 248 856 Alt: 1000m

The large triangular buttress on the right side of the corrie demarcated on either side by Raeburn's Gully and *The Black Spout*. On the left side is the fine 100m apron of the lower slabs leading to more vegetated ground above. Further right, higher up the lower reaches of *The Black Spout*, the slabs gradually steepen, merging with the steep *Black Spout Wall*. From the top of the pinnacle follow an obvious line bearing left down an easy slab to gain the col (25m). 30m of easy rocks (exposed) above lead to the plateau.

1 Pinnacle Face ★★   95m VS 4b

FA Jerry Smith & J.Dennis 4 September 1955

A major breakthrough in Cairngorm climbing, first ascended in rope soled shoes. Start about 10m above the lowest rocks at the corner of *The Black Spout*.

1 **35m 4a** Follow either of two grooves, both awkward to start. The rightmost V-groove leads to a delicate traverse left at 10m to a short chimney in the left groove. Continue up the chimney then by left-trending cracks.

2 **25m 4b** Go up to a corner and pull onto the slab on the right. Trend left up a slabby fault to belay on a large grass ledge.

3 **25m 4b** Continue left crossing the large prominent fault to twin cracks. Follow either of these continuing up the rightmost crack where the other fades then traverse rightward on flakes to belay in a niche.

4 **10m 4b** Go up the steep corner above to a ledge. Traverse easily right on grass ledges leading to a descent to **The Springboard**.

2 Pinnacle Grooves ★★   70m HVS 5a

FA Rob Archbold & Greg Strange 29 June 1975

Fine well protected climbing up the grooves to the right of *Pinnacle Face*. Start beneath the rightmost groove of *Pinnacle Face*.

1 **30m 5a** Climb the groove to grass ledges on the right at 15m. Move right then go left and up to a large downward pointing flake. Layback up

Wilson Moir on the first ascent of The Extremist. Photo Paul Allen.

its left side to step into a smooth groove on the left and climb this to a grass stance and belay.

2 **15m 5a** Continue up a groove left of the prominent overhang to a grass ledge on the right.

3 **25m** Step left then go up about 9m. Here grass ledges lead horizontally right for 15m to the start of *The Link*.

The following 5 routes all lead to a series of large vegetated ledges dubbed **The Springboard**, from where a 45m abseil can be made from a large block in the centre – slings in situ.

3 The Nihilist ★★ 45m E1 5b

FA Brian Lawrie & Dave Innes August 1976; Direct Start Guy Muhlemann & Greg Strange 28 August 1983

Superb sustained climbing following the prominent twin grooves leading to the right end of The Springboard. Start immediately left of the smooth wall near the base of 4.

1 **25m 5b** Climb up to obvious holds and make a difficult move left before swinging up to a large ledge (the *Direct Start* gains this by the prominent V-groove). Move up to a higher ledge then gain a steep narrow slab with a bulging wall above. Traverse right to an *"apparently desirable mantelshelf ledge"*. Continue traversing right to gain the main groove descending a little to a belay.

2 **20m 5a** Climb the corner.

3a The Nihilist-Extremist Combination ★★
55m E2 5b

An excellent sustained and natural line, and the fastest drying line on the pinnacle. Can readily be climbed in one long pitch.

3b The Nihilist-Existentialist Combination ★★
55m E3 5c

FA Michael Barnard & Misha Bruml 15 July 2018

Climb pitch 1 of *The Nihilist*. From the belay ledge, move up the arête and pull out right to climb the superb flake-crack of *The Existentialist*.

4 The Extremist ★★ 40m E5 6b

FA Wilson Moir, Paul Allen & Julian Lines 16 June 1996

Excellent sustained climbing. Slightly harder technically than *The Existentionalist* but better protected. Start just right of the V-groove of *Nihilist Direct Start*. 2022

1 **18m 6b** Climb an awkward corner/ramp to its top. Make hard moves up and right into the obvious niche. Continue up the crack above to belay above the overhanging wall.

2 **22m 5b** Continue in the same line up twin cracks slanting left to join and finish up the corner.

4a Radicalist ★★ 45m E6 6b

FA Gordon Lennox & Tim Rankin (both led) 14 August 2022

A right finish to *The Extremist*, climbed in one long pitch. Follow that route to the niche, then follow a handrail rightwards to cracks to pull onto a big ledge at the dirty corner of *The Nihilist*. Climb the arête on the right in a fine position to gain a higher ledge then finish direct.

5 The Existentialist ★★★★ 42m E6 6b

FA Wilson Moir & Paul Allen 8 July 1995

One of the best pitches in the Cairngorms, climbing the impressive smooth wall at the base of *The Black Spout*. Start at a small corner 10m right of *Nihilist Direct Start*. Climb boldly up into the little corner. Climb it and the crack above to a ledge. Continue up a superb flake crack until it is possible to exit by the left crack. Finish up *The Nihilist* to reach **The Springboard**. 2022

6 Extortionist ★ 35m E2 5b

FA Jules Lines & Charlie Ord 29 August 1999

Start at a right-facing corner. Climb the corner to a roof, pull left onto the wall and traverse left into the upper corner. Finish up and left out of it.

7 The Link ★ 100m VS 5a

FA Ken Grassick & Bill Brooker 16 June 1956

The time honoured upper continuation to *Pinnacle Face*. Start in the rightmost of three faults at the top of **The Springboard**.

1 **15m** Make a short awkward traverse into a vegetated V-groove which is followed to belay on the rib on the right.

2 **25m 4c** Move up the rib a short way then step back into the groove and follow it to a prominent triangular overhang. Climb up and right round a huge block to a small recess beneath an overhang. Cross this and climb a good crack slightly right to belay behind a further huge block.

3 **30m 4c** Follow the prominent right-slanting crack moving out left near its top to a steep groove and go up this to a large overhang. Pass this on the right using a 'sometimes rotating' block.

4 **30m 5a** Break through the overhang above by good cracks.

8 **Black Spout Wall** ★★★　　　　　170m E3 5c

FA Dougie Dinwoodie & Bob Smith (2 PA) 8 & 10 August 1976; FFA Brian Lawrie & Neil Morrison September 1983

Very good climbing particularly on the lower pitches, following the prominent crack-line in the pillar left of the great overhung recess in the centre of the wall. Start on a large grass ledge right of the crack.

1 **40m 5c** A very fine pitch. Shuffle left along a small ledge and climb the deep mossy crack to a pinnacle at 15m. Cross the bulge above then ascend the smooth dwindling groove to flakes and a large PR on the right. Descend to the lip of the overhang and swing left into the scoop at the base of a ramp. Go up the ramp crossing an overhang at its top to gain

Robert Durran starting up the first pitch of Black Spout Wall.
Photo Dan Moore

better holds 3m above. Move down diagonally right to a ledge and PBs directly above the initial cracks.

2 30m 5c Climb directly up the wall above, crossing a bulge to gain a groove and climb this by a thin finger-crack then the easier upper groove, which leads to an exit out left to ledges near *The Link*.

3 40m 4b Traverse right 7m, then follow slabby shelves up and right to gain the arête above the great overhangs. Go up this to a little ridge.

4 25m 5c The left-slanting 'inhospitable crack' lies to the left. Climb the long right fork of the forking system leading to the apex of the wall.

5 35m Finish up the crest to the top of the **Pinnacle**. A 60m abseil from the top of the **Pinnacle** gains *Black Spout Left Branch*.

9 **Steep Frowning Glories** ★★★   **160m E5/6 6b**

FA Wilson Moir & Neil Morrison 21 July 1996

An excellent modern companion to *Black Spout Wall* with a stunning well protected crux pitch breaching the great overhangs. Start 5m right of *Black Spout Wall*.

1 15m 5c Climb a crack up a pillar (just left of a shallow corner) to ledge.

2 20m 5c Continue up the crack to a roof. Go left and pull out onto a ledge leading left to *Black Spout Wall* PB.

3 20m 6b Go back right along the ledge and climb cracks diagonally rightward gaining the pedestal under the roof crack from the right. Climb the roof crack (full set Fs #0-4 desirable) to belay just above.

  2021

4 45m 5c Climb the continuation crack and corner and continue up to join *Black Spout Wall* at the right-slanting shelves. Go along these and the arête to belay beneath the gable wall.

5+6 60m 5c Continue up *Black Spout Wall*.

10 **Drainpipe Crack** ★★   **35m E3 5c**

FA Dougie Dinwoodie & Colin MacLean 4 August 1982

Excellent sustained jamming, though, as the name suggests, seldom dry. Start beneath the right line up the steep wall right of *Black Spout Wall*. Follow the steep crack into a recess beneath the final overhang. Cross the overhang with difficulty to finish up the right crack.

11 **The Black Spout** ★   **250m Easy**

FA John Gibson & William Douglas (winter conditions) 12 March 1893

Separating the main face from **West Buttress** is a huge scree-filled gully. Hidden from the corrie floor the Left Branch has one entertaining pitch, a traditional through route beneath a huge chokestone just above the fork. In descent either branch proves straightforward.

WEST BUTTRESS   2hr

NO 246 859 **Alt:** 980m

12 **Black Spout Buttress** ★   **250m Difficult**

FA Tom Goodeve, Willie Ling & Harold Raeburn 17 April 1908

The best easy route on the cliff though the lower buttress is very vegetated and probably better avoided by traversing in from the fork on *The Black Spout*. Start about 10m right of *The Black Spout* at the top of a grass cone. Climb the chimney fault then scramble for approx. 60m to a level arête at the end of the lower buttress. Continue along the arête then a ridge of piled blocks to a 'deceptively difficult' short chimney. Easy climbing above then leads to a short wall, which is started centrally and finished on the right by an awkward corner. Now go up a fine 10m wall on good holds then a ledge on the left to regain the crest, which leads easily to the top.

CREAG AN DUBH-LOCH
(CLIFF OF THE BLACK/DARK LOCH)

Undoubtedly the finest mountain cliff in Britain with a kilometre long, 300m expanse of impeccable granite. Superb routes of all grades from VS to E9, with the finest selection of extremes to be found anywhere.

Access: From Ballater head east out of town crossing the bridge over the River Dee then turn right and follow the B976 south for 0.7 miles/1.1km. Turn left at the Bridge of Muick up the minor C-class single track road up Glen Muick for 8 miles/13km to the car park (NO 3099 8514; 56.952341, -3.1361789) (pay & display) at the road end at the Spittal of Glenmuick.

Approach: Follow the Land Rover track past the toilets and visitor centre for 800m to a narrower track on the right crossing the outflow of Loch Muick then by the Land Rover track along the north side of the loch as far as the Glas-allt-Shiel (royal shooting lodge) – 4.5km/1 hour. Continue through the woods by a narrower track which leads after a further 3km and 250m of ascent to the Dubh-Loch. Mountain bikes can be taken to the head of Loch Muick, saving around 40 minutes.

Accommodation: There is a small draughty howff sleeping two in the boulder field directly below **The Central Slabs** and superb campsites by the golden sandy beaches around the west end of the Dubh-Loch. There is also a bothy at the rear of the Glas-allt-Shiel. Wild camping in upper Glen Muick.

Daniel Laing on the superb main pitch of Alice Springs. Photo Jules Lines

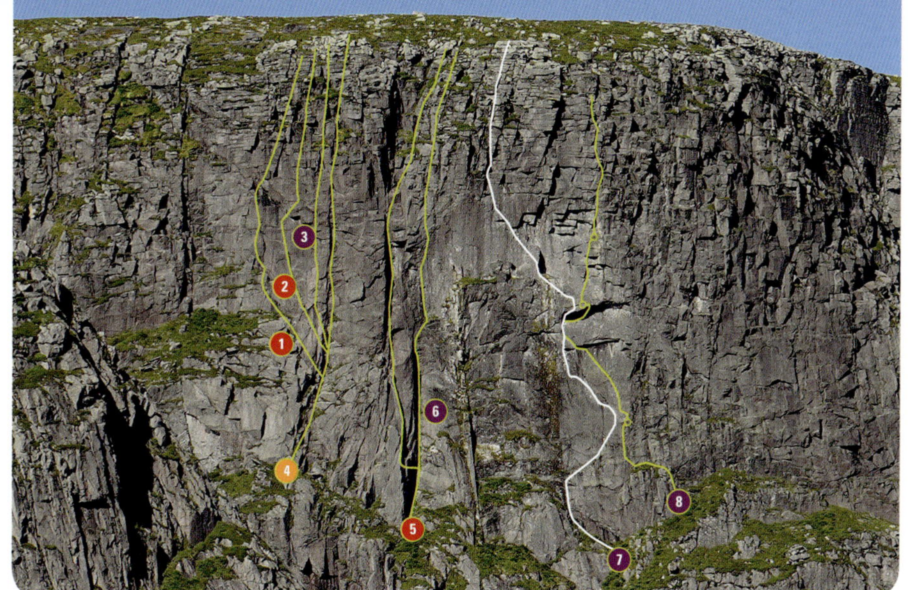

BROAD TERRACE WALL

The steep dark crag high on the left side of the cliff, steepest in its centre where the smoothest section leans out over The Grass Balcony. The huge corner fault of *The Sword of Damocles* demarcates the left edge of the steepest section, clearly visible even from the head of Loch Muick. Further left the face curves round to a slabbier section before petering out into vegetated ground above South-East Gully. This is the last section of the cliff to dry due to a large patch of snow at its top, which usually results in seepage until at least the end of June.

Approach: Gain the grassy Broad Terrace under the frontal face by zig-zagging up the left side of the broken and copiously vegetated lower crags, starting near the base of South-East Gully.

Descent: Traverse left (east) just past the loose gully bounding the left side of the crag and scramble down the vegetated rock ridge. Near the base veer left and down a slabby rock ramp (Diff) in the side wall of the gully. In wet conditions or if unsure of the line, continue traversing further east past the end of the cliffs and down the easy-angled grass slopes.

① The Eye of Allah ★★　　　　　　　　**90m E3 5c**

FA Wilson Moir & Chris Forrest 7 September 1991

　1　**20m** Start up *The Last Oasis* and follow the fault easily to belay below a steepening.

　2　**30m 5b** Climb the corner above then step left onto a rib and follow a thin crack left to a ramp. Follow the ramp easily to belay at its left end.

　3　**25m 5c** Climb a finger crack to a shelf then use a good flange to traverse right and pull into a corner. Exit right to a belay ledge below a slanting corner.

　4　**15m 5b** Climb exposed hanging slabs on the left and finish up blocks above.

② Alice Springs ★★★　　　　　　　　**95m E2 5c**

FA Murray Hamilton & Pete Whillance 29 July 1983

"A great route, though hard to get to and to catch it dry."
Excellent climbing on immaculate rock up the crack-line in the centre of the wall.

　1　**20m** Start up the fault of *The Last Oasis* and follow it easily to below a steepening.

　2　**40m 5c** Climb the corner above, as for *The Last Oasis* then step left onto the rib and follow a thin crack left to ledges. Climb the prominent finger crack to a ledge. *"A genuine contender for the best of its grade." "Absolutely mega phenomenal."*

3 35m 5b Move up to the base of a corner, step right on to a rib and reach a crack on the right. Follow this and the continuation groove above to finish.

3 The Bedouin ★★★   95m E4 6a

FA Wilson Moir & Colin Stewart 27 July 1989

The obvious line between *Alice Springs* and *The Last Oasis*. *"Amazing route. Clean, well-protected but exciting and great moves."*

1 35m 4b Climb *The Last Oasis* to the top of the second steep step.

2 45m 6a Step left on to edge and go up to enter the left-facing corner. Climb this to a small overlap. Climb the bulge (crux) and move up to an undercling. Use the short corner on the left to gain the next break and follow the twin grooves to ledge. Zig-zag up the walls above finishing through a steep slot to ledges. A fine sustained pitch on immaculate rock.

3 15m Easy rocks lead to the top.

> The less prominent leftmost of the two faults is the aptly named 4 *The Last Oasis* ★ VS 4b,4b,4c – a natural drainage line, giving good climbing when dry.

5 The Sword of Damocles ★★★   95m E2 5b

FA Graham Hunter & Doug Lang (12 PA) 20 June 1970;
FFA Dougie Dinwoodie & Bob Smith July 1977

"A really good route, the corner is amazing." The huge corner system, the most prominent line on the wall. Start at an easy chimney next to a steep wall.

1 15m 4c Climb the chimney then traverse left across the wall to a ledge at the base of the big corner.

2 35m 5a Follow the corner past a ledge on the left. Continue up the corner above to a large flake below the hanging chimney.

3 15m 5b Climb the chimney (very sustained) to large ledge at its top.

4 30m 5a Move right and follow a short corner to a white slab. Climb the chimney on the right

or left, or climb further right if damp. Finish easily up giant steps and broken ground. A useful variation if the top chimney is wet: from below the hanging chimney, cross the first bulge until below the first old bolt. Traverse the left wall to the edge, move up the arête and traverse left then up the short pale groove of *Mirage* leading to the large ledge at the top of the chimney.

6 Devolution ★★   100m E6 6b

FA Tim Rankin & Gordon Lennox June 2010

A great route with two hard contrasting pitches directly up the left side of the wall. Pitch 2 is high in the grade but well protected.

1 50m 6b Start up the chimney as for *Sword of Damocles* but where this moves left continue up the chimney to climb the fine technical groove to a belay just left of the grass terraces.

2 20m 6b Above is a stunning overhanging crack. Move up right to a groove, climb this onto the wall proper. Climb the wall and crack via technical moves to a rest at a ledge out right. Step left and continue up the steep crack in a mind blowing position to a good belay where the angle eases.

3 30m 5c Climb the groove above to where it steepens then move left across a slab to a ledge, climb the crack above and follow the line of least resistance to join and finish up *Sword of Damocles*.

7 Flodden ★★★★   120m E6 6b

FA Murray Hamilton, Kenny Spence & Rab Anderson
22 – 23 July 1983

A stunning right to left diagonal line, crossing some impressive ground with two hard contrasting pitches. The last hard pitch takes a lot of seepage and is slow to dry. Up vegetation to the base of the wall then scramble up a flake on the left and traverse right along a good ledge to belay atop a large flake. 2021

1 30m 6b *"Wild US style leaning groove grovelling."* Up an easy crack in the back of a left-facing groove to underneath the roof. Layback and jam this with difficulty (Fs #2-3) and follow this until it narrows just above a jug on the right wall. Difficult moves traversing left lead to a good flake. Up this and a groove to belay in a short right-facing groove below a big roof. *"Crack/ tape gloves and tenacity recommended."*

2 15m 6a Gain a good crack in the groove above the roof with difficulty. Up this and foot traverse right along a ledge to a thin left-trending diagonal crack. Up this and boldly up a scoop to a ledge. Belay at P at left end of ledge.

3 10m 5a Step down left onto a lower ledge, left and up a flake to belay at top of a pedestal.

4 25m 6b Move left off the ledge and up to under the rectangular roof (poor rest – extend runners under roof, otherwise excessive rope drag near end of pitch). Move left to a good rest under the next roof. Move over a flake to good nuts then step left to a good rest on a slab. Climb the slim groove above to a good nut slot, step down and move left across the slab to a good finger edge. Up the steep corner above on improving holds to belay in the base of the corner.

5 40m 5a Climb the left rib of the groove above then more easily up occasional rock and much vegetation to the plateau.

diagonal seam up the slab to a horizontal quartz vein (micro cam). Get stood on the vein and tiptoe left to gain a nubbin and make thin moves up to the roof. Traverse left under the roof until below a leaning flake-groove. Follow this to a shared belay with *Flodden* in the corner under the large roof.

4 10m 6b *The Reality Roof.* Toe traverse the slab rightwards to the point where the roof crack kicks out. Make awkward moves (the tall can reach) to gain good jams and continue round the lip in a mind bending position to gain the belay ledge above.

5 20m 6b *The Puzzle Corner.* Go straight above the belay to gain a horizontal break (RP2), reach right to a sidepull at the base of the emerging right-facing corner, place a good small wire and medium cam before making hard moves into the corner and up to a slender ledge. Make a perplexing sequence to get stood on it and continue up the corner system to a good ledge.

6 40m 6b *The Peregrine Cathedral.* Go up into the grossly overhung niche and wildly bridge your way out to better holds; continue up the overhanging groove above to reach the blocky fault line. Go up this for 5m and then swing left onto a jutting block and continue via the layback and crack system to the top.

8 Raccoon ★★★★ 120m E7 6c

FA Jules Lines & Gary Latter 8 June 2023

This exceptional route is a contender for the best E7 mountain route in the country and will require a full range of skills. An ascent will be much prized as pitch 3 is frustratingly slow to dry. Start on the left end of Broad Terrace.

1 20m 5a *The Intro.* Climb easily up left onto a smaller grass terrace (as per *Culloden*) and from its left end climb a slab and layback corner to a sloping ledge beneath the finger crack. Spike belay on the right beneath a short corner.

2 10m 6b *The Sabre Crack.* Place a high nut, step down and make hard moves left into the crack and follow it with difficulty to a toe ledge and belay.

3 20m 6c *The Perseverance Seam.* Follow the

Jules Lines on the The Perseverance Seam pitch on the first ascent of Raccoon.

THE CENTRAL SLABS

Overlapping slabs sweep the centre of the cliff for 300 metres at this point. The slabs are set at too high an angle for friction climbing; therefore the routes follow natural lines. Although the routes are slightly contrived with many of the lines interchangeable, the generally delicate nature of the climbing, with the overlaps providing variety, gives many enjoyable routes. The routes are broken into two distinct halves by a slanting mid-height terrace reducing the overall seriousness as escape is possible rightward onto **Central Gully Buttress**.
Descent: From the plateau, walk right (north-west) and go down **Central Gully**. If filled with snow early in the summer, go down the easy crest of **Central Gully Buttress**, cutting down left into the lower reaches of **Central Gully**. Near the base, a small ledge leads horizontally left (10m above a greasy ramp) to short wall of blocks just above the gully floor.

1 Dinosaur/Pink Elephant ★★★ 320m HVS 5a

FA Jim Stenhouse & Brian Lawrie (1 PA) 25 July 1964;
FFA Malcolm Nicolson & John Mothersele June 1973
FA (Pink Elephant) John Grieve & Allen Fyffe 14 June 1969

"Starts and ends well." "Fantastic top three pitches; the final crack pitch is brilliant." A good combination giving the best line on the left side of the slabs. The last (crux) pitch up the big groove is unfortunately slow to dry. Start at the lowest rocks.

> **1 25m** Follow broken cracks to a grass rake, or scramble leftward up the rake to the same point.
>
> **2 40m 4b** Follow the main crack system above to a stance 5m below the long lower overlap.
>
> **3 40m 5a** Surmount the overlap above and go up the slab over an awkward bulge. Go up slightly

higher and follow a toe traverse left (careful not to go too high). Step up left then go slightly down into the prominent shallow corner. Climb this using the left rib to the top of a large flake.

> **4+5 80m 4c** Go up a grassy niche and climb the bulging corner above. Break right over the big left-slanting overlap to reach slabs then follow the obvious line up into *Dinosaur* Gully and follow its right branch to The Terrace.
>
> **6 20m 4b** Follow the prominent corner on the right side of the Sea of Slabs to under bulges.
>
> **7 45m 4b** Continue up grooves above, break through the overhang by a short bulging slot and continue up slabs.
>
> **8 25m 5a** Follow a tapering slab leading up left to the right end of large grass ledge below the upper overhangs. Traverse right to a slabby knife edge then drop into the big upper groove.
>
> **9 45m 5a** Follow the groove, overcoming a steep step by the left wall then regain the groove. Finish up the groove or by rocks on its left side.

❷ The Blue Max ★★   330m E1 5b

FA Brian Robertson, Allen Fyffe & Bill Wilkins (2 PA/A1)
16–17 September 1967; FFA John Fraser & party 1975

"Every pitch is great; even the 4c pitches don't give it away. The moves on the crux pitch are a delight!"
Excellent sustained climbing, though the crux section is contrived. The route breaks through the main overlap at a conspicuous rockfall scar. Scramble up the grass rake to a large block.

1 **40m 5a** Follow the crack-line directly above the block over a difficult bulge. Continue up cracks then move right to belay at the base of a left-facing corner.

2 **35m 5a** Follow the corner to small ledges and go up to make a thin traverse right over a smooth slab to belay under the main overlap.

3 **25m 4c** Break through the overlap by a right traverse across the wall immediately above the rockfall scar then go up by cracks to belay.

4 **35m 5b** This is the line of *Cyclops*. Move diagonally left to below a huge diamond-shaped block in the next overlap. Climb the block by its right side and traverse the top into the corner to climb the main bulge. Follow the corner above to the upper bulge, step right and go over the bulge by a crack which twists back into the groove. Continue up the groove to belay on the rib. A fine well protected pitch.

5 **35m 4c** Continue up the rib then go straight up the succeeding crack and corner to a roof. Turn the roof on the right by cracks leading to a small ledge. Continue up walls for 10m to belay on a long grassy ledge.

6 **35m** Scramble up to the right edge of the upper slabs.

7 **40m 4c** Turn the crescent-shaped groove of *Cyclops* by a smaller groove on the right (or a wide V-groove round the right edge) to a junction below short twin grooves. Climb the left groove and continue slightly left by a short cleft to a ledge. Move slightly right, then back left and up to a grass ledge beneath a prominent nose.

8 **40m 4c** Climb up just right of the nose, then move left over a slab to the base of a pegmatite corner. Climb this until it becomes vegetated, exiting onto its right rib. Continue

up the rib to belay below grassy grooves.

9 **45m 4b** Finish up the grooves, moving onto the right wall near the top to gain broken ground.

❸ Cyclops ★★★   325m HVS 5a

FA Greg Strange & Mike Freeman 19 May 1973

"A long and varied climb with lots of interesting pitches."
The classic route on **The Central Slabs** following a direct line of cracks up the lower slabs and utilising the wall of *The Blue Max* to breach the main overlap. Start at the foot of the grass rake of *Dinosaur* at the top of a tongue of slab encircled by a ring of grass.

1 **30m 5a** Follow the left of twin parallel cracks right of a brown corner to pull onto a hanging flake from the right side. Go up the edge to a good ledge under a small overlap (crux).

2 **35m 4c** Climb the bulge above then climb the prominent crack to a scoop with a constricted groove above. Follow the groove over a long grass plug and continue to belay at top.

3 **35m 4c** Climb direct through the overlap by traversing right across the wall immediately above the rock fall scar (as for *The Blue Max*) then continue up the superb finger and hand crack to the second overlap.

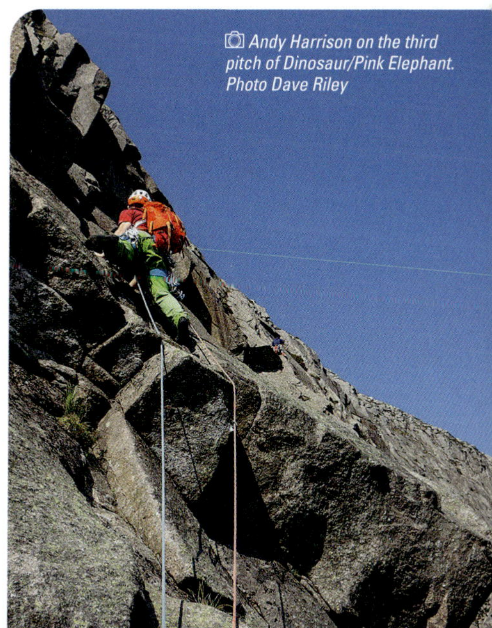

📷 *Andy Harrison on the third pitch of Dinosaur/Pink Elephant. Photo Dave Riley*

4 50m 4c Follow the continuation fault, then directly up cracks in the steeper pink water washed rock. Continue up leftwards over grassy ledges to a large grass ledge and thread belay.

5 10m Scramble up left to belay at the base of an obvious crescent-shaped groove near the right edge of the upper slabs.

6 30m 5a Climb the difficult short flared groove (crux) continuing more easily up the crescent groove above (junction with *The Blue Max*) to its top, then climb the improbable-looking short black cleft, exiting left to belay on ledge above.

7 45m 4c Move right then climb cleaned cracks slightly leftwards, then cross slab to belay at base of the pegmatite corner.

8 45m 4c Climb the corner, exiting onto the rib on the right just before it becomes vegetated. Continue directly on beautiful featured holds, then by grassy grooves to belay on the right.

9 45m 4b Finish up the grooves, moving onto the right wall near the top to gain broken ground.

4 **Black Mamba** ★★★   **335m VS 4c**
FA Allen Fyffe & John Grieve 7 June 1969

The classic on the slabs, giving excellent sustained climbing. Right of the grass ring of *Cyclops* is a grass ledge 20m up the slabs. Start from grass at a low point in the slabs.

1 20m 4c Climb a delicate shallow crack-line leading to the left end of the grass ledge.

2 45m 4c Climb the grassy crack system directly above then the shallow V-groove just right of the long corner to finish up the upper part of the corner. Belay on a ledge on the left at the top of the corner, directly beneath a prominent rock scar in the main overlap.

3 45m 4b Breach the overlap using the cracked groove on the right, then follow easier cracks above to ledges.

4 45m 4b Step round large flake and follow cracked slabs to a shallow gully containing a large pointed block. Continue leftwards

up easy slabs to turfy ground.

5 20m Climb leftwards over turfy ledges, then directly over short walls to smaller grassy ledges (The Terrace) beneath the crux pitch of *Cyclops*.

6 40m 4c Climb up and diagonally left beneath small bulges (left of the *Cyclops* groove) to gain a pink rib and slab close to *Pink Elephant*.

7 25m 4b Traverse back right to gain the cleaned cracks, (common to both *Cyclops* and *The Blue Max*) and follow them left to a grass ledge.

8 40m 4c Gain the slab above and traverse left to the short left-facing pegmatite corner. Climb the corner to exit right to the rib before it becomes vegetated. Continue up the fine rib to gain and climb grassy grooves.

9 45m 4b Finish up the grooves, moving onto the right wall near the top to gain broken ground.

5 **The Fruit Bat** ★★   **20m E5 6a**
FA Jules Lines & Kevin Smith August 1995

Very good bold slab climbing. Start 10m right of *Black Mamba*, beneath a small right-curving groove becoming an overlap. Climb to the overlap, pull over, then climb the bold blank slab above to better holds and the long grass ledge.

6 **Howff Dweller** ★★   **45m E3 6a**
FA Jules Lines & Kevin Smith August 1995

"*A really great pitch.*" The large right-facing corner curving into the overlap above the slabby wall. Climb the corner and continue under the overlap, finishing up the shallow right-facing corner and a crack in the headwall.

7 **Buddha** ★★   **40m E7 6c**
FA Julian Lines & Eddie McTavish (headpointed) 31 August 1995

About 75m up and right of *Black Mamba* is a small isolated blank-looking slab/wall topped by a right curving overlap. Start in centre of slab. Climb scoops then move right to arrange protection in cracks. Gain and climb twin converging cracks until they disappear then make desperate moves to better holds. Abseil off. Superb climbing and a soft touch — graded for an on-sight ascent. Possibly E6 6b for the tall.

CENTRAL GULLY WALL

The massive convex face forming the right side of Central Gully, sweeping round to the longer slabbier frontal face where the lines tend to follow longer easier crack lines. The wall is steepest where it overlooks the gully, split by three huge corner systems (taken by *Vertigo Wall*, *Goliath* and *The Giant* respectively). Lower down near the mouth of the gully the wall forms a maze of walls and overlapping slabs. The majority of routes finish in an area of vegetated blocks beneath the plateau. After prolonged wet spells many weeps emanate from this area for several days. If a route has wet streaks on it try something else. Many of the harder routes first climbed in the early eighties utilised pegs either for runners or belays – most of these are currently (2023) in a very poor state or missing, increasing the seriousness considerably.

Descent: Traverse left (south-east) & down **Central Gully**. There are abseil points on some of the harder routes – described within description & on topos.

1 The Shetlander ★★　　　　　60m E6 6c

FA Wilson Moir, Neil Morrison & Niall Ritchie 9 August 1995

"Extremely good climbing and safe. Can climb intro and crux in a single pitch. Probably about (cruxy) F7b."
The crack-line left of *The Wicker Man* with a very hard well protected crux.　2021

1　**15m 5b** Climb the flaky crack-line to gain a tufty break. Follow this leftward to belay at the base of a right-slanting corner.

2　**30m 6c** Go up the right-slanting corner then climb a crack leading into another corner, which leads to a resting place beneath the steep section. Continue up the crack-line to reach a semi-rest at flakes. Climb the desperate bulging groove above to a P and nut belay in a niche.

3　**15m 6a** Go up from the belay until it is possible to step right. Move up right then back left to finish up a crack.

2 The Wicker Man ★★★　　　　150m E3 6a

FA Pete Whillance & Rab Anderson 4 September 1982

"Crux pitch is brilliant. The rest are not so good, but still nae bad. Except the grassy terraces pitch..."

The conspicuous crack-line in the centre of the wall left of *Vertigo Wall*. The crux pitch is superb.

1　**30m 5a** Climb up for 6m to a grassy bay below a leftward trending groove. Follow the left wall and rib of the groove to a ledge and belay at a large perched block.

2　**35m 6a** Step right and climb twin cracks in a giant groove to a ledge below a small square-cut overhang. Climb the thin right crack into a triangular niche. Then go up the wall above to a ledge. Continue more easily up the obvious corner crack to a belay on the right. Either abseil (50m) from spike, or:

3　**25m 5b** Continue up the corner and slabby groove above exiting right at the top. Up grass ledges for 6m to belay at a large embedded flake.

4 **30m** *"The sphagnum moss pitch."* Move left around the corner and follow easy ledges leftward and then back diagonally right to reach a belay below the bulging headwall.

5 **30m 5b** Move up right onto a glacis and climb the overhang at its narrowest point to reach a right slanting groove. Climb the crack on its right to the plateau.

3 The Israelite ★★   125m E4 6a
FA Pete Whillance & John Moore 5 June 1982

The smooth orange-coloured water-worn groove just left of *Goliath*. An excellent main pitch, though unfortunately slow to dry. Adequately protected with lots of small wires. Start as for *Goliath*.

1 **25m 4b** Traverse easily right across slabs (as for *Goliath*) to a small stance and PB below the line of the groove.

2 **45m 6a** Climb the slab above then slightly leftward until a traverse right can be made to a crack leading up to an overlap. Climb the overlap on the right to enter the main groove and follow it direct to where the angle eases. Traverse right to the belay of *Goliath*.

3 **20m 5a** Up the obvious thin crack in the centre of the slab to a large ledge and belay below the huge corner.

4 **35m 5b** Climb the corner direct to the top. Scramble to the plateau.

4 Goliath ★★★   150m HVS 5a
FA Brian Findlay & Mike Rennie (4 PA); 7 June 1969;
FFA Ian Nicolson & Dave Knowles June 1970

"Tricky crux pitch until you work it out, 4 good pitches." The large slabby right-slanting corner in the upper cliff gained by a traverse in from the left. Start 30m up and left from a huge perched block.

1 **40m 4c** Move up diagonally right on small ledges then traverse right across slabs to belay below and right of two small left-facing corners in the steep wall. *"Not a pitch for the novice second."*

2 **40m 5a** Climb either of the steep corners and move up to a shelf (crux). Continue

up the fault line to belay on the right.

3 **40m 4b** Go across the slab on the left then climb to a ledge leading left to a huge slab. Climb the slab for 10m and continue to belay on a long ledge midway up the great corner.

4 **30m** Move up to a ledge above then traverse right and up to finish up wide easy cracks close to the right edge of the slab. Scramble to the plateau.

5 Goliath Eliminate ★★   160m E2 5b
FA Norrie Muir & Steve Docherty (4 PA) 22 May 1971;
FFA Brian Davison & Andy Nisbet August 1983;
Shelf Variation Jim McArtney & Brian Lawrie 1967

Good sustained climbing, though a bit of a misnomer – it shares about 3m of climbing with *Goliath*. Start 10m left of the huge poised block.

1 **20m 5b** Traverse right on a grass ledge. Hand traverse right and mantelshelf onto the block. Chimney up the back to the top of the block, move right along the shelf and up to belay on a ledge.

2 **10m 5a** Pull over the bulge above then step right and up to belay at the end of the first pitch of the normal route.

3 **30m 5a** Climb the steep corner and move up to a shelf. Move right up the slabby shelf to a ledge at the top.

4 **30m 5a** Continue up rightward to belay beneath the rock scar of *The Giant*.

5+6 **70m** Traverse left and finish as for *Goliath* (40m, 30m)

6 The Naked Ape ★★★   130m E6 6b
FA Pete Whillance, Pete Botterill, Murray Hamilton & Rab Anderson 1 August 1982

A bold and impressive route based around the big arête right of *The Giant*. Start below a groove leading directly up to the main corner system of *The Giant*.

1 **35m 5b** Climb the groove and where it forks keep right up a flake crack to reach a ledge and belay at the top of the initial ramp of *The Giant*.

2 **30m 6b** Climb the smooth groove above the belay mainly via its left rib to reach a PR on the right at 10m. Move up right onto the steep slab, traverse

Unknown climbers on the second pitch of Goliath.

right to the arête and step up to a PR. Continue traversing right along an obvious foot ledge and step up to a good ledge and belay. Pegs replaced 2023.

3 **35m 6a** Follow the leftward slanting slabby corner to a niche below a roof and place a runner in the lip. Climb delicately down leftward to a good foot hold near the arête. Step up left and climb a break in the overhang to gain a sloping ledge. Up a short steep wall (PR) to a ledge in a niche. Climb the overhanging crack above to a large ledge and spike. From here a 50m abseil on rope stretch gains easy ground on the big ramp.

4 **30m 5a** Step up right and climb cracks to enter a groove system. Follow this to a grassy terrace at its top. Scramble up right to the plateau.

7 **The Origin of Species** ★★★★　　　　**70m E6 6b**

FA Paul Thorburn & Gary Latter 22 September 1997

A long sustained and serious main pitch following the stunning blunt arête cutting through *The Naked Ape*. Take all the micro wires you can muster. Start 2m down from the start of *The Naked Ape*. 2021

1 **15m 5b** Climb the shallow groove parallel to the larger groove of *The Naked Ape* to a flake and nut belay on small ledge.

2 **40m 6b** Continue up the groove above, past a PR at a prominent rock scar to a wide crack with a

chokestone (F #4). Move up the flakes above then traverse right along horizontal break to the base of the arête (skyhooks on good flake on right). Move up the arête with committing moves to stand on the sloping shelf on the left. Follow the thin crack above with difficulty to a good hold at its top. Continue up leftward to the horizontal break of *The Naked Ape* and follow this boldly right then up to the PR on the arête. Climb the steep thin groove above, exiting onto the slab with interest. Belay in the niche beneath the right end of the roof.

3 **15m 6a** As for *The Naked Ape*. Move across the slab to a good foot hold. Pull over the roof slightly rightward then up with a hard move past PR to a good ledge above. Finish up the steep jam crack to spike and nut belay. 50m abseil reaches the ground on the stretch.

8 **The Ascent of Man** ★★★　　　　**130m E5 6b**

FA Murray Hamilton & Rab Anderson 24 July 1982

Climbs the prominent groove and crack-line up left from *Hybrid Vigour*, with a hard 3m crux section. Start directly beneath the prominent steep groove.

1 **30m 6b** Climb a crack for 3m and traverse right along a ledge to gain the lower groove. Follow this up and traverse left across a slab to regain the crack. Climb this to the break in the roof (ancient PR) and pull into the groove above, which is followed with much interest to a hanging PB.

2 **30m 6a** Step up and move across right to gain a subsidiary groove. Climb this and move across left to gain a crack which leads to a leaning wall. Step up left to climb the short crack above and hand traverse right into a groove. Up this to belay at its top.

3 **35m 5b** Climb the short corner above then up left along a slab and down into a recess occupied by a large detached block.

4 **35m 5a** Step back up right and climb cracks to enter a groove system which is followed to grassy terrace. Scramble off right.

9 Hybrid Vigour ★★★   **135m E6 6c**

*FA Rick Campbell & Paul Thorburn July 1994; Pitches 3 & 4 Rick
Campbell & Paul Thorburn (alts.) & Neil Craig 1 July 1995;
Pitch 1 gained from R & pitch 2 – Dougie Dinwoodie
& GeorgeThomson 5 July 1987; The Toyboy pitch: James
McHaffie & Ferdia Earle June 2018*

*"Ground up first ascent. 1st pitch crux is typical of this crag
in that it looks impossible until you work out what you
need to do, doing it then is often straightforward. The
'Bluebell Groove' and the pitch above are great. Pitch 3 is
a nasty piece of work and the fourth pitch was great until
it wasn't." – Rick Campbell*

An excellent varied route. The big overhanging alcove,
forming the start, provides a fantastic pitch when dry.

 2021

1 **25m 6b** Climb the groove past a peg runner (new in
2021) to a rest where the groove starts to impend.
Climb this via a hard undercut move (good cam) to
better holds in the vertical groove above. Continue
up the easing groove to a wide slab beneath another
steep groove (the Bluebell Groove). Traverse the
slab to the left edge, thread belay and possible
abseil point, common with *Perilous Journey*.

2 **25m 6c** Traverse back right to the groove (gear in the
flake in the wall above the belay), then climb the
wall just left of the groove and the groove itself,
old peg, to gain the slab on the right (possible
belay on a peg in the overlap and a cam in crack
in slab further right). Move left to gain a higher
slab under the big overlap. Gain good holds in
the lip to the right and make wild moves over
the bulge. Step left and go up a delicate groove
which rapidly eases, to belay as for *Voyage of
the Beagle*. Good wires on slab up and left.

3 **25m 6b** The original route followed the overlap
above leftwards to perched blocks. This is poorly
protected, especially for the second. Clip the belay
wires, then follow *Voyage of the Beagle* down
from the left side of the belay and across the
slab to near the arête. Follow twin cracks, (as for
Perilous Journey), up the wall above to a small
ledge (being careful of perched blocks on the
right). Belay on a peg, with good nut and spike up

on the arête on the left (possible abseil point).

4 **30m 6a** *The Toyboy* pitch. Move right past stacked
blocks (old peg runner in the break above), then
hand traverse a break rightwards into a niche in
the overhanging wall. Pull round rightwards onto
a sloping ledge at the bottom left corner of the
Cougar rockfall scar. Climb up on good positive
edges to a slot which accommodates a couple of
good microwires. Move up, then make a strange
move leftwards into the corner. Follow this to a
roof (good nut) then traverse back right to pull
steeply onto the giant blocky flakes. Follow these
leftwards to belay below a short steep corner.

5 **30m 5c** Climb the wall to slab under the final bulging
band. Move right and mantelshelf onto a block, then
up crack above to an overlap. Traverse left under
the overlap, bridging a slight gap, continuing to
gain a good rounded spike. Step up and traverse
left under a bulge until possible to move up then
left to a good ledge under a roof. From wedged
triangular block on left, abseil 20m to the abseil
spike on *The Naked Ape*, or scramble to the plateau.

10 Perilous Journey ★★★   **80m E6 6b**

FA Dougie Dinwoodie & Graeme Livingston 18 August 1983

Another fine sustained route up the steep frontal face.
Start as for *Cannibal*. Unfortunately the original start was
obliterated by the huge rockfall from *The Giant*. None of
the pegs are of any use.

1 **30m 6a** Climb the initial crack of *Cannibal* then
take the curving crack leftward into the groove
of *Hybrid Vigour*, stepping left to a thread belay.
This is a brilliant ★★★ E4 6a pitch – take a knife
to replace the tat. 30m abseil from here.

2 **20m 6b** Go up the slabby wall above to a slabby
shelf. Move up left along a creaking flake to arrange
protection in a jug. Return to the shelf and make a
hard mantelshelf move (crux) onto a smaller shelf
above. Go left and step up delicately to reach
good flake holds leading to a poor PB on the slab.

3 **30m 6b** Step down left and traverse the lip of the
slab to an *"evil sloping perch"* at the end. Swing left

round the overhanging nose and make hard moves across left and up past a PR. Climb up and step right to follow a vague groove system then twin cracks with difficulty to a PB. Good nut up left on the arête. The *Cougar* rockfall has resulted in the crack on pitch 4 now being an arête! Instead, finish up one of the other routes (*The Ascent of Man* pitch 2, *Hybrid Vigour* pitch 4, or *Voyage of the Beagle* pitch 4), or abseil off from the spike at the top of pitch 3.

11 Cannibal ★★★★   **140m E6 6b**

FA Murray Hamilton & Rab Anderson 9 June 1984

Excellent sustained climbing, taking a direct line of increasingly hard pitches through the big slab. Start up the arête right of the overhung alcove.   2021

1 **30m 5c** Climb the crack just left of the arête until moves can be made round the arête into another crack. Climb this until it fades then step left to gain a crack on the crest. Climb this steeply to a stance on the slab above.

2 **10m 6a** Move up left and gain holds which lead to a PR below the roof and good micro-cams in the roof. Pull out right until a crack above the roof leads to good wires and cams in the overlap on which to belay.

3 **30m 6b** Move up to a left-slanting corner and climb this to a roof. Traverse left and pull into the recess (PR) then pull over its right wall onto the big slab. Belay at the right end of the slab.

4 **40m 6b** Above and slightly right is a stepped corner. Climb the first step then swing right to the arête and move up until the corner can be regained. Continue up the corner and pull out right below a steep nose. Climb the wall right of the nose to gain perched block forming the top of the nose. Pull onto the top of this block and follow the hanging slab up leftwards to a stance.

5 **30m 5c** Climb the crack above, then traverse left. Continue under the overlap, bridging a slight gap, continuing to gain a good rounded spike. Step up and traverse left under a bulge until possible to move up then left to a good ledge under a roof. Scramble to the plateau.

12 Voyage of the Beagle ★★★★   **160m E5 6a**

FA Murray Hamilton & Rab Anderson 12 August 1983

An excellent sustained route, though generally on less steep ground than its companions. Start beneath a hanging left-facing groove.

1 **40m 6a** Climb a flake crack on the right up into the groove. Follow the groove initially with difficulty then more easily to a small square-cut ledge. Move left across the wall then regain the corner. Step down left to gain a thin slanting crack leading back into the fault which leads to a belay at the start of the slab.

2 **15m** Easily across slab to a small ledge at the far side. There are good wires on the upper slab a few metres up and left.

3 **40m 6a** Clip the wires mentioned above, then return to the belay. Step down to gain the lower slab. Cross the slab with some thin moves to gain the arête. Step left round the arête and straight up to enter a groove (*The Ascent of Man*) and up this to its top. Move left to gain a groove leading to the belay at the end of second pitch of *The Naked Ape*.

4+5 65m Finish up *The Naked Ape* (25m, 6a; 40m, 5c).

13 Vampire ★★   **210m E2 5b**

FA Dougie Dinwoodie & Greg Strange 8 October 1972; FFA Brian Lawrie & Andy Nisbet July 1977; Variation: Michael Barnard & Alan Hill 22 June 2014

A good sustained route with fine positions. Start 5m right of *Voyage of the Beagle*, on top of a mound with embedded blocks.

1 **30m 4c** Make a short traverse right into the often wet initial corner and follow it to a notch. Continue up left aiming for a ledge beneath an overhanging wall.

2 **30m 5b** The bulge on the right is split by an overhanging crack with a jammed flake at its base. Pull right round the bulge and ascend corner above to a good stance just above grass ledge.

3 **40m 5b** Unfortunately the big corner above is dirty (E3 5c); step back down and traverse right along ledge, then move up to a slab on the right. Climb diagonally right up slabs and a bulge to

Karen Latter on the straightforward first pitch of The Mousetrap.

A good direct line up the crack system near the left side of the front face. Reasonably quick drying. Start up left from the lowest point of the toe of the wall beneath a deep easy groove.

1 **30m 4a** Climb the groove then traverse left and climb the wide cracks to belay beneath a steepening below and left of the main crack.

2 **30m 4c** Follow the wide crack (crux – large cams useful) to small square ledge.

3 **45m 4c** Continue up recess, then twin cracks to belay beneath overlap.

4 **45m 4c** Continue up cracks above overlap, moving into another crack system just left, and up this to belay.

5 **60m** Scramble up leftwards to finish.

15 Dubh Loch Monster ★★　　　　　**250m E1 5c**

FA Ian Nicolson & Dave Knowles (1 PA) 18 June 1970;
FFA Jeff Lamb & Pete Whillance 1975

"Sustained route with interest on every pitch. Crux 5c very well protected and not too bad, but plenty of the 5a sections feel tough." "Really nice comfortable belays too."
The thin crack right of *The Mousetrap*. Very sustained at 5a with one move of 5c.

1 **30m 4b** Follow the cracked slabs to beneath a chimney break.

2 **40m 5c** Climb the chimney (crux) then continue by an awkward wall and the crack-line to beneath an overhanging notch.

3 **10m 5a** Move left and up the arête then move back right to a large ledge.

4 **20m 5b** Follow cracks over the first bulge. At the second bulge move 2m right and go up with difficulty, then cross slabs rightwards to a short left-facing corner. The belay is below and right of the biggest roof block.

5 **45m 5a** Move left to a break and up this to gain the slab above. Move left 2m and pull over a short wall into a corner. Follow the corner for 10m then the rib on the right over two bulges, turning the second on the left and move right to a ledge.

6 **50m 4b** Climb slabs to a crack at the right end of a left-slanting corner/overlap. Go

gain a blaeberry ledge. Make a toe traverse left across a tiny dwindling ledge, then up a corner to bulges. Break out left by a short wall and slab to a grass stance (poorly protected).

3a **Variation ★★ 35m** E2 5a Equally good, though bold. After the initial step around the arête onto the slab, go back up leftwards to the arête. Follow this for a few metres before moving right up a line of holds to reach thin cracks (micro-cams useful). Climb up leftwards to join the normal route at the end of its toe traverse, continuing up this to the stance.

4 **25m 4b** Move left up a slab and up a short recessed corner. Climb the slot above and up a slab to exit right.

5+6 **85m 5a** Continue trending right to finish up slabby ground.

14 The Mousetrap ★★★　　　　　**210m VS 4c**

FA Jimmy & Ronnie Marshall & Ronnie Anderson November 1959

"Fantastic and very sustained climbing. Four pitches and a 60m scramble, not a bad pitch among them!"

Callum Johnson on the second pitch of King Rat. Photo Dave Riley

up the crack then a steepening corner.
Continue in the crack to easier ground.

7 **55m** Finish up short walls and vegetation to the top.

16 Black Diamond ★★   200m E4 6a

FA Guy Robertson & Pete Macpherson 22 June 2010

Fine sustained climbing up the faint crack-line between *Dubh Loch Monster* and *King Rat*, with two good and surprisingly independent pitches. A hard crux and some bold sections. Start as for *Dubh Loch Monster*.

1 **30m 4c** Climb cracks as for *Dubh Loch Monster*
but veer right and climb a short crack to a small
ledge under a leaning wall with a short corner
defining the left end of the *King Rat* roof.

2 **30m 6a** Step up and use an undercut to gain the
leaning wall then make a desperate move right
to get established in the corner. Move up left
into then up a short groove and continue up
slightly right to a hidden jug at a steepening. Pull

directly over this and a second bulge directly
above to a good ledge and in-situ peg.

3 **30m 5c** Step left across the gap, then move up
to follow the obvious blind flake and cracks
which lead to a small niche below and left of
a cracked red wall. Climb the superb wall to a
ledge, and continue up the crack until a mossy
bulge forces a traverse right to a junction with
King Rat and belay below a steep crack.

4 **40m 5b** Climb the steep crack-line direct over
the right side of the bulging nose to a junction
with *King Rat*, and follow this to a ledge.

5+6 **70m 4c** Finish up *King Rat*.

17 King Rat ★★★   220m VS 4c with 2 PA (E1 5c free)

FA Allen Fyffe & John Bower (5 PA) 9 June 1968;
FFA Phil Thomas & Mick Fowler 23 June 1977

"Surprisingly sustained, but not too hard (other than the roof). Route finding's fine above the roof, just head up!"
The prominent crack system right of *The Mousetrap* only spoilt by inordinately hard moves through the prominent roof. Start directly below the prominent large roof at 50m, close to the left wall of the large recess at the lowest point of the wall.

1 **40m 4b** Climb steep rocks at the back of the
recess for 15m then traverse the left wall on
flakes to a ledge on the open face. Climb direct
up the crack-line to a large grass ledge.

2 **20m 5c** Move right from the ledge and up cracked
slabs to a shallow cave under the roof. Move awkwardly up left into a corner and up this to a ledge.

3 **50m 4b** Follow the cracked ribs above
until beneath a short vertical wall. Move left
up slabs to below a short leaning corner.

4 **10m 4b** Follow the corner to ledges.

5 **30m 4c** Climb a short wall to gain a slab
beneath a roof. Cross the bulge 2m left of
the roof then traverse right along a narrow
slab above the roof and up to ledges.

6+7 **70m 4b** Move into the upper crack system
and follow it to grass ledges. Scramble
away left and up to the plateau.

A good area of rock high on the right side of the cliff with plenty of good varied extremes. The wall holds the sun the longest (until around mid-day). The smooth-looking barrier wall above the slanting shelf and grassy terrace invariably provides the crux pitches of almost all the routes. This can be gained by an unpleasant VDiff. scramble (rope advised!) up the vegetated gully at the left end of the lower rocks. A much better (safer!) alternative is the first pitch of either *Sans Fer* or *Falseface*, both leading to the far right end of the terrace.

Descent: Either walk left (south-east) and down Central Gully or head out right (north-west) and down easy slopes beyond the last outcrops.

By abseil: there is a thread at the top of both *Masque* and *Anaemia* – 60m to the base of the barrier wall; and a block at the base of the smoothest section in the centre – 35m.

1 Sans Fer/Ludwig/The Snake ★★   160m E2 5c

FA Ludwig: Jules Lines & Danny Laing 22 April 2011

An excellent combination linking the easiest pitches up the left side of the barrier wall. Omitting the first pitch would give an excellent E1 5b.

1 **40m 5c** As for pitch 1 of *Sans Fer*.

2 **55m** Walk left along the slanting shelf and continue up to the next groove left of *Masque*.

3 **20m 5b** Climb the curving bow shaped groove to a ledge on the right (common with *Masque*).

4 **45m 5b** Traverse right 3m to reach a fine jam crack; climb this to join and finish up *Masque*.

2 Masque ★★   60m E2 5c

FA Murray Hamilton & Pete Whillance 28 July 1983

"*Two great pitches. Quite a physical first, followed by a more balancy second.*" The left and deeper of the two corner lines bounding the left end of the smooth barrier wall.

1 **20m 5c** Climb the leaning corner to a roof, pull out left and up crack to a ledge.

2 **40m 5b** Take the twin right-slanting cracks to a roof. Step right and follow a fault to a grassy groove to finish.

3 The Snake ★★   65m E4 6a

FA Dougie Dinwoodie & Greg Strange 5 August 1984

"*Great route - two really good pitches.*" "*Perhaps a good first E4 on the Dubh Loch – relatively tame.*" Start beneath the shallower right corner.

1 **30m 6a** Start up the corner, then move out slightly right and up to the top of a big flake. Step up past the flake and move right to flakes on the arête. Climb the flakes to an 'evil sloping shelf', step right to good protection and climb a bulging wall to another shelf. Step right again and go up a short corner past a spike to exit left at a roof to a ledge. Peg 5m up on left.

2 **35m 5b** Go up past the belay peg, traverse left round the edge then go across horizontally into a jam crack in the slabs. Climb this for 5m to join and finish up *Masque*.

4 Fer de Lance ★★★   65m E6 6b

FA Dougie Dinwoodie & Jeff Hall 10 August 1987

The wall crack at the left end of the wall gives the longest and steepest pitch on the smooth barrier wall. Initially bold, then sustained and well protected.

1 **30m 6b** Climb the sustained crack past a PR low down and up past two shake-outs to gain an awkward shelf. Climb the bulging wall above moving right and up a corner past a spike to exit left onto a ledge. PR 6m higher.

2 **35m 5c** Climb up past PR and go up diagonally right into the corner of *Slartibartfast*. Climb this past the bulge above then step left along a shelf. Climb the bulging crack above, step left and pull up onto a big flake. Move up left to easier ground, finishing up a grassy groove.

⑤ The Improbability Drive ★★★　　　**75m E6 6c**

FA Graeme Livingston & Dougie Dinwoodie 15 August 1984

Bold sustained climbing on the initial wall, leading to a difficult well protected final move. Start 10m left of *Slartibartfast*, just right of an obvious crack.

1 **20m 6c** Climb right up little foot shelves using diagonal cracks then up the wall to gain a horizontal break. Go straight up the wall to a semi resting place at the next break. Climb the twin cracks directly above using mainly the left one, with a hard move to gain a good jug at the top of the wall. PB on slab above.

2 **20m 6a** Climb directly up the bulge just left of the belay then slightly right to climb the third roof of the second pitch of *Slartibartfast* and belay on slab above.

3 **35m 5c** Continue in the same line via steep cracks to the top.

⑥ Mostly Harmless ★★★　　　**70m E5 6b**

FA Pitch 1 Jules Lines 1 August 2016 rope solo; pitches 2 & 3 Jules Lines & Daniel Laing 12 June 2015

This superb climb is equally as good as *Slartibartfast*, but will need a few days of good weather to dry it out on the upper pitches. Start to the left of *Slartibartfast* at a rectangular hole directly beneath the central crack on the wall.

1 **20m 6b** Boulder up the wall for 5m to gain the first good layback hold at the start of the crack proper; place a crucial IMP4 with difficulty in the base of the seam just to the right (crux). Continue laybacking up the crack to join *Slartibartfast* and belay on the shelf.

2 **20m 6a** Pull onto the rib on the right and follow the overhanging crack to a rest at the base of the acutely leaning corner. Climb the corner (large cam) by wide bridging, which is also required to negotiate the 'permanent' weep here. Near the top of the corner, place a high nut, step left onto the rib and climb delicately up to a shelf and flake belay. A short and tough, well protected pitch that is one of the best on the mountain.

3 **30m 5c** Climb directly behind the flake and move awkwardly rightwards onto the base of the large slabby ramp. Climb straight up a series of tiny left-facing grooves and a layback to land on a large ledge. Step right and finish up the rib.

⑦ Slartibartfast ★★★　　　**75m E5 6b**

FA Murray Hamilton, Pete Whillance & Rab Anderson 30 May 1982

Takes the most prominent crack and groove line in the centre of the wall. Start below a crack in the middle of the wall.

1 **20m 6b** A ledge on the wall just right of the crack gives access to a small groove. Climb the groove until it peters out (RP #3 out in slot on right). Move up and left (bold) to follow the crack and shallow corner, moving right at the top to belay on sloping ledges.

2 **20m 6a** Climb the obvious corner containing three small overhangs and belay on a slab above.

3 **35m 5c** Continue in the same line via steep cracks to the top.

⑧ Magrathea ★★★　　　**70m E9 7a**

FA Jules Lines & Steve Perry 5 August 2016

A stunningly minimalistic line up the blank wall to the right of *Slartibartfast*. Start at left side of the abseil block at the base of the wall.

1 **20m 7a** Climb in the line of the hairline crack (IMPs) to a poor shakeout where a cluster of

uninspiring, but 'potentially' adequate micro-wires can be placed. Continue via a desperate sequence of moves to reach a jug atop the wall. Move up onto a shelf on the right; small wires in a slot above provide the belay.

2 20m 6a Climb the beautiful leaning groove above to a large flake block.

3 30m 6a Take the short right-facing corner above and follow the line to a niche. Pull left onto a rib and continue up this to the top.

9 Anaemia ★★★   70m E5 6b

FA An Spearag: Gary Latter & Paul Thorburn 23 September 1997; Jules Lines (rope solo) 31 May 2017

This excellent varied climb takes the most natural line and is essentially a combination of *An Spearag* and *Sans Fer* with a new third pitch. It is the twin line to the *Sans Fer / Iron in the Soul* combination and is a small step up in standard. It should be nice and clean. Start at the right side of the abseil block.

1 15m 6b Climb the thin crack by mainly using good holds on the left wall. At 6m the holds run out. Stretch in a good nut, then make some tenuous moves right on sidepulls (crux with overhead protection) to gain a partial rest at the base of the flake. Climb the flake past an undercut flange with a bold finish onto a smooth shelf. Belay immediately on the right, small nut and cams. This is a less sustained method than the original, *An Spearag*, which traversed the horizontal break a little higher to gain the flake.

2 25m 6a Climb the dwindling groove above and continue through the roofed alcove to gain a crack that leads up to a block and crevasse belay.

3 30m 6a Above is a bulging crack. Go right along a shelf on undercuts and using a high pinch, pull back left into the top of the crack. Follow the line into an open V-groove, step left and continue up the easy rib to the top.

10 Sans Fer/Iron in the Soul ★★★   120m E4 6b

FA Sans Fer Murray Hamilton & Kenny Spence June 1979; FA Dougie Dinwoodie & Brian Lawrie 22 August 1984

The first route to breach the smooth barrier wall above the grassy terrace and the start of a new era. This combination links the best pitches by the most direct line. Start in the centre of the lower face 10m left of the prominent corner of *Falseface*.

1 40m 5c Climb up to a niche and slab heading for a prominent finger crack. Climb this painfully to belay on a sloping ledge at the right end of the grassy terrace. Walk left 15m to the base of the crack.

2 20m 6b Climb the crack with a difficult rounded exit to a superb incut jug at the back of the ledge. Belay just above.

3 30m 6a Climb the bulge above and go up cracks and a corner crack slightly left under the next bulge. Layback right and go over the bulge to follow a flake crack curving up and left in a slab. Go on up a short corner to belay on the ramp of *Sans Fer*.

4 30m 5a Move left up the ramp and climb a short cleaned crack then step right and follow further cracks and blocks leading to easy ground.

11 Falseface ★★   110m E2 5c

FA Graham Hunter & Doug Lang (13 PA) 7 June 1969; FFA Bob Smith & Dougie Dinwoodie July 1977

The right side of the lower wall beneath the terrace has a prominent right-facing corner. Start below and left of this.

1 20m Climb up and slightly rightward to gain a grass shelf leading into the corner.

2 20m 5c Layback strenuously up the corner to a belay at the right end of the terrace.

3 35m 5b Up the short wall above to move right into a hidden chokestone chimney and up this to a ledge on the left beneath an overlap. Gain the sloping ledge above, cross the overlap then traverse right to better holds. Move up and back left then up a series of grooves to a large ledge.

4 35m 5a Move back left and into a steep corner which is followed to near the top. Go along the right wall and up a detached flake to ledges and easy ground.

THE PASS OF BALLATER

CREAG AN T-SEABHAIG
(HAWK CRAG)

A compact selection of sunny crags situated on the pine and larch covered hillside of The Pass. Excellent granite yields a range of routes from VS upwards. Being a bit of a sun-trap, it is often possible to climb here in the winter months, apart from the depths of winter, when the sun barely rises above the neighbouring Craigendarroch hill. A popular weekend venue for Aberdonians, it is about an hour's drive from the granite city. Many more routes, crags and boulders in the area are detailed within *Ballater Rock Climbs*, available in Cairngorm Mountain Sports outdoor shop in town, or direct from juleslines@hotmail.com.

Access: The Pass lies behind the hill of Craigendarroch, on the north side of the town of Ballater itself. Travelling from the east, turn off right from the A93, 1.4 miles/2.2km before Ballater and follow the B972 for 1.1 miles/1.7km to a parking space on the north side of the road (NO 3679 9701; 57.059784, -3.0438256). From the west turn left off the main A93 road into the pass just beyond (east of) a large lay-by on the south side of the road. Follow the B972 for 0.7 miles/1.1km to a parking area directly beneath the **Central Sector** of the crag.

Approach: For the **Central Sector**, follow a steep path zig-zagging directly up from the car park. For the **Western Sector**, walk west along the track past the gate. The **Lower Tier** is visible after around 100m. For the **Middle** and **Upper Tiers**, continue for a further 200m, then follow a path just before the drystane wall, cutting out right to the respective tiers. **To minimise erosion, please try and stick to the marked paths where possible.**

WESTERN SECTOR - LOWER TIER

Altitude: 260m      5min

1 Autumn Waltz ★★　　　　　　　　**30m HS 4a**

FA Jules Lines (solo) 2018

A fine multipitch adventure on pleasant rock.

1 **20m 4a** Start at a small flake in the barrier wall and climb up and right to the rowan. Follow the handrail to its end and step down to a spacious belay. A double set of cams useful.

2 **10m 3c** Go up the clean groove and follow the smaller right continuation to the top.

1a Autumn Rib ★★   **30m HS 4a**

FA Jules Lines (solo) 2018

A single pitch variant taking the clean rib above the end of the traverse.

2 Samba Groove ★★   **15m VS 4b**

FA Jules Lines (solo) 2018

Start beneath the corner just right of 1. Mantel onto the slab and continue up into the corner to a shelf. Place a crucial small cam, rock up onto the slab and finish more easily.

3 Nerve Ending ★★   **15m VS 4c**

FA Jules Lines (solo) 2018

Varied climbing on superb rock, but a little disjointed. Climb the left wall of the flake on big holds to gain a runnel in the slab above and hence a break. The slab above is taken just left of the vein, which is followed through the bulge to the top.

WESTERN SECTOR – MIDDLE TIER

NO 3662 9694 **Altitude:** 280m   7min

Descent: Either by abseil from trees at the top, or down a worn path down either end.

1 Swivel Head ★   **8m Difficult**

FA unknown 1960s

The obvious blocky fault line. There is a Moderate descent just to the right.

2 Razor's Crack ★   **8m VS 4c**

FA Greg Strange & Dave Stuart 6 June 1971

The S-shaped crack at the left end.

3 Jumbled Blocks Crack ★   **10m Very Difficult**

FA 1960s

Also a useful descent. The prominent corner and crack near the left end, just right of the S-shaped crack. Polished.

4 Brut ★   **10m VS 5a**

FA Dougie Dinwoodie & Adhair McIvor June 1982

The corner and groove.

5. Stinker ★ — 10m E1 5b

FA Dougie Dinwoodie & Brian Lawrie August 1981

Further right is a prominent mid-height roof. This climbs the well protected crack.

6. Lime Chimney — 10m Very Difficult

FA Dave Stuart & Greg Strange 6 June 1971

The prominent short chimney, the finish is usually dirty.

7. Lucky Strike ★★ — 15m VS 5a

FA Mike Freeman, Greg Strange, Dave Stuart
& Raymond Simpson 6 June 1971

The stepped left-facing corners. Start immediately left of a rock scar. A bouldery start leads to better holds. Continue up the steep black-streaked corner to large ledge and tree. Finish up the upper corner behind the tree.

8. Flake Traverse ★ — 25m E1 5a

FA unknown

Start 3m right of 7. Follow an obvious traverse across the wall to join the final pitch of 9. Continue traversing right into the final quartzy cracks of 10 and finish up these.

9. Pretzel Logic ★★ — 20m E3 5c

FA unknown; FFA Mike McDonald & Brian Sprunt 1980;
upper corner Tony Barley & Jerry Peel 1977

Start at a short corner 4m right of 7, just right of the arête. Climb the corner until a rising traverse out right can be made to gain the crack system. Follow this to the base of the final imposing corner and finish up this in a fine position.

9a. Direct Start ★★ — 8m E3 6a

FA Graeme Livingston, Colin McLean & Alasdair 'Plod' Ross 1983

The short left-facing groove gives a problematic start to a good hold at its top. Direct up the painful quartz-filled finger crack to finish up the final corner.

10. Rattlesnake ★★ — 20m E3 6b

FA Dougie Dinwoodie & Brian Lawrie July 1981

A fine line taking the prominent quartzy cracks up the right edge of the wall. Up the initial corner with difficulty, either direct or from the right. Finish up the striking cracks

10a. Anti-Venom ★★ — 20m E1 5b

The right start to 10 avoids the 'venom' of the initial corner; good value for those not up to the original start. Start up the corner of 11 and make an ape-like traverse left along the lip of the slab to make a scrungy mantel. Pull left into the parent route.

11 Supercreeps ★   20m E1 5b

FA Alasdair 'Plod' Ross & P.Garden September 1982

The vague crack line up the left side of the slab. Start up the short left-facing groove in the alcove, pull right onto a shelf and rockover back left above the overlap (crux). Take the dwindling crack above to a small overlap and finish direct via some bold padding on some good scalloped smears.

11a Super Monsters ★★★   20m HVS 5a

FA Jules Lines (solo) 2016

The combination linking the start of 11 with the upper corner is a superb addition. A short well-protected crux. Climb the corner and step right onto the ledge. With chest high protection, make a scrungy move left to gain a ledge. Step left into the base of the corner and climb it to the top.

11a Easy Monsters ★★   10m HS 4b

Essentially the top pitch of 11a, gained by abseiling down the groove of 10.

12 Silent Spring ★★★   25m E1 5a

FA Alasdair 'Plod' Ross & P.Garden September 1982

The centre of the slab is the areas mid-grade definitive fear test. Start as for 11, and move right into the niche. Trend right to the tree, gain composure and smear out leftwards up the bold curve.

13 Medium Cool ★★★   25m VS 5a

FA Brian Lawrie & Dougie Dinwoodie 1971

A varied and exhilarating timeless classic.

1. **10m 5a** Jam and heave your way out of the slot onto the slab (crux), then traverse left to belay on a small ledge.
2. **15m 4c** Step back right and make a delightful pull through the roof and steadily up to the tree. The exciting finale up the vague crack behind the tree has only one credible runner, so just stay cool.

Miguel Alonso making short work of Rattlesnake.

WESTERN SECTOR – UPPER TIER

NO 3676 9695 **Altitude:** 300m **S** 10min

Descent: Either by abseil from trees on top, or down obvious worn paths at either end.

1 Janus ★ 10m E4 6b

FA Alasdair 'Plod' Ross & Marion Sutherland 1990

A good well protected technical pitch. Start just left of 2. Climb short crack then move left into another crack which is followed to the top.

2 Left-Hand Crack ★ 10m VS 4c

FA Stan Falconer & party June 1975

The crack on the right side of the short smooth wall, near the left end of the crag.

3 Original Route ★★ 10m VS 5a

FA Davy Duncan & party late 1960s

Further right is a corner. Gain this awkwardly then go up right to a ledge. Finish by deep cracks in the wall above.

4 Fingerwrecker ★ 10m HVS 5c

FA Brian Lawrie & Greg Strange 4 August 1979

Desperate for HVS! – probably more like E1 6a, but with sinker nuts. The obvious thin crack just left of 6. Follow the painful initial crack (not using the boulders) to gain a scoop.

5 Wrecker's Traverse ★★ 20m E3 5c

FA Alasdair 'Plod' Ross, Neil Morrison & Ged Reilly 1984; start as described: Jules Lines 2017

Start on the boulder, as for 6 (RP in a seam on the left). Pull on at undercuts and move left to gain a good pinch on the arête. Climb up to the break, gear up and sprint across the wall to a standing position on the arête. Make a thin step out left and use some flat holds to gain the top.

6 Peel's Wall ★★★★ 12m E4 6a

FA Jerry Peel & Tony Barley 1977; Right Finish: Neil Morrison (headpointed) 1 May 1999

The classic test-piece of the crag, and one of the hardest pitches in Scotland in its day. Start in the centre of the wall. Gain a good undercut flake then direct with long reaches to the horizontal break. Continue direct with an awkward move to gain a narrow ledge finishing out rightward. The *Right Finish* E4 6b moves right from the horizontal break finishing up the wall and slab just right of a blind vertical crack.

7 Smith's Arête ★★★　　　　10m E5 6a

FA Bob Smith (headpointed – 1 runner pre-placed) 1983

The prominent square-cut arête left of the central corner. Poorly protected. Climb the lower edge to the horizontal break and better protection at two-thirds height. Continue more easily directly up the final slab.

8 Little Cenotaph ★★　　　　10m HVS 5b

FA aided 1960s; FFA Guy Muhleman, Greg Strange & Rob Archbold 24 August 1975

Well protected climbing up the central square-cut corner. Wide bridging leads to a ledge on the right. Finish by the shallow corner (crux) on the left.

9 Dod's Dead Cat ★★　　　　15m E2 5c

FA Ewen Todd & Robin Patterson July 1986

Follow 8 to the horizontal break. Traverse out left along this to join 7 and finish up this.

10 Pink Wall ★　　　　10m HVS 5a

FA Greg Strange & Craig Anderson 15 October 1972

The centre of the wall right of 8 with a boulder problem start. Finish up an easy corner above the ledge.

11 Flibbertigibbet ★　　　　10m VS 5a

FA Ewen Todd (solo) June 1982

Saunter up the easy crack until level with the roof. Lean left to gain a jug at the base of the hanging flake-groove. Gather some commitment and pull into it in an exposed position; keep pedalling for the top.

12 Green Corner ★　　　　10m Very Difficult

FA unknown 1982

Reasonable climbing on good holds (some hollow) with good protection.

13 Swine Before Pearls ★★　　　　10m HVS 5a

FA Ewen Todd, Brian Lawrie & Alasdair 'Plod' Ross July 1982

Climb the left side of the arête; you will encounter small holds and little gear for a fair way. A lovely route if clean, maintaining interest throughout, especially if taking the final arête direct, if you wish to claim the generous grade.

14 Jackdaw Groove ★★　　　　10m VS 4c

FA Dave Stuart & Greg Strange 9 May 1971

A good climb up the open groove; nicely sustained with enough gear.

15 Crumbling Dice ★★　　　　10m E2 5c

FA Mike McDonald & Andy Nisbet May 1977

The set of grooves often looks dirty with the odd guano splash. Taking the first groove direct gives an interesting sequence to the half-height rest before the steep and satisfying upper groove. Make sure you leave room for your fingers in the cam slot at the crux.

16 Cookie Grooves ★★　　　　12m E2 5b

FA Jules Lines (solo) 2018

This combination sniffs out the easiest line. Climb the font of the pillar to its top, then bridge onto the arête. Climb up and make a delicate move into a delicate V-groove; continue boldly up this and pass a short crack to the top.

17 Black Custard ★★★　　　　15m E1 5b

FA John Mothersele & Mike Freeman June 1971

A fine well protected struggle up the steep left-slanting quartzy crack near the right end of the crag. Follow the crack up to a black roof and pull over this to good holds, finishing easily.

18 High Steppa ★★　　　　8m E3 6a

FA Brian Sprunt & Mike McDonald (both led) August 1980

The smooth wall is a good route to test your technical ability in relative safety. The thin crack to reach the break now has a broken hold. Above the break, do as the name suggests.

19a Second Slip ★　　　　8m E1 5c

FA Jules Lines (solo) 2018

The left finish to 19 has an involving move using an obvious sidepull. A good twin route to 18, or an easier alternative to it.

19 **First Slip** ★   8m HVS 5a

FA Alasdair 'Plod' Ross & P.Garden September 1982

Great flowing climbing up the flake system has some tough moves to gain and leave the break. Not quite E1 though.

20 **Jam Crack** ★   8m Severe 4a

FA unknown 1960s

A good jamming test. At the top of the jam crack is an awkward rollover onto a ledge. Continue up the corner and exit left, or better direct, HS 4c, up the steep top wall. Good gear, but you'll need to pull a little on some small holds.

21 **Belly Flop Finale** ★   8m VS 5a

FA Jules Lines (solo) 2018

This is so atypical of a Pass route at the grade; short and hard. Fingertip layback onto a pillar/block, then follow the slender groove in the arête. Place some micro cams in the break before enacting a finish that doesn't promote finesse.

CENTRAL SECTOR – GULLY WALL

Altitude: 280m   **w**   7min

The west-facing wall cutting up the slope, bounding the right side of the open gully.

1 **Private Parts** ★★   12m E6 6b

FA Alasdair 'Plod' Ross (redpointed) 1987

From the cutaway, go up to a PR and undercut past it to gain a shakeout on a flake, then dash up past some quartz to finish at a notch.

2 **Morphine Drip** ★★   12m E7 6c

FA Jules Lines 26 October 2016

The central line following a hairline seam 2m right of 1. Start at the diamond-shaped block and gain a slot in the wall. Make a desperate move to gain a niche hold before sprinting up to a superb finishing sequence in the scoop at the highest part of the wall. Probably F7c.

3 **Captain Copout** ★★   12m E4 6a

FA Graeme Livingston & Colin Jamieson 1984

The original line nibbling away at the line of least resistance on the wall is tougher than it looks. Climb the central weakness on the wall to a split block (high runner), step right and go up to slots in the horizontal. Continue up on edges, trending rightwards to finish in a notch.

4 Copulation ★★★   **12m E5 6a**

FA Willie Todd May 1987

A superb route on good holds; the best on the wall. It is
pumpy (no pun intended!) so important you warm up first.
Gain the roof from the left (cam in lip), and pull through
and up to quartz holds (offset cam). Move left to good
holds then up to the horizontal. Continue direct to gain
a big left sidepull (nut) and finish via quartz jugs to a
positive top out.

5 Hot Temper ★★★   **12m E5 6b**

FA Willie Todd & Graeme Livingston September 1986

The headwall finish to the now sadly defunct *Cold Rage*
gives a brilliant route. Follow quartz jugs rightwards to
a finger jug under the pod (crucial gear) and then hard
moves up into the pod followed by steady crack climbing
to the roof. Stretch in a nut, have a rest, then pull wildly
through the roof and charge up the headwall.

6 Warm Love ★★★   **20m E6 6b**

FA Jules Lines 25 March 2015

The overhanging arête gives a sensational and sustained
route. Start as for 5 and follow the quartz crimps

rightwards to the arête. Sustained slapping leads to a
shake out at the break. Continue using small opposing
sidepulls in a wild position to finish via the final layback
crack of 7.

7 Anger and Lust ★★★   **20m E2 5c**

FA Mike Freeman & Dougie Dinwoodie 1971;
FFA Bob Smith & A. Williams June1980

The prominent roof-capped corner. If you get lost on this
one, give up! Follow the corner to finish spectacularly by
undercutting left into the deep crack through the left side
of the roof (crux) leading to good finishing holds.

7a Right Finish ★★★   **6m E4 6b**

FA Mike Freeman & Dougie Dinwoodie (aid) 9 May 1971;
FFA Dougie Dinwoodie & Watty Taylor 1985

A short well protected finish. Traverse right along the
lip of the roof to a good jug. Step left and finish up the
short corner.

8 Snow Gates ★★ 20m E4 6a

FA Jules Lines 19 March 2021

The direct start to 9 and the twin right weakness avoids the loose block at the top for a safer alternative. Climb to a horizontal in the wall, make a thin move up (R 1 on left) and go up under the roof. Make a bunched pull through the roof and layback up onto the arête and an awkward mantel finish.

9 Lech Gates ★★ 20m E4 6a

FA Dougie Dinwoodie & Graeme Livingston 1983

Now harder following the removal of a shoogly chokestone jug. There is a 'creaky keyed-in' block at the top - good enough to pull on, but not advisable for gear. Follow the big holds up to the break, then go right and back left to beneath the weakness in the main roof. Finger jam up the footless slot (crux) to bigger holds, ensuring you have some strength in reserve for the final moves. The upper arête to a mantel gives a good exit.

10 Bottle of Smoke ★★ 20m E4 6a

FA Wilson Moir & Neil Morrison 1988

The photogenic arête is an exemplary route at the grade, both bold and technical. Climb the blocky crack to the top, place some big cams (thread just above) and get established between twin groove-arêtes. Climb them direct in a fine position.

11 K Nine ★★★ 20m E6 6a

FA Jules Lines & Richard Biggar 5 November 2020

This thrilling grit-like encounter just merits the grade for an on-sight. Climb the corner to a hands-off rest behind the fang. It's advisable not to place gear around the fang. (There is a R 1 on the left arête and a Camalot 0.0 under the overlap above). Now pull wildly through the overlap on flange holds and coolly paste your way up the arête to finger jugs over the top and mantel out.

12 Creak and Squeak ★★ 30m HVS 5a

FA Ian Duckworth (solo) April 1980

An excellent route now that the culprit squeaky flakes have been trundled - like a sub-extreme *Anger and Lust*.

Start directly below the corner and go up a short wall to gain its base. Climb the main corner to a shelf. The short wall above provides a sting in the tail.

CENTRAL SECTOR – SOUTH FACE

Altitude: 260m **S** 5min

Directly above the parking area.

Descent: Either by 25m abseil from trees (sling and maillon in situ on Scots pine at top of *Blutered*), or down either side of the crag after gaining a well worn path at the top.

1 The Splits ★ 35m HVS 5b

FA Brian Lawrie & Bob Smith May 1983

Climb up onto the ramp/shelf to reach a V-groove under a large roof. Start up the groove and step left to reach a flake-groove. A thin move leads to better holds and the well-protected 'splits' crux. Having reached the shelf above it is more natural to head right to climb the flake corner.

2 Giant Flake Route ★★ 30m VS 4b

FA Greg Strange & Raymond Simpson 17 September 1967

A rightward diagonal line across the front face. Start 10m right of the left edge of the front face, beneath a small larch at 10m. Climb a short difficult boulder problem wall (crux) to a good ledge at the base of the short left-facing corner. Up this onto the top of the flake. Traverse right along a large ledge and continue up another short corner on good holds continuing rightward on large flakes to finish at the large Scots pine.

3 Convoy ★★ 25m VS 4c

FA Jim McArtney, Derek Pyper & Alan Corbett 1987, FFA unknown

A direct line cutting through 2. Start 4m right of 2. Climb the groove to the large ledge on 2. Up the short corner then direct to an alcove and over the roof on good holds.

4 Stoned Immaculate ★ 15m E5 6a

FA Graeme Livingston 1985

Bold, but not too difficult for the grade. A steep start

(crux) leaves you scrabbling onto a ledge. There is good gear at the break before stepping right and head up the arête into lonely territory and a scary finishing mantel.

5 Drambuie ★★★ **15m E6 6b**

FA Dougie Dinwoodie & Alasdair 'Plod' Ross April 1985; as described: Daniel Laing & Jules Lines July 2012

Start just left of centre at some flat holds. Climb to a right sidepull at 4m; place a good IMP 2 in the slot above. Palm along the sloping handrail to a jug and gear, make a move up to gain a rail, then run it out up the slender layback to finish on finger jugs.

5a Bluter Wall ★★★ **15m E7 7a**

FA Jules Lines 22 January 2019

The centre of the wall gives immaculate climbing and required optimum January conditions. Start at a flake hold just left of 6 and make a series of desperate moves (IMP 2 or 7B+ above pads) to join 5 at the jug; finish up that route.

6 Bluter Crack ★★ **15m E4 6c**

FA Jim McArtney, Derek Pyper & Alan Corbett aid 1965; FFA Graeme Livingston 1984

The eye-catching splitter crack. Use a layaway on the right to place a R 1 to maintain the grade. Or highball above pads at 7A+.

7 Bluter Groove ★★ **20m E4 6b**

FA Jim McArtney, Derek Pyper & Alan Corbett 1965; FFA Murray Hamilton, Pete Whillance & Rab Anderson June 1982

A short well protected technical problem. Start beneath the prominent central groove. Climb the groove, initially using the right arête then bridge into the groove to reach a good jug on a block at 6m. Continue more easily above.

8 Bluter Arête ★★ **20m E8 6c**

FA Jules Lines 27 March 2021

The right arête climbed in its entirety. Climb the lower arête on both sides to a rockover on the right (6C), then easily up to the ledge. With cams placed below the flared

pod in the crack, go up onto the arête and get established on its right side, then run it out all the way to the top via a sustained set of difficult sequences. Soft at E8, but its harder and bolder than 9 which it joins for the final moves.

9 Wife Beater ★★ **20m E7 6c**

FA Tim Rankin (headpointed) May 2005

Excellent technical wall climbing with a desperate start. Boulder out the thin crack just right of the lower arête (6C+/7A), then follow 10 to the upper break. Gear up, then using an undercut make hard moves up and left and continue up the arête with a scary slap for the top jug.

10 Demon Drink ★★ **15m E5 6b**

FA Dougie Dinwoodie 4 April 1985

The centre of the wall is climbed on lovely bubbly granite. Go up the short wall via a seam and up to the break; continue directly by undercutting this to gain a flake. Climb direct to gain the next break (good gear) and forge upwards via a powerful move off a small sidepull to a protruding hold, finishing on big holds.

11 Blutered ★★★ **25m E1 5b**

FA Dave Wright, Alison Higham & Guy Muhleman July 1976

Excellent sustained climbing. Up easily to a large flake then leftward and along flared horizontal break to good holds on the arête. Step into the groove and up this by a superb hand crack. Direct up easier ground to the pine tree.

12 Doctor Dipso Direct ★★ **15m E5 6b**

FA Dougie Dinwoodie 7 April 1985; Direct: Jules Lines 2021

The headwall extension improves the outing. From the quartz nodule (micro cams above), climb direct to a good flat hold (HB 2 & cam above). Continue direct using the fat undercut, a pair of flakes and a long reach. The original (* E4 6a) moves up and right from the flat hold to better holds and an easier finish.

13 Hangover Wall ★ **15m E2 5c**

FA Andy Nisbet & Brian Lawrie 1980

Gain the quartz nodule and make a hard move up and right to gain the layback flake; power up this.

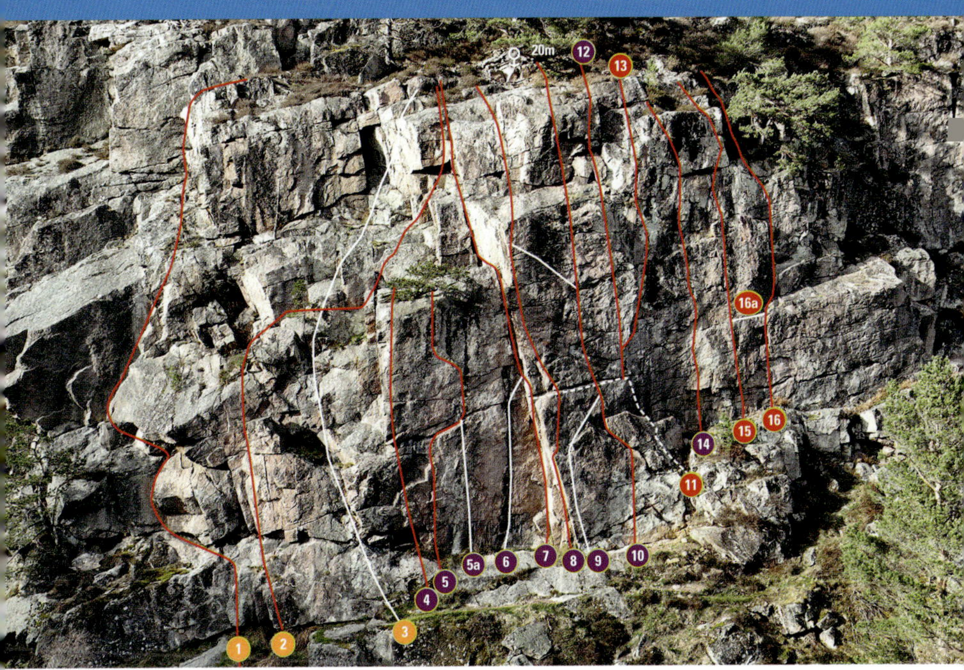

14 Sobriety ★★ 20m E5 6b

FA Jules Lines January 2017

An involving pitch up the centre of the face, at the upper limit of the grade. Make powerful bouldery moves via an undercut and blind moves to reach a break. Continue direct via a sustained sequence to gain a narrow ledge. Continue on up the crack in the upper wall veering slightly left to finish on jugs.

15 Larup Head ★★ 20m E3 5c

FA Graeme Livingston & Colin Jamieson 1985

An excellent wall climb; the lower section is bouldery with fiddly protection, making it high in the grade. Start at undercuts beneath a vague flake at 6m. Go up to the first break (R 1), then up left to the second break. Locate the quartz crimp above to gain the third break at the base of the flake line, which is followed with interest to the top.

16 Alcoholics Anonymous ★ 15m E1 5b

FA Alasdair 'Plod' Ross & Ian Davidson June 1985

Climb directly through a shallow niche to gain a ledge.

Finish more easily up the sparsely protected small corner above.

16a Alchoholic Head ★ 15m E2 5b

A popular hybrid utilising the start of 16, then traversing left to finish up 15.

17 Strawberry Ripple ★ 8m VS 5a

FA Brian Lawrie, Ged Reilly & G.Stephen June 1980

A slight route, with good technical climbing at upper limit of the grade. Climb the short discontinuous crack in the small slab on the right edge of the crack to a rounded finish.

18 Plane Slider ★★ 8m E4 5c

FA Alasdair 'Plod' Ross (solo) 1984

A unique and exciting solo, both sustained and committing, taken solely on fragile quartz razors. The climbing is maybe no more than 5b, but like many dangerous routes, the grade is for an on-sight.

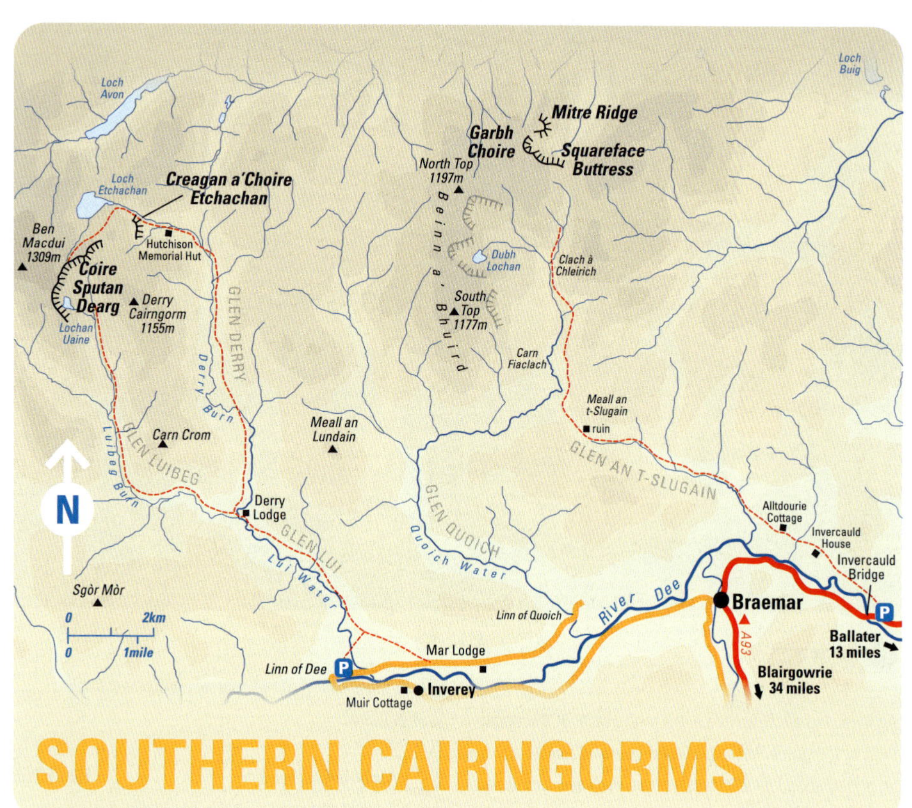

DEESIDE

Accommodation: The peculiarly self-proclaimed 'award winning Highland village' of Braemar is the main centre, with numerous hotel and B&B options. **Bunkhouse:** Gulabin Lodge Hostel, Glenshee (☎ 01250 885255; www.gulabinoutdoors.co.uk).
Youth Hostels: Braemar (☎ 01339 741659; www.hostellingscotland.org.uk); The Smugglers Hostel, Tomintoul (☎ 01807 580364; www.thesmugglershostel.co.uk).
Campsites: Braemar Caravan Park (☎ 01339 741373; www.braemarcaravanpark.co.uk); a couple of miles south of the town along the west bank of the Clunie Water, (the river running into the south end of Braemar) is a popular place to wild camp. **Club Hut:** Muir Cottage (Cairngorm Club; NO 076 896) 5 miles west of Braemar.
Amenities: There is a small supermarket in the village; outdoor shop is Braemar Mountain Sports (☎ 01339 741242; www.braemarmountainsports.com) which also does mountain bike hire and has a café.

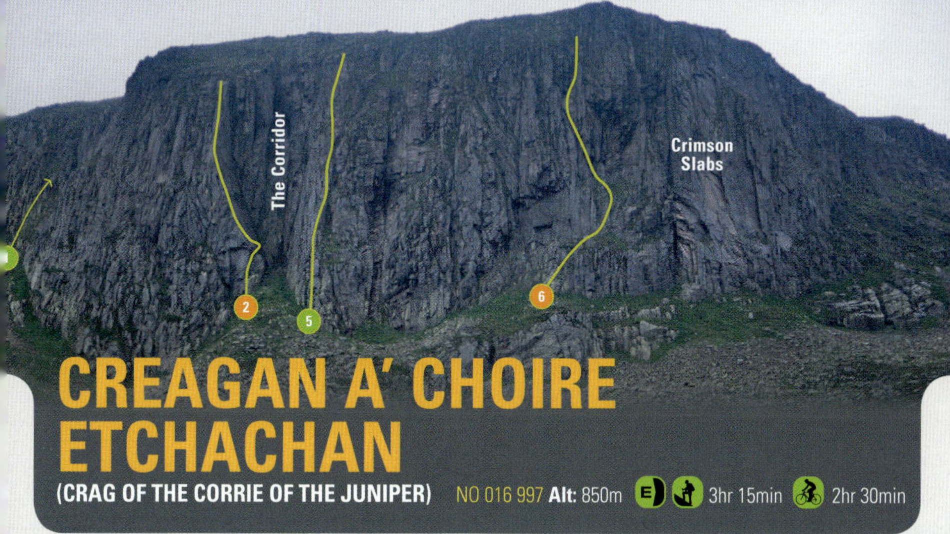

CREAGAN A' CHOIRE ETCHACHAN

(CRAG OF THE CORRIE OF THE JUNIPER) NO 016 997 **Alt:** 850m 3hr 15min 2hr 30min

A big slabby crag in an idyllic setting on the east side of Derry Cairngorm, part of the Ben Macdui massif. A visit here could also be combined with Coire Sputan Dearg, which lies less than an hour away (40 minutes from the top of the cliff).

Access: From the centre of Braemar follow the C-class road west along the south side of the River Dee for 6 miles/9.6km to Linn of Dee. Continue for a further 200m along the north side of the road back towards Braemar to a car park (pay & display) in the forest on the left.

Approach: (A) From the car park at the Linn of Dee follow the signposted path to meet up with the track from the locked Derry Gates to reach Derry Lodge in a little over 4km. Turn right at the public phone 200m beyond the lodge, cross a footbridge and head north up Glen Derry re-crossing the Derry Burn (footbridges) after 2.4km and the Glas Allt Mor after a further 3km. Continue, then cross the Coire Etchachan Burn to gain the Hutchison Hut after a further 1km. The cliff lies 750m further west up the slope. Cycles can be taken to at least the second footbridge or a couple of kms further.

(B) From the north (Coire Chas car park on Cairn Gorm). By any of the routes to the Shelter Stone near the head of Loch Avon. Continue steeply south-east up the path on the slopes of Carn Etchachan to just beyond Little Loch Etchachan. Follow the path left (east) for 200m then contour due south to gain the base of the cliff.

Accommodation: The compact Hutchison Memorial Hut (NO 023 998) beneath the cliff in Coire Etchachan makes an ideal base, sleeping about nine people.

THE BASTION

NO 016 997 **Alt:** 900m

The large buttress at the left end of the crag, between the loose and vegetated Forked Gully and the prominent slabby recess of The Corridor.

Descent: Traverse left (south) 200m beyond Forked Gully and down steep grass and scree slopes.

1 Quartzvein Edge ★★ 120m Moderate

FA Kenny Winram, G.Greig & Mac Smith 15 June 1952

A popular route following the left edge of **The Bastion** overlooking Forked Gully. Start at the base of Forked Gully above a detached block. Climb a 3m wall with inset quartz then follow the edge to a gravel patch at 35m. Layback up a left-slanting groove and continue to the lower of two shelves. Follow the shelf which develops into an open chimney ending in jumbled blocks poised over the gully. Either finish up a scree funnel or by the right side of a false tower.

2a Talisman Direct Start *   40m E1 5b

FA Andy Nisbet, Steve Kennedy & N.Mollison 13 July 1981

The prominent corner system just left of the arête below the start of the normal route. Start at the lowest point of the buttress. Ascend the corner to an overlap (PR) then follow the continuation corner on the left. A traverse back right on good holds across a steep wall leads to the arête, which is followed to the belay on *The Talisman*. Either finish up this or abseil off.

2b Talismaniac **   40m E4 6a

FA Julian Lines & Sue Harper 4 August 1996

Essentially an extended start to *Talismanic* giving a superb bold pitch. Start right of *Talisman Direct Start*. Climb the centre of the scooped wall to a bulge. Pull through this to gain an obvious quartz blotch. Move delicately right then up the arête to better holds to join the normal route.

2c Talismanic *   30m E3 5c

FA Wilson Moir & Niall Ritchie 10 June 1990

The left arête. Start at the base and pull onto the edge using big holds and go up to a perch. Continue boldly for 4m then follow cracks to the belay.

2 The Talisman ***   105m HS 4c

FA Bill Brooker & Ken Grassick 24 June 1956

Excellent varied climbing taking the crest of the wall overlooking The Corridor. Start behind a huge boulder at the base of The Corridor.

1 **35m 4b** Ascend the wide flake-crack rightward to a ledge on the right then follow the easier slab slightly leftward. An obvious traverse left leads to a fine belay on the edge.

2 **22m 4c** Traverse left to the base of a thin groove and follow this to an awkward overhanging alcove at its top. Negotiate this with interest (very well protected) to gain a fine belay stance.

3 **48m 4a** Finish up the crest, starting initially on the right wall.

Jo George starting the traverse on the first pitch of The Talisman. Photo Dave Cuthbertson, Cubby Images

3 Talking Drums ★★ 115m E2 5b

FA Colin MacLean & Andy Nisbet 4 July 1986

Sustained well protected climbing with a sensational finish up the headwall. Start at the lowest point of the buttress.

1 **40m 5b** Climb *Talisman Direct Start.*

2 **25m 4c** Follow the obvious thin diagonal crack leading right past a triangular block and through the girdling bulge. Move slightly back left and up to a poor belay under the headwall.

3 **25m 5a** Gain the base of a shallow groove via two small scoops and follow the groove to a good foothold at its top. Move right and up on hidden holds to a hand traverse and mantelshelf leading right to a PB.

4 **25m 5b** Make a delicate move up to gain a small ledge and PR then swing down right and across the wall to reach the base of a corner crack. Follow this to finish.

JUNIPER BUTTRESS

The buttress right of the large recess of The Corridor.

4 The Hex ★ 80m HS 4b

FA Greg Strange, Mike Forbes & Brian Lawrie 10 July 1971

A short but steep route on excellent clean rough rock. Start at a shallow pink bay.

1 **40m 4b** Climb a vague rock fault to a system of grass cracks and belay at top of cracks beneath a bulge.

2 **40m 4b** Step left and follow the continuation fault to a point where the cracks become grass choked. Step right to the crest and into *Pikestaff* and finish up the last 6m of that route.

5 Pikestaff ★ 120m Very Difficult

FA Tom Patey, Bill Brooker & John Hay 7 August 1954

Follows the left rib. Go up The Corridor flank of the rib to a point just beyond a ferny trough which climbs to a large grass platform on the buttress. Start beside a curious little pocket in the rock. Climb over short walls and then up alongside the trough to within 5m of its top. Slant up the Corridor wall for 18m by cracks until a break on the right leads to a crest. Follow the rib directly ahead. First comes a smooth nose, then easier rocks to a pointed belay. Take off from a flake on the left (crux) and finish the pitch by moving right. A final 35m pitch leads to easy ground. Scramble to the plateau.

THE MEADOW FACE

6 Carmine Groove ★★ 140m VS 4b

FA Greg Strange, Dave Stuart & Dougie Dinwoodie 17 July 1971

The obvious red groove left of the overhangs on the upper section of the cliff. Slow to dry, though it provides two superb pitches on steep clean rock. Start directly below a recent rock scar in the lower overhangs.

1+2 60m Scramble up right to the upper meadow.

3 **20m 4b** Start up the groove above the left edge of the meadow, moving right to a flake, then back left to re-enter the groove. Follow this to belay below an overhang.

4 **25m 4b** Step left onto a steep red slab and climb directly up the corner to a roof. Exit right to a small ledge and follow a thin crack in the slab to ledges.

5 **35m –** Move right and follow a narrow grey rib soon leading to easier ground.

THE CRIMSON SLABS

The very fine sweep of slabs bounding the right end of the cliff.

7 Scalpel Direct ★★ 120m E1 5b

FA Dougie Dinwoodie & Greg Strange 11 July 1977;
Direct Finish Andy Nisbet & Alasdair 'Plod' Ross 20 October 1985

The very shallow tilted corner towards the left edge of the slabs. Fairly bold in its upper reaches. Start 4m right of the left-bounding Red Chimney.

1 **40m 4c** Ascend the arête overlooking corners on the left to a ledge. Step into the edge of the overlap then direct over a small bulge. Step right to a small stance.

2 **45m 5b** Go up the shallow corner until it finishes at an overlap. Step right and place good runners in *King Crimson* then return to make a rising left traverse (crux) to the arête of a curving corner. Continue up cracks just right of the arête leading to easier ground.

3 Either continue up broken rock (35m), or move right to *Djibangi* and descend via the terrace or by abseil.

8 In the Pink/King Crimson ★★ 125m E3 5c

FA King Crimson: Alasdair 'Plod' Ross & Greg Strange 25 August 1984; In the Pink: Andy Nisbet & Alasdair 'Plod' Ross 27 October 1985

An excellent combination with the main pitch on immaculate pink rock midway between the corners of *Scalpel* and *Djibangi*. Start in the alcove as for *Djibangi*.

1 **35m 4c** Climb a flake crack midway between *Djibangi* and *Sgian Dubh* then move left round a bulge and up to join *Djibangi*. Climb its small corner to a small ledge at its top.

2 **45m 5c** Go out left to a prominent jug, then climb the slab to a shallow flake with unexpected but crucial runners. Climb straight up the pink rock to gain a tiny foot ledge and arrange runners under a shallow curving overlap. Move up left to grey rock and continue to beneath an obvious notch in the main overlap. Move right then cross the overlap just right of a small corner forming the notch. Continue directly, pulling left into a small groove (F #0) and follow this to broken ground beneath an area of clean cracked slabs.

3 **45m 5a** Climb directly up the slabs by cracks passing through a break in the overlap and right of a large block. Scramble to finish.

9 Djibangi ★★ 135m VS 4c

FA John Hay, Ronnie Wiseman & Allan Cowie (2 PA) 22 July 1956; FFA early sixties

The leftmost and less prominent of the two corner lines on the slabs. The first pitch of *Sgian Dubh* provides a better start, more consistent in standard with the main pitch. Start at the base of a prominent pink right-facing corner in a grassy alcove beneath the line.

1 **20m** Climb diagonally left to a large platform on the edge of the slabs.

2 25m Trend back right and ascend a small corner in the centre of the slabs then easier to the base of the main corner.

3 30m 4c The fine corner, stepping right to a P and NB on the rib at the top. Either make two abseils (40m; 30m, or one 60m) from PBs or continue up the inferior upper slabs.

4+5 60m Move back left into the corner and climb it then move out left and follow the obvious groove leading to a huge block. Climb round this and continue up a rib to easy ground.

10 Sgian Dubh ★★   105m HVS 5a

FA Andy Nisbet & Mary Bridges May 1978

The right arête of the main *Djibangi* corner. Start at the left side grassy alcove as for *Djibangi*.

1 30m 4c Follow a series of shallow left-facing corners leading up to the base of the main *Djibangi* corner.

2 35m 5a Ascend the arête overlooking the corner making a short deviation out right in the mid section. Abseil as for *Djibangi*, or:

3 40m Grassy grooves lead out rightward to the terrace.

11 Stiletto ★★   110m E2 5c

FA Mike Forbes & Mike Rennie (3 PA) 17 August 1966;
FFA Dougie Dinwoodie & Adair McIvor 1976

Sustained well protected climbing tackling the impressive vertical crack slicing into the slabs, midway between *Djibangi* and *The Dagger*.

1 25m 4b Go up the lower continuation of the crack to belay beneath the crack

2 40m 5c Cruise up the crack.

3 45m Traverse right into *The Dagger* and follow it easily to the terrace.

12 The Dagger ★★   130m HVS 5a

FA Tom Patey & John Hay (4 PA) 4 September 1955

The great sweeping corner dominating the right side of the slabs. The original line on the slabs with a classic main pitch. Start in the grassy alcove as for *Djibangi*.

1 25m 4b Traverse right and up a little for 5m then move back left into a short corner. Continue slightly left again to gain easy ground leading rightward to the base of the main corner.

2 35m 5a Go up the corner to a hanging belay on a huge spike beneath an overhang.

3+4 70m Move left below a bulge to avoid the overhang then move into the grassy groove above and scramble to a large platform beneath the final slab (terrace descends right from here). Climb the slab by a crack slanting right then direct to finish.

13 Scabbard ★★★   106m VS 4c

FA Mike Rennie & Mike Forbes September 1966

The wonderfully situated arête overlooking *The Dagger* corner – very quick drying. Start by a right-facing clean–cut corner directly below the line.

1 28m 4c Ascend the corner then continue directly by cracks to belay at the base of the arête.

2 28m 4c Go up the prominent finger crack and continuation cracks to a small overlap. Continue up the edge to the spike belay at the top of the corner.

3 50m 4b Move right and down a short way, and pull out right to an edge. Follow cracks and blocks to gain the terrace at its highest point.

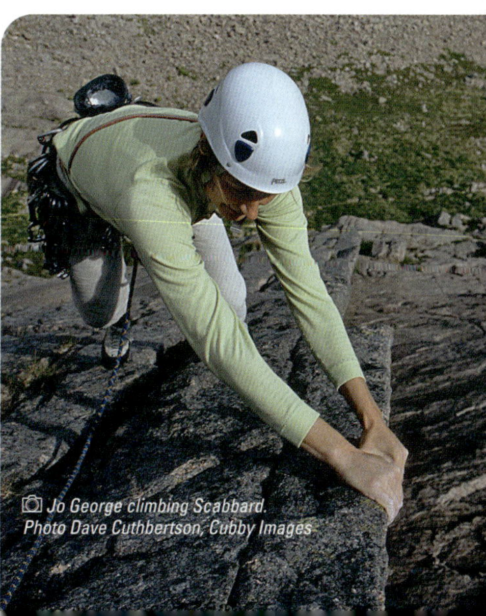

Jo George climbing Scabbard.
Photo Dave Cuthbertson, Cubby Images

BEN MACDUI
(HILL OF MACDUFF)

COIRE SPUTAN DEARG
(CORRIE OF THE RED SPOUTS)

NO 003 991 **Alt:** 1090m   3hr 30min   2hr 30min

This coire lies high on the east side of Ben Macdui, at 1309m Britain's second highest mountain. The slabby rock is rough, clean and very quick to dry. With short routes and easy descents it is often possible to climb several routes in a day. A pleasant open friendly atmosphere makes this one of the best lower grade venues in the whole of the Cairngorms.

Access: From the centre of Braemar follow the C-class road west along the south side of the River Dee for 6 miles/9.6km to Linn of Dee. Continue for a further 200m along the north side of the road back towards Braemar to a car park (pay & display) in the forest on the left.

Approach: (A) From the car park at the Linn of Dee follow the signposted path to meet up with the track from the locked Derry Gates to reach Derry Lodge in a little over 4km. Cross the footbridge over the Derry Burn and follow it downstream to pick up the track heading west up Glen Luibeg. After 2.5km break off right up a path up the right (east) bank of Luibeg Burn for 2km to a fork. Continue up the right fork then in the same direction for a further 3km to the cliff (3 hours). Cycles can be taken as far as the Luibeg Burn (3 km from the cliff).

(B) From the Hutchison Hut head up the side of the stream in the grassy bay to the left of the cliff and over the col between the summit and Derry Cairngorm (1 hour).

(C) From the north (Coire Cas car park on Cairn Gorm) the fastest approach is to follow the path up the ridge on the west side of Coire an Lochain as far as Lochan Buidhe, then contour round the north-eastern slopes of Ben Macdui, heading south-east across the upper reaches of Garbh Uisge Mor. Descend down the scree-filled Glissade Gully (NO 002 990) beneath the south end of a tiny lochan to arrive in the centre of the corrie (2½ hours).

Accommodation: The closest base is the Hutchison Memorial Hut (NO 023 998) in a wonderful situation beneath Creagan a' Choire Etchachan. Further away, though still a useful base is Bob Scott's Bothy (NO 042 932) just south of Derry Lodge. Good camping spots are numerous, including just beyond Derry Lodge; Robbers Copse (NO 015 938) – the stand of Caledonian pines just beyond the ford over the Luibeg Burn; the flat grassy area just beyond the confluence of the Allt Carn a' Mhaim 1.5km further up the burn; or under the cliff itself. There are also some draughty howffs in the corrie, best suited for dry sunny weather.

Descent: The scree-filled Glissade Gully in the centre of the corrie between **Grey Man's Crag** and *Snake Ridge* provides the easiest descent.

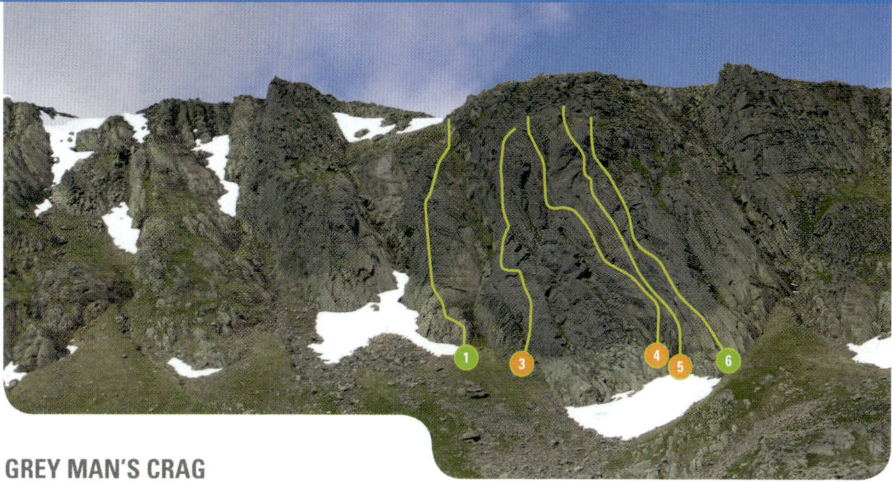

GREY MAN'S CRAG

① Crystal Ridge ★★   110m Difficult

FA Bob Still & Elizabeth Lawrence1 September 1948

The great slabby ridge abutting the left side of the steep upper left side of the main buttress.

1 **15m** Scramble up to a large platform beneath the main slab.

2 **50m** Continue, keeping close to the crest of the slab.

3 **45m** Continue in the same line to reach more broken ground. Scrambling remains.

② Zircon ★   120m E2 5c

FA Simon Richardson & John Ashbridge 7 June 1992

Good sustained climbing up the wall left of *Amethyst Pillar*. Start 10m below a chimney.

1 **10m 4c** Step onto the slab on the right and move up to a large ledge.

2 **35m 5c** Follow *Amethyst Pillar* for a few metres, then the crack system trending left. Pass below a long narrow rectangular roof. Climb the steep crack system 3m to its left and exit strenuously on big holds to reach a ledge.

3 **25m 5a** Continue up the left-facing chimney groove directly above.

4 **20m 5c** Gain a sloping triangular ledge from the right and follow a line of flakes diagonally left across the wall to a steep finish on the left arête.

5 **30m 4a** Finish up easy cracks and grooves.

③ Amethyst Pillar ★★★   100m HVS 5a

FA Final crack: Jim McArtney & James Stenhouse 13 July 1964; Lower section: Rob Archbold & Dave Nichols 10 June 1979

Superb steep climbing up the left side of the buttress. Start at the base of the prominent left-slanting chimney fault.

1 **40m 4c** Move out right onto the steep slabs, starting about 5m below a 'curious round niche'. Climb up by cracks, pulling left over a bulge to the obvious hanging corner. Step left beneath a small roof, step down left and traverse left across steep wall by a flake ledge to belay at its end.

2 **25m 4c** Above is a shallow cracked groove with a bulge. Climb this and exit right onto the crest of the ramp.

3 **35m 5a** Move up right and climb a prominent vertical crack above a pedestal using a jammed flake on the initial bulge. Finish up slabs.

④ Grey Slab ★★★   117m HS 4b

FA M.Higgins, John Innes & Brian Lawrie 14 September 1963

A brilliant route following the prominent left trending ramp in the centre of the buttress. Start by scrambling up broken ground to a good platform 10m beneath the corner.

1 **12m 4a** Climb the cracked groove to gain the right end of the ramp. P & NB.

2 **20m 4b** Climb the ramp to a grass ledge.

3 **40m 4a** Continue in the same line to an overhang,

turn this on the left and climb to a ledge. Climb the slab or the corner on the right until forced to move left at the top. P & NB on small ledge.

4 45m 4a Step down and move into the chimney on the left and finish easily up this. Alternatively, the short slab above the belay (4c) can be climbed to gain the chimney higher up.

5 Ferlas Mor ★★   **122m HS 4b**

FA John Mothersele, Greg Strange, Dave Stuart, Brian Lawrie & Dougie Dinwoodie 22 May 1971

Ferlas Mor is the big grey man, a spectre reputed to haunt the slopes of Ben Macdui, first witnessed by Norman Collie in 1891.
A fine companion to *Grey Slab*, taking a direct line up the centre of the crag.

1 12m As for *Grey Slab*, or climb the fine long layback crack on the right (20m VS 5a).

2 30m Swing up right and climb grooves and ribs on the edge overlooking *Grey Slab* to a basalt fault. Climb this to a ledge below the crux of *Hanging Dyke*.

3 36m Move left and continue up grooves to a small stance level with the top of *Grey Slab*.

4 44m Go up and right and follow the crest via a short chimney and easy rib to top.

6 Hanging Dyke ★★   **105m Very Difficult**

FA Allan Parker & John Young 29 March 1949

Start to the right of the lowest rocks at the foot of a broad slab. The dyke goes up the centre of the slab.

1 30m Follow the dyke to a small ledge then climb a grass-filled crack to a good stance beneath a crack-seamed slab.

2 30m Climb a wide slab inclining left by a series of parallel cracks to a sloping corner. The dyke steepens, forming a rib to the left of a groove. Climb the rib for 18m on small holds (crux), until the dyke falls back into a chimney. Spike and NB on ledge on right.

3 45m Follow the continuation of the dyke up left on the crest then by the left wall of a grassy groove. Scramble out right, then up a wide easy grassy gully to finish.

Karen Latter on the superb first pitch of Amethyst Pillar.

Karen Latter moving up the ramp on the fine second pitch of the classic Grey Slab.

SNAKE RIDGE

There are two ridges right of Glissade Gully, this is the rightmost one.

7 Snake Ridge ★★   130m Severe 4a

FA Bill Brooker, Doug Sutherland & Charlie Hutcheon 25 June 1949

An excellent climb. One of the original classics of the corrie. *Snake Ridge* is the second of the long ridges to the right of Glissade Gully, so named from *"its fancied resemblance as seen from the top, to a snake, head down"*. For easier recognition its lower rocks fan out into three ridges *"giving a fair impression of inverted Prince of Wales feathers"*. The left side of the ridge is low and angles easily into a grassy gully running alongside.

1+2 60m Start on the left *"feather"* and follow the crest directly for two pitches to a platform and belay below a step on the ridge.

3 15m Climb this on the left to a stance and spike belay on the right.

4 25m 4a The crux pitch follows. Go up a short wall and use good holds to pull up into a groove. Continue rightward up the groove to the crest, or (easier) leave the groove and climb to a good hold on the left.

5 30m Further climbing leads to the broken upper buttress.

THE CENTRAL BUTTRESSES

Three buttresses close together, right of *Snake Ridge*.

8 The Black Tower ★★   110m Severe 4a

FA Tom Patey, 'Goggs' Leslie & Mike Taylor 21 April 1952

Fine climbing up the central buttress, unfortunately marred by a horribly wet and slimy approach pitch. Start to the left of the buttress.

Starting 15m above the lowest rocks follow a prominent groove on the left flank of the buttress. This leads in 25m to easier ground. Climb broken slabs on the left to a platform 9m above, and at the foot of a steep 6m groove, close to the true crest of the tower. The groove, *"entered from the right by a Severe movement"*, gives access to a platform and block. The steep slab round the corner on the left is climbed by a 6m crack to a short arête. Traverse delicately for 5m across the slab on the right until it is possible to regain the crest and the summit of the tower by a short crack. The tower is connected to the plateau by a shattered arête. Clean, rough rock throughout.

9 Flake Buttress ★   110m Moderate

FA John Tewnion, Ernie Smith, Mac Smith, Kenny Winram 22 May 1949

Good varied climbing up the right buttress. More difficult variations have been made on the lower section. Start just right of the lowest rocks.

1 18m Follow a grassy depression rightward to a stance on the broken crest.

2+3 92m Continue in the same line until a flake crack leads back left to a short right-angled corner. Swing up left on good holds. Go up a chimney on the right and then by easy ledges to a stance and belay below an overhang on the crest. Behind a huge flake on the left climb a vertical crack on grand holds and continue up slabby blocks to a gap. Ignore the easy ground on the right and take the arête straight ahead to the plateau.

SPIDER BUTTRESS

The slabby buttress high up in the coire, between *Flake Buttress* and **The Main Spout**.

 **Bolero** ★★   **50m E5 6b**

FA Wilson Moir & Tim Rankin 27 July 1997

1 **30m 6b** Start up the initial corner of the big low angled corner, then go right to climb the small left-facing corner, pulling out right at its top to the PR on *The Skater* (the PR is a joke but other gear can be arranged). Step up then reach left to climb the arête (crux) to a ledge. Continue up the easier arête above to a belay.

2 **20m 5c** Traverse 5m right and pull over the overhang via a flake layaway. Continue up the hanging slab to the top.

11 **The Skater** ★   **45m E4 6a**

FA Colin MacLean & Andy Nisbet 26 July 1984

Ascends the left edge of the buttress overlooking the large easy-angled corner. Start at the base of this.

1 **30m 6a** Ascend a shallow corner just right of the arête of the large corner exiting right. Move diagonally left across the slab to a jug on the arête and poor PR (good F). Climb the slab on the right to the prominent corner at the left end of the cigar-shaped overhang. Climb the corner to a belay.

2 **15m 4b** Continue up and slightly left over easy blocks, finishing by a short overhanging crack

12 **The Fly** ★★   **45m VS 4c**

FA Dougie Dinwoodie & Brian Lawrie 19 June 1971

A good well protected pitch up the steep cracked wall near the centre. Climb a crack leftward then move up right to gain a horizontal crack. Traverse this across the wall and climb easily up left to the prominent deep S-shaped crack. Climb the crack then either the slab or the deep flake crack on its right to finish.

13 **Flying Saucers** ★   **55m E4 5c**

FA Wilson Moir & Tim Rankin 27 July 1997

1 **35m 5c** Start up the initial crack of *The Fly* to gain the left arête of the wall. Climb the arête to its top and arrange bombproof runners. Continue up the unprotected scooped slab-rib which runs up rightward to a horizontal break. Step right and ascend a short crack leading to a belay.

2 **20m 4c** Go directly up from the belay to climb flakes up the slab above.

THE MAIN SPOUT – TERMINAL BUTTRESS

The largest piece of rock on the right side of the base of **The Main Spout**.

Descent: Down the grassy shelf slanting down leftward into **The Main Spout**.

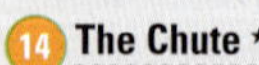 **The Chute** ★   **45m HVS 5a**

FA John Ingram, Brian Findlay, Greg Strange & Dave Stuart 13 June 1970

The leftmost of a series of grooves and corners on the wall left of the descent route. Follow the corner to a *"curious jammed block"* and go behind the block. Step left and climb the left-slanting crack to a shallow groove left of a prominent overhanging prow. Follow the groove to a good platform. Scramble to finish.

15 **Terminal Wall** ★★   **70m HS 4b**

FA Brian Lawrie & Jim McArtney 8 September 1963

Fine exposed climbing following a series of cracks and grooves up the left edge of the wall at the right side of the corrie. Start just right of a short gully and prominent fault. Climb straight up over an overhang then left to a stance on the edge. Move right, climb cracks above then a fault to a sloping ledge leading right. From the end of this ledge go straight up and make an awkward move left round a corner on a hidden foot hold. Finish by cracks and split blocks.

BEINN A'BHUIRD (TABLE HILL)

A vast mountain, with its extensive summit plateau containing a series of fine corries cutting into its eastern slopes, together with the even remoter Garbh Choire at the head of An Slochd Mor (the great pit).

Access: Turn north off the A93 (signposted Keiloch), 2.8 miles/4.5km east of Braemar, 0.2 miles/0.3km east of Invercauld Bridge over the Dee. Park in the small car park up on the right (pay & display) (NO 1882 9132; 57.005943, -3.338172).

Approach: Mountain bikes recommended. Follow single track road (signposted Water of Quoich) north-west through the forest past Invercauld House. Take the right fork 500m beyond the farm at Alltdourie (signposted Slugain) through a plantation and onto a new track up Gleann an t-Slugain which reverts to a stalkers path at the ruined Slugain Lodge. Leave bikes here. Continue until the path splits 500m beyond. Take the right fork and follow it north, passing the huge boulder of Clach a' Chleirich, then the stream which leads north in a further 2km to the col of The Sneck (the notch), 15km.

On foot: If the Dee is running low, ford it just downstream from Braemar Castle and head north to pick up the track at the bottom of Gleann an t-Slugain. This reduces the approach on foot by around half an hour.

Accommodation: There is a small howff sleeping two or three people at NO 098 995 at the base of Dividing Buttress, under the second largest boulder immediately beneath the lowest slabs (200m west of Dubh Lochain). Otherwise, excellent wild camping in any of the corries or the upper reaches of Glen Quoich.

GARBH CHOIRE 
(ROUGH COIRE)
4hr   3hr

A fantastically remote corrie, with grand views north over the head of An Slochd Mor.

Descent: From the top of the path at The Sneck, head west up the rim for a few hundred metres to drop down to an obvious grassy hollow on the right. Contour west from here to pick up a path which descends diagonally across scree slopes into the base of the corrie. A prominent narrow scree fan a few hundred metres north of the top of *Mitre Ridge* provides a convenient descent from the other end of the corrie.

SQUAREFACE BUTTRESS

NJ 111 013 **Alt:** 950m

High up on the left side of the corrie, ending abruptly on the plateau, and roughly midway between The Sneck and *Mitre Ridge* is a fine steep buttress with a superb slabby west wall. Very quick drying.

Approach: Easiest from the plateau. Follow a path west round the rim for about 250m beyond the top of the buttress. Just beyond a shallow gully, follow a vague path diagonally down to the base. From the corrie floor, follow the diagonal scree and boulder-filled rake tortuously up to the base.

 **Squareface ★★★★**   **90m Very Difficult**

FA Tom Patey & Mike Taylor July 1953

The best route of its grade in the entire massif, well worth the long approach. Start at a large flat area beneath the centre of the slab.

1 **40m** Move slightly left and follow a crack/groove over a small bulge at mid-height to large platform. Continue up the ridge to belay on a ledge on the left side of the arête.

2 **25m** Traverse right for 9m past the first prominent crack then direct up the slab to another obvious traverse line leading back left to a fine belay ledge

on the arête. Thread and nut belay.

3 25m Follow the obvious line up rightwards to a good horizontal crack. Climb the obvious wide fissure, then rightwards to a fine wide left-facing layback flake (the fissure continues at Severe). Climb the flake to its top then move up rightwards pleasantly. A stunning pitch.

2 Angel's Edgeway ★★   **65m VS 4b**

FA pitch 2 W.Gault & A.Kane 1959; complete Brian Findlay & Greg Strange 17 June 1989

Good situations and exposure, cutting through *Squareface*. Start right of *Squareface*, left of a corner crack.

1 30m 4a Ascend slabs to where they steepen. Move right and climb the crack through a bulge which leads to the belay on *Squareface*.

2 15m 4b Step out right and follow a line just right of the arête to share another belay with *Squareface*. A fine exposed pitch.

3 20m 4a Follow *Squareface* to the good horizontal crack. Traverse left and layback the prominent hollow flake, then easy ground to finish.

 *Unknown climbers on the finely positioned top pitch of Squareface.*

MITRE RIDGE NJ 107 014 **Alt:** 950m 

1 Mitre Ridge ★★   **220m HS 4b**

FA Sandy Wedderburn, Pat Baird & Jock Leslie 4 July 1933; Variation Start: Charlie Ludwig & Donald Dawson September 1933

One of the great classic Cairngorm ridges, despite some vegetation and loose rock. After the first pitch (avoidable), the climbing is Very Difficult. Start beneath large groove midway between the lowest rocks and the right edge of the ridge.

1a 35m 4a Climb slabs then the groove and a short bulging wall to belay.

1b 25m The first pitch can be avoided by following a line of weakness (Moderate) leftward from the ridge to belay at the top of the original first pitch.

2–6 Follow the general line of a rising shelf round to beneath a deep chimney on the west face. Follow the chimney and enter a shallow gully which leads in 12m to a shoulder on the ridge. Make a delicate traverse right then climb direct to the base of the steep wall beneath the first tower. Gain the large grass platform above either directly by 5m inset right-angled corner or, slightly easier, by moving leftward over a slab and climbing a splintered chimney. Climb the wall and continue to the col between the first and second towers (junction with *Cumming-Crofton Route*). Either turn the second tower on the left or direct by a steep crack to finish along a narrow arête and final tower.

THE WEST WALL 

Steep and impressive, but well endowed with holds.

2 The Fundamentalist ★★   **175m E2 5c**

FA Simon Richardson & Iain Small 15 July 2005

An excellent route taking the right edge of the front face of *Mitre Ridge*. Pitch 3 is sensationally exposed. Start directly below the well defined pillar split by a crack that lies just left of the initial corner of *Cumming-Crofton Route*.

1 30m 5a Scramble up to the crack and climb it through

an overlap to a good platform on top of the pillar.

2 30m 5b The route continues up the slim hanging groove in the edge (between the corners). Move right, then back left into the hanging groove and climb it to exit on easier ground on the front face of *Mitre Ridge*. Continue easily up the edge where *Mitre Link Variation* joins from the right.

3 35m 5c Go up to the foot of the First Tower and climb the right edge on hidden holds.

4 20m Scramble along the ridge to the notch below the Second Tower.

5 20m 5c Climb the wall on the front face of the tower past a prominent protruding flake. Strenuous and awkward to protect.

6 40m Scramble to the plateau.

3 Cumming-Crofton Route ★★ 160m Severe 4b
FA Stephen Cumming & John W.Crofton 4 July 1933;
Pitch 5: Jim Bell May 1935

A classic and the most impressive of all the pre-war routes in the Cairngorms. Start directly beneath the large open corner bounding the left end of a steep clean wall. Scramble up slightly left to a small platform.

1 25m 4b Climb the obvious open chimney passing a steep projecting flake with interest (crux) at mid-height. Belay in a cleft at the top.

2 15m 4a Move up to ledge above then climb diagonally rightward up a smooth slabby groove until it ends at a bulge. Move up then step left round the rib to belay at base of corner.

3 20m 4a Climb the flaky corner to a good ledge on the right.

4 45m 4a Climb the wide crack in the left wall then the general line of the corner. Cross a loose gully to large blocks then up to belay on the crest of the ridge.

5 25m 4a Step off the top of the large block and up to a ledge above. Traverse a wide flake right round the edge. Step down into corner and up this, stepping left up a flake on good holds to the top of the First Tower. Continue along a level section to a large block belay.

6 30m Continue along the airy crest, via a good ledge down on the left side to drop down into a gap. Finish easily on the left of the final wall. Thread belay.

4 The Spear of Destiny ★★ 50m E5 6a
FA Pete Benson & Guy Robertson 26 August 2007

The smooth slab and bulging wall left of *The Empty Quarter*, heading for then through the obvious arrow-shaped niche. Superb sustained climbing on beautiful rock. Start below a big open groove 10m down from *The Empty Quarter*.

1 15m Climb the easy groove, or cracks on its left wall, to belay comfortably on a perched block thread belay.

2 35m 6a Step left onto the slab directly above the belay and move left to reasonable holds by a good diagonal crack (protection). Stand up on these, tip-toe a little further, then step left to where a series of moves leads directly to the obvious jug. Continue up to another diagonal crack, climb past this, then traverse rightward into the niche. Turn this on the left, step back right, then climb sustained cracks into the obvious fault and easy ground. Abseil off (in situ thread on left).

5 The Empty Quarter ★★ 95m E3 5c
FA Dougie Dinwoodie & Greg Strange 6 August 1983

An excellent well protected pitch up the shallow vertical corner at the right side of the wall.

1 40m 5c Scramble up to a ledge then move right onto the wall and ascend to the corner. Climb a good crack leftward until it is possible to move up to a small foot ledge beneath an overlap. Surmount this and head straight up to the ledge before traversing left easily and finishing up the groove to the terrace.

2+3 55m 4c From the middle of the terrace, climb the wall above, then continue straight to easy ground below the Second Tower.

6 Bounty Hunter ★ 60m E2 5b
FA Guy Robertson & Pete Benson 7 September 2007

The fine crack system right of *The Empty Quarter*.

1 35m 5b Climb the crack over an overhang, then trend left past another. Continue directly up the wall above to a ledge and belay below a tower.

2 25m 5b Start up a groove in the right arête, swing left onto the front face at an obvious break, then climb directly to easier ground which leads to a platform. From here either abseil off or scramble up left to the col behind the First Tower.

THE SLOCHD WALL (THE PIT)

7 Slochd Wall ★★★ 110m HVS 5a

FA Mike Rennie & Greg Strange (A3) 30 August 1969;
FFA Brian Lawrie & Andy Nisbet 3 July 1979

Excellent clean climbing taking the large vertical corner above the left end of the roof system. Start at the foot of North-West Gully.

1 15m 4a Climb the gully to the first depression. Traverse onto the wall on the left via a grass shelf to an obvious belay stance.

2 30m 5a Go up a crack rightwards for 5m then follow the slab slightly leftwards at first, then heading up to and following the left edge of a hanging corner to reach a guarding overlap. Move up and right into another hanging corner and follow this for 8m to a belay ledge on the right.

3 15m 5a Step back left into the corner and follow it leftwards under a roof (usually wet) to gain the continuation corner, follow this for 10m to a belay ledge on the right.

4 20m 5a Go up the rib at the left end of the ledge for a short way to swing left to regain the corner. Climb this, bypassing the overhang on the left. Trend diagonally left to an arête and belay just above.

5 30m Continue up the edge to finish by a crack.

8 Freebird ★★★ 100m E4 6a

FA Guy Robertson & Pete Benson 7 September 2007

Brilliant climbing up the right arête of *Slochd Wall* corner; start as for that route.

1 30m 6a Climb *Slochd Wall* to just past the bolt, step right under the obvious slot and make hard moves up right to a rest on the

arête. Continue directly through bulges and up a slab (crux), to belay just left of overhangs.

2 30m 6a Pull right through the overhangs into a hanging groove, step right again, then go back left and climb cracks up the wall in a superb position. Belay at the right end of the second big ledge.

3 40m 5a Climb cracks in the right side of the upper wall.

9 The Primate ★★ 110m E1 5b

FA John Anderson & Andy Nisbet 9 July 1979;
Pitch 1 Greg Strange & R.Ross 8 July 1984

The wide crack splitting the right side of the roof system. A fine companion to *Slochd Wall*. Start as for that route.

1 15m As for *Slochd Wall*.

2 25m 4c Follow the obvious right-slanting crack to gain cracks which lead to a ledge below and right of the roof crack.

3 25m 5b Traverse left and follow the crack with a *"fascinating sequence of moves"* through the roof. Continue up the crack to a grass terrace beneath the final wall.

4 45m Climb short walls to a left-slanting groove in the headwall, then the groove leading almost to the left edge to finish up a crack.

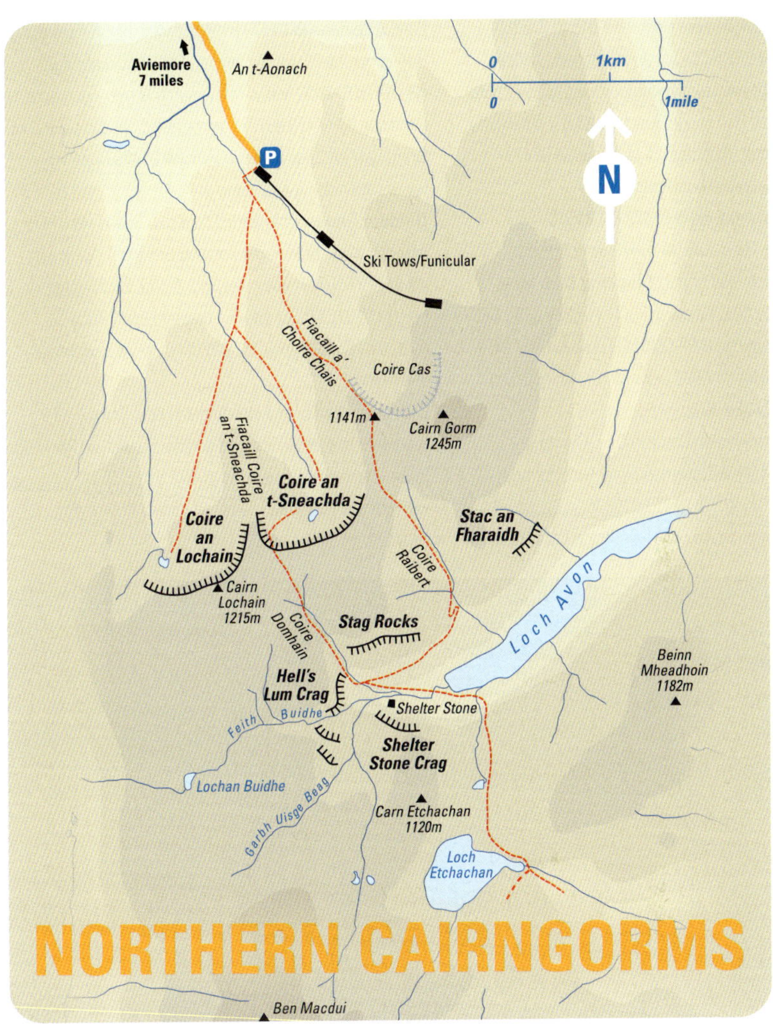

Amenities & Accommodation: Aviemore (the Las Vegas of the Highlands!) is the most convenient base though it has all the charm of a new town designed by wee men in suits sitting at a drawing board. There are a multitude of eating and drinking establishments here, though much better towns can be found in all directions (anywhere away from Aviemore!). There is also a bar in Glenmore Lodge (☎ 01479 861256; www.glenmorelodge.org.uk) where a variety of self-catering chalets and B&B accommodation are also available. There is a large Tesco supermarket in the centre of Aviemore and an Aldi supermarket in the retail park 0.3mile/0.5km further on, heading north through town. **TIC** in Aviemore (☎ 01479 810930; www.visitscotland.com).

Bunkhouses: Newtonmore Hostel (☎ 01540 673360; www.newtonmorehostel.co.uk) and Craigower Lodge (☎ 01540 210000; www.activeoutdoorpursuits.com) both Newtonmore; Insh Hall Lodge, Kincraig (☎ 01540 651272; www.lochinsh.com); Aviemore Bunkhouse (☎ 01479 811181; www.aviemore-bunkhouse.com).

Youth Hostels: Aviemore (☎ 01479 810345) & Cairngorm Lodge, Glenmore (☎ 01479 861238) both www.hostellingscotland.org.uk. **Climbing Club huts:** Milehouse, 1 mile from Kincraig (LSCC; NH 839 043); Mill Cottage in Feshiebridge (Mill Cottage Fund; NH 847 047); and the Raeburn Hut on A889 between Dalwhinnie and Laggan (SMC; NN 636 909).

Campsites: Invernahavon Caravan Site, Glentruim (☎ 01540 673534; www.invernahavon.com); Dalraddy Holiday Park, Loch Insh (☎ 01479 810330; www.campingaviemore.co.uk); High Range Caravan Park, Aviemore (☎ 01479 810636; www.highrange.co.uk); Rothiemurchus Camp & Caravan Park, Coylumbridge (☎ 01479 812800; www.rothiemurchus.net); Glenmore Campsite (☎ 01479 861271; www.glenmore-campsite.com); Grantown-on-Spey Caravan Park (☎ 01479 872474; www.caravanscotland.com).

Outdoor Shops: Cairngorm Mountain Sports (☎ 01479 810903; www.braemarmountainsports.com); Nevisport (☎ 01479 810227; www.nevisport.com); Ellis Brigham (☎ 01479 810175; www.ellis-brigham.com); Tiso, Aviemore Retail Park (01479 788840; www.tiso.com) all in Aviemore.

Climbing Walls: small leading and bouldering wall in Glenmore Lodge; leading and bouldering wall at The Ledge, Telford Retail Park, Inverness (07873 818 868; www.theledgeclimbing.com) including a fitness gym and café.

Prore (page 446), Coire an Lochain. Scott Muir climbing. Photo Dave Cuthbertson, Cubby Images

NORTHERN CORRIES OF CAIRN GORM STRATH SPEY

Clearly visible from the A9 are three great corries scooped into the northern slopes of Cairn Gorm. The east most is Coire Cas, of interest purely to the skier. Further west lie the two most accessible cliffs in the Cairngorms, Coire an t-Sneachda and Coire an Lochain.

Due to the large amounts of broken rock in the corries they are better known as superb and very popular winter venues. Nevertheless, they are of some interest to the rock climber with a few particular routes as good as anything of their grade in the area.

COIRE AN T-SNEACHDA
(CORRIE OF THE SNOW)

The central of the three great northern corries (korin'tray-achk) lies midway between the ridges of the Fiacaill a' Choire Chais and the Fiacaill Coire an t-Sneachda.

Approach: From the Coire Cas car park, cut down right past the bottom T-bar to follow a signposted path heading diagonally right (south-west) then head up the left (east) side of the Allt Coire an t-Sneachda (stream). Continue along the path up into the corrie.

Descent: By the Goat Track, a well-worn path right of the main mass of the crag. This follows a zig-zagging line, steeply at first, then heads diagonally right (east) under the cliffs.

ALADDIN'S BUTTRESS

The large buttress left of the centre of the corrie, bounded on its left side by the scree-filled zig-zag gully of **Aladdin's Couloir.**

1 The Magic Crack ★★★★ **100m HVS 5a**

FA Greg Strange, Mungo Ross, Jim Wyness & Dougie Dinwoodie 16 May 1981; Start as described: Allen Fyffe & Martin Bagness July 1984

A contender for the best HVS in the Gorms, with a superb final pitch up an immaculate finger crack. Unfortunately winter ascents since the late nineties have greatly scarred the route. Start at a deep left-facing corner by a huge beak of rock.

1 **35m 4c** Follow the corner then the broad blunt rib above to large spike belay on platform.

2 **25m 4c** Follow the rib to gain the thin crack on the right, which leads up into corners.

3 **40m 5a** Move right and follow the immaculate finger crack. Higher up cross an overlap and finish up the cracked wall leading to easier ground. It is also possible to abseil (slings in situ) back down the line in 2 rope lengths.

2 Damnation ★   90m E1 5b

FA Dave Sharp & Bob Taplin (1 PA) 1969; FFA unknown

A good main pitch up the rightmost corner. Start up from the lowest rocks.

1 **45m** Follow easy cracked slabs and corners to a huge spike belay at the base of a pale corner.

2 **45m 5b** Climb the initial corner then the main one to finish up the cracked wall on the left.

3 Pygmy Ridge ★★   90m Moderate

FA Harold Raeburn, William Garden, George Almand, A.Roth 1 April 1904

High up in the centre of the corrie is a prominent ridge. Either scramble up much broken ground to the base of the route, or gain the base by scrambling or abseiling down its right (west) side from the corrie rim. Follow the well defined rib, crossing a horizontal rib.

FLUTED BUTTRESS

NH 992 029 **Alt:** 1080m

The broken buttress high on the right side of the corrie overlooking the Goat Track.

4 Fingers Ridge Direct ★   110m VS 4c

FA Allen Fyffe & Martin Bagness 9 July 1984

This is the direct line up the slabby rib culminating in some distinctive pinnacles just short of the plateau. Pleasant climbing though escapable in places. Start in the middle of the ridge.

1 **35m 4b** Climb directly up pink slabs, cross an awkward bulge and belay in the open groove.

2 **20m 4b** Work up and right to a stance by Red Gully.

3 **30m 4c** Follow the diagonal crack in the fine slab to ledges.

4 **25m** Go up the next diagonal crack to join the normal route.

Scott Muir and Jo George on the fantastic final pitch of The Magic Crack. Photo Dave Cuthbertson, Cubby Images

The westmost of the three great northern corries lies west of the Fiacaill Coire an t-Sneachda, beneath the summit of Cairn Lochan, the most westerly of the Cairn Gorm tops.

Approach: From the Coire Cas car park, cut down right past the bottom T-bar to follow a path heading diagonally right (south-west) to cross the Allt Coire an t-Sneachda (stream) after about 10 minutes. Continue along the path, branching off left after 75m, to follow a narrow path up into the coire, skirting round the right side of the boulder field to arrive at the lochan. Head steeply up the hillside to the base of the chosen cliff.

The buttresses are numbered left to right.

Descent: Down either end of the coire. For the western descent, from the top of *Savage Slit* continue west along the rim of the coire for 400m, then down gravelly ground, skirting back round on grass right (east) to the base of **No. 4 Buttress**.

NO.1 BUTTRESS

NH 985 027 **Alt:** 1030m  1hr 30min

The buttress on the left side of the corrie, with a steep right side rising out of The Vent, a short well defined gully funnelling into easy angled open ground above.

1 **Ventriloquist** ★★　　　　　　　　**80m HVS 5a**

FA John Lyall & Ben Kellet 2 August 1990

A scrappy start but improves with height, taking cracks near the left side of the front face. Start beneath a chimney crack 5m left of *Ventricle*.

1 **30m 4c** Climb the chimney crack then move right and follow a crack-line up into a recess. Pull out right to beneath a wide crack.

2 **20m 5a** Climb the crack with difficulty then trend up right by two short corners to a ledge.

3 **30m 4c** Move up left to a line of flakes and traverse right to a thin crack. Move up the crack, bypassing the steepest section by a short detour on the right. Above, cross the 'crevasse', finishing up a deep crack.

② Cardiac Arête ★★   85m HVS 5a

FA Lord John Mackenzie & Andy Nisbet 13 July 2006

"Brilliant and varied climbing." The arête is impressive but has good holds and protection. Start as for *Ventricle*.

1 **20m 5a** Climb the overhanging crack of *Ventricle*, then right a short way. Move left to below an overhanging groove. Continue easily left to a smaller groove.

2 **20m 5a** Climb this groove for 5m, then make an impressive but easy traverse right to the upper section of the groove. Climb this to the big sloping rock ledge. Move right to the arête.

3 **35m 5a** Step round the arête into the left of two shallow grooves (*Ventricle* climbs the right one). Climb this to near its top, then step back on to the arête. Climb immediately right of the arête, then the arête itself to a ledge. Make a move up the wall 3m left of the arête and traverse back to the arête. Go up to a bulge again passed 3m on the left followed by an immediate return to the arête. The bulge direct is 5b. Climb the arête to a ledge.

4 **10m 4a** Finish up the wall above, as for *Ventricle*.

③ Ventricle ★★   95m E1 5b

FA John Cunningham & George Shields (1 PA) summer 1968; FFA unknown

The line of cracks and grooves on the left wall of The Vent. Start near where the face changes aspect.

1 **15m 5b** Climb the initial overhanging crack (crux), move right along a ledge and climb a wall near

📷 *Gary Latter starting up the finely positioned Prore.*
Photo Karen Latter

the right edge to belay in a small mossy recess.

2 **15m 5b** Climb directly above the belay heading for prominent groove, with an awkward move to gain the belay ledge.

3 **20m 4c** Follow the right most of the two shallow grooves to reach a steep groove. Climb this until a traverse can be made to a block belay at the top of a groove.

4 **45m 4b** Follow the wide crack in the groove, finishing up the steep wall to ledges.

④ Daddy Longlegs ★★   70m HVS 5a

FA George Shields & Brian Hall (11 PA) 25 August 1968; FFA Brian Davison & Andy Nisbet 29 August 1983

The steep groove and cracks high on the right side of the buttress. Start 15m beneath the chokestone in The Vent.

1 **35m 5a** Climb the groove, step right into another groove and up this past an overhang to ledges.

2 **10m** Scramble up left.

3 **25m 4c** Follow two vertical cracks in the wall right of the wide corner crack of *Ventricle*.

NO. 4 BUTTRESS

1a Direct Finish ★★★   25m HVS 5a

FA Michael Barnard & John MacLeod 27 August 2016

A finely positioned and generally well protected pitch. Take a full set of cams up to 4". Where the normal route goes left through the twin roofs, instead climb steeply up the top wall, moving left to use the left edge.

2 Savage Slit ★★★   75m Severe 4a

FA Richard Frere & Kenneth Robertson 17 July 1945

"It appeared to be very deeply cut in places, and to penetrate into the depths of the mountain." – Frere.
A compelling line tackling the unmistakable wide chimney slot in the huge left-facing corner up the centre of the buttress. Start directly below the main corner.

1 **10m** Climb easily up to the base of the corner.
2 **28m 4a** The corner to ledge at base of the wider chimney.
3 **12m 4a** Continue to good ledge at top of the chimney.
4 **25m** Finish up the gully above, with a short steep rock step at the top to gain easy ground.

NH 9836 0257 **Alt:** 1125m   1hr 30min

Bounding the right edge of the corrie, down and right of the two branches of Y Gully is a fine buttress with a number of excellent prominent corner lines.

Approach: Follow a path heading steeply up right, then back left under the base of the buttress.

Descent: Abseil 50m down line of 3 *Prore*, or 55m directly down 4 *Fallout Corner* (free hanging in lower section).

1 Bulgy ★★   60m VS 4b

FA George Shields & Robert Doig (1 PA) 21 August 1968

The left arête of *Savage Slit* finishing up the wide crack through the prominent twin roofs.

1 **24m 4b** Climb easily to the base of the arête, then move up onto the arête, then a shallow groove to belay in the corner beneath a steepening.
2 **28m 4b** Move up right and climb the corner to prominent wide crack through the upper roof. Pull left round this and up easier ground to large block belay.
3 **8m –** Traverse rightwards and down to gain the abseil point on *Savage Slit*.

Tom Challands & Craig Wilson on Fall-out Corner (page 446).

③ Prore ★★　　　　　　　　　　**80m VS 4c**

FA G.Bradshaw & Bob Taplin 5 July 1969

Excellently situated climbing up the prominent curving right arête of 2, starting and finishing up that route.

1　**10m** Climb easily up to the base of the corner.

2　**45m 4c** Climb the right side of the arête to belay on large ledge at the top.

3　**25m** Finish as for *Savage Slit*.

④ Fall-Out Corner ★★★　　　　　**90m VS 4b**

FA Tom Patey, Robin Ford & Mary Stewart (1 PA) 17 May 1964

Well protected and low in the grade. On the right side of the pillar is an equally fine right-facing corner – the line. Start beneath the corner.

1　**8m** Go up to belay beneath the roof blocking the corner.

2　**28m 4b** Cross the roof and climb the excellent sustained corner to a good ledge on the right.

3　**24m 4a** Continue up the corner, to belay on block strewn terrace.

4　**30m** Drop down into the 'crevasse' and move left, finishing up the final pitch of *Savage Slit*.

THE LOCH AVON BASIN

Loch Avon (pronounced 'aan) lies in a deep 300m hollow between Cairn Gorm and Ben Macdui. The western end of the loch is overlooked by a variety of contrasting cliffs, dominated by the imposing **Shelter Stone Crag** at the head of the corrie. **Hell's Lum Crag** sits glistening opposite while further east not far below the Cairn Gorm plateau are **Stag Rocks** and **Stac an Fharaidh**.

Access: From the A9 take the turn-off just south of Aviemore and follow the B9152 north (parallel to the A9) for 1 mile/1.6km to turn off right along the B970 (on the southern fringes of the town) which passes through the Glen More Forest Park before climbing steeply up to the large extensive car park (NH 990 060) 630m/2000 feet up the northern slopes of Cairn Gorm (7 miles /11km).

Approach: From the Coire Cas car park at the funicular follow the track up the left side of the railway passing underneath it at the first station. Continue up the wide vehicle track for about 10 minutes past a couple of zig-zags to gain a well constructed path (marked Fiacaill path) leading up onto the ridge, Fiacaill a' Choire Chais (tooth of steep corrie). One hour of ascent to cairn at top of ridge (spot height 1141).

Accommodation: Clach Dhion (stone of the shelter), a renowned howff is situated low down in the corrie at the base of the boulder field beneath the **Shelter Stone Crag**, near the stream at NJ 002 016. There is a cairn on top. Less popular and not quite so claustrophobic howffs exist further up the boulder field. There are also many superb camping spots by the stream and around the head of the loch and on some beautiful golden sandy beaches along the shores of the loch.

Bouldering: There is some excellent bouldering around the base of the boulder field, including a finely situated long 6a arête above a pool just above the path.

HELL'S LUM CRAG  NH 995 017 **Alt:** 920m 1hr 30min

The extensive crags on the south-east flank of Cairn Lochan at the head of the Loch Avon basin, sandwiched between the streams of the Feith Bhuidhe and the Allt Coire Domhain. Many of the routes on **The Frontal Face** take a lot of drainage, though many of the other lines dry quickly.

Approach: From top of Fiacaill a' Choire Chais (spot height 1141) follow a path on the right which contours initially south then south-west across the hillside (not marked on 1:25,000 map) for a little less than 2km to gain a well worn path down the east side of the Allt Coire Domhain. The cliff comes into view on the right just beneath the steepest section of the path. Cross the burn level with the base of the cliff, just beneath a large boulder on the path.

THE GREY BUTTRESS

A very quick drying buttress due to its detached nature lying between **Hell's Lum** and *Deep-Cut Chimney*. The front face is steeper lower down, lying back in the upper section where the lines become less defined.

1 Chariots of Fire ★★   45m E4 6a

FA John Lyall, Andy Nisbet & Jonathan Preston 4 August 1991

A sensational route with flaky rock adding to the excitement. Although overhanging all the way the crux is short and well protected. Gain the start by a 40m abseil from the 'viewing block', a large flat rock from where one can lie and peer into The Lum.

1 **15m 5b** Climb out rightward into a shallow corner, actually a big flake. Follow this briefly then go rightward again into another corner. Go up this to belay below a roof.

2 **15m 6a** Make a high traverse left onto the overhanging wall then pull up to good holds (crux). Traverse left to a spike then go up to a line of flakes leading leftward to a small ledge.

3 **15m 5a** Continue up the flakes then straight up over the bulge to the viewing block.

2 The Seventh Circle ★★★   50m E6 6b

FA Iain Small & Gary Latter 10 August 2007

The stunning slim groove and hanging crack in the wall just below the main pitch in The Lum. Gain the base by a 60m abseil from 8m left of 1 abseil point (starting down front face).

1 **23m 6b** Climb the groove, moving out left then back into the groove, past a PR. Pull over small overlap, stepping left to a good rest. Return and make difficult moves up and right to get established in the crack, which leads with difficulty to good ledge.

Iain Small on the first ascent of The Seventh Circle.

2 27m 6b Traverse left into good flake crack and follow this past projecting block to rest on wall on right. Step left and pull over the roof with difficulty leading to better holds above. Continue more easily up crack to good ledge above. Scramble out rightwards to finish.

3 Good Intentions ★★ — 125m VS 5b

FA Bill March & M.McArthur (1PA) 12 September 1969

Finely situated and very quick drying; very well protected on the harder sections.

1 25m 4a Ascend green rock over a wedged block to a ledge below and right of the ramp.

2 25m 4c Move diagonally left through a bulging recess, then follow left-slanting ramp to a corner.

3 10m 5b Climb the corner, hard to start, to a ledge.

4 35m 4a Move diagonally left to a ledge below a wall with a prominent V-groove overlooking the Lum.

5 30m Climb the groove past blocks at the top.

4 The Exorcist ★★ — 100m E1 5b

FA Ado Liddell & Bob Smith June 1975

An excellent crux pitch. Start at an isolated right-facing corner, about 10m left of *Hell's Lump*.

1 20m Climb corner to grassy ledges, then scramble up left to beneath an overhung recess.

2 35m 5b Climb wide crack in the recess, then head diagonally right to gain the corner via a shallow groove. Climb the corner, which has a prominent crack in its right wall, then break right to a ledge on the rib.

3 45m The rib above to easy ground.

5 Evil Spirits ★★ — 110m E2 5c

FA Allen Fyffe & Bob Barton 26 June 1986

The left of a pair of right-slanting ramps. Start just right of *The Exorcist*.

1 15m 5a Climb green rock just right of the corner to a horizontal crack, go right to a slim corner then continue to the large flake.

2 35m 5c Climb into a deep crack, as for *Hell's Lump*, then directly into the corner above. Climb this to the

roof, traverse left, then go up the next corner which leads to a ledge above the corner on *The Exorcist*.

3 15m 5a Go diagonally left, then descend to obvious overlap and pink rock. Follow this diagonally left to its end and go up to belay above the crux corner of *Good Intentions*.

4 45m As for *Good Intentions* - climb the rib above to easy ground.

6 Hell's Lump ★★ — 100m VS 4c

FA Jimmy Marshall & James Stenhouse September 1961

The rightmost of twin right-slanting ramps. Start just left of the left-facing corner.

1 40m 4c Climb a narrow dyke up the slab to gain a large flake at 15m. Continue by the groove and crack above for 12m. Move up rightward to turn the square roof by a layback crack on the right and gain a grassy bay.

2 30m 4a Move up left to the recess. Follow cracks and corners slightly rightwards to below a wall adjoining *Deep Cut Chimney*.

3 30m 4c Finish up the wall via a tricky central bulge.

7 Devil's Advocate ★★ — 50m E4 6a

FA Jules Lines & Tang Hui Li 23 June 1999

A slightly contrived line but with superb climbing and amazingly quick drying. Start at a thin vertical crack in a slim green buttress just left of the obvious *Direct Start* to *Deep-Cut Chimney*. Climb the crack to a small foot ledge at 4m. Step up and make a very thin move up and left to better holds. Place a runner in the crack above, step down and move right onto a slab. Make awkward moves up and slightly left to below twin hairline cracks. Climb between these (marginal RPs) to the top. F #3 for belay.

THE FRONTAL FACE

The main highest section of the crag, gradually steepening as height is gained.

8   **Deep Cut Chimney ★**   **150m Very Difficult**

FA Ian Brooker & Marie Newbigging September 1950

A good climb up the impressive slit with a spectacular
and unexpected finish. Though vegetated in its lower
part the walls are close enough to back and foot and
avoid the greenery. Start either directly below or by the
easy terraced fault cleaving the smooth lower slabs.
Once in the chimney proper there are a number of
pitches to be overcome, mostly fern and grass-grown,
hence climb them back and foot leaving the herbage
untouched. About 50m from the top the rock scenery
becomes quite remarkable; the chimney cuts far into the
cliff and chokestones are jammed well out between the
walls forming a tunnel. Back and foot outwards below
the final overhang to reach a crazy pile of boulders
wedged in the outer jaws. The finish comes with
startling suddenness. *"…the final pitch looked like a*
dead end. To save us all going up the green, dark, greasy,
wet, sodden, mossy, moist, slippery, scary, rory blumner
piece of sugar honey ice tea that the last pitch was, I
escaped into the daylight and climbed the last pitch on
real life dry rock (novel!)."

9   **Salamander ★★★**   **155m HVS 5a**

FA Dougie Dinwoodie & John Tweddle 18 September 1971

A really good route, with a particularly good, though bold,
crux pitch. Start below an obvious right-facing corner.

1   **40m 4b** Climb directly and delicately up the
corner and follow this to a ledge. Continue
upwards passing a prominent overlap by a
bulging slab on the left to reach a platform.

2   **20m** Follow easier rocks to a glacis.

3   **20m 5a** Climb twin cracks up steep slabs
left of an obvious corner to a platform.

4 20m 4b Climb directly by a black bulge just right of a short corner and go up slabs trending rightward to an obvious shallow scoop.

5 30m 4b Continue up slabs to reach a shallow groove. Follow this to break through the upper overhangs by a striking chimney slit. Thread belay.

6 25m Continue direct through an open funnel to easy ground.

10 The Bats, The Bats ★★   50m HVS 5a

FA Blair & Allen Fyffe 14 August 2003

A good pitch on immaculate rock following the crack lines in the clean pink slab. Climb the clean crack into the short shallow right-facing corner. From the top of the corner trend left to gain the prominent crack and climb it to where it disappears then move right to a square nose. From the right side of the nose follow another crack to the fault. Descend down the fault.

11 Hellfire Corner ★★   175m VS 4b

FA George Annand & Ronnie Sellars 14 September 1958

Good climbing up the large corner to the left of *The Clean Sweep*. Slow to dry. Start at a crack at the lowest point of the slab.

1 50m 4a Climb the crack and left-facing corner leading to the large diagonal fault.

2 45m 4a Ascend the short deep left-facing corner leading up into the main corner system which is followed to a belay beneath a large overhang.

3 20m 4b Continue in the same line up a depression and corner to an awkward move onto a platform beneath the main corner.

4 20m 4b Continue up the steepening corner, chimneying through the overlap in a fine position to move out right to a large ledge.

5 40m Finish more easily up the right-facing corner left of the big fault above.

12 The Clean Sweep ★★★   185m VS 4c

FA Robin Smith & Graham Tiso September 1961

The crag classic. Start down and left of the prominent left-slanting diagonal fault leading into *Deep-Cut Chimney*

beneath a left-facing corner at the right side of the slab.

1 30m 4c Follow the corner for a short way to gain a groove. Climb this to gain cracks on the crest of the green whale-back leading to a belay beneath the fault.

2 45m 4a Climb the slab above the fault moving left and up corners to a huge block below the pink corner.

3 45m 4c Climb the superb corner then the continuation fault to a large ledge on the right underneath an overhang. A fine sustained pitch – very well protected.

4 40m 4b Continue up the round grey edge by cracks, corners and bulges.

5 25m – Continue more easily above.

13 Prince of Darkness ★★   150m E1 5b

FA Bob Barton & Allen Fyffe 10 August 1984

Good climbing up the big red slab in the centre of the cliff. Start just right of the base of the diagonal fault.

1 25m 4a Climb the crack in a green slab to grooves. Belay below an obvious single crack splitting the slab.

2 45m 4b Follow cracks and corners to below the prominent red slab.

3 30m 4c Climb cracks in the big red slab to a good ledge.

4 20m 5b Climb the right-facing corner with overlaps to a ledge on the left. An excellent pitch.

5 30m 4c Return right into the continuation corner and crack leading to the top.

14 Auld Nick ★★★   160m HS 4b

FA George Brown, Ian Houston & Ian Small 11 October 1963

Excellent sustained climbing up the right side of the slabs. Start at a prominent crack in the slab 20m down left from *The Escalator* depression.

1 45m 4a Follow the crack to belay beneath the first overlap.

2 45m 4b Continue up the right-facing corner, then the crack in the slab to the next bulge. Move right along horizontal crack then diagonally up right.

3 20m 4b Ascend the left-facing corner over a series of steps to belay beneath a right-tapering roof.

then make a hard move to easier ground.
An easier rib now leads to the top.

4 30m 4a Pass the roof on the right then continue leftward, passing a block to a large ledge. Continue up the thin crack above the centre of the ledge to belay beneath large grey block.

5 20m Climb either side of the block then scramble to the top.

15 The Devil's Alternative ★★★ 160m E1 5a

FA Allen Fyffe & Bob Barton 1 August 1981

Interesting climbing on excellent rock following a direct line up the right side of the slabs. Start at a greenish buttress just left of the depression of *The Escalator*.

1 45m 4b Climb the buttress by shallow cracks to a huge terrace.

2 45m 5a Twin cracks rise above the right side of the overlap. Gain these from a scoop on their right and climb them to the next overlap (*Auld Nick* crosses here). Work left across the overlap to the glacis.

3 25m 5a Above is a stepped wall. Zig-zag up this to gain a short left-leaning corner, above which moves up and right lead to the next glacis.

4 45m 5a Climb into a niche in the grey wall above, go leftward to a horizontal crack

16 The Escalator ★★ 150m Moderate

FA Graeme Nicol, Tom Patey & Miss E.Davidson 30 September 1955

The prominent fault towards the right end of the cliffs. Clean sound rock but often a watercourse. Easy scrambling up a gully leads to a large platform below the steeper section where an easy shelf leads off to the right. Pleasant climbing up pink water-worn rock leads to a finish just left of the top of the watercourse.

17 Sneer ★ 120m Very Difficult

FA Alasdair 'Bugs' McKeith (solo) 27 June 1963

A varied route on very clean rock up the overlapping slab at the right side of the cliff. Climb the obvious large open corner about 30m right of *The Escalator* to easier rock. Go up rightward by easy grooves finishing up the right of two obvious cracks through three overlaps on the often wet upper slabs.

18 Highway to Hell ★★ 20m E2 5b

FA Jules Lines (on-sight solo) August 1995

The slabby right-angled arête at the far right end of the cliff, climbed boldly on its right side.

STAG ROCKS  1hr 45min

This is the collective name for the long face of dark rough granite crags on the southern slopes of Cairn Gorm. Quick drying, the cliffs rise in two sections separated by the open scree-filled Diagonal Gully.

Approach: From top of Fiacaill a' Choire Chais (spot height 1141), head down the path close to the left (east) bank of the Allt Coire Raibert and traverse right (west) to the base of the crag. In clear weather the prominent Y-shaped gully (NH 000 022) can be descended to gain the base of *Afterthought Arête*. The long open Diagonal Gully (with scree chute almost reaching the loch — NH 001 022) gives the fastest approach to **The Right Section**, though care is required especially on the steep grassy upper section. Alternatively, for **The Left Section**, follow the **Hell's Lum** approach to pick up a vague path contouring left underneath the steep clean **Cascade Wall**. Cross a couple of wet shelves then regain the path which leads across the Y-shaped Gully to the base of *Afterthought Arête*. 35 – 40 minutes from ·1141.
Descent: As for the approach.

THE LEFT SECTION   NH 001 021 **Alt:** 950m 

1 Afterthought Arête ★★★   180m Moderate
FA Ronnie Sellars & Mac Smith September 1956

The excellent exposed ridge defining the left edge of **Stag Rocks**, bounding the right side of the wide scree-filled Y-shaped gully. Avoid the initial steep start (Severe 4a) by moving in from the right further up the slope. *"After this the rocks develop into an excellent, steep knife-edge for 75m and maintain their interest to the top."* Climbed in three full rope lengths. Finish more easily over broken boulders.

2 Serrated Rib ★   150m Moderate
FA Jock Nimlin (solo) July 1930

The rib on the right overlooking Diagonal Gully. Climbed in three rope lengths, the middle pitch mainly heather.

THE RIGHT SECTION 

NJ 001 022 **Alt:** 950m

3 **Final Selection** ★★★　　　　　　**50m Difficult**

FA Ronnie Sellars & Mac Smith November 1956

The last defined arête near the top of Diagonal Gully
gives a very fine route. The right side of the arête
is steep, the left a cracked slab angling into a large
right-facing groove.

1　**40m** Start up the large groove then break out right
　　onto the arête. Continue up the edge and cracks
　　on the left of the edge to a platform on the right
　　just below the level of a prominent overhang.

2　**10m** Finish easily up the corner on the right.

The arête can be followed more directly by stepping
right from the base of the initial groove, finishing
steeply on good holds over the small overlap above
the belay ledge at Severe 4a.

4 **Purge** ★　　　　　　**55m Very Difficult**

FA B & D.Taplin 7 August 1969

The groove down to the right of *Final Selection*.

1　**40m** Climb up into the groove and follow it to
　　belay at the right end of the ledge at its top.

2　**15m** Move slightly right then trend left to finish up
　　the edge of the final easy groove on *Final Selection*.

*Colleen Maclellan pulling through the
crux bulge on Final Selection Direct.*

STAC AN FHARAIDH
(PRECIPICE OF THE LADDER)

NJ 014 029 **Alt:** 900m   1hr 40min

These glaciated slabs lie on the southern flank of Cairn Gorm overlooking Loch Avon. The slabs are divided into an **East** and **West Flank** by a broad gully and are broken by overlaps or steep walls which provide the cruxes of most of the routes. On the west flank the top overlaps extend into a steep wall providing climbing of a different character.

Approach: From the top of the Fiacaill a' Choire Chais head south-east skirting round the east side of .1082 and down the west (right) bank of the Feith Bhuidhe (yellow bog-stream) that drains south-east into the loch. The best line lies about 50m right (west) of the stream, involves some easy-angled scrambling and is almost always very wet and slippery. Continue down grass slopes to skirt round beneath the base of the crag. A slightly longer approach is to descend easier slopes 300m east (left) of the Feith Bhuidhe then contour back west below some broken rocks to the crag, re-crossing the stream just before the crag.

Descent: As for the approach described above.

EAST FLANK

1 Whispers ★★   130m VS 4c

FA John Cunningham & George Shields Summer 1969

On the extreme left of the right slab are two cracks – the right is the route. Very quick drying – one of the first routes to come into condition early in the spring.

1 **45m 4a** Climb the second crack system from the left. Go straight up cracks (heathery, sometimes wet) to belay at an easing in the angle.

2 **40m 4b** Continue up the same crack system, at one point quite near left edge and up to belay at the base of huge triangular flake.

3 **45m 4c** Move up to the top of the flake then traverse right past PR for 6m. Move up rightward through a bulge and make thin moves onto the slab above. Climb past a flake-crack onto the ledge above then traverse diagonally leftward and straight up to belay on a grass ledge.

2 Bellows ★★   140m HVS 5a

FA Rab Carrington & Jimmy Marshall 5 July 1970

Good climbing up a crack-line 12m right of and parallel to *Whispers*. Start directly beneath the prominent right-facing corner beneath the first overlap system.

1 **25m** Climb easy vegetated crack to belay at the base of a steep grassy corner.

2 **40m 4c** Climb the corner then step right to a good foot hold on the lip. Move right into the crack system and follow this up the slab then an easy enormous flat scoop to belay in the short left-facing corner crack below the overlap.

3 **35m 5a** Climb a series of sloping ledges up leftward to a crack then move up and left over an overlap and follow cracks to belay at the base of the huge triangular flake.

4 **40m 4c** Continue up the wide crack to a PR then traverse right. Move diagonally right through the crux bulge of *Whispers*. Continue up past a ledge then boldly up a thin rounded crack finishing more easily trending left.

3 Pushover ★★   140m HVS 5a

FA John Cunningham & Glenmore Lodge party Summer 1969

A good line up the centre of the slab though slow to dry. Start at the crack at the left side of the large boulder.

1 **45m 4c** Follow the sustained crack taking the right fork of the Y at 10m to beneath the bow-shaped overlap. Sustained and poorly protected.

2 **12m 5a** Pull onto the overlap and move left until the upper slab can be gained. Move up this to a ledge.

3 **40m 4c** Trend up and leftward to a steeper wall which is climbed by a series of cracks to below mossy blocks.

4 **43m 4b** Climb over the blocks and up slabs to a chimney to finish up the wall above.

4 Rockover ★★   130m E1 5b

FA Ron Kenyon & B.Cosby 23 July 1994

A good quick drying route.

1 **45m 5b** Climb the slab, bold, to gain an obvious
   groove line and follow this to near the left end of
   the initial left-slanting crack of pitch 2 of *Pipit Slab*.

2 **25m 5b** Move up, then right onto the slab above
   (as for *Pipit Slab*), traverse halfway along the
   slab and reach over the overlap to gain a faint
   ledge. Rockover to gain the slab (crux). Move
   up then left to a mucky groove, and pass this by
   the slab on the left to gain a belay just above.

3 **40m 4c** Continue right up a slab to an overlap
   with an obvious crack above. Climb the crack
   to a ledge. Climb up on the right and move
   left slightly to a belay. A bold pitch.

4 **20m 4a** Continue up to the right
   to finish as for *Pipit Slab*.

5 Pipit Slab ★★   135m Severe 4b

FA John Cunningham & Bill March 14 June 1970;
Direct Andy Nisbet & party 1987

The right crack-line. Fine climbing though quite run-out at
times. First 5m is the crux. Start 3m right of the boulder.

1 **45m 4b** Climb either of the twin grassy cracks
   which converge at 15m crack then continue in
   the same line to a ledge and chokestone belays.

2 **30m 4a** Trend left up a easy crack to a grassy ledge
   below a wall. Now step up right and traverse right
   along a ledge to regain the line (can be climbed direct
   up the wide crack at HS 4b). Continue up thin slab
   bearing slightly left to small stance and nut belay.

3 **40m** Climb directly up a shallow corner above some
   overlaps to belay in a small wet recess. This pitch
   is slow to dry but a faster drying *Direct Variation*,
   HVS 4c ascends the smooth slab forming a vague
   rib just left of the corner above the overlaps.
   Perfect rock but very sparsely protected.

4 **20m** Up a recess onto a slab, which leads to short
   overhanging wall. Climb this to the top of the crag.

6 Linden ★★   70m Severe 4a

FA Bill March, L.Rae & S.Matthewson 4 June 1970

An excellent little route up the prominent sloping ramp
high on the right side of the slab. Start in the large
grassy bay on the right where a narrow grass strip runs
up leftward to a wide right-facing flake crack.

1 **25m 4a** Climb a short crack 6m right of the corner,
   move left then follow thin flakes up the blunt arête
   (poorly protected). Step right past a large pocket
   then up to belay in cracks at the base of the ramp.

2 **45m 4a** Climb the ramp then the flake crack
   and a short steep corner to a glacis. Move
   out right up a slab to a P and nut belay at the
   right edge of the slabs. Scrambling remains.

SHELTER STONE CRAG

NJ 002 013 **Alt:** 850m  2hr

With a series of excellent routes from E1 – E8 the **Shelter Stone Crag** is far and away the best cliff in the northern Cairngorms for the extreme climber. As the most accessible high standard cliff in the Cairngorms it is consequently relatively popular. The 'big three' routes on **The Bastion** often have several teams on them on most summer weekends, even in poor weather.

Approach: From the Choire Chais car park at the funicular follow the track up the left side of the railway passing underneath it at the first station. Continue up the track to gain a well constructed path (marked Fiacaill path) leading up onto the ridge (Fiacaill a'Choire Chais – tooth of steep corrie) (1hr of ascent to top of ridge). Walk due south and down Coire Raibert (Robert's Corrie) by a path on the left bank of the burn. Cut across the stream bed a couple of hundred metres short of Loch Avon and follow a worn path west to the head of the loch. From here cross the stream and follow a path which weaves through the massive boulder field up to the base of the crag. If going in for a day trip and climbing on **The Bastion** (or intending to top out from **The Central Slabs**) it is worth walking round the rim of the corrie and gearing up at the top of the cliff, thus avoiding climbing a further 300m back out of the corrie floor.

Dan Moore on the crack for thin fingers pitch of The Needle (page 467). Photo Martin Stephens

Paul Thorburn on the third pitch of The Harp (page 458).

THE CENTRAL SLABS

In many ways the showpiece of the crag, this 100m sweep of steepening slabs lies sandwiched between two large areas of unpleasant grass and broken ground – dubbed the **Low Ledge** and **High Ledge** respectively.

Approach: A short awkward corner near the left end provides access to the **Low Ledge** – an extensive long right-slanting grass terrace at the base of the slabs proper.

Descent: Either continue by much vegetated scrambling above the slabs and traverse right on the plateau to descend as for the routes on **The Bastion**, or abseil. In situ anchors are in place at the top of *The Run of the Arrow* (P and nut) down to *Thor* dièdre then from 2 PRs to Low Ledge, or from 2 PRs and long sling 10m up and left from the top of *The Pin* then by thread on belay again to Low Ledge. Scramble back right (facing out) along this and down awkward short groove to base of cliff.

Photo David Riley

LOWER TIER

 **Freya** ★★   **45m E3 5c**

FA Jules Lines & Lawrence Hughes 4 July 2001

"A good approach to the terrace, with over 40m of pretty sustained climbing. Plenty of gear where you need it." Climbs good clean hairline cracks in a slab to the right of the obvious grey right-facing corner. Start at the toe of the slab 8m right of the corner. Climb directly up the slab on reasonable holds for 15m to reach a scoopy groove on the left, just right of the corner. Make thin moves up and right then direct, passing a couple of overlaps to gain the ledges beneath the main routes.

MAIN FACE

 **The Harp** ★★★   **80m E3 6a**

FA Pete Whillance & Ray Parker 27 July 1983

Excellent well-protected climbing on the third pitch. Start down and left of the lower left-facing corner of *Thor*.

1 **15m** Move leftward across slab to an old PB in a recess left of the main slabs.

2 **15m 6a** Move right and up to a weakness in the overlap. Through this (crux) and easily up a slab to belay 5m above.

3 **50m 5c** Up crack/groove and move out right to join the left-trending diagonal fault of *Snipers*, which provides a more logical finish. Alternatively, move out rightward to belay as for *The Run of the Arrow* (30m) finishing up the final pitch of that route (25m 5b).

3 **Immortals** ★   **35m E5 6a**

FA Jules Lines & Rick Campbell 20 July 1996

"Good, but very bold after the roof." A good alternate start to many of the routes on the slab. Start at the left end of the lower slabs at a right facing corner leading into an arching roof. Climb the corner easily to the roof, pull over and follow a crack-line upwards to its termination (RP #1 at its top). Move right to gain better holds, step up above a small overlap and delicately traverse right to gain the huge loose flake/block on the initial corner of *Thor*.

4 **Aphrodite** ★★★   **95m E7 6b**

FA Rick Campbell & Alistair Moses July 1990

A long sustained and serious main pitch, following a superb uncompromising line up the left side of the slabs. Start at the base of the left-facing corner leading up to the base of the *Thor* dièdre.

1 **35m 5a** Climb the corner, moving out left round a prominent overhang. Continue up, teetering past a large loose flake near the top to belay beneath the main roof.

2 **10m 6b** Cross the overlap past a peg (replaced 2014) with difficulty to a ledge above. Follow the thin crack leftward to a good nut placement. Return to belay at P and nut placement just above the lip of the roof. An additional 10m of rope required to arrange a satisfactory belay for the next pitch.

3 **50m 6b** Move up and right into a weakness leading into *The Run of the Arrow*. Follow that route to the crucial protection then move up and right and up again to a tiny foot ledge. Continue directly up to good foot holds in the scoop where *Cupid's Bow* traverses in from the right. Follow *Cupid's Bow* and its final flake crack to the peg belay.

5 **Quiver** ★★   **95m E4 6a**

FA Jules Lines & Steve Perry (on-sight) 28 June 2018

Good climbing based on the crack-line to the left of *The Run of the Arrow*. *"The first hard pitch is good and well protected; top pitch basically unprotected. Pretty exciting!"*

1 **40m 5a** As for *The Run of the Arrow*.

2 **30m 6a** Climb round the rib as for *The Run of the Arrow* to the RPs. Make a thin move left to a sloping foothold in the scoop. Continue on more positive holds until you can step left onto a sloping ledge. Continue up the vague crack-line in the rib until a step right reaches a small wire belay on *The Run of the Arrow*.

3 **25m 6a** Step back left into the line and continue up the blind flake-line via a difficult mantel and better holds to the top.

6 The Run of the Arrow ★★★ **105m E6 6b**

FA Pete Whillance & Tony Furnis 24 July 1982

The main pitch follows the faint fading crack up the slab down and left of *Cupid's Bow*. It is very bold, with two sustained hard sections. The route has become a grade harder (much more serious) since the removal of the hammered nuts (which were poor anyway) just before the last protection.

1 **40m 5a** Climb easily up rightward to a grassy platform at 10m (possible belay) above a small niche. Move left around the rib and follow a crack up into the base of the *Thor* dièdre. Climb the dièdre for 6m to belay at 2 PRs.

2 **40m 6b** Move up the short corner above the belay and pull left onto the slab. Out left on good holds to thin twin parallel cracks (many small nuts). Up these with difficulty to an easy middle section leading to a large flat hold. Step right precariously to place protection (F #0 under overlap, Wallnut #9 on side or old style R #7 and good small nut in between; F #Z2 also). Make hard delicate moves left and up leading to better holds and continue leftward to belay.

3 **25m 5b** Follow the obvious rightward slanting line for 12m to a flake crack. Finish up this.

7 Cupid's Bow ★★★★ **95m E5 6b**

FA Dougie Dinwoodie & Dick Renshaw 27 May 1978 (4PA); FFA Murray Hamilton & Rab Anderson 4 June 1982

Superb climbing with an excellent sustained main pitch up the bow-shaped corner above the *Thor* dièdre. Start at the base of the left-facing corner of *Thor*.

1 **10m** Climb easily up rightward to belay on a grassy platform above a small niche.

2 **40m 5b** Go straight up the cracks on the left then traverse right into a shallow corner. Go up this then move left onto the rib and traverse into the *Thor* dièdre. Go up this for a short way to take a stance in slings beside a ledge of sorts.

3 45m 6b Move up the dièdre then swing out left onto a good ledge below the bow. Climb the corner with difficulty (crucial F #0) until it is possible to gain the left rib. Move up this past an awkward bulging section at the top of the corner. Continue up the continuation of the bow until it kinks to the right. Step left onto a difficult slabby grey wall and climb this to traverse right by small ledges to beneath a prominent flake-crack in the headwall. Finish by moving left up this.

7a Cupid's True Bow ★★★★　　　80m E6 6b

FA Jules Lines & Caelan Barnes 27 June 2018

Follows the bow feature in its entirety, offering one of the best and most varied pitches on the slabs.

1 40m 5b As per the original line.

2 40m 6b Follow the original line to the kink in the bow where the original steps delicately left onto the *Aphrodite* slab. Climb rightwards along the flake of the bow to near its end where an obvious hold provides access to the slab above. Continue up the slab on distant positive holds to reach a thin smear move, stepping into a large scallop to gain a ledge and into the groove. Pull out left onto a shelf and continue left for 5m to belay on medium/large cams. The abseil point is 5m further left.

8 Thor ★★★　　　100m E5 6b

FA Mike Rennie & Greg Strange (A2) 7 September 1968;
FFA Rick Campbell & Neil Craig 15 July 1989

One of the most striking natural features on any crag in Scotland, following an exquisite sweeping line across the slabs. Start down and left of the huge dièdre.

1 30m 5a Climb up into the left-facing corner. At the prominent stepped roof move left onto the slab on the left and up this. Near the top step right across large perched flake to peg belay just above.

2 15m 5b Go under the overhang and up the diedre to peg belay on *Cupid's Bow*.

3 20m 6b Continue up the diedre, crossing an alcove to peg belay (back up with cams).

4 35m 6b Continue up the corner with a long

reach to gain a crescent-shaped crack, hand traverse this and the lip of the continuation overlap to reach a rising flake line. Follow this more easily to join and finish up *The Pin*.

9 Sweet Granite Kiss ★★★　　　100m E7 6b

FA Jules Lines & Steve Perry 13 July 2018

A fantastic direct line up the slabs, amalgamating sections of older routes with new sections. Pitch 1 is *Cupid's Bow Variation Start*.

1 20m 5c Scramble up to below the crack in the steep wall to its right. Climb it up into the corner and continue for 5m to a ledge on the left.

2 30m 6b Go back into the corner, climb it to its top and continue over an overlap to follow a thin crack to join the diagonal of *Missing Link*. Follow *Missing Link* for 10m to a small overlap/flake feature, gear up here and make a hard rockover onto it and make harrowingly thin moves to gain the left flake that leads into the *Thor* belay.

3 50m 6b A superb slab pitch. Climb *Thor* to the crescent shaped crack and from its left end, climb thinly up onto a small foothold. A flake jug is agonisingly out of reach. Make a harrowing move, or jump, to gain it (crux). (Very tall people might reach the jug, making the route significantly easier and E6 overall). Continue up to the overlap on *Cupid's True Bow*, gain the hold above the overlap and boldly climb the slab into the corner before stepping out left onto the shelf. Having arranged gear on the shelf, climb the rib above into the diagonal crack and slab above to reach a thin ledge (possible belay, 40m). Continue to the *Icon* belay at the apex of the slabs.

10 The Missing Link ★★　　　115m E4 5c

FA Dave Cuthbertson & Derek Jamieson summer 1981

A fine bold diagonal pitch across the slab, with the technical crux leaving the *Thor* diedre.

1 35m 5a As for *Aphrodite*.

2 15m 5b Up the dièdre to the hanging stance below ledge common to *Cupid's Bow*.

3 40m 5c Traverse right and follow the long

narrow overlap to gain a hollow sounding flake. Traverse this to its end and pull into *The Pin*. Climb up a short way to belay.

4 25m 5a Finish up *The Pin*.

 **10a** **True Link** ★★   **15m E3 5c**

FA Graham Tyldesley & Ross Bingham 9 July 2022

Follow the continuation line with a well-protected, reachy sting in the tail to a PB.

11 **Athene** ★★★   **105m E7 6c**

FA Pitch 1 Andy Cunningham & Allen Fyffe summer 1990; Pitch 2 Rick Campbell & Gary Latter 29 August 1993; Pitch 3 Rick Campbell (previously top-roped) 23 July 1994; as described Jules Lines & Paul Thorburn 1 August 1999

Magnificent sustained climbing with two hard contrasting pitches. Start as for *Cupid's Bow* to belay under a steep wall.

1 30m 5c Move up the initial wall then follow thin tapering ramp up rightward to good ledge at the base of the left-facing corner.

Jules Lines starting the tenuous traverse on the crux pitch of The Realm of the Senses. Photo Dave Cuthbertson, Cubby Images

2 30m 6c *The Realm of the Senses*. Very sustained with three separate hard sections – by far the safest of the E7 pitches on the slabs. Follow the tapering left-facing corner with increasing difficulty to good runners at the overlap. Move left with increasing difficulty to join *The Missing Link*. Follow this to gain a standing position on the hollow flakes. Place crucial runners above the overlap, move back down and traverse left above the overlap to gain a flake leading to the *Thor* belay.

3 45m 6b The *L'Elisir d'Amore* pitch. A run out pitch. Follow *Thor* for 10m to a flake pocket. Traverse up and left to gain a small ledge, crucial small wires and micro cam. Go up the tiny flake to its top and then boldly make a few thin moves left into a toe

pocket in the red streak. Smear up and left to a line of shallow pockets that lead left into the corner (RPs and micro cam). Climb out of the corner using the left wall then straight up on reasonable holds and follow the holds curving rightward until they run out. Make a thin move to gain a flake groove on the right and continue up it to the top. Climb the diagonal flake up left to the *Icon* belay.

⑫ Icon of Lust ★★★★   **115m E8 6c**
*FA p.1 Jules Lines & Lawrence Hughes (on-sight) August 2001;
p2&3 Jules Lines & Paul Thorburn 30 June 2003*

A monumental voyage up the full height of the slabs taking a direct line through *The Realm of the Senses*. Start 20m down and left from the initial crack of *The Pin*.

Bob Duncan nearing the top of the first pitch of *The Pin*.

1 35m 6a The serious pitch! Climb a cracked groove to a ledge at 10m. Pull onto a bleached slab and climb it veering right into a vague left-facing corner at its top (good gear). Go straight up the steep wall on small positive edges to pockets over the top, step left and pull onto the ramp and follow this rightward to the belay.

2 25m 6c A brilliant pitch, desperate and bold. Follow *The Realm of the Senses* groove to the overlap, step right and pull through the overlap with disbelief, sketch up the slab to a weird pocket (crucial micro cams), move diagonally leftward, very thin, to join *The Missing Link* amidst its crux. Gain good holds and gear then traverse down and left above the overlap to gain a flake leading to the *Thor* belay.

3 55m 6b Extremely bold, a cool head required. Climb *Thor* to the crescent crack, step up into pockets and make thin moves up and left to a jug in the red streak. A precarious stretch left enables a micro cam to be placed and potential small wires. Climb the streak to a pocket, step right and up the right edge of the streak past a flat hold into a scoop (frightening). Climb a vertical wall (RP/micro cam) into a further scoop, move diagonally left along flakes and onto the slab above. Continue up the slab to a grassy ledge and possible belay. Continue up the wall above to the apex of the slabs, PB.

13 The Pin ★★★   **70m E2 5b**

FA Rab Carrington & Jimmy Gardner August 1968 (2 PA); FFA Ben Campbell-Kelly & Mike Kosterlitz, early 70s

Good sustained climbing up the prominent vertical crack-line near the right edge of the slabs. Start directly beneath the crack.

1 35m 5b Follow the crack (belay possible at 20m) through an awkward bulge high up then slightly right and direct to a good thread belay.

2 35m 5b Continue in the same line up the steep wall above the belay then the fine crack above.

THE BASTION

This steepening 250m tapering wedge dominates the head of the Loch Avon basin.

Descent: From the plateau head out right (west) beyond the broken scree-filled Pinnacle Gully then down the mainly grassy buttress just before the stream. Cut back right across the top of the boulder field to regain the base.

1 The Shard ★★   **280m E5 6a**

FA Callum Johnson & Guy Robertson 4 July 2018

A superb route combining thin slabs low down with strenuous cracks high up. Well protected apart from a serious section near the start of the second pitch. Start at the very lowest point of the crag.

1 35m 5a From just up and right, step left to gain and follow a crack-line which leads up slightly left to an overlap and hollow blocks. Pull over, move left, then climb the slab using the obvious thin curving crack. Head up left to the grass filled groove and belay by a huge block on the left.

2 45m 6a Climb the groove, then go right and up to the overlap and good protection. Traverse delicately left to gain good holds, stand on the highest of these, then make thin moves directly (serious) to gain a horizontal break. Step left at the next overlap, then pull over with difficulty to gain and follow the flake-groove to the next overlap and the superb thin crack-line which is followed to a ledge.

3 60m 4c Climb the easier continuation groove to join *Haystack* and follow this to its belay below the *Steeple* fault.

4 50m 6a Start up *Haystack/Steeple*, but quickly step left onto the right rib of the *Citadel* fault, then up to gain and follow the overhanging groove and crack-line between big overhangs. Above the overhangs, continue up easier ground to join *Citadel* where it comes back right. Belay below the cracked nose with a chimney just up on the left.

5 **40m 6a** Step up right then pull left into a slight groove to gain a sloping ledge. Pull up then steeply out right into the overhanging finger crack and follow this with hard moves near its top just before it lies back. Continue up the superbly-positioned crack to join *Citadel* where it comes in from the right. Climb over the bulge (as for *Citadel*) and belay at the foot on the long, right-trending shallow groove.

6 **50m 5c** Climb the sustained groove, then move up slightly left into a final short corner.

② **The Heel Stone ★★★**   **260m E5 6a**

*FA Guy Robertson & Dave Cowan 11 August 2018; pitch 1
Michael Barnard & Alan Hill 21 July 2013*

A scintillating line up the centre. Superb sustained climbing, much easier than it looks, with a memorable climax up the wall left of *Haystack*'s top crack. Start at the base of the *Steeple* corners.

1 **50m 5c** Climb the short corner (as for *Haystack* pitch 2) and pull left onto the slabs, then continue direct up thin cracks on the right with increasingly difficulty to a tricky section gaining a groove. Climb this past a bulge and pull left to belay.

2 **20m 5c** Pull directly up to gain the right-curving crack and follow this boldly right to protection. Make thin moves back left to a crack in the bulge which gain a groove and easier ground and a belay on the right.

3 **40m 4c** Climb short walls and grassy ledges up to the grassy terrace. Belay on the right.

4 **40m 6a** The crux - a truly stunning pitch. A few metres right of *Steeple/Haystack*, an obvious line of sloping holds leads temptingly up rightwards into the heart of the smooth wall. Follow these holds with no protection to below a scary mantelshelf onto two jugs at 10m. Place a crucial skyhook out left (better than it sounds) & a poor RP1 directly below. Mantel up off the two good holds, then follow a finger flake out right to a thin crack

which leads up to an overlap. Balance/shuffle left under this to gain an overhanging crack and follow this to a junction with *Steeple*. Continue up leftwards across the *Steeple* ramp to the next ramp (*Haystack*) which is followed to a pull up left onto a commodious ledge and belay. *"Bold but climbing is steady and positive with the gear improving with height as the climbing gets a bit trickier."*

5 **40m 5c** Climb a short tapering groove with difficulty to a jug, then swing left onto the slab. Move up then climb cracks in the left side of the pillar above to a junction with *Citadel* which is followed to below the headwall.

6 **25m 6a** Gain the top of the huge flake on the left, then climb the flake/crack in the wall above to a strenuous pull into the niche. The wall above is climbed directly before a step right to a jug at the top - a sensational and well-protected pitch. *"Great, really out there in a stunning position!"*

7 **45m 5b** Go easily up right to the final wall cracks of Steeple which is then followed to the top.

③ **Haystack ★★★**   **275m E3 5c**

*FA Rab Carrington & Ian Nicolson (1 PA) summer 1971;
FFA unknown*

Excellent climbing heading for the stunning steep wide crack on the upper left side of the front face. Start 10m up from the toe of the buttress beneath the leftmost of two right-facing corner systems.

1 **25m 5b** Follow the right-facing corner to the base of a short corner.

2 **25m 4b** Climb the short corner then exit left onto slabs. Continue up and left to a good grass ledge beneath a shallow corner.

3 **40m 5a** Climb the corner then move out right and cross the overlap above by a prominent crack. Move left and follow a right-slanting corner to terraces.

4 **40m 4b** Follow short walls and grassy ledges leftward to beneath the steep wall.

5 **45m 5c** Follow the steep line of weakness (common to *Steeple*) but continue direct through a slight break then move right up a ramp (above the

Photo Dan Moore

Steeple ramp) to spike belays on a pinnacle.

6 **40m 5a** Make delicate moves left onto a ledge
and follow pleasant cracks left to a break in
the arête. Follow an initially steep crack to
gain a ledge beneath an overhanging wall.

7 **30m 5b** Climb the impressive wide overhanging
crack, more difficult above the prominent flake.

8 **30m 5a** Move right and follow a short vertical
crack in the wall above finishing more easily.

4 Steeple ★★★★   240m E2 5c

*FA Pitches 1-6 Kenny Spence & Mike Watson (2 PA); Pitches 7–9
John Porteous, Kenny Spence & Mike Watson August 1968;
FFA Jeff Lamb & Pete Whillance May 1975*

Excellent sustained climbing linking the lower and upper
corners by a good natural line. Start beneath a prominent
right-facing corner system (the rightmost of two) up right
from the toe of the buttress. This is left of the steep
grassy gully which splits the lower section of the face.

Scramble 6m up the gully and belay at the start of a
large corner.

1 **35m 5a** Climb the corner passing two small
overlaps to belay beyond the second.

2 **25m 5a** Continue up the corner above,
exiting left at the top to block belay.

3 **40m 4b** Climb by short walls and grass ledges
to a terrace below the steep crux wall.

4 **35m 5c** Climb the steep line of weakness
and move right with difficulty then more
easily up the rightward-slanting ramp.

5 **45m 5a** Climb the obvious line of layback
cracks up and right to the foot of the
wonderfully situated upper corner.

6 **40m 5b** Climb the corner past a niche
to belay on ledges on the left.

7 **20m 5b** Finish by a thin crack on the right, which
leads by ledges and cracks to the top of the crag.

⑤ The Heel Stone/The Spire ★★★★   270m E4 6a

FA Murray Hamilton & Rab Anderson 5 May 1982;
pitches 2 & 6 Guy Robertson & Dave Cowan 11 August 2018

An excellent line featuring a lot of technical and sustained climbing. In essence, *The Spire* is two independent pitches above the grass terrace at one third height. This combination links all the best pitches at the grade.

1-3 110m 5c,5c,4c As for *The Heel Stone*.

4 45m 5c Climb the steep shallow groove, pull right over the bulge at the top and continue up to the *Steeple/The Needle* belay ledge. Climb directly above to belay at a large pointed block. A superb pitch. *"Awesome weird positions, upwards caterpillar backstrokes, lying down in the groove with a double knee scum. Glad I wore the high friction jeans."*

5 45m 6a Above is an obvious ramp. Gain the ramp with difficulty and follow it and cracks above to belay to the left of the large blocks at the base of the *Steeple* corner.

6-7 70m 6a,5b As for *The Heel Stone*.

⑥ The Needle ★★★★   260m E1 5b

FA Robin Smith & Davy Agnew 8 June 1962

The crag classic, with excellent varied climbing – the first extreme climbed in the Cairngorms. It follows an inspired line up the centre of the cliff, heading for the prominent chimney splitting the upper cliff. Start directly below the upper chimney, about 30m up right from the right-facing corner of *Steeple*.

1 30m 4b Climb straight up a slab with a step left onto a nose at 20m then direct to a ledge and block belay.

2 50m 5b Go up, first right then left then by a line of twin zig-zag cracks to a short steep wall at 40m. Pull up rightward onto the steep rib and up to belay below the grass terrace.

3 25m Cross the terrace, straight up the slab and flake crack on the left to the top of a huge block.

4 30m 5b Go up left for 6m then step right and up a flake crack to traverse left along a thin ledge leading to a bulging crack. Climb the crack to belay on ledges.

5 15m 4c From the left end of the ledge move up then go diagonally rightward up a slabby ramp to ledges. Continue right to belay.

6 35m 5a Go diagonally up and right along the 'crack for thin fingers'. Above this break out right then up and left by blocks and ledges.

7 20m 4c Go up grooves to the foot of the chimney-crack splitting the final rocks of the crag.

8 35m 5a Climb the chimney-crack (the Needle Crack) to a ledge and loose blocks.

9 20m Continue by the line of the chimney to thread a pile of chokestones and emerge on plateau.

⑦ Stone Bastion ★★   255m E4 6a

FA lower pitches Rab Anderson & Chris Anderson 17 May 1992; top pitches – Rab & Chris Anderson 23 July 1994 (reached on abseil, and redpointed)

"A bit disjointed low down but builds to a great top section with a superb top pitch." "Low end E4, perhaps even E3. Brilliant top two pitches can be led in one big pitch." The highlight is the fine sustained and strenuous

Gary Latter on the second pitch of Steeple. Photo Karen Latter

top pitch, tackling the big corner right of *The Needle*. Start just up right from *The Needle*.

1 **50m 5a** Climb easily to a small overlap. Cross this to gain an obvious groove line slanting up left. Follow this to a junction with *The Needle* just below its stepped corners, then traverse right beneath a wall to gain another obvious left-slanting corner which is climbed to a small ledge.

2 **50m 5c** Continue up the corner a short way, pull up a flange on the right to reach a ledge, then awkwardly pull up right to another ledge beneath a leaning wall. Step round the edge on the right and go easily up a ramp to thread belay on the terrace, as for *The Needle*.

3 **25m 5b** Follow the fault up right for a short way, then traverse left and along beneath a leaning wall to gain the top of a huge block beside *The Needle* belay. Step right, pull onto the wall and gain a ledge, traverse right, then go up to another ledge.

4 **35m 5b** Move up a short way, then go left into a corner/groove and pull up into the recess beneath a roof. Exit the recess up and out left, then go up a short corner to reach easy ground leading up right to below the 'crack for thin fingers' of *The Needle*.

5 **15m 5a** Climb the 'crack for thin fingers' to the groove above.

6 **25m 4b** Exit from the groove and follow the grassy shelf up right beneath the upper bastion. Continue up a slabby groove to some blocks.

7 **15m 5c** Go up the groove on the right past loose rock, then move left into a corner. Climb the corner past a dubious flake to a good sloping shelf.

8 **30m 6a** Climb the corner above and swing out right onto a sloping ledge occupied by some blocks (the hold at the top of the corner is best avoided). Climb the crack in the right wall of the next corner until able to bridge into the corner, then continue up the fine wall on horizontal breaks. Belay in a recess.

9 **10m** Pull out right and scramble to the top.

 **The Camel** ★★   **250m E6 6a**

FA Iain Small & Tess Fryer 5 August 2006; Pitches 1, 2, 6 Iain Small & Chris Cartwright July 2005

Sustained, independent climbing threading around the line of *The Needle*; leads to a final crux pitch tackling the exposed headwall right of *The Needle* crack. A double set of small cams is useful. *"Great climbing throughout, but top pitch needs a clean."* Start right of *Stone Bastion* at grassy ledges.

1 **50m 5c** Climb easily up right to an overlap and gain a grassy fault above. Step left onto the slab and follow a seam leftwards to its end (bold). Move up to a steep wall, take the left-facing inset corner in the arête, continue up the rib then step right onto a slab and traverse rightwards to a good ledge.

2 **35m 5b** Move up left-trending ramps heading for a steep alcove right of the *The Needle* rib. Climb the steep wall (stubby spike runner), to the terrace.

3 **45m 5c** From the left end of a long narrow ledge above, make tricky moves up thin flakes then swing right to a grassy ledge. Step back left into a groove leading to a steep wall and take a crack on perfect spaced slots to a footledge. Move out left onto a rib, finishing up its left side to a square-cut ledge.

4 **20m 4c** Move right and up to a bigger ledge and follow the flake-crack, then a dirty groove onto a big ledge below a leaning wall.

5 **30m 5c** Start up the 'crack for thin fingers' and make a long reach left for a handrail flake leading to a ledge. Take the corner at its right end and climb easier ground up left to the foot of the *Steeple* corner.

6 **25m 5c** Follow a diagonal crack in the fine right wall (awkward start) to gain a small corner leading to a narrow ledge below *The Needle* crack.

7 **45m 6a** Sustained and intricate face climbing aiming for the niche at the source of the pink weep. Step down and right off the ledge and pass the rounded spike. Continue in the same line with occasional deviations out right. Move right into a haven in the niche. Move out right to finish up the headwall on excellent breaks with a blocky exit.

Harrison Connie nearing the top of the second pitch of *Steeple* (page 466). Photo Gavin McColl

EASTER ROSS

Situated north-west from Inverness, this area often benefits from a generally drier climate than the main climbing areas further west.

Accommodation: Riverside Chalets Caravan Park, Contin (☎ 07862 322990; www.lochness-chalets.co.uk); hotels and B&Bs in most local towns.

TIC: Inverness (☎ 01463 252401; www.visitscotland.com).

Amenities: Small supermarket and petrol station at Contin on A835.

MOY ROCK

NH 4953 5493 **Alt:** 100m  4-7min

A very convenient collection of conglomerate open quick drying sport crags overlooking the main A835 Tore-Ullapool road, just east of Contin. The crag is a real sun trap with climbing possible most of the year. With a good spread of grades, the crag is popular.

Access: Leave the A9 at Tore roundabout and follow the A835 west. Continue for 3.4 miles/5.4km beyond Maryburgh roundabout to park considerately on the verge by a gated forestry road immediately beneath the crags (NH 4957 5482; 57.558361, -4.5159978), 2.2miles/3.5km east from Contin. There is further parking in a long layby on the south side of the road, 250m to the east.

Approach: Follow the forestry track for 30m, then head steeply up left on a path to arrive at the base of **Chimney Zone**. From here paths traverse horizontally beneath the base.

Nesting Restrictions: Birds often nest from March-June on **Chimney Zone** and **Raven's Wall**. Adhere to any restrictions posted by the local ranger at the base of the crags. Check the current status at www.mountaineering.scot/access/birds-and-nesting/disturbance.

RAVEN'S WALL

NH 4947 5493

Stays dry, even in heavy rain.

1 Moy Racer ★   14m F6a
FA Andy Nisbet 9 September 2012

2 The Herring ★★   15m F6c+
FA Ian Taylor & Andy Wilby 2008

3 Moy del Sol ★   17m F6c+
FA Jamie Love & Frank Kremer 5 September 2021

4 Fighting off the Vultures ★★★   22m F6a+
FA Tess Fryer & Andy Wilby 1 November 2011

5 The Old Man of Moy ★   18m F6a+
FA Andy Nisbet July 2011

6 Moy Bueno ★   14m F6b
FA Andy Wilby July 2011

7 The Harling Direct ★★   20m F6b+
FA Andy Tibbs 30 March 2023

8 Pebbledash ★★   20m F6b
FA Andy Nisbet 24 October 2011

9 Corvus ★★   22m F5+
FA Andy Nisbet July 2011

10 Shaggy ★★   25m F6c
FA Andy Tibbs 27 May 2023

11 Scoopy Doo ★   22m F6a+
FA Martin Hind & Lord John Mackenzie 16 October 2011

12 Raven's Nest ★   27m F6a+
FA Ray Wilby August 2011

13 Black Streak ★★   25m F6a+
FA Ray Wilby July 2011

14 Pyramid ★   25m F6a+
FA Ray Wilby August 2011

Unknown climber on the fine steady Conglomarete.

THE BAY

NH 4944 5493

The leftmost crag, just left of **Raven's Wall**.

1 Wasted Grit　　　　　　　　　**12m F6b**

FA Jamie Skelton & Morag Eagleson 26 February 2022

2 The Flying Pebble ★　　　　　**18m F6b+**

FA Jamie Skelton & Morag Eagleson 26 February 2022

3 Vangelis ★　　　　　　　　　**20m F6a+**

FA Frank Kremer 7 November 2021

4 Eldorado ★★　　　　　　　　**24m F6a+**

FA Andy Nisbet 4 September 2013

5 Holly Tree Groove ★★　　　　**22m F6a**

FA Sandy Allan & Andy Nisbet 22 November 2011

6 Conglomarete ★★　　　　　　**20m F4+**

FA Sandy Allan & Andy Nisbet 22 November 2011

7 Conglomarete Cliff ★　　　　**26m F4+**

FA Frank Kremer 14 September 2023

BIG FLAT WALL

NH 4952 5493

The showpiece sector, with lots of good sustained routes in the mid grades.

1 Match if you are Weak ★★　　**20m F6c+**

FA Andy Wilby 2008

2 The Dark Side ★★★　　　　　**20m F6c**

FA Andy Wilby 16 December 2007

3 Little Teaser ★★★　　　　　**20m F6b+**

FA Andy Wilby 2007

4 Little Squeezer ★　　　　　　**20m F6c+**

FA Andy Tibbs & Ian Taylor June 2016

5 The Diagonal ★★★　　　　　**25m F6c+**

6 Pulling on Pebbles ★★　　　　**20m F7a+**

FA Andy Wilby 2008

7 Cobbledegook ★★　　　　　　**25m F7a+**

FA Ian Taylor February 2017

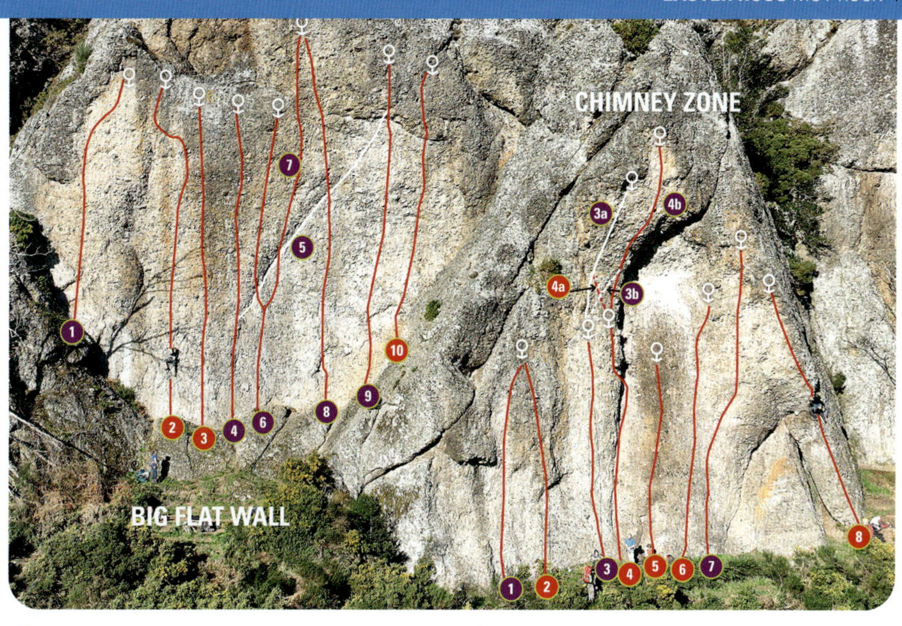

8	**The Ticks Ate all the Midges** ★★★	**20m F7a**

FA Andy Wilby 2007

9 Cloak and Dagger ★★   **20m F6c+**

FA Andy Wilby 2007

10 Kitemark ★★   **20m F6c**

FA Ian Taylor June 2015

CHIMNEY ZONE

NH 4953 5492

The first crag encountered, at the top of the steep path.

1 One Man went to Moy ★   **10m F7a**

FA Neil Shepherd 27 August 2011

2 Burning Barrels ★   **10m F6c**

FA Neil Shepherd 27 August 2011

3 The Seer ★★   **12m F7a+**

FA Andy Wilby 2007

3a The Seer Left Continuation   **18m F7b**

FA Andy Wilby 2011

3b The Fear ★★   **20m F7b**

FA Alan Cassidy & Andy Wilby 25 March 2012

4 The Adder Stone ★   **14m F6a+**

FA Ian Taylor November 2017

4a The Adder Stone into The Seer Left Continuation ★★   **20m F6b+**

FA Morag Eagleson & Jamie Skelton 18 June 2021

4b The Adder Stone into The Fear ★   **20m F7a**

FA Ian Taylor November 2017

5 Guano Apes   **12m F6a**

FA Frank Kremer 6 June 2023

6 Scat, maan   **14m F6b**

FA Jamie Love 1 August 2023

7 Constant Flux ★★   **15m F6c+**

FA Ian Taylor November 2009

8 Priscilla the Squirrel ★   **15m F6b+**

FA Tess Fryer & Ian Taylor 24 February 2023

THE SLABS

Just up right from **Chimney Zone**, a pair of easy-angled broken pillars host the easiest routes at Moy.

1 Easy Slab ★★ 25m F4+

FA Andy Wilby 2008

2 Ankle Biter's Delight ★ 20m F4+

FA Gary Kinsey & James Edwards 29 September 2011

3 Ephemeral Artery ★ 23m F4+

FA Gary Kinsey & Christine Paterson 29 September 2011

4 Venus Return ★★ 24m F4+

FA Gary Kinsey & James Edwards 29 September 2011

OAK TREE AREA

Just right from **The Slabs**.

1 Moy Soldiers ★ 17m F5+

FA Ian Taylor & Lord John Mackenzie 2011

2 Pebble Party ★ 17m F6a+

FA Will Wilkinson 25 August 2008

3 L-Plate ★ 17m F5+

FA Andy Nisbet April 2011

4 The Fly ★★ 20m F6a+

FA Andy Wilby 2008

5 Magnifascent ★★ 23m F6a+

FA (trad) Lord John Mackenzie, Richard McHardy & M.Birch 31 May 1978; sport: Ray Wilby March 2014

6 The Silver Fox ★★ 20m F7a

FA Tess Fryer 2010

7 The Clansman ★★★ 25m F7a+

FA Andy Wilby 2008

ROUTE INDEX